REVIEWS in MINERALOGY
Volume 15

MATHEMATICAL CRYSTALLOGRAPHY

An INTRODUCTION to the MATHEMATICAL FOUNDATIONS of CRYSTALLOGRAPHY

The Authors

M. B. BOISEN, JR.
Department of Mathematics
Virginia Polytechnic Institute & State University
Blacksburg, Virginia 24061

G. V. GIBBS
Department of Geological Sciences
Virginia Polytechnic Institute & State University
Blacksburg, Virginia 24061

Series Editor: PAUL H. RIBBE
Virginia Polytechnic Institute & State University

MINERALOGICAL SOCIETY OF AMERICA

Printed by

BOOKCRAFTERS, Inc.

(formerly: SHORT COURSE NOTES)

ISSN 0275-0279

Volume 15: MATHEMATICAL CRYSTALLOGRAPHY
An Introduction to the Mathematical
Foundations of Crystallography

ISBN 0-939950-19-7

* * * * * * * * * *

ADDITIONAL COPIES of this volume as well as
those listed below may be obtained from

Mineralogical Society of America

2000 Florida Avenue, N.W., Washington, D.C. 20009

Volume		Pages
1	SULFIDE MINERALOGY, P.H. Ribbe, editor (1974)	284
2	FELDSPAR MINERALOGY, P.H. Ribbe, editor (1975; revised 1983)	363
3	OXIDE MINERALS, Douglas Rumble, III, editor (1976)	502
4	MINERALOGY and GEOLOGY of NATURAL ZEOLITES, F.A. Mumpton, editor (1977)	232
5	ORTHOSILICATES, P.H. Ribbe, editor (1980; revised 1982)	450
6	MARINE MINERALS, R.G. Burns, editor (1979)	380
7	PYROXENES, C.T. Prewitt, editor (1980)	525
8	KINETICS of GEOCHEMICAL PROCESSES, A.C. Lasaga and R.J. Kirkpatrick, editors (1981)	391
9A	AMPHIBOLES and OTHER HYDROUS PYRIBOLES - MINERALOGY, D.R. Veblen, editor (1981)	372
9B	AMPHIBOLES: PETROLOGY and EXPERIMENTAL PHASE RELATIONS, D.R. Veblen and P.H. Ribbe, editors (1982)	390
10	CHARACTERIZATION of METAMORPHISM through MINERAL EQUILIBRIA, J.M. Ferry, editor (1982)	397
11	CARBONATES: MINERALOGY and CHEMISTRY, R.J. Reeder, editor (1983)	394
12	FLUID INCLUSIONS, Edwin Roedder, author (1984)	644
13	MICAS, S.W. Bailey, editor (1984)	584
14	MICROSCOPIC to MACROSCOPIC: Atomic Environments to Mineral Thermodynamics, S.W. Kieffer & A. Navrotsky (1985)	428

to our wives

HELEN and NANCY

whose patience and support is
sincerely appreciated

REVIEWS in MINERALOGY

FOREWORD TO VOLUME 15

MATHEMATICAL CRYSTALLOGRAPHY represents a new direction for the *Reviews in Mineralogy* series. This text book is not a review volume in any sense of the term, but in fact it is, as its subtitle suggests, "An Introduction to the Mathematical Foundations of Crystallography." Written by a mathematician, M.B. Boisen, Jr., and a mineralogist, G.V. Gibbs, Volume 15 was carefully prepared and illustrated over a period of several years. It contains numerous worked examples, in addition to problem sets (many with answers) for the reader to solve.

The book was first introduced at a Short Course of the same title in conjunction with the annual meetings of the Mineralogical Society of America and the Geological Society of America at Orlando, Florida, October 24-27, 1985. Boisen and Gibbs instructed 35 participants with the assistance of Karen L. Geisinger, (Pennsylvania State University), James W. Downs (Ohio State University), and Bryan C. Chakoumakos (University of New Mexico), who led the computer-based laboratory sessions.

Paul H. Ribbe
Series Editor
Blacksburg, VA
9/13/85 - a Friday

MATHEMATICAL CRYSTALLOGRAPHY

PREFACE

This book is written with two goals in mind. The first is to derive the 32 crystallographic point groups, the 14 Bravais lattice types and the 230 crystallographic space group types. The second is to develop the mathematical tools necessary for these derivations in such a manner as to lay the mathematical foundation needed to solve numerous basic problems in crystallography and to avoid extraneous discourses. To demonstrate how these tools can be employed, a large number of examples are solved and problems are given. The book is, by and large, self-contained. In particular, topics usually omitted from the traditional courses in mathematics that are essential to the study of crystallography are discussed. For example, the techniques needed to work in vector spaces with non-cartesian bases are developed. Unlike the traditional group-theoretical approach, isomorphism is not the essential ingredient in crystallographic classification schemes. Because alternative classification schemes must be used, the notions of equivalence relations and classes which are fundamental to such schemes are defined, discussed and illustrated. For example, we will find that the classification of the crystallographic space groups into the traditional 230 types is defined in terms of their matrix representations. Therefore, the derivation of these groups from the point groups will be conducted using the 37 distinct matrix groups rather than the 32 point groups they represent.

We have been greatly influenced by two beautiful books. Hermann Weyl's book entitled *Symmetry* based on his lectures at Princeton University gives a wonderful development of the point groups as well as an elegant exposition of symmetry in art and nature. Fredrik W. H. Zachariasen's book entitled *Theory of X-ray Diffraction in Crystals* presents important insights on the derivation of the Bravais lattice types and the crystallographic space groups. These two books provided the basis for many of the ideas developed in this book.

The theorems, examples, definitions and corollaries are labelled sequentially as a group whereas the problems are labelled separately as a group as are the equations. The manner in which these are labelled is self-explanatory. For example, T4.15 refers to Theorem (T) 15 in Chapter 4 while DA1.1 refers to Definition (D) 1 in Appendix (A) 1.

We have strived to write this book so that it is self-teaching. The reader is encouraged to attempt to solve the examples before appealing to the solution presented and to work all of the problems.

ACKNOWLEDGEMENTS

We wish to thank Virginia Chapman, a lady of exceptional talent and ingenuity, for her enthusiastic and tireless contribution to this project. In particular, her preparation of the GML files used to produce this book is greatly appreciated. We thank John C. Groen for his dedication and his meticulous preparation of the many illustrations in this book. Margie Strickler is gratefully acknowledged for her preparation of the GML files for the appendices. It is also a pleasure to thank Karen L. Geisinger for painstakingly preparing the stereoscopic pair diagrams of the *G*-equivalent ellipsoids for the 32 crystallographic point groups *G*. We thank Don Bloss and Hans Wondratschek for beneficial discussions, Sharon Chang for important technical advice on the drafting of the figures and her assistant, Melody L. Watson, for her draftwork. We also thank Bryan C. Chakoumakos, Department of Geology, University of New Mexico, Albuquerque, New Mexico; James W. Downs, Department of Geology and Mineralogy, Columbus, Ohio; Karen L. Geisinger, Department of Geoscience, The Pennsylvania State University, University Park, Pennsylvania and Neil E. Johnson, Department of Geological Sciences, VPI&SU, Blacksburg, Virginia and David R. Veblen, Department of Earth and Planetary Sciences, The Johns Hopkins University, Baltimore, Maryland for their reading of the earlier drafts and useful remarks. However, they are neither responsible for any errors that may be present in the book nor for the point of view we have taken in this project. Finally, we gratefully acknowledge the Series Editor Paul H. Ribbe for his helpful comments and criticisms and the National Science Foundation Grant EAR-8218743 for partial support of this project.

EXPLANATION OF SYMBOLS

SYMBOL	DESCRIPTION
S	Geometric three-dimensional space.
P	A primitive lattice or the basis for a primitive lattice.
$[\mathbf{r}]_D$	The triple representation of $\mathbf{r}$ with respect to the basis D.
L_D	The lattice generated by D.
$D^* = \{\mathbf{a}^*,\mathbf{b}^*,\mathbf{c}^*\}$	The reciprocal lattice of D.
$M_D(\alpha)$	The 3×3 matrix representation with respect to the basis D of α when α is a point isometry and of the linear component of α when α is an isometry.
$M_D(G)$	The set of all $M_D(\alpha)$ where $\alpha \in G$.
I	The set of all isometries.
Γ	The set of all translations.
$D(\mathbf{o})$	The basis $\{\tau_x(\mathbf{o}),\tau_y(\mathbf{o}),\tau_z(\mathbf{o})\}$ of S where $D = \{\tau_x,\tau_y,\tau_z\}$ is a basis for Γ.
$R_{D(\mathbf{o})}(G)$	The set of all $R_{D(\mathbf{o})}(\alpha)$ where $\alpha \in G$.
$\{M \mid \mathbf{t}\}$	The Seitz notation for the 4×4 matrix representation of an isometry.
$\mathrm{orb}_T(\mathbf{o})$	The orbit of $\mathbf{o}$ under the translations of T.
$\Lambda_{\mathbf{o}}(\alpha)$	The linear component of α.
$\Lambda_{\mathbf{o}}(G)$	The set of $\Lambda_{\mathbf{o}}(\alpha)$ where $\alpha \in G$.
$T(G)$	The set of all translations in G.
T_L	The set of translations associated with the lattice L.
$\mathrm{tr}(M)$	The trace of the matrix M.
$\det(M) = \|M\|$	The determinant of the matrix M.
$\#(H)$	The number elements in H.
$o(\alpha)$	The order of the isometry α.
$[uvw]_n$	An nth-turn whose axis is along $u\mathbf{a} + v\mathbf{b} + w\mathbf{c}$ where $D = \{\mathbf{a},\mathbf{b},\mathbf{c}\}$ is a given basis.
$[uvw]_{n_m}$	An nth-turn screw about the vector $u\mathbf{a} + v\mathbf{b} + w\mathbf{c}$ with a translation of mr/n where $\mathbf{r}$ is the shortest nonzero vector in the lattice in the $[uvw]$ direction.
i	The inversion.
Basic conventions	Points in S are denoted by lower case letters, vectors and their endpoints by bold-faced lower case letters, lengths of vectors by italics, sets by capital italics and matrices by capital letters.

MATHEMATICAL CRYSTALLOGRAPHY

CONTENTS

Chapter 3. POINT ISOMETRIES - VEHICLES FOR DESCRIBING SYMMETRY

Chapter 4. THE MONAXIAL CRYSTALLOGRAPHIC POINT GROUPS

Chapter 5. THE POLYAXIAL CRYSTALLOGRAPHIC POINT GROUPS

Chapter 6. THE BRAVAIS LATTICE TYPES

Chapter 7. THE CRYSTALLOGRAPHIC SPACE GROUPS

Appendix 1. MAPPINGS

Appendix 2. MATRIX METHODS

Appendix 3. CONSTRUCTION AND INTERPRETATION OF MATRICES REPRESENTING POINT ISOMETRIES

CHAPTER 1

MODELING SYMMETRICAL PATTERNS AND GEOMETRIES OF MOLECULES AND CRYSTALS

"When certain causes produce certain effects, the symmetry elements of the causes ought to reappear in the effects produced". -- Pierre Curie

INTRODUCTION

Symmetrical Patterns in Molecular Structures: The *minimum energy principle* states that the atoms in an aggregate of matter strive to adopt an arrangement wherein the total energy of the resulting configuration is minimized. When such a condition is realized, the atoms in the aggregate, whether large or small in number, are characteristically repeated at regular intervals in a symmetrical pattern. In recent years, molecular orbital methods have had great success in finding minimum total energy structures for small aggregates (molecules), using various algorithms for optimizing molecular geometry (cf. Hehre *et al.*, in press and references therein). Not only do these calculations reproduce known molecular structures within the experimental error, but they also show that when the total energy is minimized certain atoms in the molecule are repeated at regular intervals in a symmetrical pattern about a point, line or a plane.

(E1.1) Example - Repetition of a pattern at regular intervals about a point and a line: Monosilicic acid, H_4SiO_4, is an example of a small molecule whose atoms are repeated at regular intervals in a symmetrical pattern about a point and a line (Figure 1.1). An optimization of the geometry of this molecule using molecular orbital methods shows that its total energy is minimized when an SiOH group of the molecule is repeated at regular intervals of 90° about a point and at regular intervals of 180° about a line to give the molecular structure displayed in Figure 1.1. The repetition of the group at regular intervals about a line is called rotational symmetry, whereas that about a point is called rotoinversion symmetry. Rotations and rotoinversions and how they can be used to define the symmetry of this molecule will be examined in Chapters 3 and 4.

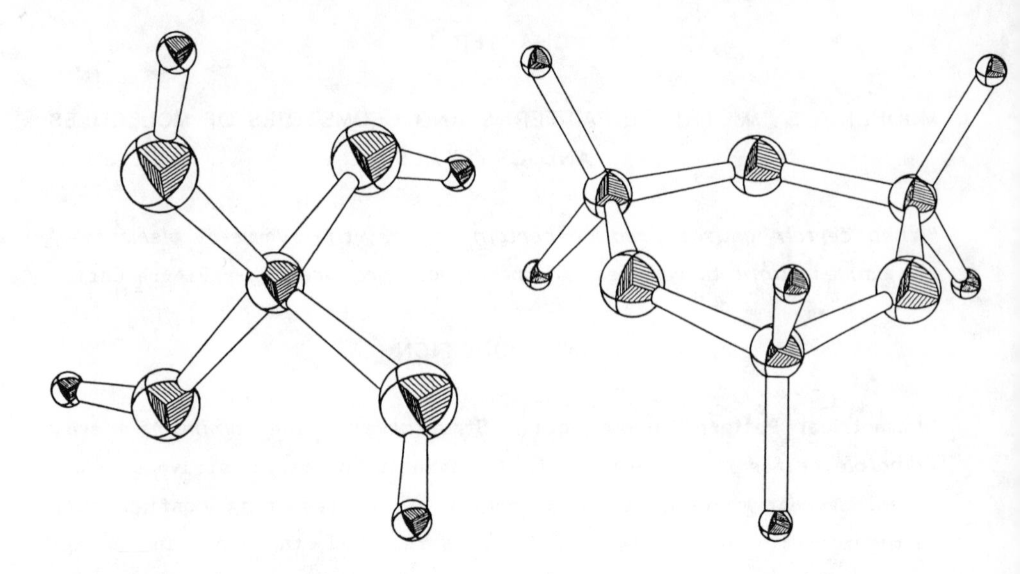

Figure 1.1 (to the left): A drawing of the molecular structure of monosilicic acid, H_4SiO_4, as determined in near Hartree-Fock molecular orbital calculations (Gibbs et al., 1981; O'Keeffe and Gibbs, 1984). The intermediate-sized sphere centering the molecule represents Si, the four largest spheres represent O, and the smallest spheres represent H. The sizes of the spheres in this drawing or in any other drawing in this book are not intended to mimic the actual sizes of atoms in molecules or crystals.

Figure 1.2 (to the right): A drawing of the structure of a tricyclosiloxane molecule composed of three "tetrahedral" $Si(OH)_2$ groups bonded together into a 6-membered ring. The large spheres represent O, the intermediate-sized spheres Si, and the small ones H. The structure of the molecule was determined by O'Keeffe and Gibbs (1984), using molecular orbital methods.

(E1.2) Example - Repetition of a pattern at regular intervals about lines and across planes: Tricyclosiloxane, $H_6Si_3O_3$, is an example of a molecule whose atoms are repeated at regular intervals in a symmetric pattern about lines and across planes (Figure 1.2). The total energy of this molecule is minimized when a "tetrahedral" SiO_2H_2 group is repeated at regular intervals about a line to form a planar 6-membered ring of three Si and three O atoms. Imagine a line drawn perpendicular to the plane of the ring and passing through its center. Note that the atoms of the molecule are repeated in a symmetrical pattern about this line at regular intervals of 120°. Also note that the hydrogen atoms of the molecule are repeated across the plane of the ring. The regular repetition of the structure across a plane is called reflection symmetry. There are other lines and planes in the molecule about which the atoms of the molecule are repeated, but a study of these elements will be deferred to Chapter 5.

Symmetrical Patterns in Crystals: When a large aggregate of atoms (typically ~10^{24}) strives to adopt a configuration wherein the total energy is minimized, we may again find that some atomic pattern of the aggregate is similarly repeated at regular intervals about points and lines or across planes. But, unlike a molecule, the pattern may also be repeated at regular intervals along straight lines to produce a periodic pattern of atoms in three dimensions. Such a three-dimensional aggregate of atoms is said to be a ***crystal*** or a ***crystalline solid***, whereas the actual atom arrangement of the solid is referred to as its ***crystal structure***. In this book, we shall only be concerned with ***ideal crystals***. Unlike real crystals which contain a variety of flaws and irregularities, ideal crystals are envisaged as being perfect in every respect. Also the atoms in such a solid are envisaged to repeat indefinitely at regular intervals in three dimensions. In addition, the positions of the atoms in such a crystal are envisaged to be static and to be specified exactly by a set of atomic coordinates. Thus, when we make reference to a crystal or a crystalline solid, we shall be referring to an ideal crystal. The repetition of a structure at regular intervals along straight lines is called translational symmetry, a subject discussed in Chapters 6 and 7.

(E1.3) Example - Repetition of a pattern at regular intervals along a straight line: As stated above, a characteristic feature of a crystalline solid is that some atomic pattern in a large aggregate of atoms is repeated at regular intervals along straight lines. Consider the string of silicate tetrahedra in Figure 1.3 isolated from the crystal structure of α-quartz, a common mineral of SiO_2 composition comprising about 12 percent of the continental crust. Fix your attention on any two adjacent tetrahedra in the string and imagine that this pair of tetrahedra is repeated at regular intervals indefinitely in the direction of the string. Whenever some atomic pattern in a crystal (like the pair of tetrahedra in the string) is repeated along some line at regular intervals, it is convenient to represent such a periodic pattern by a directed line segment (a vector) parallel to the string and whose length equals the repeat unit of the pattern. Such a vector (labelled **a** in this case), drawn next to the string of tetrahedra in Figure 1.3, is called a translation vector.

(E1.4) Example - The crystal structure of α-quartz: A drawing of the α-quartz crystal structure is presented in Figure 1.4. As observed for

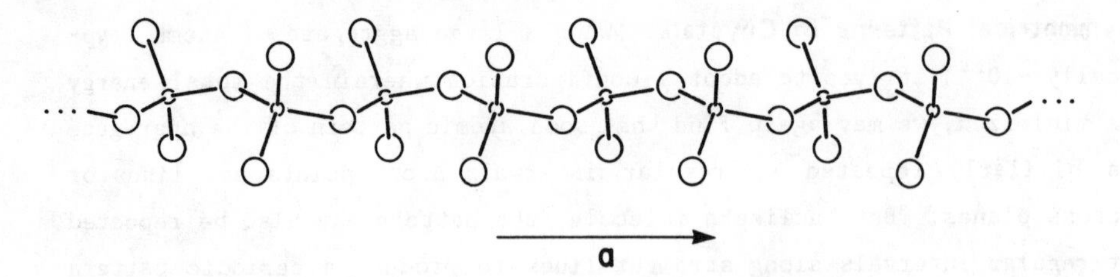

Figure 1.3: A string of silicate tetrahedra isolated from the crystal structure of α-quartz. Each silicon atom (small sphere) in the string is bonded to four oxygen atoms (large spheres) disposed at the corners of a SiO_4 silicate tetrahedron. The lines connecting the atoms in the string represent the bonds between Si and O. Direct your attention on any two adjacent tetrahedra in the string and note that this pair of tetrahedra is repeated at regular intervals so that the repeat unit is given, for example, by the distance between Si atoms in alternate silicate tetrahedra. The repeat along the string is represented by a parallel vector **a** whose magnitude, *a*, equals the repeat unit along the string. The ellipses, (...), at both ends of the string indicates that this sequence of silicate tetrahedra repeats indefinitely in the direction of the string, even though the sequence is terminated in the figure.

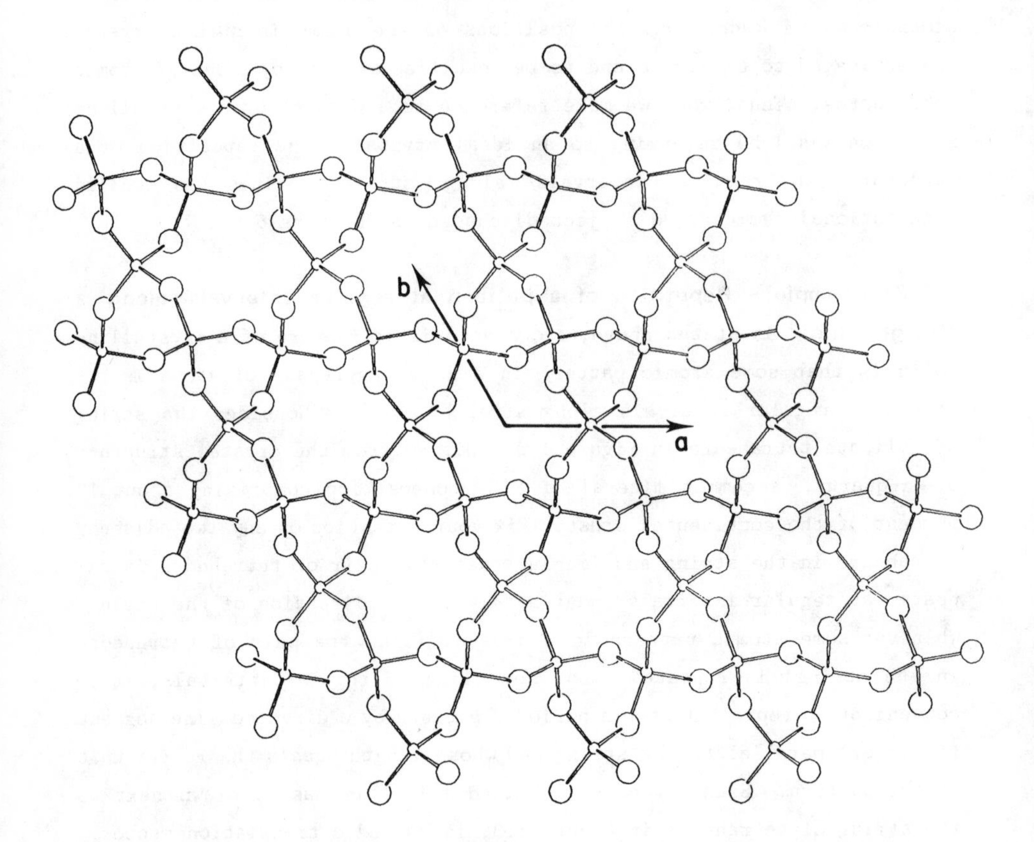

Figure 1.4: The crystal structure of α-quartz projected down a direction in the crystal along which the pattern of atoms displayed in the drawing is repeated. The SiO_4 groups in the structure share corners and form spirals of tetrahedra that advance toward the reader and that are linked laterally to form a continuous framework of corner silicate tetrahedra. The vector **a** represents the repeat unit of the structure in the direction of **a**; **b** represents the repeat unit in the direction of **b**.

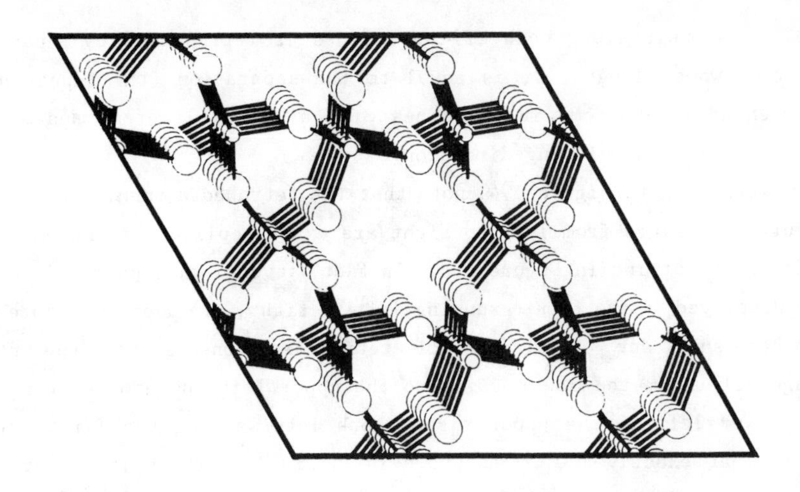

Figure 1.5: A view of the α-quartz structure tilted about 2° off the viewing direction in Figure 1.4. The repeat unit along the viewing direction is represented by the vector **c** whose magnitude equals the separation between equivalent atoms in the lines of atoms paralleling **c**. Because crystals consist of periodic three-dimensional patterns of atoms disposed along well-defined lines, whenever a crystal like α-quartz is viewed along one of these directions, the arrangement will be simplified by the fact that only the atoms in the repeating unit will be seen as in Figure 1.4. Also, the shorter the repeat unit along such a viewing direction, the simplier the arrangement because the repeat unit must involve fewer atoms. When the crystal is tilted off this direction, then the view becomes much more complicated with the uncovering of the many atoms that lie beneath the atoms in the repeating unit. Although the repeat pattern of the structure displayed in Figure 1.4 and 1.5 is finite, the pattern is assumed to continue uninterrupted in the direction of **a**, **b** and **c** indefinitely in a ideal crystal.

monosilicic acid, each Si atom in the structure is bonded to four nearest neighbor O atoms disposed at the corners of a tetrahedron. In addition, as each O atom is bonded to two nearest neighbor Si atoms, the structure can be viewed as a framework structure of corner sharing silicate tetrahedra.

As the structure in Figure 1.4 is viewed down one of the lines along which an atomic pattern is repeated, the atoms along this line are one on top of the other so that only the atoms in one repeat unit along the line are visible. Thus beneath each Si atom displayed in the figure, there exists a line of Si atoms equally spaced at regular intervals that extends indefinitely. Likewise beneath each oxygen atom in the figure, there exists a comparable line of oxygen atoms also equally spaced at the same regular intervals. However, when the structure in Figure 1.4 is tilted off the viewing direction by about 2°, the repeating pattern of atoms along this direction is exposed as in Figure 1.5. By convention, the repeat unit along each of these lines is represented by a parallel vector **c** whose length, *c*, is equal to the separation (the repeat unit) between adjacent atoms in the lines of equivalent Si atoms and O atoms that parallel the viewing direction.

Returning to Figure 1.4, note that the tetrahedra running across the figure in strings from left to right are exact replicas of the ones comprising the string in Figure 1.3. In fact, several such parallel strings are displayed in the figure running left to right. In addition, note that another such set of strings of tetrahedra runs across the figure diagonally from the lower right to the upper left and another runs from the lower left to the upper right, both intersecting the first set of strings at exactly 120°. As shown in Figure 1.4, the unit repeat along the left-to-right trending set of strings is represented by the translation vector **a**, whereas that along the diagonal set of strings running from the lower right to the upper left is represented by the translation vector **b**. In addition to defining the repeat unit along strings of tetrahedra, the pattern at both ends of these vectors is exactly the same regardless of their location in the structure provided the vectors have not been rotated from their original orientations. This property can be illustrated by laying a sheet of tracing paper on Figure 1.4, tracing **a** and **b** on it and then sliding (translating) the sheet over the figure making sure that the vectors on the sheet are kept parallel with those on the drawing. Note that the pattern at the end points of both **a** and **b** is an exact copy of that at their origin regardless of the placement of the vectors on the drawing, provided that the vectors on the sheet are kept parallel with those on the figure.

The lengths (magnitudes) *a*, *b* and *c* of the translation vectors $D = \{\mathbf{a},\mathbf{b},\mathbf{c}\}$ and the interaxial angles between them denoted $\alpha = \angle(\mathbf{b}:\mathbf{c})$, $\beta = \angle(\mathbf{c}:\mathbf{a})$ and $\gamma = \angle(\mathbf{a}:\mathbf{b})$ can be determined in an X-ray diffraction experiment when a crystal or crystal powder is bathed in an X-ray beam and its diffraction pattern is recorded on a film or strip chart. By measuring the positions of the spots on the film or the peaks on the chart, the lengths and the interaxial angles of the translation vectors can be determined.

A set of X-ray data measured for a powdered α-quartz crystal is given in Table A2.2. A least-squares refinement of the data (see EA2.12 and PA2.4) shows that $a = b = 4.914$A, $c = 5.409$A, $\alpha = \beta = 90°$ and $\gamma = 120°$ where 1A = 10^{-8} cm.

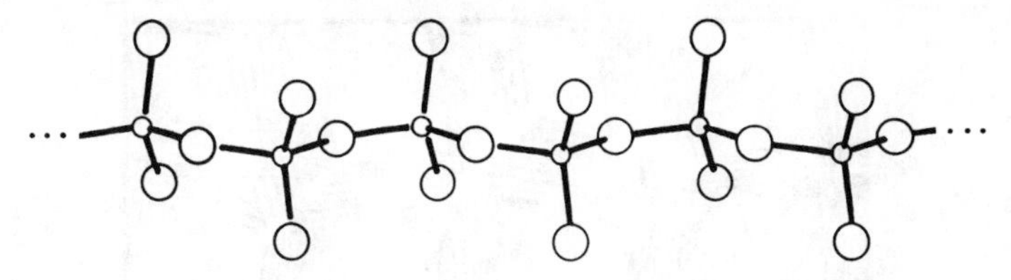

Figure 1.6: A string of silicate tetrahedra isolated from the α-cristobalite structure.

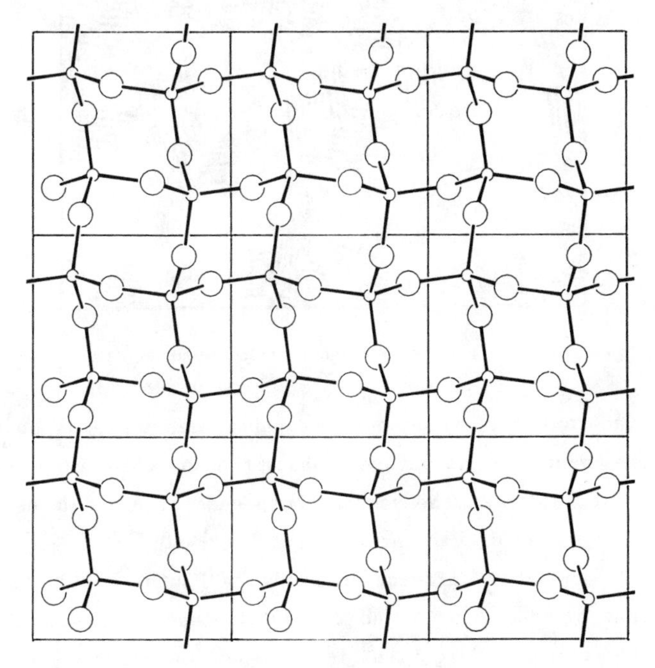

Figure 1.7: A projection of the α-cristobalite structure viewed down **c**, the repeat unit of the structure perpendicular to the drawing. The structure consists of corner sharing silicate tetrahedra that form spirals that are linked laterally into a framework of SiO_2 composition.

(P1.1) Problem: Study the string of silicate tetrahedra in Figure 1.6 isolated from the structure of α-cristobalite, a rare polymorph of SiO_2 found in highly silicic volcanic rocks such as rhyolite.

(1) Determine the number of silicate tetrahedra in the repeat unit.

(2) Draw a translation vector **a** alongside the string. Examine the structure of α-cristobalite displayed in Figure 1.7 where it is viewed down an important repeat direction and find two repeat directions in the plane that match that in Figure 1.6. Draw the vectors **a** and **b** to represent the repeat units along these directions as was done for the α-quartz structure in Figure 1.4 and measure the interaxial angle $\gamma = \angle(\mathbf{a}:\mathbf{b})$. As done in E1.4,

Figure 1.8 A view of the α-cristobalite structure tilted about 2° off the viewing direction in Figure 1.7.

trace **a** and **b** on a sheet of paper and note, as observed for the α-quartz structure, that the pattern of atoms in α-cristobalite is the same at both ends of **a** and **b** as the sheet is translated over the drawing.

An X-ray diffraction study by Pluth *et al.* (1985) shows that $a = b = 4.971$A and $\gamma = 90°$. The repeat unit, **c**, along the viewing direction is exposed in Figure 1.8 where the structure has been tilted 2°. The study also shows that $c = 6.928$A and $\alpha = \beta = 90°$.

A MATHEMATICAL DESCRIPTION OF THE GEOMETRIES OF MOLECULES AND CRYSTALS

It is clear from the discussion, examples and problems presented above that there are numerous geometrical features of molecules and crystals that must be described in order to understand their atomic patterns and properties. This cannot be done with pictures or illustrations alone, because accurate calculations are necessary for a description of their symmetry and structure together with their bond lengths and angles. To accomplish this we shall create a mathematical model of the real world that will enable us to describe all of the geometrical features of a given crystal with respect to its natural frame of reference and translation vectors. For example, we have seen in our discussion of the silica polymorphs α-quartz and α-cristobalite that ideal crystals have special directions along which atoms are repeated over and over again indefinitely in periodic patterns. The mathematical model will be "adjustable" in the sense that its principle directions can be chosen so that they are along special directions of the object under study. We will also see that the collection of all the motions that send a crystal into self-coincidence will be used to describe its symmetry. The model will be such that even very complicated motions can be easily described using matrices. We will begin the development of the model by examining the geometry of the real world.

Geometric three-dimensional space: The three-dimensional real world in which molecules and crystals actually occur will be called the *geometric three-dimensional space* and will be denoted by S. We will view S as a set of points such that the distance between any two points can be calculated, the angle between any two intersecting lines can be measured and such that we can tell when two lines are parallel. All that we will add to S as we go along is a frame of reference that will enable us to model S algebraically. The first step in imposing a frame of reference on S is to establish a point o to be the origin. Any point in S can be chosen as the origin but, as we shall see later, there are certain simplifications that arise when the origin is chosen on a point, line or plane about which the pattern of a molecule or a crystal is repeated. Once the origin is determined, S can be described by the set of all vectors emanating from o to each point in S. Such vectors will be denoted by bold-faced letters. The magnitude or length of a vector **r** is the distance from the origin to its end point. The magnitude of **r** will usually be denoted by an italics r or, sometimes, when an ambiguity is possible, by $\|\mathbf{r}\|$. The absolute value of a scalar x will be denoted by $|x|$. Since the vector emanating from o to o is of length zero, it is called the zero vector and is denoted **0**. For the most part we shall not distinguish between a vector emanating from the origin and its end point. Consequently, unless otherwise stated, we shall let **r** denote both the vector from the origin to the point r and the point as well.

To impose some organization on this set of vectors, we choose a set of three noncoplanar, nonzero vectors $D = \{\mathbf{a},\mathbf{b},\mathbf{c}\}$. Using the notions

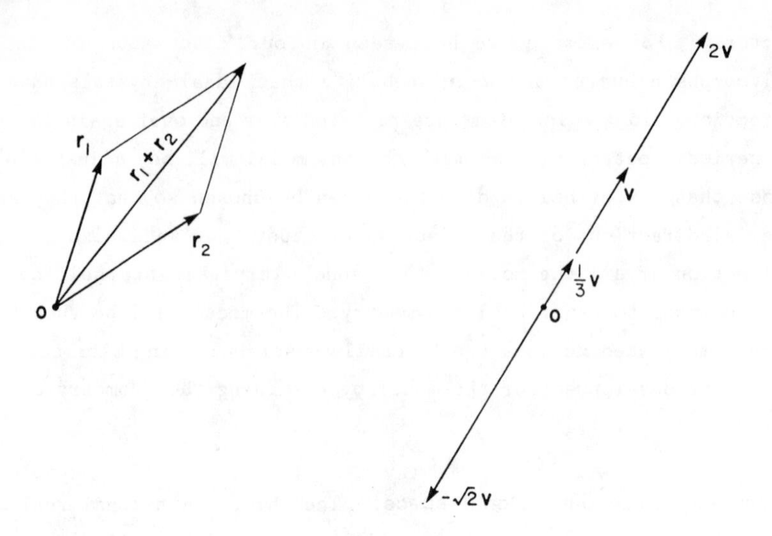

Figure 1.9 (to the left): The geometric addition of any two vectors $\mathbf{r}_1$, $\mathbf{r}_2 \in S$ radiating from the origin, O of S. Note that $\mathbf{r}_1$ and $\mathbf{r}_2$ form the sides of a parallelogram and that their sum $\mathbf{r}_1 + \mathbf{r}_2$ forms its diagonal. An alternate way of obtaining $\mathbf{r}_1 + \mathbf{r}_2$ is to place a parallel copy of $\mathbf{r}_1$ at the end point of $\mathbf{r}_2$. In this way, the vector sum $\mathbf{r}_1 + \mathbf{r}_2$ is given by the vector that radiates from O to the end point of the parallel copy of $\mathbf{r}_2$.

Figure 1.10 (to the right): The multiplication of a vector $\mathbf{v}$ by scalars 2, 1/3 and $-\sqrt{2}$ generates the vectors $2\mathbf{v}$, $(1/3)\mathbf{v}$ and $-\sqrt{2}\mathbf{v}$. We note that every point on a line in S coincident with $\mathbf{v}$ occurs at the end points of the vectors contained in the set $\{x\mathbf{v} \mid x \in R\}$.

of vector addition and scalar multiplication, we shall see that each vector in S can be expressed in a compact manner in terms of **a**, **b** and **c**. Hence, the placement of the origin and the choice of $D = \{\mathbf{a},\mathbf{b},\mathbf{c}\}$ will establish the frame of reference for the model and therefore will be selected to reflect the properties of the crystal under study. To see how this is accomplished, we shall explore the important operations of vector addition and scalar multiplication defined on S.

Vector addition and scalar multiplication: The addition operation on S is defined by geometric addition. By geometric addition (or *vector addition*), we mean that the sum of two vectors $\mathbf{r}_1$ and $\mathbf{r}_2$, denoted by $\mathbf{r}_1 + \mathbf{r}_2$, is the vector that is the diagonal of a parallelogram constructed with sides $\mathbf{r}_1$ and $\mathbf{r}_2$ (Figure 1.9). In addition to the operation of vector addition, there is another important operation defined between real numbers and vectors in S called scalar multiplication. The set of all real numbers (scalars) is denoted by R and consists of the set Z of all integers, the set Q of all quotients a/b of integers ($b \neq 0$) (Q is called the set of rational numbers) and the set of irrational numbers such as $\sqrt{2}$, π, etc. The set R does not, however, contain numbers involving $i = \sqrt{-1}$. If $\mathbf{r} \in S$ (read "$\mathbf{r}$ is an element of the set S") and $x \in R$, then $x\mathbf{r}$ is defined to be a vector pointing along the same line as $\mathbf{r}$ with a length $|x|$ times that of $\mathbf{r}$ and with direction the same as that of $\mathbf{r}$ when $x > 0$ and the opposite of $\mathbf{r}$ when $x < 0$. The multiplication of a scalar x times the vector $\mathbf{r}$ is what is meant by *scalar multiplication*. For example, if the vector $\mathbf{v}$ is multiplied by 2, then the resulting vector $2\mathbf{v}$ is a vector in the direction of $\mathbf{v}$ that has a length twice that of $\mathbf{v}$ while $(1/3)\mathbf{v}$ is a vector in the direction of $\mathbf{v}$ with a length one third that of $\mathbf{v}$ (Figure 1.10). We define $0\mathbf{r}$ to be the zero vector $\mathbf{0}$. If x is a negative number, then $x\mathbf{r}$ is a vector pointing in the opposite direction from $\mathbf{r}$ with a length $|x|$ times that of $\mathbf{r}$. For example, $(-\sqrt{2})\mathbf{v}$ is a vector pointing in the opposite direction as $\mathbf{v}$ with a length $|-\sqrt{2}| = \sqrt{2}$ times that of $\mathbf{v}$. Thus, given any real number x and any vector $\mathbf{r} \in S$, scalar multiplication assigns to x and $\mathbf{r}$ a uniquely determined vector $x\mathbf{r} \in S$.

Recall that S can be viewed as the set of all vectors emanating from a chosen origin $\mathbf{0}$. If $D = \{\mathbf{a},\mathbf{b},\mathbf{c}\}$ is a set of three noncoplanar vectors in S, then for any vector $\mathbf{r}$ in S there exists a unique set of real numbers x, y and z such that $\mathbf{r} = x\mathbf{a} + y\mathbf{b} + z\mathbf{c}$. An expression like $x\mathbf{a} + y\mathbf{b} + z\mathbf{c}$ is called a linear combination of $\{\mathbf{a},\mathbf{b},\mathbf{c}\}$ and the set $D = \{\mathbf{a},\mathbf{b},\mathbf{c}\}$ is called a basis. We shall study the nature of bases more thoroughly later. Using set notation, these statements about S can be written concisely as

$$S = \{x\mathbf{a} + y\mathbf{b} + z\mathbf{c} \mid x,y,z \in R\} .$$

This expression is read "S is the set of all vectors $x\mathbf{a} + y\mathbf{b} + z\mathbf{c}$ such that x, y and z are elements of R, the real numbers."

Triples: Since each vector $\mathbf{r}$ in S can be expressed uniquely as a linear combination $\mathbf{r} = x\mathbf{a} + y\mathbf{b} + z\mathbf{c}$, $\mathbf{r}$ can be unambiguously determined with respect to the basis $D = \{\mathbf{a},\mathbf{b},\mathbf{c}\}$ by the three scalars x, y and z. When we write these scalars in a vertical column enclosed in brackets as

$$\begin{bmatrix} x \\ y \\ z \end{bmatrix},$$

they form a *triple*. The set of all such triples is denoted by R^3. That is,

$$R^3 = \{ \begin{bmatrix} x \\ y \\ z \end{bmatrix} \mid x,y,z \in R\} . \tag{1.1}$$

Hence each vector $\mathbf{r}$ in S can be assigned to a unique triple in R^3 with respect to the basis D. Since this triple representative of $\mathbf{r}$ is dependent on which basis D is chosen, we denote it by $[\mathbf{r}]_D$. That is

$$[\mathbf{r}]_D = \begin{bmatrix} x \\ y \\ z \end{bmatrix} \quad \text{if and only if} \quad \mathbf{r} = x\mathbf{a} + y\mathbf{b} + z\mathbf{c} ,$$

where $D = \{\mathbf{a},\mathbf{b},\mathbf{c}\}$.

The decision as to which basis is to be used is usually made by taking into account the natural geometry of the crystal. For example, if a crystal like α-quartz is to be studied, then it would be convenient for each atom in the structure to have a triple that defines its position in the crystal relative to three noncoplanar vectors like **a**, **b** and **c** which radiate from a common origin, **0**, and which lie along well-defined directions along which the structure is repeated in a relatively short interval. When referring to molecular and crystal structures, it is customary to speak of the coordinates of a point rather than its triple. In this book, we shall use the words "triples" and "coordinates" of a point interchangeably.

Space lattice: If $D = \{\mathbf{a},\mathbf{b},\mathbf{c}\}$ is a set of three noncoplanar nonzero vectors in S, then the subset of vectors (points)

$$L_D = \{u\mathbf{a} + v\mathbf{b} + w\mathbf{c} \mid u,v,w \in Z\}$$

defines an important set of points L_D called a *space lattice*. The vectors comprising L_D are special in the sense that each is of the form

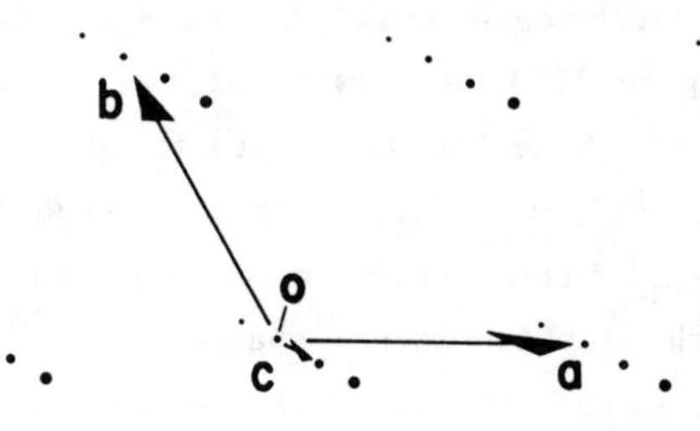

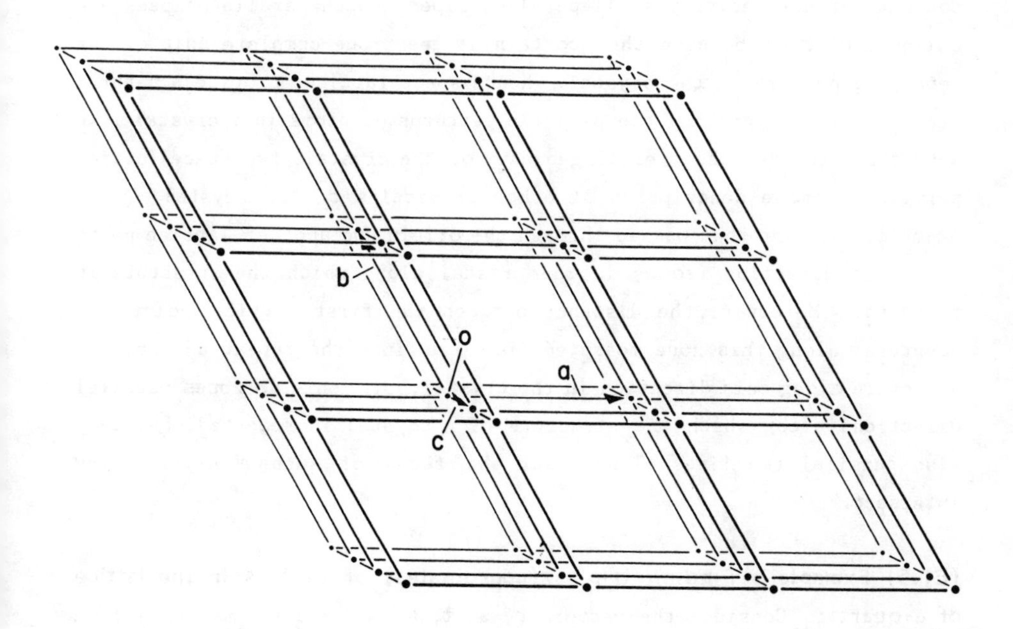

Figure 1.11 (top): The points of the space lattice L_D generated by the vectors $D = \{a,b,c\}$ of α-quartz. The pattern of points in this figure is assumed to continue indefinitely in all directions.

Figure 1.12 (bottom): A line representation of the space lattice in Figure 1.11 constructed by passing lines parallel to **a**, **b** and **c**, respectively, through each point of the figure.

$u\mathbf{a} + v\mathbf{b} + w\mathbf{c}$ where u, v and w are integers. A drawing of the space lattice associated with the α-quartz structure is displayed in Figure 1.11. The vectors **a**, **b** and **c** were chosen in this case to be three of the shortest repeat units in the structure. The lattice can be seen more easily by drawing lines that are parallel to the vectors **a**, **b** and **c**, respectively, through each point in the space lattice as shown in Figure 1.12. In Figure 1.13, a drawing of the α-quartz structure is placed in the lattice. Note how the contents of each of the resulting parallelepipeds is exactly the same. This is true regardless of the placement of the origin of the lattice in the α-quartz structure, provided the orientation of the lattice is not changed. Also the atomic pattern at each lattice point in the structure will be exactly the same regardless of the choice of the origin. Hence, the whole structure of an α-quartz crystal can be constructed by starting with any one of these parallelepipeds and by simply repeating it indefinitely over and over again to fill space. This illustrates how space lattices enable us to give a precise description to a theoretically infinite crystal by the contents of a relatively small parallelepiped. Such parallelepipeds are called *unit cells* because they contain at least one complete unit of the repeating pattern of the crystal. Hence space lattices provide a natural framework for describing the periodic patterns of atoms in a crystal like α-quartz. Besides the overall geometry of the crystal, the space lattice permits a simple description of other properties of the crystal. For example, any line that passes through the origin and another lattice point defines a direction (*zone*) in the crystal along which the structure is repeated. Moreover, the distance between the first lattice point encountered along this zone measured from **0** defines the repeat unit of the structure along this direction in the crystal. Not only do zones parallel directions along which the structure of a crystal is repeated, but they also parallel the lines along which the faces of such a crystal may intersect.

(E1.5) Example - Finding triples (coordinates) of vectors in the lattice of α-quartz: Consider the vectors **r**, **s**, **t**, **u**, **v** and **w** emanating from the origin of the space lattice L_D associated with the α-quartz structure (Figure 1.14) and find $[\mathbf{r}]_D$, $[\mathbf{u}]_D$ and $[\mathbf{w}]_D$.

Solution: An examination of Figure 1.14 shows that $\mathbf{r} = 2\mathbf{a} + 1\mathbf{b} + 2\mathbf{c}$, $\mathbf{u} = 1\mathbf{a} + 2\mathbf{b} + 0\mathbf{c}$ and $\mathbf{w} = -1\mathbf{a} - 1\mathbf{b} - 1\mathbf{c}$ are vectors contained in L_D, thus

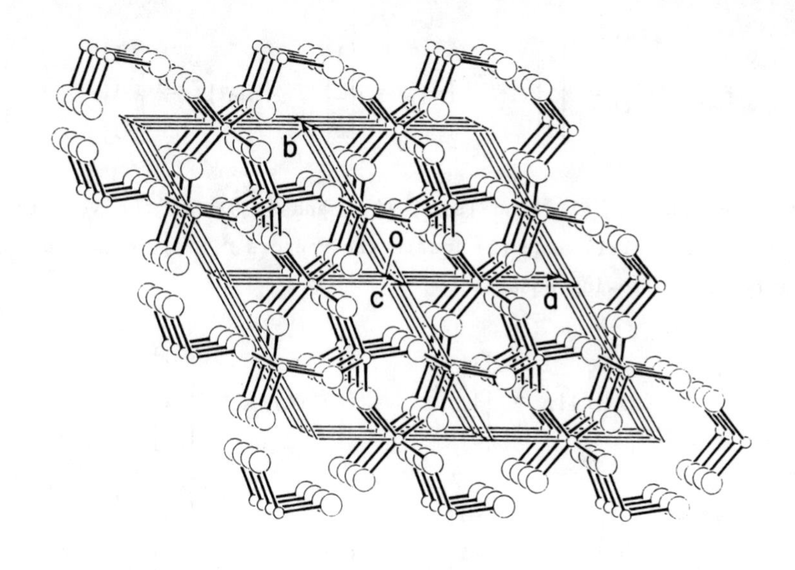

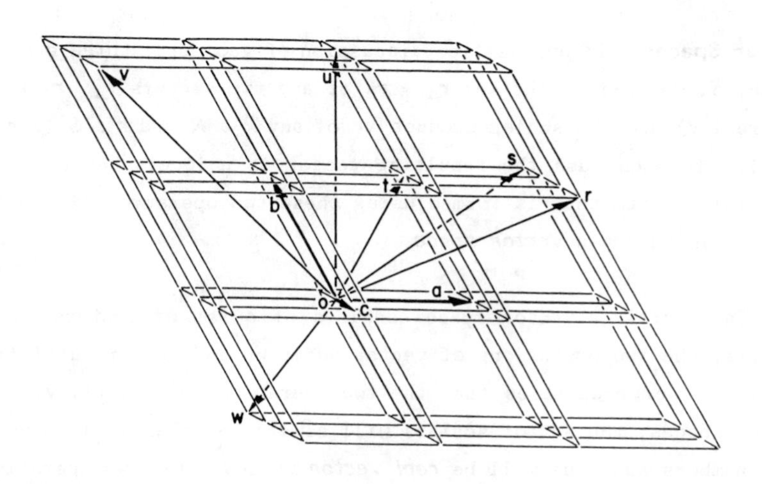

Figure 1.13 (top): A drawing of the crystal structure of α-quartz and its lattice representation. Note that each parallelepiped of the lattice partitions the α-quartz structure into an indefinite number of disjoint identically constituted unit cell volumes each of which contains at least one complete and representative unit of the repeating pattern of the structure.

Figure 1.14 (bottom): The space lattice, L_D, of α-quartz generated by $D = \{\mathbf{a},\mathbf{b},\mathbf{c}\}$. A vector $\mathbf{p}$ is contained in L_D if there exists three integers p_1, p_2 and p_3 such that $\mathbf{p} = p_1\mathbf{a} + p_2\mathbf{b} + p_3\mathbf{c}$. Thus, the vectors $\mathbf{r},\mathbf{s},\ldots,\mathbf{w}$ are each contained in L_D because the coefficients of each are integers.

$$[\mathbf{r}]_D = \begin{bmatrix} 2 \\ 1 \\ 2 \end{bmatrix}, \qquad [\mathbf{u}]_D = \begin{bmatrix} 1 \\ 2 \\ 0 \end{bmatrix}, \qquad [\mathbf{w}]_D = \begin{bmatrix} -1 \\ -1 \\ -1 \end{bmatrix}.$$

□

(P1.2) Problem: (1) Find $[\mathbf{s}]_D$, $[\mathbf{t}]_D$ and $[\mathbf{v}]_D$ for the vectors $\mathbf{s}$, $\mathbf{t}$, $\mathbf{v} \in L_D$ (Figure 1.14) and (2) locate the vectors $\mathbf{r}_1$, $\mathbf{r}_2$, $\mathbf{r}_3$, $\mathbf{r}_4 \in L_D$ that satisfy the following equalities:

$$\text{(a)}\ [\mathbf{r}_1]_D = \begin{bmatrix} 1 \\ 1 \\ 0 \end{bmatrix} \qquad \text{(c)}\ [\mathbf{r}_3]_D = \begin{bmatrix} -1 \\ 2 \\ -1 \end{bmatrix}$$

$$\text{(b)}\ [\mathbf{r}_2]_D = \begin{bmatrix} -1 \\ 1 \\ 0 \end{bmatrix} \qquad \text{(d)}\ [\mathbf{r}_4]_D = \begin{bmatrix} 1 \\ -1 \\ 1 \end{bmatrix}.$$

Vector Spaces: In our earlier discussion of geometric three-dimensional space, S, we defined the sum $\mathbf{r}_1 + \mathbf{r}_2$ of any two vectors $\mathbf{r}_1$, $\mathbf{r}_2 \in S$ (see Figure 1.9) and the scalar product $x\mathbf{r}$ of any $x \in R$ and $\mathbf{r} \in S$ (see Figure 1.10). In each case the resulting vector is uniquely determined. The algebraic system that is formed using these two operations is an example of what is called a ***vector space***.

In general, a vector space consists of a set of vectors, a set of scalars, the two operations of vector addition and scalar multiplication and a list of properties that the two operations must satisfy. For our vector spaces, the set of scalars will always be taken to be the set of real numbers and thus will be ***real vector spaces***. The two operations can be defined in various ways. In earlier discussions, we gave a geometric definition of these operations. Of course, if vectors do not have a geometric interpretation, other definitions must be devised. The required properties of a vector space are listed in a formal definition given below. Notice that each of these properties is satisfied by S.

(D1.6) Definition of a vector space: Let V denote a nonempty set and let R denote the set of real numbers. Consider an operation of vector addition denoted by + that combines any two elements in V to yield an element in V, and the operation of scalar multiplication that combines an element in R with an element in V to yield an element in V denoted by juxtaposition (i.e., if $x \varepsilon R$ and $\mathbf{r} \varepsilon V$ then the scalar product of x with $\mathbf{r}$ is denoted $x\mathbf{r}$). Then V is a *vector space* if for all $\mathbf{r}, \mathbf{s}, \mathbf{t} \varepsilon V$ and $x, y \varepsilon R$ the following rules hold:

(1) $\mathbf{r} + \mathbf{s} = \mathbf{s} + \mathbf{r}$ (Commutative Law);

(2) $\mathbf{r} + (\mathbf{s} + \mathbf{t}) = (\mathbf{r} + \mathbf{s}) + \mathbf{t}$ (Associative Law);

(3) There exists a vector $\mathbf{u} \varepsilon V$ such that $\mathbf{v} + \mathbf{u} = \mathbf{v}$ for all $\mathbf{v} \varepsilon V$ (The existence of an identity element $\mathbf{u}$);

(4) Corresponding to each vector $\mathbf{v} \varepsilon V$, there exists a vector $\mathbf{w}$ such that $\mathbf{v} + \mathbf{w} = \mathbf{u}$ where $\mathbf{u}$ is the identity element (The existence of an inverse $\mathbf{w}$ for each $\mathbf{v} \varepsilon V$);

(5) $x(\mathbf{r} + \mathbf{s}) = x\mathbf{r} + x\mathbf{s}$ (Left Distributive Law);

(6) $(x + y)\mathbf{r} = x\mathbf{r} + y\mathbf{r}$ (Right Distributive Law);

(7) $(xy)\mathbf{r} = x(y\mathbf{r})$ (Associative Law);

(8) $1\mathbf{r} = \mathbf{r}$ (Unit Element).

Besides the vector space S of all geometric vectors, there are many other important vector spaces. For example, the set R^3 defined in (1.1) with operations

$$\begin{bmatrix} x_1 \\ y_1 \\ z_1 \end{bmatrix} + \begin{bmatrix} x_2 \\ y_2 \\ z_2 \end{bmatrix} = \begin{bmatrix} x_1 + x_2 \\ y_1 + y_2 \\ z_1 + z_2 \end{bmatrix} \quad \text{and} \quad x \begin{bmatrix} x_1 \\ y_1 \\ z_1 \end{bmatrix} = \begin{bmatrix} xx_1 \\ xy_1 \\ xz_1 \end{bmatrix}$$

for all $x, x_1, y_1, z_1, x_2, y_2, z_2 \varepsilon R$ is a real vector space. Note that equality in R^3 is defined to be

$$\begin{bmatrix} x_1 \\ y_1 \\ z_1 \end{bmatrix} = \begin{bmatrix} x_2 \\ y_2 \\ z_2 \end{bmatrix}$$

if and only if $x_1 = x_2$, $y_1 = y_2$, and $z_1 = z_2$.

(P1.3) Problem: Show that R^3 satisfies the properties of D1.6 and qualifies as a vector space.

Vector Space Bases: We observed that if **a**, **b** and **c** are noncoplanar vectors in S, then each vector $\mathbf{v} \varepsilon S$ can be expressed as a linear combination

$$\mathbf{v} = x\mathbf{a} + y\mathbf{b} + z\mathbf{c} \text{ where } x,y,z \ \varepsilon \ R \ . \tag{1.2}$$

When each vector in a vector space can be expressed as a linear combination of a set of vectors, $D = \{\mathbf{a},\mathbf{b},\mathbf{c}\}$, we say that D *spans* the vector space.

If $\mathbf{v} = x\mathbf{a} + y\mathbf{b} + z\mathbf{c}$, where **a**, **b**, and **c** are noncoplanar, then the distance that **v** lies from the plane defined by **a** and **b** is completely determined by z. Hence z is uniquely determined. Similarly, x and y are uniquely determined. By uniquely determined we mean that if $\mathbf{v} = x_1\mathbf{a} + y_1\mathbf{b} + z_1\mathbf{c}$ and $\mathbf{v} = x_2\mathbf{a} + y_2\mathbf{b} + z_2\mathbf{c}$, then $x_1 = x_2$, $y_1 = y_2$ and $z_1 = z_2$. If D is a set of vectors in a vector space such that each vector that is a linear combination of D can be written in only one way as a linear combination of D, then D is said to be *linearly independent*. Consequently, a set of three noncoplanar vectors $D = \{\mathbf{a},\mathbf{b},\mathbf{c}\}$ both spans S and is linearly independent. Any set of vectors in a vector space that both spans the vector space and is linearly independent is called a *basis* for the vector space. Thus, any three noncoplanar vectors in S qualifies as a basis. However, unless otherwise stated, the bases we use will be assumed to be right-handed (see Appendix 4). For example, given a, b, c, α, β and γ, two choices of $\{\mathbf{a},\mathbf{b},\mathbf{c}\}$ are possible - one left-handed and one right-handed. In this book a, b, c, α, β, γ will refer unambiguously to the right-handed one. A commonly used basis for S is $C = \{\mathbf{i},\mathbf{j},\mathbf{k}\}$ where **i**, **j** and **k** are mutually perpendicular unit length vectors. Usually C is referred to as a *cartesian basis*. In vector spaces other than S, it may be difficult to visualize the set of vectors that forms a basis. The following example illustrates how to determine whether a set of vectors in R^3 is a basis or not.

(E1.7) Example - A demonstration that a set of vectors qualifies as a basis for R^3: Show that the set of vectors

$$D = \left\{ \begin{bmatrix} 1 \\ 1 \\ 0 \end{bmatrix}, \begin{bmatrix} -1 \\ 1 \\ 0 \end{bmatrix}, \begin{bmatrix} 0 \\ 0 \\ 1 \end{bmatrix} \right\} .$$

forms a basis for R^3.

Solution: To show that D is a basis, we must show that given any vector $\mathbf{v} \in R^3$, $\mathbf{v}$ can be expressed in exactly one way as a linear combination of the vectors in D. Hence, we shall show that there exists a unique solution for x, y and z in the equation

$$\mathbf{v} = \begin{bmatrix} v_1 \\ v_2 \\ v_3 \end{bmatrix} = x \begin{bmatrix} 1 \\ 1 \\ 0 \end{bmatrix} + y \begin{bmatrix} -1 \\ 1 \\ 0 \end{bmatrix} + z \begin{bmatrix} 0 \\ 0 \\ 1 \end{bmatrix} . \tag{1.3}$$

This equation is equivalent to the system of equations

$$\begin{aligned} x - y &= v_1 \\ x + y &= v_2 \\ z &= v_3 . \end{aligned} \tag{1.4}$$

This system can be written in matrix form (see Appendix 2) as

$$\begin{bmatrix} 1 & -1 & 0 \\ 1 & 1 & 0 \\ 0 & 0 & 1 \end{bmatrix} \begin{bmatrix} x \\ y \\ z \end{bmatrix} = \begin{bmatrix} v_1 \\ v_2 \\ v_3 \end{bmatrix} . \tag{1.5}$$

Using any of the methods in Appendix 2, it is found that the system has the unique solution $x = v_1/2 + v_2/2$, $y = -v_1/2 + v_2/2$, $z = v_3$. Since (1.3) has a unique solution for each choice of the vector $\mathbf{v}$, D is a basis for R^3.

□

(P1.4) Problem: Let D denote the basis given in E1.7. Using the method described in the solution to E1.7, write

$$\begin{bmatrix} 2 \\ -6 \\ 5 \end{bmatrix}$$

as a linear combination of the vectors in D.

(P1.5) Problem: Show that

$$D = \left\{ \begin{bmatrix} 1 \\ 1 \\ 1 \end{bmatrix}, \begin{bmatrix} 1 \\ -1 \\ -1 \end{bmatrix}, \begin{bmatrix} -1 \\ -1 \\ 0 \end{bmatrix} \right\}$$

is a basis of R^3.

As demonstrated in E1.7, to determine whether $D = \{\mathbf{v}_1, \mathbf{v}_2, \ldots, \mathbf{v}_n\}$ is a basis for R^3, one must show that the matrix equation $A\mathbf{x} = \mathbf{v}$ has a unique solution for all $\mathbf{v}$ where the columns of A are the triples $[\mathbf{v}_1]_D$, $[\mathbf{v}_2]_D, \ldots, [\mathbf{v}_n]_D$. If $n < 3$, then A has fewer columns than rows implying that for some $\mathbf{v} \in R^3$, $A\mathbf{x} = \mathbf{v}$ has no solution (see Appendix 2). If $n > 3$, then A has more columns than rows implying that solutions are not unique. Hence, unless $n = 3$ there is no possibility that D is a basis. Furthermore, if $n = 3$, $A\mathbf{x} = \mathbf{v}$ has a unique solution if and only if $\det(A) \neq 0$ (that is, when A^{-1} exists (see Appendix 2)). In summary, we have the following theorem:

(T1.8) Theorem: All bases for R^3 have 3 vectors. Furthermore, $D = \{\mathbf{v}_1, \mathbf{v}_2, \mathbf{v}_3\}$ is a basis for R^3 if and only if $\det(A) \neq 0$ where A is the 3×3 matrix with columns $[\mathbf{v}_1]_D$, $[\mathbf{v}_2]_D$ and $[\mathbf{v}_3]_D$.

The first statement in Theorem T1.8 can be generalized to any vector space that has a finite set of vectors as a basis. That is, if V is a vector space with a basis consisting of n vectors, then every basis for V must have exactly n vectors. In this case we say that the ***dimension of V is n***. Hence S and R^3 both have dimension 3.

(P1.6) Problem: Determine which of the following sets of vectors qualify as a basis for R^3.

(a) $D = \left\{ \begin{bmatrix} 2 \\ 1 \\ 0 \end{bmatrix}, \begin{bmatrix} -1 \\ -1 \\ 0 \end{bmatrix}, \begin{bmatrix} -1 \\ -1 \\ -1 \end{bmatrix} \right\}$ (b) $D = \left\{ \begin{bmatrix} 1 \\ 0 \\ 0 \end{bmatrix}, \begin{bmatrix} 0 \\ 1 \\ 0 \end{bmatrix}, \begin{bmatrix} 0 \\ 0 \\ 1 \end{bmatrix} \right\}$

(c) $D = \left\{ \begin{bmatrix} 2 \\ 1 \\ 0 \end{bmatrix}, \begin{bmatrix} -1 \\ 1 \\ 0 \end{bmatrix}, \begin{bmatrix} 0 \\ 0 \\ 1 \end{bmatrix} \right\}$ (d) $D = \left\{ \begin{bmatrix} 1 \\ -1 \\ 0 \end{bmatrix}, \begin{bmatrix} -2 \\ 2 \\ 0 \end{bmatrix}, \begin{bmatrix} 0 \\ 0 \\ 1 \end{bmatrix} \right\}$.

The one-to-one correspondence between S and R^3: Let $D = \{\mathbf{a},\mathbf{b},\mathbf{c}\}$ denote a basis for S and let $\mathbf{w}$ denote any vector in S. Then

$$[\mathbf{w}]_D = \begin{bmatrix} w_1 \\ w_2 \\ w_2 \end{bmatrix},$$

where the unique representation of $\mathbf{w}$ as a linear combination of the vectors of D is $\mathbf{w} = w_1\mathbf{a} + w_2\mathbf{b} + w_3\mathbf{c}$. Hence corresponding to each vector in S, there is a unique triple $[\mathbf{w}]_D$ in R^3. Conversely, each vector in R^3 is the triple of some linear combination of D and hence gives rise to a unique vector in S. Consequently, we have the one-to-one correspondence between S and R^3 given by

$$\mathbf{w} \longleftrightarrow [\mathbf{w}]_D ,$$

or written another way:

$$w_1\mathbf{a} + w_2\mathbf{b} + w_3\mathbf{c} \longleftrightarrow \begin{bmatrix} w_1 \\ w_2 \\ w_3 \end{bmatrix}.$$

This one-to-one correspondence is the fundamental reason why R^3 can be used to model three-dimensional geometric space S.

We have already seen that S and R^3 are both three-dimensional vector spaces. As such, each has a vector addition and scalar product defined on it. One important property of the $\mathbf{w} \leftrightarrow [\mathbf{w}]_D$ correspondence is that it "preserves" these vector space operations. For example, if $\mathbf{u}$, $\mathbf{v} \in S$, then $\mathbf{u} + \mathbf{v}$ corresponds to $[\mathbf{u}]_D + [\mathbf{v}]_D$. Hence, to add two vectors in S we first find their corresponding triples, add the triples and then convert that sum into its corresponding vector in S. This eliminates the necessity of using the clumsy parallelogram rule for adding geometric vectors. Similarly, if $\mathbf{u} \in S$ and $x \in R$, then $x\mathbf{u}$ corresponds to $x[\mathbf{u}]_D$. To formalize and justify these remarks we have the following theorem.

(T1.9) Theorem: Let $D = \{\mathbf{a},\mathbf{b},\mathbf{c}\}$ denote a basis of S. Then the following two statements are true:

(1) $[\mathbf{u} + \mathbf{v}]_D = [\mathbf{u}]_D + [\mathbf{v}]_D$ for all $\mathbf{u}, \mathbf{v} \in S$;

(2) $[x\mathbf{u}]_D = x[\mathbf{u}]_D$ for all $x \in R$ and $\mathbf{u} \in S$.

Proof: Let $\mathbf{u}, \mathbf{v} \in S$. Since D is a basis of S, there exist real numbers u_1, u_2, u_3, v_1, v_2 and v_3 such that $\mathbf{u} = u_1\mathbf{a} + u_2\mathbf{b} + u_3\mathbf{c}$ and $\mathbf{v} = v_1\mathbf{a} + v_2\mathbf{b} + v_3\mathbf{c}$. Hence

$$\begin{aligned} \mathbf{u} + \mathbf{v} &= (u_1\mathbf{a} + u_2\mathbf{b} + u_3\mathbf{c}) + (v_1\mathbf{a} + v_2\mathbf{b} + v_3\mathbf{c}) \\ &= (u_1 + v_1)\mathbf{a} + (u_2 + v_2)\mathbf{b} + (u_3 + v_3)\mathbf{c}\ . \end{aligned} \tag{1.6}$$

In R^3 we have

$$[u]_D = \begin{bmatrix} u_1 \\ u_2 \\ u_3 \end{bmatrix}, \quad [v]_D = \begin{bmatrix} v_1 \\ v_2 \\ v_3 \end{bmatrix} \quad \text{and} \quad [u]_D + [v]_D = \begin{bmatrix} u_1 + v_1 \\ u_2 + v_2 \\ u_3 + v_3 \end{bmatrix}.$$

But according to equation (1.6), this last triple is also $[\mathbf{u} + \mathbf{v}]_D$. Hence $[\mathbf{u} + \mathbf{v}]_D = [\mathbf{u}]_D + [\mathbf{v}]_D$. □

(P1.7) Problem: Prove part (2) of T1.9.

(P1.8) Problem - Calculating vectors in the lattice of α-quartz: In E1.5 we found the triple corresponding to several vectors labeled in Figure 1.14. Estimate $\mathbf{r} + \mathbf{u}$ using the parallelogram rule and then determine $[\mathbf{r} + \mathbf{u}]_D$ from the figure. Now calculate $[\mathbf{r} + \mathbf{u}]_D$ using Theorem T1.9 and compare your answers. Use the theorem to calculate each of the following:

(a) $[-\mathbf{r}]_D$

(b) $[6\mathbf{r} + 2\mathbf{u}]_D$

(c) $[3\mathbf{r} - 5\mathbf{u}]_D$.

Theorem T1.9 shows that, as vector spaces, S and R^3 are identical. The mathematical statement of this fact is that S and R^3 are isomorphic and that the mapping of S to R^3 that takes $\mathbf{w}$ to $[\mathbf{w}]_D$ is an isomorphism.

If we choose an origin $\mathbf{0}$ and a basis $D = \{\mathbf{a},\mathbf{b},\mathbf{c}\}$ in S, then as observed earlier the vectors $\mathbf{0}$, $\mathbf{a}$, $\mathbf{b}$ and $\mathbf{c}$ are fundamental to establishing a frame of reference on S. To find the triples that correspond to these vectors, we observe that $\mathbf{0} = 0\mathbf{a} + 0\mathbf{b} + 0\mathbf{c}$, $\mathbf{a} = 1\mathbf{a} + 0\mathbf{b} + 0\mathbf{c}$, $\mathbf{b} = 0\mathbf{a} + 1\mathbf{b} + 0\mathbf{c}$ and $\mathbf{c} = 0\mathbf{a} + 0\mathbf{b} + 1\mathbf{c}$. Hence

$$[\mathbf{0}]_D = \begin{bmatrix} 0 \\ 0 \\ 0 \end{bmatrix}, \quad [\mathbf{a}]_D = \begin{bmatrix} 1 \\ 0 \\ 0 \end{bmatrix}, \quad [\mathbf{b}]_D = \begin{bmatrix} 0 \\ 1 \\ 0 \end{bmatrix}, \quad [\mathbf{c}]_D = \begin{bmatrix} 0 \\ 0 \\ 1 \end{bmatrix}.$$

In particular, the basis of R^3 that corresponds to D is

$$\left\{ \begin{bmatrix} 1 \\ 0 \\ 0 \end{bmatrix}, \begin{bmatrix} 0 \\ 1 \\ 0 \end{bmatrix}, \begin{bmatrix} 0 \\ 0 \\ 1 \end{bmatrix} \right\}. \tag{1.7}$$

Consequently, no matter how peculiar the geometric relationship between **a**, **b** and **c** may be, the corresponding triples in R^3 are very simple. This simplicity is a great help in conveniently describing features in S with respect to the established frame of reference. For example, the set of all vectors in the space lattice defined by D is represented in R^3 as the set Z^3 of all triples with integer entries. However, caution must be exercised when lengths of vectors and angles between vectors are considered since, for example, in spite of the great similarity of the three vectors in (1.7), their corresponding vectors **a**, **b** and **c** may vary greatly in length. We shall discuss how to overcome this problem later in the chapter.

Vectors in S that lie in the direction of **a** can be written in the simple form $x\mathbf{a}$ for some $x \in R$. Similarly, vectors along **b** and **c** can be written in the form $y\mathbf{b}$ and $z\mathbf{c}$, respectively, where $y, z \in R$. Since $D = \{\mathbf{a},\mathbf{b},\mathbf{c}\}$ forms a basis, any vector $\mathbf{v} \in S$ can be written as a linear combination $\mathbf{v} = x\mathbf{a} + y\mathbf{b} + z\mathbf{c}$. Hence, geometrically $\mathbf{v}$ can be pictured as the sum of three vectors each in the direction of a basis vector (see Figure 1.15(a)). As the correspondence between S and R^3 suggests, we can decompose a vector in R^3 in a similar manner. Since

$$[x\mathbf{a}]_D = \begin{bmatrix} x \\ 0 \\ 0 \end{bmatrix}, \quad [y\mathbf{b}]_D = \begin{bmatrix} 0 \\ y \\ 0 \end{bmatrix}, \quad [z\mathbf{c}]_D = \begin{bmatrix} 0 \\ 0 \\ z \end{bmatrix},$$

and since these are scalar multiples of the basis vectors in (1.7), we have the correspondence pictured in Figure 1.15(b).

Coordinate Axes: Let $D = \{\mathbf{a},\mathbf{b},\mathbf{c}\}$ denote a basis for S. As in the case of α-quartz, **a**, **b** and **c** will usually be chosen so that each lies in an

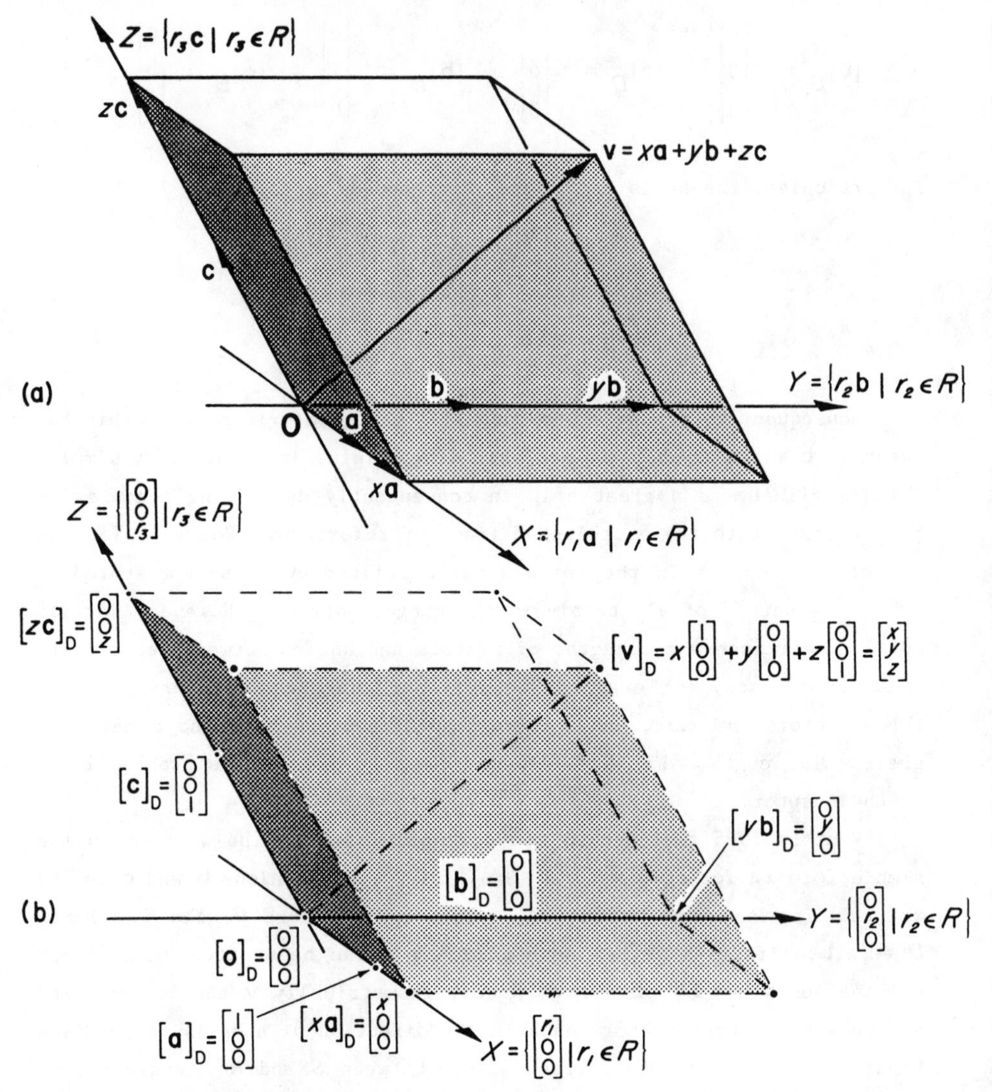

Figure 1.15. A graphical representation of the basis vectors and coordinate axes in (a) geometric space S and (b) R^3. In (a), basis vectors **a**, **b** and **c** are directed along coordinate axes X, Y and Z, respectively. In (b), the triple representatives of these vectors are displayed with $[\mathbf{a}]_D$, $[\mathbf{b}]_D$ and $[\mathbf{c}]_D$ placed along X, Y and Z. For each point $\mathbf{v} \in S$, there exists three real numbers x, y and z such that $\mathbf{v}$ can be written as $\mathbf{v} = x\mathbf{a} + y\mathbf{b} + z\mathbf{c}$. Likewise, for each vector $\mathbf{v} \in S$, there exists a triple representative of $\mathbf{v}$, $[\mathbf{v}]_D \in R^3$ such that

$$[\mathbf{v}]_D = x[\mathbf{a}]_D + y[\mathbf{b}]_D + z[\mathbf{c}]_D$$

The purpose of this figure is to show the correspondence that exists between vectors $v \in S$ and $[\mathbf{v}]_D \in R^3$. No geometrical significance should be attached to Figure 1.15(b) other than as a model for Figure 1.15(a). The vectors in S have length and direction and an angle can be defined between any two nonzero vectors in S. On the other hand, as R^3 is just a set of triples, there is no intrinsic meaning to the notions of the length of a vector or the angles between vectors in R^3.

important direction relative to the structure of the crystal under study. Since these directions will be important to us, we call the set of all points lying along them coordinate axes. For example, the X-axis is defined to be the set of points lying on a line passing through **0** and including **a**. That is, the X-axis is the set of vectors $\{r_1\mathbf{a} \mid r_1 \varepsilon R\}$ and is shown in Figure 1.15(a). The Y-axis is defined to be $\{r_2\mathbf{b} \mid r_2 \varepsilon R\}$ and the Z-axis is defined to be $\{r_3\mathbf{c} \mid r_3 \varepsilon R\}$. As the X-axis consists of the vectors $\mathbf{r} = r_1\mathbf{a} + r_2\mathbf{b} + r_3\mathbf{c}$ where $r_2 = r_3 = 0$, the triples for the points on the X-axis in R^3 are (see Figure 1.15(b)):

$$X = \left\{ \begin{bmatrix} r_1 \\ 0 \\ 0 \end{bmatrix} \mid r_1 \varepsilon R \right\} .$$

Similarly, the triples for the points on the Y- and Z-axes in R^3 are

$$Y = \left\{ \begin{bmatrix} 0 \\ r_2 \\ 0 \end{bmatrix} \mid r_2 \varepsilon R \right\} \text{ and } Z = \left\{ \begin{bmatrix} 0 \\ 0 \\ r_3 \end{bmatrix} \mid r_3 \varepsilon R \right\} ,$$

respectively. By convention, the interaxial angles between these coordinate axes are denoted $\alpha = \angle(Y:Z)$, $\beta = \angle(Z:X)$ and $\gamma = \angle(X:Y)$.

LENGTHS AND ANGLES

Inner Product: In geometric space S, the length of a given vector or the angle between two vectors can actually be measured. However, a vector in R^3 has no intrinsic length nor is there a natural way to define the angle between two vectors in R^3. Since the objective is to model the real world by R^3, a method must be devised so that the length of a geometric vector **v** and the angle between such vectors can be calculated from the coordinates in R^3. Consequently, the lengths assigned to vectors in R^3 must be based on information about the geometry of the basis D. The necessary information is best described by the inner product (this product is also referred to as the *dot product*). In geometric space, the *inner product* of two vectors **v** and **w** is defined to be

$$\mathbf{v} \bullet \mathbf{w} = vw\cos\theta , \tag{1.8}$$

where θ is the angle between **v** and **w** such that $0° \leq \theta \leq 180°$ and where

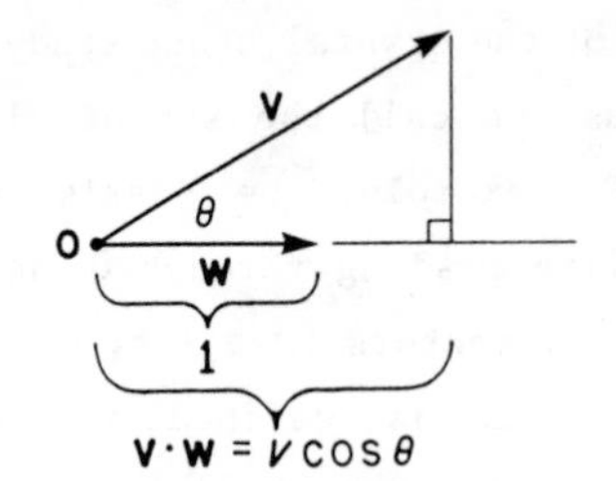

Figure 1.16: The projected length of a vector $\mathbf{v}$ on a unit vector $\mathbf{w}$ is given by $v\cos\theta$, where $\theta = \angle(\mathbf{v}:\mathbf{w})$.

v and w denote the magnitudes of $\mathbf{v}$ and $\mathbf{w}$, respectively.

Some useful properties of the inner product include the following. For all vectors $\mathbf{u}, \mathbf{v}, \mathbf{w} \in S$ and all real numbers $r \in R$:

(1) $\mathbf{v} \bullet \mathbf{v} \geq 0$ with equality only when $\mathbf{v} = 0$;

(2) $\mathbf{u} \bullet (\mathbf{v} + \mathbf{w}) = (\mathbf{u} \bullet \mathbf{v}) + (\mathbf{u} \bullet \mathbf{w})$ (Left Distributive Law);

(3) $(\mathbf{u} + \mathbf{v}) \bullet \mathbf{w} = (\mathbf{u} \bullet \mathbf{w}) + (\mathbf{v} \bullet \mathbf{w})$ (Right Distributive Law);

(4) $\mathbf{u} \bullet \mathbf{v} = \mathbf{v} \bullet \mathbf{u}$ (Commutative Law);

(5) $r(\mathbf{u} \bullet \mathbf{v}) = (r\mathbf{u}) \bullet \mathbf{v} = \mathbf{u} \bullet (r\mathbf{v})$. (1.9)

If $\mathbf{u} \bullet \mathbf{v} = 0$, with $u \neq 0$ and $v \neq 0$, then $\mathbf{u}$ and $\mathbf{v}$ are perpendicular. Also, note that $\mathbf{v} \bullet \mathbf{v} = v^2$ because $\cos(0°) = 1$.

Suppose that $\mathbf{w}$ is a unit length vector. Then $\mathbf{v} \bullet \mathbf{w} = v\cos\theta$ is the projection of $\mathbf{v}$ onto the direction of $\mathbf{w}$ (see Figure 1.16). In general $(\mathbf{v} \bullet \mathbf{w})/w$ is the projection of $\mathbf{v}$ onto the direction of $\mathbf{w}$.

Metrical Matrix: Suppose that $D = \{\mathbf{a},\mathbf{b},\mathbf{c}\}$ is a basis for S and suppose that $\mathbf{v}, \mathbf{w} \in S$ such that

$$\mathbf{v} = v_1\mathbf{a} + v_2\mathbf{b} + v_3\mathbf{c} \quad \text{and} \quad \mathbf{w} = w_1\mathbf{a} + w_2\mathbf{b} + w_3\mathbf{c} \ .$$

Then $\mathbf{v} \bullet \mathbf{w}$ can be found as follows

$$\begin{aligned} \mathbf{v} \bullet \mathbf{w} &= (v_1\mathbf{a} + v_2\mathbf{b} + v_3\mathbf{c}) \bullet (w_1\mathbf{a} + w_2\mathbf{b} + w_3\mathbf{c}) \\ &+ [(v_1\mathbf{a} + v_2\mathbf{b} + v_3\mathbf{c}) \bullet w_1\mathbf{a}] + [(v_1\mathbf{a} + v_2\mathbf{b} + v_3\mathbf{c}) \bullet w_2\mathbf{b}] \\ &+ [(v_1\mathbf{a} + v_2\mathbf{b} + v_3\mathbf{c}) \bullet w_3\mathbf{c}] \ . \end{aligned}$$

(Property 2 of (1.9))

Hence,

$$\begin{aligned} \mathbf{v} \bullet \mathbf{w} &= [(v_1\mathbf{a}) \bullet (w_1\mathbf{a})] + [(v_2\mathbf{b}) \bullet (w_1\mathbf{a})] + [(v_3\mathbf{c}) \bullet (w_1\mathbf{a})] \\ &+ [(v_1\mathbf{a}) \bullet (w_2\mathbf{b})] + [(v_2\mathbf{b}) \bullet (w_2\mathbf{b})] + [(v_3\mathbf{c}) \bullet (w_2\mathbf{b})] \\ &+ [(v_1\mathbf{a}) \bullet (w_3\mathbf{c})] + [(v_2\mathbf{b}) \bullet (w_3\mathbf{c})] + [(v_3\mathbf{c}) \bullet (w_3\mathbf{c})] \ . \end{aligned}$$

(Property 3 of (1.9))

Therefore,

$$
\begin{aligned}
\mathbf{v} \bullet \mathbf{w} &= v_1w_1(\mathbf{a} \bullet \mathbf{a}) + v_2w_1(\mathbf{b} \bullet \mathbf{a}) + v_3w_1(\mathbf{c} \bullet \mathbf{a}) \\
&+ v_1w_2(\mathbf{a} \bullet \mathbf{b}) + v_2w_2(\mathbf{b} \bullet \mathbf{b}) + v_3w_2(\mathbf{c} \bullet \mathbf{b}) \\
&+ v_1w_3(\mathbf{a} \bullet \mathbf{c}) + v_2w_3(\mathbf{b} \bullet \mathbf{c}) + v_3w_3(\mathbf{c} \bullet \mathbf{c}) \,.
\end{aligned} \tag{1.10}
$$

(Property 4 of (1.9))

Hence the inner product of any two vectors can be calculated by evaluating the inner products of the basis vectors with each other. Equation (1.10) is somewhat unwieldy but recasting it in matrix form, it can be simplified (see Appendix 2 for matrix review) and written as

$$
\mathbf{v} \bullet \mathbf{w} = [v_1 v_2 v_3] \begin{bmatrix} w_1\mathbf{a} \bullet \mathbf{a} & w_2\mathbf{a} \bullet \mathbf{b} & w_3\mathbf{a} \bullet \mathbf{c} \\ w_1\mathbf{b} \bullet \mathbf{a} & w_2\mathbf{b} \bullet \mathbf{b} & w_3\mathbf{b} \bullet \mathbf{c} \\ w_1\mathbf{c} \bullet \mathbf{a} & w_2\mathbf{c} \bullet \mathbf{b} & w_3\mathbf{c} \bullet \mathbf{c} \end{bmatrix} ,
$$

which yields

$$
\mathbf{v} \bullet \mathbf{w} = [v_1 v_2 v_3] \begin{bmatrix} \mathbf{a} \bullet \mathbf{a} & \mathbf{a} \bullet \mathbf{b} & \mathbf{a} \bullet \mathbf{c} \\ \mathbf{b} \bullet \mathbf{a} & \mathbf{b} \bullet \mathbf{b} & \mathbf{b} \bullet \mathbf{c} \\ \mathbf{c} \bullet \mathbf{a} & \mathbf{c} \bullet \mathbf{b} & \mathbf{c} \bullet \mathbf{c} \end{bmatrix} \begin{bmatrix} w_1 \\ w_2 \\ w_3 \end{bmatrix} . \tag{1.11}
$$

The 3×3 matrix in (1.11) is called the *metrical matrix* (also called the *metric tensor*) and is denoted as

$$
G = \begin{bmatrix} g_{11} & g_{12} & g_{13} \\ g_{21} & g_{22} & g_{23} \\ g_{31} & g_{32} & g_{33} \end{bmatrix} = \begin{bmatrix} \mathbf{a} \bullet \mathbf{a} & \mathbf{a} \bullet \mathbf{b} & \mathbf{a} \bullet \mathbf{c} \\ \mathbf{b} \bullet \mathbf{a} & \mathbf{b} \bullet \mathbf{b} & \mathbf{b} \bullet \mathbf{c} \\ \mathbf{c} \bullet \mathbf{a} & \mathbf{c} \bullet \mathbf{b} & \mathbf{c} \bullet \mathbf{c} \end{bmatrix} . \tag{1.12}
$$

Note that since the inner product is commutative, G is a *symmetric* matrix. That is, $g_{ij} = g_{ji}$ for all i and j. With this notation, the inner product in (1.11) can be written compactly as

$$
\mathbf{v} \bullet \mathbf{w} = [\mathbf{v}]_D^t G [\mathbf{w}]_D , \tag{1.13}
$$

where $[\mathbf{v}]_D^t$ is the transpose $[v_1 v_2 v_3]$ of $[v]_D$. A special case of the inner product is when the basis vectors form a cartesian basis yielding

$$
G = \begin{bmatrix} \mathbf{i} \bullet \mathbf{i} & \mathbf{i} \bullet \mathbf{j} & \mathbf{i} \bullet \mathbf{k} \\ \mathbf{j} \bullet \mathbf{i} & \mathbf{j} \bullet \mathbf{j} & \mathbf{j} \bullet \mathbf{k} \\ \mathbf{k} \bullet \mathbf{i} & \mathbf{k} \bullet \mathbf{j} & \mathbf{k} \bullet \mathbf{k} \end{bmatrix} = \begin{bmatrix} 1 & 0 & 0 \\ 0 & 1 & 0 \\ 0 & 0 & 1 \end{bmatrix} = I_3 .
$$

Thus for a cartesian basis C, (1.13) becomes the more familiar inner product

$$\mathbf{v} \bullet \mathbf{w} = [\mathbf{v}]_C^t I_3 [\mathbf{W}]_C = [\mathbf{v}]_C^t [\mathbf{w}]_C \quad (1.14)$$

$$= [v_1 v_2 v_3] \begin{bmatrix} w_1 \\ w_2 \\ w_3 \end{bmatrix}$$

$$= v_1 w_1 + v_2 w_2 + v_3 w_3 \ .$$

(E1.10) Example - Calculation of bond lengths and angles for H_4SiO_4, a molecule with an atom at the origin: A structural analysis of the monosilicic acid molecule (Figure 1.1), using molecular orbital methods yields the cartesian coordinates of the Si, O and H atoms listed in Table 1.1. Calculate the bond lengths between Si and O_3 and between Si and O_4, denoted $R(SiO_3)$ and $R(SiO_4)$, respectively, and the angle between the SiO_3 and SiO_4 bonds denoted $<(O_3SiO_4)$.

Table 1.1: The atomic coordinates for H_4SiO_4.

Atom	x	y	z	Atom	x	y	z
Si	0.0	0.0	0.0	H_1	1.875	−0.237	1.089
O_1	1.281	0.466	0.877	H_2	−0.237	−1.875	−1.089
O_2	0.466	−1.281	−0.877	H_3	−1.875	0.237	1.089
O_3	−1.281	−0.466	0.877	H_4	0.237	1.875	−1.089
O_4	−0.466	1.281	−0.877				

Solution: In this example, the basis vectors are cartesian with unit lengths of 1A, i.e., $\|\mathbf{i}\| = \|\mathbf{j}\| = \|\mathbf{k}\| = 1A$. Because the Si atom was chosen at the origin, the distance between Si and O_3 is found by evaluating the length of the vector $\mathbf{r}_3 = -1.281\mathbf{i} - 0.466\mathbf{j} + 0.877\mathbf{k}$ that radiates from Si to O_3. Forming the inner product $\mathbf{r}_3 \bullet \mathbf{r}_3$ (Equation 1.14), we get

$$\mathbf{r}_3 \bullet \mathbf{r}_3 = r_3^2 = [\mathbf{r}_3]_C^t [\mathbf{r}_3]_C$$

$$= [-1.281 \quad -0.466 \quad 0.877] \begin{bmatrix} -1.281 \\ -0.466 \\ 0.877 \end{bmatrix} = 2.627A^2 \ .$$

By taking the square root of this number, the length of $\mathbf{r}_3$ is found, i.e., r_3 = 1.621A and so R(SiO_3) = 1.621A.

To determine θ_{34} = <(O_3SiO_4), the angle between the SiO_3 and SiO_4 bonds, we calculate $\mathbf{r}_3 \bullet \mathbf{r}_4$ where

$$\mathbf{r}_4 = -0.466\mathbf{i} + 1.281\mathbf{j} - 0.877\mathbf{k}$$

is the vector that radiates from Si to O_4 and solve for θ_{34} in (see (1.8))

$$\mathbf{r}_3 \bullet \mathbf{r}_4 = r_3 r_4 \cos\theta_{34}$$

Hence

$$\cos\theta_{34} = \mathbf{r}_3 \bullet \mathbf{r}_4/(r_3 r_4)$$

$$= ([\mathbf{r}_3]_C^t[\mathbf{r}_4]_C)/(([\mathbf{r}_3]_C^t[\mathbf{r}_3]_C)^{\frac{1}{2}}([\mathbf{r}_4]_C^t[\mathbf{r}_4]_C)^{\frac{1}{2}}) ,$$

where

$$[\mathbf{r}_3]_C^t[\mathbf{r}_4]_C = [-1.281 \quad -0.466 \quad 0.877]\begin{bmatrix}-0.466\\1.281\\-0.877\end{bmatrix} = -0.769A^2 ,$$

and

$$[\mathbf{r}_4]_C^t[\mathbf{r}_4]_C = [-0.466 \quad 1.281 \quad -0.877]\begin{bmatrix}-0.466\\1.281\\-0.877\end{bmatrix} = 2.627A^2 .$$

Note that the distance between Si and O_4 is equal to r_4 = $(2.627)^{\frac{1}{2}}$ = 1.621A, which is also equal to r_3. Thus, $\cos\theta_{34}$ = -0.769/((1.621)(1.621)) = -0.293 and the angle between the SiO_3 and SiO_4 bonds is <(O_3SiO_4) = 107.0°. □

(P1.9) Problem: Calculate R(SiO_1) and R(SiO_2) and <(O_1SiO_2) for H_4SiO_4, using the coordinates given in Table 1.1.

(E1.12) Example - A calculation of bond lengths for $H_6Si_3O_3$, a molecule lacking an atom at the origin: A structure analysis of $H_6Si_3O_3$ by O'Keeffe and Gibbs (1984) yielded the cartesian coordinates in Table 1.2. In this example, we will calculate the SiO bond lengths R(Si_1O_1) and R(Si_2O_1).

Table 1.2: Coordinates of the atoms for tricyclosiloxane, $H_6Si_3O_3$.

Atom	x	y	z	Atom	x	y	z
Si_1	1.7668	0.0	0.0	H_1	2.5940	0.0	1.2098
Si_2	-0.8833	1.5301	0.0	H_2	2.5940	0.0	-1.2098
Si_3	-0.8833	-1.5301	0.0	H_3	-1.2970	2.2465	1.2098
O_1	0.7456	1.2914	0.0	H_4	-1.2970	2.2465	-1.2098
O_2	-1.4912	0.0	0.0	H_5	-1.2970	-2.2465	1.2098
O_3	0.7456	-1.2914	0.0	H_6	-1.2970	-2.2465	-1.2098

Solution: The length of the Si_1O_1 bond equals the magnitude of a vector denoted $\mathbf{r}_1$ drawn from O_1 to Si_1. This vector is found by simply subtracting from the vector $\mathbf{v}_1 = 1.7668\mathbf{i}$ (a vector drawn from the origin $\mathbf{0}$ to Si_1) the vector $\mathbf{v}_2 = 0.7456\mathbf{i} + 1.2914\mathbf{j}$ (a vector drawn from $\mathbf{0}$ to O_1), i.e.,

$$\mathbf{r}_1 = (1.7668 - 0.7456)\mathbf{i} - 1.2914\mathbf{j}$$
$$= 1.0212\mathbf{i} - 1.2914\mathbf{j} \ .$$

Next, evaluating the inner product

$$\mathbf{r}_1 \bullet \mathbf{r}_1 = [1.0212 \quad -1.2914 \quad 0.0] \begin{bmatrix} 1.0212 \\ -1.2914 \\ 0.0 \end{bmatrix} = 2.711A^2$$

and taking the square root of the result, we find that $R(Si_1O_1) = 1.646A$. The distance between Si_2 and O_1 is found by calculating the length of $\mathbf{r}_2$, a vector that radiates from O_1 to Si_2. The vector $\mathbf{r}_2$ is found by subtracting $\mathbf{v}_2$ from $\mathbf{v}_3 = -0.8833\mathbf{i} + 1.5301\mathbf{j}$ (a vector taken to radiate from $\mathbf{0}$ to Si_2); i.e.,

$$\mathbf{r}_2 = (-0.8833 - 0.7456)\mathbf{i} + (1.5301 - 1.2914)\mathbf{j}$$
$$= -1.6289\mathbf{i} + 0.2387\mathbf{j} \ .$$

Evaluating $(\mathbf{r}_2 \bullet \mathbf{r}_2)^{\frac{1}{2}}$, we find that $R(Si_2O_1)$ also equals 1.646A.

If you had difficulty following the vector assignment made in this problem, use the coordinates in Table 1.2 and make a careful drawing of

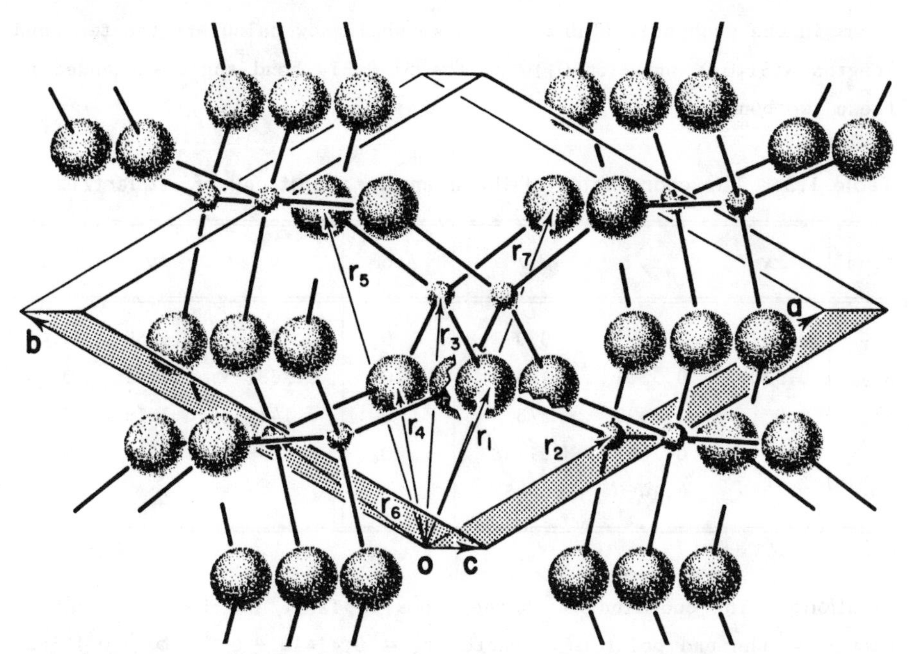

Figure 1.17: A drawing of a block of the α-quartz structure. The parallelepiped in the drawing outlined by $D = \{\mathbf{a},\mathbf{b},\mathbf{c}\}$ contains a total of three Si atoms and six oxygen atoms for a total of three SiO_2 formula units per unit cell volume. The coordinates of the atoms in the unit cell are given in Table 1.3. The oxygen atoms O_1, O_3, O_4, O_6 are located at the end points of $\mathbf{r}_1$, $\mathbf{r}_7$, $\mathbf{r}_4$, $\mathbf{r}_5$, respectively, and the Si atoms Si_1, Si_2, Si_3 are located at the end points of $\mathbf{r}_2$, $\mathbf{r}_3$ and $\mathbf{r}_6$, respectively.

the structure of the molecule viewed down the Z-axis perpendicular to the ring. Construct the line segments $\mathbf{v}_1$, $\mathbf{v}_2$ and $\mathbf{v}_3$ directed from the origin, $\mathbf{O}$, to Si_1, O_1 and Si_2, respectively, and then construct $\mathbf{r}_1$ directed from O_1 to Si_1 and $\mathbf{r}_2$ directed from O_1 to Si_2. With this construction, it should be apparent that $\mathbf{r}_1 = \mathbf{v}_1 - \mathbf{v}_2$ and $\mathbf{r}_2 = \mathbf{v}_3 - \mathbf{v}_2$□

(P1.10) Problem: Calculate $\angle(Si_1O_1Si_2)$, $R(Si_2H_3)$, $R(Si_2H_4)$ and $\angle(H_3Si_2H_4)$ for tricyclosiloxane, using the cartesian coordinates given in Table 1.2.

(E1.13) Example - Calculation of bond distances and angles for α-quartz: Earlier in this chapter we observed that the atoms in a crystal like α-quartz are repeated along numerous directions to form a periodic pattern in three dimensions. We also observed that the repeat units along three of these directions can be represented by three vectors $D = \{\mathbf{a},\mathbf{b},\mathbf{c}\}$ that outline a unit cell that contains at least one complete unit of the repeating pattern of the structure (Figure 1.17). A structural analysis of α-quartz by Levien *et al.* (1980) yielded the atomic coordinates of the atoms in one such cell (Table 1.3). We shall now calculate the two bond lengths $R(Si_1O_1)$ and $R(Si_2O_1)$ and the $Si_1O_1Si_2$ bond angle subtended by these two bonds.

Table 1.3: The coordinates of the atoms in a unit cell of α-quartz.

Atom	*x*	*y*	*z*	Atom	*x*	*y*	*z*
Si_1	0.4699	0.0	0.0	O_3	0.8540	0.5859	0.4521
Si_2	0.5301	0.5301	1/3	O_4	0.2681	0.4141	0.5479
Si_3	0.0	0.4699	2/3	O_5	0.1460	0.7319	0.8812
O_1	0.4141	0.2681	0.1188	O_6	0.5859	0.8540	0.2145
O_2	0.7319	0.1460	0.7855				

Solution: The coordinates of the atoms in Table 1.3 indicate that O_1 occurs at the end point of a vector $\mathbf{r}_1 = 0.4141\mathbf{a} + 0.2681\mathbf{b} + 0.1188\mathbf{c}$, that Si_1 occurs at the end of $\mathbf{r}_2 = 0.4699\mathbf{a}$ and that Si_2 occurs at the end of $\mathbf{r}_3 = 0.5301\mathbf{a} + 0.5301\mathbf{b} + (1/3)\mathbf{c}$ (see Figure 1.17). $R(Si_1O_1)$ is equal to the length of the vector

$$\mathbf{r}_2 - \mathbf{r}_1 = 0.0558\mathbf{a} - 0.2681\mathbf{b} - 0.1188\mathbf{c}$$

and $R(Si_2O_1)$ is equal to the length of the vector

$$\mathbf{r}_3 - \mathbf{r}_1 = 0.1160\mathbf{a} + 0.2620\mathbf{b} + 0.2145\mathbf{c}\ .$$

To calculate these bond lengths, the metrical matrix must be constructed (Equation 1.12). Since $\alpha = \beta = 90°$, $\mathbf{a} \cdot \mathbf{c} = \mathbf{b} \cdot \mathbf{c} = \mathbf{c} \cdot \mathbf{a} = \mathbf{c} \cdot \mathbf{b} = 0$. Also, since $a = b = 4.914$A, $c = 5.409$A and $\gamma = 120°$, the remaining inner products are

$$\begin{aligned} \mathbf{b} \cdot \mathbf{a} &= \mathbf{a} \cdot \mathbf{b} = ab\cos(120°) = -12.0737 \\ \mathbf{a} \cdot \mathbf{a} &= a^2 = 24.1474 \\ \mathbf{b} \cdot \mathbf{b} &= b^2 = 24.1474 \\ \mathbf{c} \cdot \mathbf{c} &= c^2 = 29.2573\ . \end{aligned}$$

Hence, the metrical matrix G for α-quartz is

$$G = \begin{bmatrix} 24.1474 & -12.0737 & 0 \\ -12.0737 & 24.1474 & 0 \\ 0 & 0 & 29.2573 \end{bmatrix} .$$

When $\mathbf{v}$ and $\mathbf{w}$ in Equation 1.13 are both replaced by $\mathbf{r}_2 - \mathbf{r}_1$, we have

$$R(Si_1O_1)^2 = (\mathbf{r}_2 - \mathbf{r}_1) \bullet (\mathbf{r}_2 - \mathbf{r}_1) = [\mathbf{r}_2 - \mathbf{r}_1]_D^t G[\mathbf{r}_2 - \mathbf{r}_1]_D$$

$$= [0.0558 \quad -0.2681 \quad -0.1188]G \begin{bmatrix} 0.0558 \\ -0.2681 \\ -0.1188 \end{bmatrix}$$

$$= [0.0558 \quad -0.2681 \quad -0.1188] \begin{bmatrix} 4.5844 \\ -7.1476 \\ -3.4758 \end{bmatrix} = 2.5850A^2 ,$$

and so $R(Si_1O_1) = 1.608A$.

Likewise, replacing $\mathbf{v}$ and $\mathbf{w}$ in Equation 1.13 by $\mathbf{r}_3 - \mathbf{r}_1$, we obtain

$$R(Si_2O_1)^2 = (\mathbf{r}_3 - \mathbf{r}_1) \bullet (\mathbf{r}_3 - \mathbf{r}_1) = [\mathbf{r}_3 - \mathbf{r}_1]_D^t G[\mathbf{r}_3 - \mathbf{r}_1]_D$$

$$= [0.1160 \quad 0.2620 \quad 0.2145]G \begin{bmatrix} 0.1160 \\ 0.2620 \\ 0.2145 \end{bmatrix} = 2.595A^2 ,$$

and so $R(Si_2O_1) = 1.611A$.

To calculate the angle $\theta = \angle(Si_1O_1Si_2)$, the inner product $(\mathbf{r}_2 - \mathbf{r}_1) \bullet (\mathbf{r}_3 - \mathbf{r}_1)$ is required because

$$(\mathbf{r}_2 - \mathbf{r}_1) \bullet (\mathbf{r}_3 - \mathbf{r}_1) = R(Si_1O_1)R(Si_2O_1)\cos\theta .$$

Since

$$(\mathbf{r}_2 - \mathbf{r}_1) \bullet (\mathbf{r}_3 - \mathbf{r}_1) = [\mathbf{r}_2 - \mathbf{r}_1]_D^t G[\mathbf{r}_3 - \mathbf{r}_1]$$

$$= [0.0558 \quad -0.2681 \quad -0.1188]G \begin{bmatrix} 0.1160 \\ 0.2620 \\ 0.2145 \end{bmatrix} = -2.087A^2 ,$$

$$\cos\theta = -2.087/(R(Si_1O_1)R(Si_2O_1)) = -2.087/((1.608)(1.611)) ,$$

and

$$\theta = \cos^{-1}[-2.087/((1.608)(1.611))] = 143.7° . \qquad \square$$

(P1.11) Problem: Find four oxygen atoms in the unit cell of α-quartz that are each bonded to Si_2, i.e., the four oxygen atoms nearest to Si_2. Calculate the distances between Si_2 and each of these oxygen atoms and the six bond angles, $<(OSi_2O)$, subtended by the bonds of the SiO_4 group. Note in Figure 1.17, Si_2, O_3, O_4 and O_6 are located at the end points of $\mathbf{r}_3$, $\mathbf{r}_7$, $\mathbf{r}_4$ and $\mathbf{r}_5$, respectively.

(P1.12) Problem: Calculate the bridging $<(Si_3O_4Si_2)$ angle in α-quartz (Si_3 is located at the end point of $\mathbf{r}_6$).

In the computation of the interatomic separations and angles in α-quartz, we note that particular values are encountered again and again. This is by no means an accident of nature but a consequence of the fact that the atoms in the mineral strive to adopt a minimum energy configuration. When this condition is realized, a pattern of atoms is repeated at regular intervals in a symmetrical pattern as discussed in the introduction. As a consequence, particular bond lengths and angles in the aggregate are repeated as observed not only in the case for α-quartz but also for the molecules H_4SiO_4 and $H_6Si_3O_3$.

(P1.13) Problem: Anorthite is an important feldspar mineral of $CaAl_2Si_2O_8$ composition found in volcanic ejecta and contact metamorphic rocks. Wainwright and Starkey (1971) studied its structure using a triclinic basis $D = \{\mathbf{a},\mathbf{b},\mathbf{c}\}$ where $a = 8.173$A, $b = 12.869$A and $c = 14.165$A, $\alpha = 93.11°$, $\beta = 115.91°$ and $\gamma = 91.26°$ and determined that an oxygen atom occurs at the end point of $\mathbf{r}_1$, a silicon atom at the end point of $\mathbf{r}_2$, and an aluminum atom at the end point of $\mathbf{r}_3$, where

$$\mathbf{r}_1 = 0.3419\mathbf{a} + 0.3587\mathbf{b} + 0.1333\mathbf{c}\ ,$$
$$\mathbf{r}_2 = 0.5041\mathbf{a} + 0.3204\mathbf{b} + 0.1099\mathbf{c}\ ,$$
$$\mathbf{r}_3 = 0.1852\mathbf{a} + 0.3775\mathbf{b} + 0.1816\mathbf{c}\ .$$

Find the lengths of the SiO and AlO bonds and the SiOAl angle between them.

Cross Product: A second way to multiply vectors is to use the cross product. Unlike the inner product which yields a scalar, the cross product of two vectors yields a vector. Let $\mathbf{v}$ and $\mathbf{w}$ denote vectors in

S and let θ denote the angle between them such that $0° \le \theta \le 180°$. Let $\mathbf{N}$ denote the unit vector perpendicular to $\mathbf{v}$ and $\mathbf{w}$ so that $\{\mathbf{v},\mathbf{w},\mathbf{N}\}$ forms a right-handed system. Then the *cross product* $\mathbf{v} \times \mathbf{w}$ is defined to be

$$\mathbf{v} \times \mathbf{w} = (vw\sin\theta)\mathbf{N} . \tag{1.15}$$

Since the parallelogram with sides $\mathbf{v}$ and $\mathbf{w}$ has base v and height $w\sin\theta$, the area of the parallelogram is $vw\sin\theta$. Hence $\mathbf{v} \times \mathbf{w}$ is a vector whose length equals the area of the parallelogram with sides $\mathbf{v}$ and $\mathbf{w}$ such that $\{\mathbf{v},\mathbf{w},(\mathbf{v} \times \mathbf{w})\}$ forms a right-handed system. Note that one apparent flaw in our definition of the cross product is that if $\mathbf{v}$ and $\mathbf{w}$ are collinear then $\mathbf{N}$ is not uniquely defined. However, in this case, $\sin\theta = 0$ and so $(\mathbf{v} \times \mathbf{w}) = \mathbf{0}$ and the direction of $\mathbf{N}$ is of no importance. The cross product has the following properties for all $\mathbf{u}$, $\mathbf{v}$ and $\mathbf{w} \in S$ and for all $r \in \mathrm{R}$:

(1) $\mathbf{u} \times (\mathbf{v} + \mathbf{w}) = (\mathbf{u} \times \mathbf{v}) + (\mathbf{u} \times \mathbf{w})$ (Left Distributive Law);
(2) $(\mathbf{u} + \mathbf{v}) \times \mathbf{w} = (\mathbf{u} \times \mathbf{w}) + (\mathbf{v} \times \mathbf{w})$ (Right Distributive Law);
(3) $(\mathbf{u} \times \mathbf{v}) = -(\mathbf{v} \times \mathbf{u})$ (Anticommutative Law); (1.16)
(4) $r(\mathbf{u} \times \mathbf{v}) = (r\mathbf{u}) \times \mathbf{v} = \mathbf{u} \times (r\mathbf{v})$.

These properties are proved in any standard calculus text. In this section, we shall present a method for calculating the cross product $\mathbf{v} \times \mathbf{w}$ in the case where $\mathbf{v}$ and $\mathbf{w}$ are expressed with respect to a cartesian coordinate system $C = \{\mathbf{i},\mathbf{j},\mathbf{k}\}$. This method will not be justified here as it is discussed in the standard calculus texts. In Chapter 2 (2.23) we shall give a more general formula. If

$$\mathbf{v} = v_1\mathbf{i} + v_2\mathbf{j} + v_3\mathbf{k} \quad \text{and} \quad \mathbf{w} = w_1\mathbf{i} + w_2\mathbf{j} + w_3\mathbf{k}$$

are vectors in S, then

$$\mathbf{v} \times \mathbf{w} = \begin{vmatrix} v_2 & w_2 \\ v_3 & w_3 \end{vmatrix} \mathbf{i} - \begin{vmatrix} v_1 & w_1 \\ v_3 & w_3 \end{vmatrix} \mathbf{j} + \begin{vmatrix} v_1 & w_1 \\ v_2 & w_2 \end{vmatrix} \mathbf{k} ,$$

which in determinant form equals

$$\mathbf{v} \times \mathbf{w} = \begin{vmatrix} \mathbf{i} & v_1 & w_1 \\ \mathbf{j} & v_2 & w_2 \\ \mathbf{k} & v_3 & w_3 \end{vmatrix} .$$

Since $\mathbf{v} \times \mathbf{w}$ is a vector perpendicular to $\mathbf{v}$ and $\mathbf{w}$, the cross product can be used to find a vector normal to a face or to a plane of atoms in a crystal.

(E1.14) Example - The tilt angle of a carbonate group in aragonite: The carbonate CO_3 groups in the mineral aragonite, $CaCO_3$, are arranged in corrugated hexagonal close-packed layers perpendicular to **c** (Villiers, 1971). The corrugated nature of the layer is related to a tilting of the groups out of the layer to accommodate the bonding requirements of the Ca atom. The coordinates of the three oxygen atoms comprising a carbonate group in one such layer are given in Table 1.4.

Table 1.4: Coordinates of a carbonate group in aragonite.

Atom	x	y	z
C	0.25	0.7622	−0.0862
O_1	0.25	0.9225	−0.0962
O_2	0.4736	0.6810	−0.0862
O_3	0.0264	0.6810	−0.0862

Calculate a vector perpendicular to the plane defined by the three oxygen atoms of the carbonate group and determine the tilt angle, i.e., the angle between the perpendicular to the plane and **c** given the cell dimensions $a = 4.96$A, $b = 7.97$A, $c = 5.74$A, $\alpha = \beta = \gamma = 90°$.

Solution: Choosing a cartesian basis with **i**, **j**, **k** directed along the coordinate axes X, Y, Z, respectively, we have that $\mathbf{a} = a\mathbf{i} = 4.96\mathbf{i}$, $\mathbf{b} = b\mathbf{j} = 7.97\mathbf{j}$, $\mathbf{c} = c\mathbf{k} = 5.74\mathbf{k}$. Then the vectors $\mathbf{v}_1$ and $\mathbf{v}_2$ from O_2 to O_3 and from O_2 to O_1, respectively, defined in terms of a cartesian basis are

$$\mathbf{v}_1 = (-0.4472)(4.96\mathbf{i}) = -2.218\mathbf{i}$$

$$\mathbf{v}_2 = (-0.2236)(4.96\mathbf{i}) + (0.2415)(7.97\mathbf{j}) + (-0.0100)(5.74\mathbf{k})$$
$$= -1.1091\mathbf{i} + 1.9248\mathbf{j} - 0.0574\mathbf{k} \ .$$

Hence,

$$\mathbf{v}_2 \times \mathbf{v}_1 = \begin{vmatrix} \mathbf{i} & -1.1091 & -2.218 \\ \mathbf{j} & 1.9248 & 0 \\ \mathbf{k} & -0.0574 & 0 \end{vmatrix}$$

$$= 0.1273\mathbf{j} + 4.2692\mathbf{k}$$

is a vector perpendicular to the plane of the three oxygen atoms. To find the tilt angle between $\mathbf{c}$ and $\mathbf{v} = \mathbf{v}_2 \times \mathbf{v}_1$, we evaluate

$$\cos^{-1}((\mathbf{v} \bullet \mathbf{k})/v) = \cos^{-1}(0.9996) = 1.7° \ .$$

Hence, the carbonate group is tilted out of the close-packed layer 1.7°. □

(P1.14) Problem: The structure of amphibole is based on double chains of corner sharing silicate tetrahedra that are interlocked in layers paralleling **b** and **c**. The chains are not planar but are curved away from the **b**,**c**-plane to relieve strain ascribed to a dimensional misfit between octahedral and tetrahedral layers. The coordinates of the oxygen atoms O_4, O_5, and O_6, defining the base of the Si_2 tetrahedron of the chain in protoamphibole (Gibbs, 1969) are given in Table 1.5. If the chains were planar, the plane defined by the three oxygen atoms would be perpendicular to **a**. Calculate the angle between the perpendicular and **a** to obtain an estimate of the angle of misfit given that a = 9.335, b = 17.880, c = 5.287A, $\alpha = \beta = \gamma = 90°$.

Table 1.5: Coordinates of the oxygen atoms comprising the base of a silicate group in protoamphibole.

Atom	x	y	z
O_4	0.3764	0.2494	0.6873
O_5	0.3479	0.1214	0.4276
O_6	0.3509	0.1302	0.9311

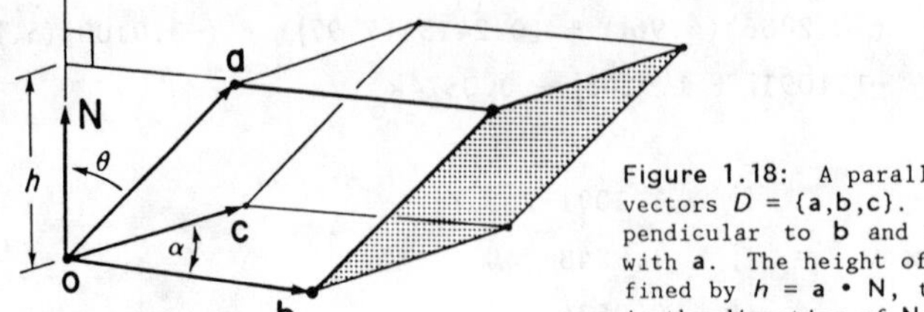

Figure 1.18: A parallelepiped outlined by basis vectors $D = \{\mathbf{a},\mathbf{b},\mathbf{c}\}$. The unit vector **N** is perpendicular to **b** and **c** and makes an angle of θ with **a**. The height of the parallelepiped is defined by $h = \mathbf{a} \cdot \mathbf{N}$, the scalar component of **a** in the direction of **N**.

Triple scalar product: A useful combination of the inner and cross products is the *triple scalar product* $\mathbf{u} \cdot (\mathbf{v} \times \mathbf{w})$. Let us obtain a formula for the product. If $\mathbf{u} = u_1\mathbf{i} + u_2\mathbf{j} + u_3\mathbf{k}$, $\mathbf{v} = v_1\mathbf{i} + v_2\mathbf{j} + v_3\mathbf{k}$ and $\mathbf{w} = w_1\mathbf{i} + w_2\mathbf{j} + w_3\mathbf{k}$, then

$$\mathbf{u} \cdot (\mathbf{v} \times \mathbf{w}) =$$

$$(u_1\mathbf{i} + u_2\mathbf{j} + u_3\mathbf{k}) \cdot \left(\begin{vmatrix} v_2 & w_2 \\ v_3 & w_3 \end{vmatrix} \mathbf{i} - \begin{vmatrix} v_1 & w_1 \\ v_2 & w_3 \end{vmatrix} \mathbf{j} + \begin{vmatrix} v_1 & w_1 \\ v_2 & w_2 \end{vmatrix} \mathbf{k} \right)$$

$$= u_1 \begin{vmatrix} v_2 & w_2 \\ v_3 & w_3 \end{vmatrix} - u_2 \begin{vmatrix} v_1 & w_1 \\ v_2 & w_3 \end{vmatrix} + u_3 \begin{vmatrix} v_1 & w_1 \\ v_2 & w_2 \end{vmatrix}$$

This can be written more concisely as

$$\mathbf{u} \cdot (\mathbf{v} \times \mathbf{w}) = \begin{vmatrix} u_1 & v_1 & w_1 \\ u_2 & v_2 & w_2 \\ u_3 & v_3 & w_3 \end{vmatrix} .$$

Thus the triple scalar product is a scalar that equals the determinant of the coordinates of three vectors written in terms of a cartesian basis. Note that that the columns of the determinant comprise the coordinates of **u**, **v** and **w** written in terms of the cartesian basis.

The triple scalar product has an important geometrical meaning with respect to a (right-handed) basis $D = \{\mathbf{a},\mathbf{b},\mathbf{c}\}$. Consider the height h of the end point of the vector **a** above the base of the parallelepiped outlined by **a**, **b** and **c** (Figure 1.18). As the volume v of such a parallelepiped is equal to the area of its base A times its height, h, we have that $V = hA$. Since $A = \|\mathbf{b} \times \mathbf{c}\| = bc\sin\alpha$ and $h = \mathbf{a} \bullet \mathbf{N}$ where $\mathbf{N}$ is a unit vector perpendicular to the $\mathbf{b}$, $\mathbf{c}$ plane,

$$
\begin{aligned}
V = \; & \mathbf{a} \bullet \mathbf{N}\|\mathbf{b} \times \mathbf{c}\| \\
& \mathbf{a} \bullet \|\mathbf{b} \times \mathbf{c}\|\mathbf{N} \\
& \mathbf{a} \bullet \mathbf{b} \times \mathbf{c} \, .
\end{aligned}
\tag{1.17}
$$

Hence the volume of a parallelepiped with adjacent edges $\mathbf{a}$, $\mathbf{b}$ and $\mathbf{c}$ is given by the triple scalar product $\mathbf{a} \bullet (\mathbf{b} \times \mathbf{c})$. Note that had D been a left-handed basis, then $\mathbf{a} \bullet (\mathbf{b} \times \mathbf{c})$ would have been negative and so, in that case, $V = -\mathbf{a} \bullet (\mathbf{b} \times \mathbf{c})$. In fact the triple scalar product can be used to test whether D is right-handed or left-handed since D is right-handed if and only if $\mathbf{a} \bullet (\mathbf{b} \times \mathbf{c}) > 0$.

We shall now show that $\mathbf{a} \bullet (\mathbf{b} \times \mathbf{c}) = \mathbf{b} \bullet (\mathbf{c} \times \mathbf{a}) = \mathbf{c} \bullet (\mathbf{a} \times \mathbf{b})$. If $\mathbf{a} = a_1\mathbf{i} + a_2\mathbf{j} + a_3\mathbf{k}$, $\mathbf{b} = b_1\mathbf{i} + b_2\mathbf{j} + b_3\mathbf{k}$ and $\mathbf{c} = c_1\mathbf{i} + c_2\mathbf{j} + c_3\mathbf{k}$, then

$$
\mathbf{a} \bullet (\mathbf{b} \times \mathbf{c}) = \begin{vmatrix} a_1 & b_1 & c_1 \\ a_2 & b_2 & c_2 \\ a_3 & b_3 & c_3 \end{vmatrix} . \tag{1.18}
$$

Using a theorem of determinants which states that an even number of column interchanges of a determinant leaves its sign unchanged, we observe that both

$$
\mathbf{b} \bullet (\mathbf{c} \times \mathbf{a}) = \begin{vmatrix} b_1 & c_1 & a_1 \\ b_2 & c_2 & a_2 \\ b_3 & c_3 & a_3 \end{vmatrix}
$$

and

$$
\mathbf{c} \bullet (\mathbf{a} \times \mathbf{b}) = \begin{vmatrix} c_1 & a_1 & b_1 \\ c_2 & a_2 & b_2 \\ c_3 & a_3 & b_3 \end{vmatrix}
$$

equal $\mathbf{a} \bullet (\mathbf{b} \times \mathbf{c})$.

(E1.15) Example - Calculation of the unit cell volume of coesite: The cell dimensions of coesite, a polymorph of silica stable at high pressures, are $a = 7.135$A, $b = 12.372$A, $c = 7.173$A, $\beta = 120.36°$, $\alpha = \gamma = 90°$ (Gibbs *et al.*, 1977). Find its unit cell volume, V.

Solution: We begin by setting up a cartesian basis $C = \{\mathbf{i},\mathbf{j},\mathbf{k}\}$ that relates in a simple fashion to the basis $\{\mathbf{a},\mathbf{b},\mathbf{c}\}$. Since $\mathbf{b}$ and $\mathbf{c}$ are perpendicular to one another, we place $\mathbf{j}$ along $\mathbf{b}$ and $\mathbf{k}$ along $\mathbf{c}$. Since $\mathbf{a}$ and $\mathbf{b}$ are perpendicular, $\mathbf{a}$ will then lie in the $\mathbf{i},\mathbf{k}$ plane. Since β is the angle between $\mathbf{a}$ and $\mathbf{k}$, we have the following linear combinations:

$$\begin{aligned}\mathbf{a} &= a\sin\beta\mathbf{i} + a\cos\beta\mathbf{k}\\ \mathbf{b} &= b\mathbf{j}\\ \mathbf{c} &= c\mathbf{k}\ .\end{aligned}$$

The volume of the unit cell is given by the triple scalar product

$$V = \mathbf{a} \bullet (\mathbf{b} \times \mathbf{c}) = \begin{vmatrix} a\sin\beta & 0 & 0 \\ 0 & b & 0 \\ a\cos\beta & 0 & c \end{vmatrix} = abc\sin\beta\ .$$

Using the cell dimension given for coesite, we get

$$V = (7.135\text{A})(12.372\text{A})(7.173\text{A})\sin(120.36°) = 546.4\text{A}^3\ . \qquad \square$$

(P1.15) Problem: Given the basis vectors for α-quartz, place $\mathbf{k}$ along $\mathbf{c}$ and $\mathbf{j}$ along $\mathbf{b}$ and show that

$$\begin{aligned}\mathbf{a} &= (\sqrt{3}a/2)\mathbf{i} - (a/2)\mathbf{j}\\ \mathbf{b} &= b\mathbf{j}\\ \mathbf{c} &= c\mathbf{k}\ .\end{aligned}$$

Then show that $V = (\sqrt{3}/2)a^2c$ and that the unit volume of α-quartz is 113.1A^3

CHAPTER 2

SOME GEOMETRICAL ASPECTS OF CRYSTALS

"Symmetry becomes the norm to which all things approximate; it is the fundamental blueprint which nothing is expected to follow to the letter." -- M. Senechal

INTRODUCTION

It was observed in Chapter 1 that when a large aggregate of atoms adopts a structure wherein the total energy is minimized, that the atoms in the aggregate are characteristically ordered into a regular and symmetric pattern referred to as a crystal. Such crystals occur in nature as minerals, constituting the bulk of the rocks of the earth's crust and mantle and the metals of the inner core. The mineral grains in these systems are intimately and firmly intergrown with surrounding minerals and as such, they are almost always irregular in shape and lack faces. When conditions are right and minerals are allowed to grow slowly in a cavity of a rock or vein, they may form as polyhedra of varying degrees of regularity bounded by smooth faces that meet in edges and vertices.

More than half a century ago, Max von Laue (1912) and his research assistants Walter Friedrich and Paul Knipping undertook a series of epoch-making diffraction experiments in the Mineralogy Department at the University of Munich and found that crystals by virtue of their regular and periodic structure and their minute interatomic spacing ($\sim10^{-8}$cm) behave as three-dimensional diffraction gratings for X-rays. The following year W.H. and W.L. Bragg (1913) demonstrated that each potential face on such a crystal parallels a stack of planes in the lattice used to represent its three dimensional periodic structure. In this chapter, the properties of the lattice and its planes together with those of the reciprocal lattice will be developed. Theorems will be proved for transforming atomic coordinates, face poles, etc. This will be followed by a variety of problems, in addition to solved examples worked out in detail, to illustrate the theorems and concepts presented in this chapter and Chapter 1.

EQUATION OF PLANES AND LATTICE PLANES

Lattice Planes: A plane passing through any three non-collinear lattice points in a lattice is called a *lattice plane*. Let $D = \{\mathbf{a},\mathbf{b},\mathbf{c}\}$ denote a basis for the lattice and let

$$\mathbf{p} = p_1\mathbf{a} + p_2\mathbf{b} + p_3\mathbf{c} \quad , \quad \mathbf{q} = q_1\mathbf{a} + q_2\mathbf{b} + q_3\mathbf{c} \quad , \quad \mathbf{r} = r_1\mathbf{a} + r_2\mathbf{b} + r_3\mathbf{c}$$

denote lattice vectors radiating from the origin to these points where $p_i, q_i, r_i \in Z$. Let P denote the set of lattice points on this plane. If we define $\mathbf{s} = \mathbf{q} - \mathbf{p}$ and $\mathbf{t} = \mathbf{r} - \mathbf{p}$, then a set of lattice points in P is given by

$$P = \{\mathbf{p} + n\mathbf{s} + m\mathbf{t} \mid n,m \in Z\} \quad . \tag{2.1}$$

This is illustrated in Figure 2.1. From Equation (2.1) we see that there are an infinite number of lattice points in such a plane. If $\mathbf{u} = u_1\mathbf{a} + u_2\mathbf{b} + u_3\mathbf{c}$, where $u_i \in Z$, is any lattice vector whose terminus is not on P, then

$$U = \{\mathbf{u} + n\mathbf{s} + m\mathbf{t} \mid n,m \in Z\}$$

lies on a lattice plane parallel to P passing through the terminus of $\mathbf{u}$. Therefore, every lattice point lies on a lattice plane parallel to P (Figure 2.1); thus, the whole lattice can be viewed as a stack of lattice planes parallel to any given lattice plane.

The equation of a plane: A method other than (2.1) that can be used to describe a plane in S is to locate a vector $\mathbf{s}$ perpendicular to the plane and a vector $\mathbf{p}$ whose terminus is on the plane. Then a vector $\mathbf{x}$ will have its terminus on the plane (Figure 2.2) if and only if

$$\mathbf{s} \cdot (\mathbf{x} - \mathbf{p}) = 0 \quad ,$$

or expressed another way,

$$\mathbf{s} \cdot \mathbf{x} = \mathbf{s} \cdot \mathbf{p} \quad . \tag{2.2}$$

Then (2.2) can be expressed as

$$[\mathbf{s}]_D^t G[\mathbf{x}]_D = \mathbf{s} \cdot \mathbf{p} \quad , \tag{2.3}$$

where G is the metrical matrix for the basis D.

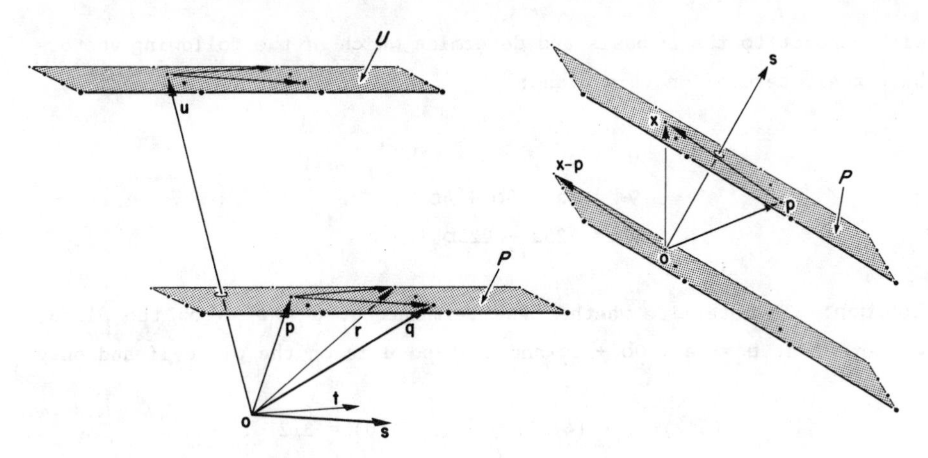

Figure 2.1 (to the left): A drawing of two parallel lattice planes in a space lattice L_D. The end points of the noncollinear vectors **p**, **q** and **r** define the plane

$$P = \{\mathbf{p} + n(\mathbf{q} - \mathbf{p}) + m(\mathbf{r} - \mathbf{p}) \mid n,m \in Z\}$$

of lattice points. The second plane

$$U = \{\mathbf{u} + n(\mathbf{q} - \mathbf{p}) + m(\mathbf{r} - \mathbf{p}) \mid n,m \in Z\}$$

parallels P and is located at the end point of the lattice vector **u**. Note that both P and U are planes in L_D that extend indefinitely in two dimensions.

Figure 2.2 (to the right): A drawing of a lattice plane, P, passing through the end points of vectors **p** and **x** and a parallel lattice plane passing through the origin, O. A vector **s** is perpendicular to P only when the inner product of **s** and the vector $(\mathbf{x} - \mathbf{p})$ vanishes for all $\mathbf{x}, \mathbf{p} \in P$.

Letting $[\mathbf{x}]_D = [x,y,z]^t$, multiplying (2.3) out and observing that $\mathbf{s} \bullet \mathbf{p}$ is a scalar, we have the equation

$$t_1x + t_2y + t_3z = w \tag{2.4}$$

where t_1, t_2, t_3, $w \in \mathbf{R}$ and x, y and z are indeterminates. Equation (2.4) completely describes the plane because a vector

$$\mathbf{v} = v_1\mathbf{a} + v_2\mathbf{b} + v_3\mathbf{c}$$

has its terminus on the plane if and only if $x = v_1$, $y = v_2$ and $z = v_3$ satisfies (2.4).

(E2.1) Example - Verification that the termini of a set of vectors lie on a plane: Consider the plane defined by the equation

$$(3/2)x + (4/3)y - (2/5)z = 3/2$$

with respect to the D basis and determine which of the following vectors have their termini on this plane:

$$\begin{aligned} \mathbf{u} &= \mathbf{a} \ , \\ \mathbf{v} &= 2\mathbf{a} - 3\mathbf{b} + 4\mathbf{c} \ , \\ \mathbf{w} &= 225\mathbf{a} - 225\mathbf{b} + 90\mathbf{c} \ . \end{aligned}$$

Solution: To determine whether the terminus of $\mathbf{u} = \mathbf{a}$ is on the plane, we note that $\mathbf{u} = 1\mathbf{a} + 0\mathbf{b} + 0\mathbf{c}$ and that so $\mathbf{u}$ is on the plane if and only if

$$(3/2)(1) + (4/3)(0) - (2/5)(0) = 3/2 \ .$$

Since this equality is true, we conclude that the terminus of $\mathbf{u}$ lies on the plane. Similarly the terminus of $\mathbf{v}$ is on the plane if and only if

$$(3/2)(2) + (4/3)(-3) - (2/5)(4) = 3/2 \ .$$

In this case, the equality is false and thus the terminus of $\mathbf{v}$ is not on the plane. The demonstration that $\mathbf{w}$ is on the plane is left as an exercise for the reader. □

Equation (2.4) enables us to easily determine many of the properties of the plane it describes. For example, the point at which the plane crosses the X-axis is the solution of the form $\mathbf{v} = v_1\mathbf{a} + 0\mathbf{b} + 0\mathbf{c}$. Hence, setting $x = v_1$, and $y = z = 0$ in (2.4), we obtain $t_1 v_1 = w$. If $t_1 \neq 0$, this equation has the solution w/t_1 and hence the plane intersects the X-axis at $(w/t_1)\mathbf{a}$. If $t_1 = 0$, then either $w = 0$, in which case the X-axis lies in the plane or $w \neq 0$, in which case there is no solution and so the X-axis parallels the plane and the plane does not intersect the X-axis anywhere. A similar analysis holds for the Y-axis and the Z-axis. Thus the plane $t_1 x + t_2 y + t_3 z = w$ intercepts the X-axis at $(w/t_1)\mathbf{a}$, the Y-axis at $(w/t_2)\mathbf{b}$ and the Z-axis at $(w/t_3)\mathbf{c}$ when t_1, t_2 and t_3 are non-zero.

Miller Indices: In Appendix 5, lattice planes are discussed and the facts we cite here are established. The equation of a lattice plane (2.4) can be expressed in the form

$$hx + ky + \ell z = m \ , \tag{2.5}$$

where h, k, ℓ, $m \in Z$, not all of h, k and ℓ are zero, and the largest common integer factor of h, k and ℓ is 1. Conversely, any such equation defines a lattice plane. Note that (2.5) results from (2.3) if

$$[\mathbf{s}]_D^t = [hk\ell]G^{-1} .$$

Hence h, k and ℓ determine a vector **s** perpendicular to the plane. Consequently, two equations of the form (2.5) with different m values but the same h, k and ℓ values will define parallel planes. If $m = 0$, the plane passes through the origin. Amongst the lattice planes not passing through the origin, the two closest to the origin occur when $m = \pm 1$, the next two closest when $m = \pm 2$, etc. Furthermore, the planes are equally spaced. This can be seen by observing that a plane of the form (2.5) crosses the X-axis at $(m/h)\mathbf{a}$. (If $h = 0$, do the same analysis using k or ℓ, whichever is not zero). Hence, any two adjacent planes cross the X-axis at points a distance of $(1/h)a$ apart and so they are equally spaced. However, note that, unless **a** is perpendicular to the plane, $(1/h)a$ is not the actual distance or d-spacing between the planes. We shall calculate this distance at the end of this section. The set of three integers denoted by $(hk\ell)$ is called the *Miller indices* of the face of the crystal that parallels these planes. By convention Miller indices are enclosed between parenthesis and a negative index is denoted by a minus sign set over the index. Thus $(\bar{h}k\bar{\ell})$ specifies a stack of parallel planes whose first plane ($m = 1$) intercepts X, Y and Z at $-\mathbf{a}/h$, $\mathbf{b}/k$ and $-\mathbf{c}/\ell$ (when h, k and ℓ are nonzero).

Figure 2.3 shows a stack of planes with indices (312) intercepting the parallelepiped (unit cell) outlined by the basis vectors of α-quartz. Note that the planes divide **a**, **b** and **c** into 3, 1 and 2 parts, respectively, and that the $m = 1$ plane intercepts X, Y, and Z at $\mathbf{a}/3$, $\mathbf{b}/1$, and $\mathbf{c}/2$, respectively. The $m = 1$ plane with Miller indices (001) parallels both the X and Y axis and intercepts Z at **c**. Likewise, the one for (100) parallels Y and Z and intercepts X at **a** and the one for (010) parallels X and Z and intercepts Y at **b**.

(E2.2) Example - Determination of a, b, c given the indices of a plane and its intercepts on X, Y, Z: Suppose a plane in the mineral anorthite

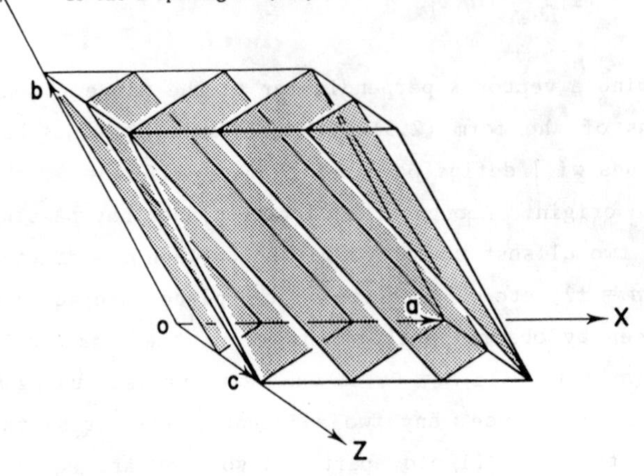

Figure 2.3: A stack of parallel lattice planes (lattice points are not shown) with Miller indices (312) intercepting the basis vectors $D = \{\mathbf{a},\mathbf{b},\mathbf{c}\}$ outlining a unit cell of the α-quartz structure. The equation of each plane in the stack is $3x + y + 2z = m$, where $m \in Z$. Each plane intercepts the coordinate axes X, Y, Z at $m\mathbf{a}/3$, $m\mathbf{b}/1$, $m\mathbf{c}/2$, respectively; The $m = 1$ plane intercepts X, Y, Z at $\mathbf{a}/3$, $\mathbf{b}$, $\mathbf{c}/2$, the $m = -1$ plane intercepts X, Y, Z at $-\mathbf{a}/3$, $-\mathbf{b}$, $-\mathbf{c}/2$, etc. The separation between each plane in the stack is denoted as the d-spacing of (312).

(see P1.13) with Miller indices (132) and $m = 1$ intercepts X at 8.173A, Y at 4.290A and Z at 7.082A. Find the lengths of **a**, **b** and **c**.

Solution: We are given that $a/h = 8.173\text{A}$, $b/k = 4.290\text{A}$ and $c/\ell = 7.082\text{A}$ where $h = 1$, $k = 3$ and $\ell = 2$. Thus, the lengths of **a**, **b**, and **c** are $a = 8.173\text{A}$, $b = 3 \times 4.290\text{A} = 12.87\text{A}$, and $c = 2 \times 7.082\text{A} = 14.164\text{A}$. □

d-spacings: The *d-spacing*, the distance between adjacent planes with given Miller indices $(hk\ell)$, is equal to the distance between the origin (which lies on the $m = 0$ plane) and the $m = 1$ plane. The equation for the $m = 1$ plane is

$$hx + ky + \ell z = 1 \quad . \tag{2.6}$$

We can see from (2.3) that the **s** vector giving rise to (2.6) is such that **s** is perpendicular to the plane and $\mathbf{s} \bullet \mathbf{p} = 1$ for any vector **p** whose terminus is on the $m = 1$ plane. Since $1 = \mathbf{s} \bullet \mathbf{p} = sp\cos\theta$ where θ is the angle between **s** and **p**, and since $p\cos\theta$ is the projection of **p** onto **s** (see Chapter 1), we see that $1/s = p\cos\theta$ is the projection of **p** onto **s**. But the projection of **p** onto the direction perpendicular to the plane is the distance from the origin to the plane. Therefore, $1/s$ is the distance from the origin to the plane. From (2.3) we can show (details are given in Appendix 5) that

$$s^2 = [hk\ell]G^{-1}\begin{bmatrix} h \\ k \\ \ell \end{bmatrix} .$$

Hence, in summary, **s** yields (2.6) from (2.3) if and only if

(1) **s** is perpendicular to the $(hk\ell)$ planes, and

(2.7)

(2) $\mathbf{s} \bullet \mathbf{p} = 1$ for all **p** whose termini lie on the plane of (2.6).

Furthermore, in this case the d-spacing for the $(hk\ell)$ planes is $1/s$ and

$$s^2 = [hk\ell]G^{-1}\begin{bmatrix} h \\ k \\ \ell \end{bmatrix} .$$

(E2.3) Example - Calculation of the *d*-spacing for a plane in a crystal: With the cell dimensions given in P1.13, compute the d-spacing for the (312) plane of anorthite.

Solution: To find the d-spacing of this plane, first we use the cell dimensions of anorthite to compute the metrical matrix G for D and then we invert it and find

$$G^{-1} = \begin{bmatrix} 66.7979 & -2.3146 & -50.5929 \\ -2.3146 & 165.6112 & -9.8993 \\ -50.5924 & -9.8993 & 200.6472 \end{bmatrix}^{-1} = \begin{bmatrix} 0.01855 & 0.00054 & 0.00470 \\ 0.00054 & 0.00607 & 0.00044 \\ 0.00470 & 0.00044 & 0.00619 \end{bmatrix} .$$

Then we evaluate the product

$$s^2 = [312]G^{-1}\begin{bmatrix} 3 \\ 1 \\ 2 \end{bmatrix} = 0.2595 .$$

Since $d = 1/s$, we see that the d-spacing of (312) is $d = 1.963$A. □

RECIPROCAL BASIS VECTORS

We observed in the last section that each potential face on a crystal parallels a stack of equi-spaced parallel lattice planes with a common interplanar spacing, d. The Braggs discovered in their pioneering experiments on rock salt (NaCl) that X-ray diffraction by such a crystal can be considered as reflections from these planes. They also found for a given wavelength λ that X-ray beams are diffracted from these planes by constructive interference, but only at definite angles defined by the Bragg equation, $n\lambda = 2d\sin\theta$, where $n \in Z$, d is the interplanar spacing, and θ is the reflection angle. Recording these beams on a photographic film or with an electronic counting device and measuring the positions of each, the directions perpendicular to the planes and the interplanar spacings d for the crystal can be measured. From such measurements, the cell dimensions of the crystal can be determined (*cf.* PA2.4).

In the previous section we saw that corresponding to each stack of parallel lattice planes in the lattice generated by $D = \{\mathbf{a},\mathbf{b},\mathbf{c}\}$ there is a vector $\mathbf{s}$ such that when $\mathbf{s}$ is substituted into (2.3), we obtain

$$hx + ky + \ell z = 1 \quad , \tag{2.8}$$

where $(hk\ell)$ are the Miller indices of the stack of planes. By (2.7), $\mathbf{s}$ is perpendicular to the planes and $\mathbf{s} \cdot \mathbf{p} = 1$ for any $\mathbf{p}$ whose terminus is on the plane of (2.8). In general, such an $\mathbf{s}$ is not in the lattice generated by D. Our goal in this section is to create a *reciprocal basis* $D^* = \{\mathbf{a}^*,\mathbf{b}^*,\mathbf{c}^*\}$ such that, for any set of Miller indices $(hk\ell)$, $\mathbf{s}$ is expressed as

$$\mathbf{s} = h\mathbf{a}^* + k\mathbf{b}^* + \ell\mathbf{c}^* .$$

That is,

$$[\mathbf{s}]_{D^*} = \begin{bmatrix} h \\ k \\ \ell \end{bmatrix} \quad . \tag{2.9}$$

When we find such a basis, we will be able to easily locate and describe all of the lattice planes in the lattice generated by D. To discover D^* we first consider the planes (100). In this case,

$$[\mathbf{s}]_{D^*} = \begin{bmatrix} 1 \\ 0 \\ 0 \end{bmatrix} ,$$

and so $\mathbf{s} = \mathbf{a}^*$. Hence we want $\mathbf{a}^*$ to satisfy (2.7) with respect to (100). Therefore $\mathbf{a}^*$ is perpendicular to the (100) planes and, since the terminus of $\mathbf{a}$ is on the plane $1x + 0y + 0z = 1$, $\mathbf{a}^* \bullet \mathbf{a} = 1$. Note that the (100) planes parallel the plane in which $\mathbf{b}$ and $\mathbf{c}$ lie. Since $\mathbf{b} \times \mathbf{c}$ is a vector perpendicular to the (100) planes, $\mathbf{a}^* = r(\mathbf{b} \times \mathbf{c})$ for some real number r. Using the fact that $\mathbf{a}^* \bullet \mathbf{a} = 1$, we obtain

$$r(\mathbf{b} \times \mathbf{c}) \bullet \mathbf{a} = 1 \quad ,$$

and so

$$r = 1/[\mathbf{a} \bullet \mathbf{b} \times \mathbf{c}] \quad .$$

Consequently,

$$\mathbf{a}^* = (\mathbf{b} \times \mathbf{c})/[\mathbf{a} \bullet \mathbf{b} \times \mathbf{c}] \quad .$$

By a similar argument,

$$\mathbf{b}^* = (\mathbf{c} \times \mathbf{a})/[\mathbf{a} \bullet \mathbf{b} \times \mathbf{c}]$$

$$\mathbf{c}^* = (\mathbf{a} \times \mathbf{b})/[\mathbf{a} \bullet \mathbf{b} \times \mathbf{c}] \quad .$$

(P2.1) Problem: Explain why $\mathbf{b}^*$ is perpendicular to the (010) planes and $\mathbf{b}^* \bullet \mathbf{b} = 1$. Then show that $\mathbf{b}^* = r(\mathbf{a} \times \mathbf{c})$ where $r \in R$ and that

$$\mathbf{b}^* = (\mathbf{c} \times \mathbf{a})/[\mathbf{a} \bullet \mathbf{b} \times \mathbf{c}] \quad .$$

(P2.2) Problem: Show, as in P2.1, that

$$\mathbf{c}^* = (\mathbf{a} \times \mathbf{b})/[\mathbf{a} \bullet \mathbf{b} \times \mathbf{c}] \quad .$$

Note that by the way D^* was selected,

$$\mathbf{a}^* \bullet \mathbf{a} = \mathbf{b}^* \bullet \mathbf{b} = \mathbf{c}^* \bullet \mathbf{c} = 1 \quad ,$$

and

$$\mathbf{a}^* \bullet \mathbf{b} = \mathbf{a}^* \bullet \mathbf{c} = \mathbf{b}^* \bullet \mathbf{a} = \mathbf{b}^* \bullet \mathbf{c} = \mathbf{c}^* \bullet \mathbf{a} = \mathbf{c}^* \bullet \mathbf{b} = 0 \quad . \qquad (2.10)$$

Also, note that since $\mathbf{a}^*$, $\mathbf{b}^*$ and $\mathbf{c}^*$ are perpendicular to the planes (100), (010) and (001), respectively, and since these planes do not intersect in parallel lines, $\mathbf{a}^*$, $\mathbf{b}^*$, and $\mathbf{c}^*$ are not coplanar. Thus, D^* qualifies as a basis with $\mathbf{a}^*$, $\mathbf{b}^*$, $\mathbf{c}^*$ directed along the coordinate axes X^*, Y^*, Z^*, respectively, with interaxial angles $\alpha^* = \angle(Y^*:Z^*)$, $\beta^* = \angle(Z^*:X^*)$, $\gamma^* = \angle(X^*:Y^*)$.

(T2.4) Theorem: If **s** satisfies (2.7) for the Miller indices ($hk\ell$), then

$$[\mathbf{s}]_{D^*} = \begin{bmatrix} h \\ k \\ \ell \end{bmatrix} .$$

Proof: Since **s** satisifies (2.7), $\mathbf{s} \cdot \mathbf{p} = 1$ for all **p** whose termini are on the plane

$$hx + ky + \ell z = 1 .$$

Assume that $h \neq 0$, $k \neq 0$ and $\ell \neq 0$. Recall that this plane crosses the X-axis at $\mathbf{a}/h$, the Y-axis at $\mathbf{b}/k$ and the Z-axis at $\mathbf{c}/\ell$. Hence, $\mathbf{s} \cdot (\mathbf{a}/h) = \mathbf{s} \cdot (\mathbf{b}/k) = \mathbf{s} \cdot (\mathbf{c}/\ell) = 1$. Since $D^* = \{\mathbf{a}^*, \mathbf{b}^*, \mathbf{c}^*\}$ is a basis, there exists real numbers t_1, t_2, t_3 such that

$$\mathbf{s} = t_1\mathbf{a}^* + t_2\mathbf{b}^* + t_3\mathbf{c}^* . \tag{2.11}$$

Hence,

$$1 = \mathbf{s} \cdot (\mathbf{a}/h) = (1/h)(t_1\mathbf{a}^* + t_2\mathbf{b}^* + t_3\mathbf{c}^*) \cdot \mathbf{a}$$

$$= (1/h)(t_1\mathbf{a}^* \cdot \mathbf{a} + t_2\mathbf{b}^* \cdot \mathbf{a} + t_3\mathbf{c}^* \cdot \mathbf{a})$$

$$= (1/h)(t_1) \qquad \text{(applying (2.10))} .$$

Therefore, $t_1 = h$. Similarly, $t_2 = k$ and $t_3 = \ell$. Therefore,

$$\mathbf{s} = h\mathbf{a}^* + k\mathbf{b}^* + \ell\mathbf{c}^* ,$$

and so

$$[\mathbf{s}]_{D^*} = \begin{bmatrix} h \\ k \\ \ell \end{bmatrix} .$$

When some of h, k and ℓ are zero, a similar argument can be employed. For example, if $h = 0$ then **a** parallels the plane and so $\mathbf{s} \cdot \mathbf{a} = 0$. But $\mathbf{s} \cdot \mathbf{a} = 0$ implies that $t_1 = 0$. We took care of the case when two of h, k and ℓ are zero when we defined $\mathbf{a}^*$, $\mathbf{b}^*$ and $\mathbf{c}^*$. □

(P2.3) Problem: In the $h \neq 0$, $k \neq 0$ and $\ell \neq 0$ case, show that for (2.11)

$t_2 = k$ and $t_3 = \ell$.

(P2.4) Problem: In the $h = 0$, $k \neq 0$, $\ell \neq 0$ case, show that

$$[\mathbf{s}]_{D^*} = \begin{bmatrix} 0 \\ k \\ \ell \end{bmatrix} .$$

(P2.5) Problem: In the $h \neq 0$, $k = 0$, $\ell \neq 0$ case, show that

$$[\mathbf{s}]_{D^*} = \begin{bmatrix} h \\ 0 \\ \ell \end{bmatrix} .$$

Direct and reciprocal lattices: In Chapter 1, we defined a lattice L_D as a subset of points in S such that each vector $\mathbf{r} \in L_D$ can be expressed as an integral combination of D. In the context of D and D^*, D is called the *direct basis*. Hence L_D is called the *direct lattice*. With the introduction of the reciprocal basis D^*, a second subset of points in S

$$L_{D^*} = \{h\mathbf{a}^* + k\mathbf{b}^* + \ell\mathbf{c}^* \mid h,k,\ell \in Z\}$$

known as the *reciprocal lattice* can be defined. Because of the orientation of the lattice vectors in L_{D^*} relative to the planes of L_D, the reciprocal lattice has found important applications in X-ray crystallography in the study of the diffraction symmetry of crystals, crystal structure analysis and in solid state chemistry in the study of bonding (Slater, 1972). When an X-ray beam is diffracted by the structure of a crystal, the positions of the diffracted beams forming the diffraction record provide a map of the reciprocal lattice of the crystal. By analyzing the pattern, not only can the size, shape and orientation of the unit cell be found, but also the symmetry, the lattice type and certain space group symmetry operations of the crystal may be deduced (Buerger, 1942).

***D* and *D** compared:** The relationship between D and D^* is an important one that goes well beyond the task of finding the lattice planes in the lattice generated by D. In this section we shall establish a number of results that will prove to be useful when dealing with S described in terms of D or D^*.

The strategy we employed to solve for t_1, t_2, and t_3 in (2.11) can be applied in a more general setting. Let $\mathbf{r} \in S$. Then there exist real numbers r_1, r_2, and r_3 such that if

$$\mathbf{r} = r_1\mathbf{a} + r_2\mathbf{b} + r_3\mathbf{c} \ ,$$

then

$$\mathbf{r} \cdot \mathbf{a}^* = (r_1\mathbf{a} \cdot \mathbf{a}^*) + (r_2\mathbf{b} \cdot \mathbf{a}^*) + (r_3\mathbf{c} \cdot \mathbf{a}^*)$$

$$= r_1 \qquad \text{(applying (2.10))} \quad .$$

Similarly, $\mathbf{r} \cdot \mathbf{b}^* = r_2$ and $\mathbf{r} \cdot \mathbf{c}^* = r_3$.

(P2.6) Problem: Show that $\mathbf{r} \cdot \mathbf{b}^* = r_2$ and $\mathbf{r} \cdot \mathbf{c}^* = r_3$.

Hence, we have the important equation

$$\mathbf{r} = (\mathbf{r} \cdot \mathbf{a}^*)\mathbf{a} + (\mathbf{r} \cdot \mathbf{b}^*)\mathbf{b} + (\mathbf{r} \cdot \mathbf{c}^*)\mathbf{c} \quad . \tag{2.12}$$

Similarly,

$$\mathbf{r} = (\mathbf{r} \cdot \mathbf{a})\mathbf{a}^* + (\mathbf{r} \cdot \mathbf{b})\mathbf{b}^* + (\mathbf{r} \cdot \mathbf{c})\mathbf{c}^* \quad . \tag{2.13}$$

(P2.7) Problem: Show that (2.13) is a true statement.

From (2.12) and (2.13), we see that the following theorem is true.

(T2.5) Theorem: Let D denote a basis and D^* its reciprocal basis. If $\mathbf{r} \in S$, then

$$[\mathbf{r}]_D = \begin{bmatrix} \mathbf{r} \cdot \mathbf{a}^* \\ \mathbf{r} \cdot \mathbf{b}^* \\ \mathbf{r} \cdot \mathbf{c}^* \end{bmatrix} \qquad \text{and} \qquad [\mathbf{r}]_{D^*} = \begin{bmatrix} \mathbf{r} \cdot \mathbf{a} \\ \mathbf{r} \cdot \mathbf{b} \\ \mathbf{r} \cdot \mathbf{c} \end{bmatrix} \quad .$$

Suppose that $[\mathbf{r}]_D = [r_1, r_2, r_3]^t$ is known and we wish to find $[\mathbf{r}]_{D^*}$. By (T2.5), $[\mathbf{r}]_{D^*} = [\mathbf{r} \cdot \mathbf{a}, \mathbf{r} \cdot \mathbf{b}, \mathbf{r} \cdot \mathbf{c}]^t$. But

$$\mathbf{r} \cdot \mathbf{a} = (r_1\mathbf{a} + r_2\mathbf{b} + r_3\mathbf{c}) \cdot \mathbf{a}$$
$$= r_1(\mathbf{a} \cdot \mathbf{a}) + r_2(\mathbf{b} \cdot \mathbf{a}) + r_3(\mathbf{c} \cdot \mathbf{a}) \quad .$$

Similarly,

$$\mathbf{r} \cdot \mathbf{b} = r_1(\mathbf{a} \cdot \mathbf{b}) + r_2(\mathbf{b} \cdot \mathbf{b}) + r_3(\mathbf{c} \cdot \mathbf{b})$$
$$\mathbf{r} \cdot \mathbf{c} = r_1(\mathbf{a} \cdot \mathbf{c}) + r_2(\mathbf{b} \cdot \mathbf{c}) + r_3(\mathbf{c} \cdot \mathbf{c}) \quad .$$

Writing this in matrix form we have

$$[\mathbf{r}]_{D^*} = \begin{bmatrix} \mathbf{a} \cdot \mathbf{a} & \mathbf{a} \cdot \mathbf{b} & \mathbf{a} \cdot \mathbf{c} \\ \mathbf{b} \cdot \mathbf{a} & \mathbf{b} \cdot \mathbf{b} & \mathbf{b} \cdot \mathbf{c} \\ \mathbf{c} \cdot \mathbf{a} & \mathbf{c} \cdot \mathbf{b} & \mathbf{c} \cdot \mathbf{c} \end{bmatrix} \begin{bmatrix} r_1 \\ r_2 \\ r_3 \end{bmatrix} , \tag{2.14}$$

where we have used the fact that for any two vectors $\mathbf{u}$ and $\mathbf{v}$, $\mathbf{u} \cdot \mathbf{v} = \mathbf{v} \cdot \mathbf{u}$. The 3×3 matrix in (2.14) is the metrical matrix G for D introduced in Chapter 1. Hence,

$$[\mathbf{r}]_{D^*} = G[\mathbf{r}]_D \tag{2.15}$$

for all $r \varepsilon S$. We have found a matrix, G, that allows us to change from the notation of one basis, D, to another, D^*. We shall see later that given any two bases D_1 and D_2 there always exists a matrix T such that

$$T[\mathbf{r}]_{D_1} = [\mathbf{r}]_{D_2}$$

for all $\mathbf{r} \varepsilon S$ and we will present a technique for finding T.

In view of the symmetry of the equation in (2.14) it is reasonable to expect that D is the reciprocal basis of D^*. The following theorem shows that this is the case.

(T2.6) Theorem: If $D^* = \{\mathbf{a}^*,\mathbf{b}^*,\mathbf{c}^*\}$ is the reciprocal basis with respect to $D = \{\mathbf{a},\mathbf{b},\mathbf{c}\}$, then D is the reciprocal basis with respect to D^*.

Proof: Let the reciprocal basis with respect to D^* be denoted by $(D^*)^* = \{(\mathbf{a}^*)^*, (\mathbf{b}^*)^*, (\mathbf{c}^*)^*\}$. Using Equation (2.12) to write $(\mathbf{a}^*)^*$ in terms of D, we have

$$(\mathbf{a}^*)^* = ((\mathbf{a}^*)^* \cdot \mathbf{a}^*)\mathbf{a} + ((\mathbf{a}^*)^* \cdot \mathbf{b}^*)\mathbf{b} + ((\mathbf{a}^*)^* \cdot \mathbf{c}^*)\mathbf{c} \quad . \tag{2.16}$$

Since $(D^*)^*$ is the reciprocal basis with respect to D^*, equations of (2.10) yield

$$\mathbf{a}^* \bullet (\mathbf{a}^*)^* = 1$$

and

$$\mathbf{b}^* \bullet (\mathbf{a}^*)^* = \mathbf{c}^* \bullet (\mathbf{a}^*)^* = 0 \quad .$$

Since the inner product is commutative, (2.16) becomes $(\mathbf{a}^*)^* = \mathbf{a}$. Similarly $(\mathbf{b}^*)^* = \mathbf{b}$ and $(\mathbf{c}^*)^* = \mathbf{c}$. Hence $(D^*)^* = \{\mathbf{a},\mathbf{b},\mathbf{c}\}$. □

In view of T2.6 and the definition of a reciprocal basis we have shown that

$$\mathbf{a} = (\mathbf{b}^* \times \mathbf{c}^*)/[\mathbf{a}^* \bullet (\mathbf{b}^* \times \mathbf{c}^*)]$$
$$\mathbf{b} = (\mathbf{c}^* \times \mathbf{a}^*)/[\mathbf{a}^* \bullet (\mathbf{b}^* \times \mathbf{c}^*)]$$
$$\mathbf{c} = (\mathbf{a}^* \times \mathbf{b}^*)/[\mathbf{a}^* \bullet (\mathbf{b}^* \times \mathbf{c}^*)] \quad .$$

Let H denote the metrical matrix for D^*. By (2.15),

$$H[\mathbf{r}]_{D^*} = [\mathbf{r}]_{(D^*)^*}$$

for all $\mathbf{r} \in S$. But by T2.6, $(D^*)^* = D$. Hence

$$H[\mathbf{r}]_{D^*} = [\mathbf{r}]_D$$

for all $\mathbf{r} \in S$. Also, by (2.15), $G[\mathbf{r}]_D = [\mathbf{r}]_{D^*}$ for all $\mathbf{r} \in S$ where G is the metrical matrix for D. Hence $[\mathbf{r}]_D = G^{-1}[\mathbf{r}]_{D^*}$ (see Appendix 2 for a discussion on the inverse of a matrix and why, since $\det(G) \neq 0$, G^{-1} exists). Therefore, $H[\mathbf{r}]_{D^*} = G^{-1}[\mathbf{r}]_{D^*}$ for all $\mathbf{r} \in S$. Since the equality holds for all $\mathbf{r} \in S$, $H = G^{-1}$. Hence we have proved the following result.

(C2.7) Corollary: If G is the metrical matrix for D, then G^{-1} is the metrical matrix for D^*.

The fact that the metrical matrix of D^* is the "reciprocal" of the metrical matrix of D is consistent with calling D^* the reciprocal basis of D. The metrical matrix of D^* is denoted G^*. The complete geometry of D^* is described in terms of that of D in the following theorem.

(T2.8) Theorem: Let $D = \{\mathbf{a},\mathbf{b},\mathbf{c}\}$ denote a basis of S and let $D^* = \{\mathbf{a}^*,\mathbf{b}^*,\mathbf{c}^*\}$ denote its reciprocal basis. Then the geometry of D^* can be written in terms of that of D in the following way:

$$
\begin{aligned}
a^* &= (bc/v)\sin\alpha \quad ; \\
b^* &= (ac/v)\sin\beta \quad ; \\
c^* &= (ab/v)\sin\gamma \quad ; \\
\cos\alpha^* &= (\cos\beta\cos\gamma - \cos\alpha)/\sin\beta\sin\gamma \quad ; \\
\cos\beta^* &= (\cos\alpha\cos\gamma - \cos\beta)/\sin\alpha\sin\gamma \quad ; \\
\cos\gamma^* &= (\cos\alpha\cos\beta - \cos\gamma)/\sin\alpha\sin\beta \quad ;
\end{aligned}
$$

where $v = \mathbf{a} \cdot (\mathbf{b} \times \mathbf{c})$.

Proof: Since $\|\mathbf{b} \times \mathbf{c}\| = bc\sin\alpha$ and $\mathbf{a} \cdot (\mathbf{b} \times \mathbf{c}) = v$ (see (1.15) and (1.17)),

$$a^* = (\|\mathbf{b} \times \mathbf{c}\|)/(\mathbf{a} \cdot (\mathbf{b} \times \mathbf{c})) = (bc/v)\sin\alpha \quad .$$

The expressions for b^* and c^* can be similarly established. To calculate $\cos\alpha^*$, we shall use the fact that $\mathbf{b}^* \cdot \mathbf{c}^* = b^*c^*\cos\alpha^*$ and the vector identity (White, 1960, p. 76)

$$(\mathbf{x} \times \mathbf{y}) \cdot (\mathbf{z} \times \mathbf{w}) = \begin{vmatrix} \mathbf{x} \cdot \mathbf{z} & \mathbf{x} \cdot \mathbf{w} \\ \mathbf{y} \cdot \mathbf{z} & \mathbf{y} \cdot \mathbf{w} \end{vmatrix} . \tag{2.17}$$

Hence

$$\cos\alpha^* = (\mathbf{b}^* \cdot \mathbf{c}^*)/(b^*c^*) = (v^2\mathbf{b}^* \cdot \mathbf{c}^*)/(a^2bc\sin\beta\sin\gamma) \tag{2.18}$$

But

$$
\begin{aligned}
\mathbf{b}^* \cdot \mathbf{c}^* &= (1/v^2)(\mathbf{c} \times \mathbf{a}) \cdot (\mathbf{a} \times \mathbf{b}) \\
&= (1/v^2)\begin{vmatrix} \mathbf{c} \cdot \mathbf{a} & \mathbf{c} \cdot \mathbf{b} \\ \mathbf{a} \cdot \mathbf{a} & \mathbf{a} \cdot \mathbf{b} \end{vmatrix} = (1/v^2)\begin{vmatrix} ac\cos\beta & cb\cos\alpha \\ a^2 & ab\cos\gamma \end{vmatrix} \\
&= (a^2\, bc/v^2)(\cos\beta\cos\gamma - \cos\alpha) \quad .
\end{aligned} \tag{2.19}
$$

Substituting (2.19) into (2.18) we obtain

$$\cos\alpha^* = (\cos\beta\cos\gamma - \cos\alpha)/\sin\beta\sin\gamma \ .$$

The other parts of the theorem are proved in a perfectly similar manner.□

Because of the duality between D and D^* discussed in T2.6, any expression like $\cos\alpha^* = (\cos\beta\cos\gamma - \cos\alpha)/\sin\beta\sin\gamma$ can be used to write the equality

$$\cos\alpha = (\cos\beta^*\cos\gamma^* - \cos\alpha^*)/\sin\beta^*\sin\gamma^* \quad .$$

Similarly,

$$\cos\beta = (\cos\alpha^*\cos\gamma^* - \cos\beta^*)/\sin\alpha^*\sin\gamma^*$$

and

$$\cos\gamma = (\cos\alpha^*\cos\beta^* - \cos\gamma^*)/\sin\alpha^*\sin\beta^* \quad .$$

CHANGE OF BASIS

We learned in Chapter 1 that there are innumerable directions in a crystal along which a representative pattern of the structure is repeated at regular intervals. Furthermore, not only do these directions define a natural coordinate system along which three non-coplanar coordinate axes can be placed, but the repeat unit along each of these directions defines a natural set of basis vectors $D = \{\mathbf{a},\mathbf{b},\mathbf{c}\}$. Thus, for a given crystal, innumerable sets of axes, basis vectors and planes can be chosen, depending on the problem to be studied. In addition, as there are innumerable ways of describing such a crystal, it is essential to be able to change its geometric framework with respect to the observations being made. Moreover, if we wish to study crystal structure relationships (Smith, 1982), polymorphism, phase transformations (Hazen and Finger, 1982) or defects (Lasaga and Kirkpatrick, 1981), it is imperative that these features be related to a single set of basis vectors. In particular, if data are provided by one investigator describing some feature of a crystal relative with respect to one basis and if another investigator provides a similar set of data on another such crystal but defined with respect to a second basis, it is important to express both data sets with respect to a common basis in order to make comparisons and to draw conclusions. In other words, we must be able to make a change of basis.

In the last section we saw how, given a basis $D = \{\mathbf{a},\mathbf{b},\mathbf{c}\}$ and its reciprocal basis $D^* = \{\mathbf{a}^*,\mathbf{b}^*,\mathbf{c}^*\}$, we can easily translate geometrical information given in terms of D into the language of D^*. For example, if $\mathbf{r} \in S$ and if we know $[\mathbf{r}]_D$, then we can find $[\mathbf{r}]_{D^*}$ by, see (2.15),

$$G[\mathbf{r}]_D = [\mathbf{r}]_{D^*} \quad ,$$

where G is the metrical matrix of D. In this section, we will see how to translate geometrical information from a basis $D_1 = \{\mathbf{a}_1, \mathbf{b}_1, \mathbf{c}_1\}$ to a basis $D_2 = \{\mathbf{a}_2, \mathbf{b}_2, \mathbf{c}_2\}$.

Just as in the special case when $D_1 = D$ and $D_2 = D^*$, a matrix will be found that will convert $[\mathbf{r}]_{D_1}$ into $[\mathbf{r}]_{D_2}$ for all $\mathbf{r} \in S$ and this matrix will be used to calculate the metrical matrix of D_2 from that of D_1.

(D2.9) Definition: The change of basis matrix from the basis D_1 to the basis D_2 is the matrix T such that

$$T[\mathbf{r}]_{D_1} = [\mathbf{r}]_{D_2}$$

for all $\mathbf{r} \in S$.

Since we know what job T is supposed to perform, we can find T by observing that

$$\begin{bmatrix} t_{11} & t_{12} & t_{13} \\ t_{21} & t_{22} & t_{23} \\ t_{31} & t_{32} & t_{33} \end{bmatrix} \begin{bmatrix} 1 \\ 0 \\ 0 \end{bmatrix} = \begin{bmatrix} t_{11} \\ t_{21} \\ t_{31} \end{bmatrix} = \text{1st column of T}$$

$$\begin{bmatrix} t_{11} & t_{12} & t_{13} \\ t_{21} & t_{22} & t_{23} \\ t_{31} & t_{32} & t_{33} \end{bmatrix} \begin{bmatrix} 0 \\ 1 \\ 0 \end{bmatrix} = \begin{bmatrix} t_{12} \\ t_{22} \\ t_{32} \end{bmatrix} = \text{2nd column of T} \qquad (2.20)$$

$$\begin{bmatrix} t_{11} & t_{12} & t_{13} \\ t_{21} & t_{22} & t_{23} \\ t_{31} & t_{32} & t_{33} \end{bmatrix} \begin{bmatrix} 0 \\ 0 \\ 1 \end{bmatrix} = \begin{bmatrix} t_{13} \\ t_{23} \\ t_{33} \end{bmatrix} = \text{3rd column of T} \quad .$$

Since $[\mathbf{a}_1]_{D_1} = [100]^t$, $[\mathbf{b}_1]_{D_1} = [010]^t$, and $[\mathbf{c}_1]_{D_1} = [001]^t$, and since $T[\mathbf{a}_1]_{D_1} = [\mathbf{a}_1]_{D_2}$, $T[\mathbf{b}_1]_{D_1} = [\mathbf{b}_1]_{D_2}$ and $T[\mathbf{c}_1]_{D_1} = [\mathbf{c}_1]_{D_2}$ we have $[\mathbf{a}_1]_{D_2}$ = 1st column of T, $[\mathbf{b}_1]_{D_2}$ = 2nd column of T and $[\mathbf{c}_1]_{D_2}$ = 3rd column of T. Now let $\mathbf{r}$ denote an arbitrary vector in S. Since D_1 is a basis, there exists r_1, r_2, $r_3 \in R$ such that $\mathbf{r} = r_1\mathbf{a}_1 + r_2\mathbf{b}_1 + r_3\mathbf{c}_1$. By T1.9, we have $[\mathbf{r}]_D = r_1[\mathbf{a}_1]_D + r_2[\mathbf{b}_1]_D + r_3[\mathbf{c}_1]_D$. But then

$$
\begin{aligned}
T[\mathbf{r}]_D &= T(r_1[\mathbf{a}_1]_{D_1} + r_2[\mathbf{b}_1]_{D_1} + r_3[\mathbf{c}_1]_{D_1}) \\
&= r_1 T[\mathbf{a}_1]_{D_1} + r_2 T[\mathbf{b}_1]_{D_1} + r_3 T[\mathbf{c}_1]_{D_1} \\
&= r_1[\mathbf{a}_1]_{D_2} + r_2[\mathbf{b}_1]_{D_2} + r_3[\mathbf{c}_1]_{D_2} \\
&= [r_1\mathbf{a}_1 + r_2\mathbf{b}_1 + r_3\mathbf{c}_1]_{D_2} \\
&= [\mathbf{r}]_{D_2} \ .
\end{aligned}
$$

Therefore, the fact that T "works" for $[\mathbf{a}_1]_{D_1}$, $[\mathbf{b}_1]_{D_1}$ and $[\mathbf{c}_1]_{D_1}$ implies that it "works" for all $[\mathbf{r}]_{D_1}$, $\mathbf{r} \in S$. Consequently, we have established the following theorem.

(T2.10) Theorem: The change of basis matrix T from $D_1 = \{\mathbf{a}_1, \mathbf{b}_1, \mathbf{c}_1\}$ to $D_2 = \{\mathbf{a}_2, \mathbf{b}_2, \mathbf{c}_2\}$ is

$$
T = \begin{bmatrix} | & | & | \\ [\mathbf{a}_1]_{D_2} & [\mathbf{b}_1]_{D_2} & [\mathbf{c}_1]_{D_2} \\ | & | & | \end{bmatrix} .
$$

If T is the change of basis matrix from D_1 to D_2, then it is straightforward to show that T^{-1} exists (since the columns of T are linearly independent). Furthermore, by multiplying both sides of $T[\mathbf{r}]_{D_1} = [\mathbf{r}]_{D_2}$ by T^{-1}, we have

$$
[\mathbf{r}]_{D_1} = T^{-1}[\mathbf{r}]_{D_2} \ .
$$

Therefore, T^{-1} is the change of basis matrix from D_2 to D_1. Applying T2.10 with the roles of D_1 and D_2 switched, we have the following corollary.

(C2.11) Corollary: If T is the change of basis matrix from D_1 to D_2, then T^{-1} is the change of basis matrix from D_2 to D_1 and

$$
T^{-1} = \begin{bmatrix} | & | & | \\ [\mathbf{a}_2]_{D_1} & [\mathbf{b}_2]_{D_1} & [\mathbf{c}_2]_{D_1} \\ | & | & | \end{bmatrix} .
$$

As we shall see in a later section on applications, it is often easy to find by inspection the change of basis matrix in one direction, say from D_2 to D_1. The change of basis matrix in the other direction is then found by inverting the first matrix.

Now that we have a procedure for finding the change of basis matrix from D_1 to D_2, we need to discover how to find the metrical matrix G_2 of D_2 from the metrical matrix G_1 of D_1. We begin by obtaining a useful expression for T using the reciprocal bases $D_1^* = \{\mathbf{a}_1^*, \mathbf{b}_1^*, \mathbf{c}_1^*\}$ and $D_2^* = \{\mathbf{a}_2^*, \mathbf{b}_2^*, \mathbf{c}_2^*\}$. By T2.5,

$$[\mathbf{a}_1]_{D_2} = \begin{bmatrix} \mathbf{a}_1 \bullet \mathbf{a}_2^* \\ \mathbf{a}_1 \bullet \mathbf{b}_2^* \\ \mathbf{a}_1 \bullet \mathbf{c}_2^* \end{bmatrix}, \quad [\mathbf{b}_1]_{D_2} = \begin{bmatrix} \mathbf{b}_1 \bullet \mathbf{a}_2^* \\ \mathbf{b}_1 \bullet \mathbf{b}_2^* \\ \mathbf{b}_1 \bullet \mathbf{c}_2^* \end{bmatrix}, \quad [\mathbf{c}_1]_{D_2} = \begin{bmatrix} \mathbf{c}_1 \bullet \mathbf{a}_2^* \\ \mathbf{c}_1 \bullet \mathbf{b}_2^* \\ \mathbf{c}_1 \bullet \mathbf{c}_2^* \end{bmatrix} .$$

Hence,

$$T = \begin{bmatrix} \mathbf{a}_1 \bullet \mathbf{a}_2^* & \mathbf{b}_1 \bullet \mathbf{a}_2^* & \mathbf{c}_1 \bullet \mathbf{a}_2^* \\ \mathbf{a}_1 \bullet \mathbf{b}_2^* & \mathbf{b}_1 \bullet \mathbf{b}_2^* & \mathbf{c}_1 \bullet \mathbf{b}_2^* \\ \mathbf{a}_1 \bullet \mathbf{c}_2^* & \mathbf{b}_1 \bullet \mathbf{c}_2^* & \mathbf{c}_1 \bullet \mathbf{c}_2^* \end{bmatrix} . \qquad (2.21)$$

We now have the following equations:

$$\begin{aligned} T[\mathbf{r}]_{D_1} &= [\mathbf{r}]_{D_2} \\ G_2[\mathbf{r}]_{D_2} &= [\mathbf{r}]_{D_2^*} \\ G_1[\mathbf{r}]_{D_1} &= [\mathbf{r}]_{D_1^*} \quad . \end{aligned}$$

Defining S to be the change of basis matrix from D_1^* to D_2^*, we also have

$$S[\mathbf{r}]_{D_1^*} = [\mathbf{r}]_{D_2^*} \quad .$$

These relationships are summarized in the following circuit diagram:

$$\begin{array}{ccc} [\mathbf{r}]_{D_1} & \overset{T}{\rightarrow} & [\mathbf{r}]_{D_2} \\ G_1 \downarrow & & \downarrow G_2 \\ [\mathbf{r}]_{D_1^*} & \overset{S}{\rightarrow} & [\mathbf{r}]_{D_2^*} \end{array} .$$

According to this diagram,

$$SG_1T^{-1}[\mathbf{r}]_{D_2} = G_2[\mathbf{r}]_{D_2}$$

for all $\mathbf{r} \in S$. Hence, $SG_1T^{-1} = G_2$. Consequently we need to calculate S. By T2.6, D_2 is the reciprocal basis of D_2^*. Hence,

$$[\mathbf{a}_1^*]_{D_2^*} = \begin{bmatrix} \mathbf{a}_1^* \bullet (\mathbf{a}_2^*)^* \\ \mathbf{a}_1^* \bullet (\mathbf{b}_2^*)^* \\ \mathbf{a}_1^* \bullet (\mathbf{c}_2^*)^* \end{bmatrix} = \begin{bmatrix} \mathbf{a}_1^* \bullet \mathbf{a}_2 \\ \mathbf{a}_1^* \bullet \mathbf{b}_2 \\ \mathbf{a}_1^* \bullet \mathbf{c}_2 \end{bmatrix} .$$

Applying this observation to each of the columns of the change of basis matrix from D_1^* to D_2^* written in the form of (2.21), we obtain

$$S = \begin{bmatrix} \mathbf{a}_1^* \bullet \mathbf{a}_2 & \mathbf{b}_1^* \bullet \mathbf{a}_2 & \mathbf{c}_1^* \bullet \mathbf{a}_2 \\ \mathbf{a}_1^* \bullet \mathbf{b}_2 & \mathbf{b}_1^* \bullet \mathbf{b}_2 & \mathbf{c}_1^* \bullet \mathbf{b}_2 \\ \mathbf{a}_1^* \bullet \mathbf{c}_2 & \mathbf{b}_1^* \bullet \mathbf{c}_2 & \mathbf{c}_1^* \bullet \mathbf{c}_2 \end{bmatrix} .$$

Consider,

$$S^t = \begin{bmatrix} \mathbf{a}_2 \bullet \mathbf{a}_1^* & \mathbf{b}_2 \bullet \mathbf{a}_1^* & \mathbf{c}_2 \bullet \mathbf{a}_1^* \\ \mathbf{a}_2 \bullet \mathbf{b}_1^* & \mathbf{b}_2 \bullet \mathbf{b}_1^* & \mathbf{c}_2 \bullet \mathbf{b}_1^* \\ \mathbf{a}_2 \bullet \mathbf{c}_1^* & \mathbf{b}_2 \bullet \mathbf{c}_1^* & \mathbf{c}_2 \bullet \mathbf{c}_1^* \end{bmatrix} .$$

According to the analogous statement to (2.21), S^t is the change of basis matrix from D_2 to D_1. That is, $S^t = T^{-1}$. Therefore, $S = (T^{-1})^t$. Since $(T^{-1})^t = (T^t)^{-1}$, we write T^{-t}. Hence we have the following theorem.

(T2.12) Theorem: Let D_1 and D_2 denote bases, T the change of basis matrix from D_1 to D_2 and let G_1 denote the metrical matrix for D_1. Then the relationship between G_1 and G_2, the metrical matrix for D_2, is given by

$$G_1 = T^t G_2 T \quad \text{and} \quad G_2 = T^{-t} G_1 T^{-1} \quad .$$

We summarize these results in the following circuit diagram.

$$\begin{array}{ccccc} & [\mathbf{r}]_{D_1} & \xrightarrow{T} & [\mathbf{r}]_{D_2} & \\ T^t G_2 T = G_1 & \downarrow & & \uparrow & G_2 = T^{-t} G_1 T^{-1} \\ & [\mathbf{r}]_{D_1^*} & \xrightarrow{T^{-t}} & [\mathbf{r}]_{D_2^*} & \end{array} \quad . \qquad (2.22)$$

□

(T2.13) Theorem: Let $D = \{\mathbf{a},\mathbf{b},\mathbf{c}\}$ denote a basis with metrical matrix G. Then the volumne v of the parallelpiped outlined by D is such that

$$v^2 = \det(G) \quad .$$

Proof: Let C denote a cartesian basis and let T denote the change of basis matrix from C to D. By (1.18) we know that $v = \det(M)$ where

$$M = \left[\begin{array}{c|c|c} [\mathbf{a}]_C & [\mathbf{b}]_C & [\mathbf{c}]_C \end{array} \right] \quad .$$

Note that

$$\begin{aligned} TM &= [T[\mathbf{a}]_C \mid T[\mathbf{b}]_C \mid T[\mathbf{c}]_C] \\ &= [[\mathbf{a}]_D \mid [\mathbf{b}]_D \mid [\mathbf{c}]_D] \\ &= I_3 \quad . \end{aligned}$$

Hence $\det(TM) = 1$. Therefore,

$$\begin{aligned} 1 &= \det(TM) \\ &= \det(T)\det(M) \qquad \text{(by (A2.8))} \\ &= \det(T)v \quad . \end{aligned}$$

Consequently, $\det(T) = 1/v$. Since the metrical matrix for C is I_3, by T2.12,

$$I_3 = T^t G T \quad .$$

Since $\det(T^t) = \det(T)$,

$$1 = \det(T^t)\det(G)\det(T)$$
$$= (1/v^2)\det(G) \quad .$$

Therefore, $\det(G) = v^2$. □

We are now in a position to show how the cross product is calculated when the vectors are expressed with respect to an arbitrary basis D. In Chapter 1, we discussed how to calculate $\mathbf{v} \times \mathbf{w}$ with respect to a cartesian basis.

(T2.14) Theorem: Let $D = \{\mathbf{a},\mathbf{b},\mathbf{c}\}$ denote a basis with metrical matrix G and let $\mathbf{r}$ and $\mathbf{s}$ denote vectors in S. Then

$$\mathbf{r} \times \mathbf{s} = \det(G)^{-\frac{1}{2}} \begin{vmatrix} \mathbf{a} & & \\ \mathbf{b} & G[\mathbf{r}]_D & G[\mathbf{s}]_D \\ \mathbf{c} & & \end{vmatrix} \tag{2.23}$$

Proof: Let r_1, r_2, r_3, s_1, s_2 and $s_3 \in R$ be such that

$$\mathbf{r} = r_1\mathbf{a}^* + r_2\mathbf{b}^* + r_3\mathbf{c}^* \quad ,$$

and

$$\mathbf{s} = s_1\mathbf{a}^* + s_2\mathbf{b}^* + s_3\mathbf{c}^* \quad ,$$

where $D^* = \{\mathbf{a}^*,\mathbf{b}^*,\mathbf{c}^*\}$ is the reciprocal basis for D. Then

$$\mathbf{r} \times \mathbf{s} = (r_1\mathbf{a}^* + r_2\mathbf{b}^* + r_3\mathbf{c}^*) \times (s_1\mathbf{a}^* + s_2\mathbf{b}^* + s_3\mathbf{c}^*)$$
$$= (r_2s_3 - r_3s_2)(\mathbf{b}^* \times \mathbf{c}^*) + (r_3s_1 - r_1s_3)\mathbf{c}^* \times \mathbf{a}^*$$
$$+ (r_1s_2 - s_1r_2)\mathbf{a}^* \times \mathbf{b}^* \quad ,$$

where we have used the properties of the cross product discussed in Chapter 1. By T2.6,

$$\mathbf{r} \times s = v^*[(r_2s_3 - r_3s_2)\mathbf{a} + (r_3s_1 - r_1s_3)\mathbf{b} + (r_1s_2 - s_1r_2)\mathbf{c}] \quad ,$$

where $v^* = \mathbf{a}^* \bullet (\mathbf{b}^* \times \mathbf{c}^*)$. Hence,

$$\mathbf{r} \times \mathbf{s} = v^* \begin{vmatrix} \mathbf{a} & r_1 & s_1 \\ \mathbf{b} & r_2 & s_2 \\ \mathbf{c} & r_3 & s_3 \end{vmatrix} .$$

By T2.13, $(v^*)^2 = \det(G^*)$. By C2.7, $G^* = G^{-1}$ and so

$$\begin{aligned} v^* &= (\det(G^*))^{\frac{1}{2}} \\ &= (\det(G^{-1}))^{\frac{1}{2}} \\ &= (\det(G))^{-\frac{1}{2}} \quad ; \end{aligned}$$

thus the theorem is established. □

Let $D = \{\mathbf{a},\mathbf{b},\mathbf{c}\}$ denote a basis with metrical matrix G. By T2.13,

$$v^2 = \det(G) = \begin{vmatrix} a^2 & ab\cos\gamma & ac\cos\beta \\ ab\cos\gamma & b^2 & bc\cos\alpha \\ ac\cos\beta & bc\cos\alpha & c^2 \end{vmatrix} ,$$

and so we find that

$$v = abc(1 - \cos^2\alpha - \cos^2\beta - \cos^2\gamma + 2\cos\alpha\cos\beta\cos\gamma)^{\frac{1}{2}} .$$

(P2.8) Problem: Show that $(v^*)^2 = \det(G^*)$ where v^* is the volume of the parallelepiped outlined $D^* = \{\mathbf{a}^*,\mathbf{b}^*,\mathbf{c}^*\}$ and that

$$v^* = a^*b^*c^*(1 - \cos^2\alpha^* - \cos^2\beta^* - \cos^2\gamma^* + 2\cos\alpha^*\cos\beta^*\cos\gamma^*)^{\frac{1}{2}} .$$

By C2.7, G^{-1} is the metrical matrix for D^*. This reciprocal relationship between the metrical matrices of D and D^* yields the reciprocal volume relationship

$$v^2 = \det(G) = 1/\det(G^{-1}) = 1/v^{*2}$$

and therefore $v = 1/v^*$.

Zones: Suppose the nonparallel lattice planes P_1 and P_2 have Miller indices $(h_1\ k_1\ \ell_1)$ and $(h_2\ k_2\ \ell_2)$, respectively, then the zone axis defined by them is in the direction of a vector parallel to the intersection of P_1 and P_2. Hence if

$$\mathbf{s}_1 = h_1\mathbf{a}^* + k_1\mathbf{b}^* + \ell_1\mathbf{c}^* ,$$

and

$$\mathbf{s}_2 = h_2\mathbf{a}^* + k_2\mathbf{b}^* + \ell_2\mathbf{c}^* ,$$

then the zone axis is in the direction of $\mathbf{s}_1 \times \mathbf{s}_2$.

(E2.15) Example - The vector product of two non-collinear reciprocal lattice vectors: Show that $\det(G)^{\frac{1}{2}}(\mathbf{s}_1 \times \mathbf{s}_2)$ is in the direct lattice L_D where G is the metrical matrix for D.

Solution: By T2.14,

$$\det(G)^{\frac{1}{2}}(\mathbf{s}_1 \times \mathbf{s}_2) = \begin{vmatrix} \mathbf{a} & & \\ \mathbf{b} & G[\mathbf{s}_1]_D & G[\mathbf{s}_2]_D \\ \mathbf{c} & & \end{vmatrix}$$

$$= \begin{vmatrix} \mathbf{a} & h_1 & h_2 \\ \mathbf{b} & k_1 & k_2 \\ \mathbf{c} & \ell_1 & \ell_2 \end{vmatrix} .$$

Hence $\det(G)^{\frac{1}{2}}(\mathbf{s}_1 \times \mathbf{s}_2) = u\mathbf{a} + v\mathbf{b} + w\mathbf{c}$ where

$$u = \begin{vmatrix} k_1 & k_2 \\ \ell_1 & \ell_2 \end{vmatrix} , \quad v = -\begin{vmatrix} h_1 & h_2 \\ \ell_1 & \ell_2 \end{vmatrix} , \quad w = \begin{vmatrix} h_1 & h_2 \\ k_1 & k_2 \end{vmatrix} .$$

Since h_1, k_1, ℓ_1, h_2, k_2 and ℓ_2 are all integers, so are u, v and w. Hence $\det(G)^{\frac{1}{2}}(\mathbf{s}_1 \times \mathbf{s}_2) \in L_D$. □

The u, v and w in this example define the zone axis. To distinguish such a set of integers from a set defining a crystal plane $(hk\ell)$, the integers for the zone are enclosed between a pair of brackets $[uvw]$.

(E2.16) Example - Finding a vector perpendicular to the plane defined by two non-collinear vectors in a crystal: The lunar mineral pyroxferroite $(Fe,Ca)SiO_3$ is triclinic with cell dimensions a = 6.621A,

b = 7.551A, c = 17.381A, α = 114.27°, β = 82.68°, γ = 94.58° (Burnham, 1971). Calculate the cross product of the following two vectors

$$\mathbf{r}_{34} = -0.0963\mathbf{a} + 0.1243\mathbf{b} + 0.2018\mathbf{c}$$
$$\mathbf{r}_{54} = 0.1084\mathbf{a} - 0.0880\mathbf{b} + 0.2947\mathbf{c}$$

to find a vector perpendicular to their plane.

Solution: Calculating G with the cell dimension of pyroxferroite, we find

$$G = \begin{bmatrix} 43.8376 & -3.9922 & 14.6624 \\ -3.9922 & 57.0176 & -53.9461 \\ 14.6624 & -53.9461 & 302.0992 \end{bmatrix}$$

and $(\det G)^{\frac{1}{2}} = 785.3466$. Hence,

$$\mathbf{r}_{34} \times \mathbf{r}_{54} = (785.3466)^{-1} \begin{vmatrix} \mathbf{a} & -1.7589 & 9.4243 \\ \mathbf{b} & -3.4146 & -21.3482 \\ \mathbf{c} & 52.8461 & 95.3653 \end{vmatrix}$$

$$= 1.0219\mathbf{a} + 0.8478\mathbf{b} + 0.0888\mathbf{c}$$

is a vector perpendicular to the plane of $\mathbf{r}_{34}$ and $\mathbf{r}_{54}$.

(P2.9) Problem: Calculate the cross product $\mathbf{k} \times \mathbf{i}$ for pyroxferroite given that

$$\mathbf{k} = 0.1166\mathbf{a} + 0.0968\mathbf{b} + 0.0101\mathbf{c}$$

and

$$\mathbf{i} = -0.0286\mathbf{a} + 0.0369\mathbf{b} + 0.0599\mathbf{c} \quad .$$

APPLICATIONS

In the previous section, we introduced a method for finding the entries of a transformation matrix T. In practice, however, the entries of such a matrix may be found by making a careful drawing of a crystal structure or its lattice representation that shows how one set of basis vectors is related to another set. Then by examining how the vectors of one basis can be expressed as a linear combination of the other, the coefficients of the second basis vectors are used as columns in the ma-

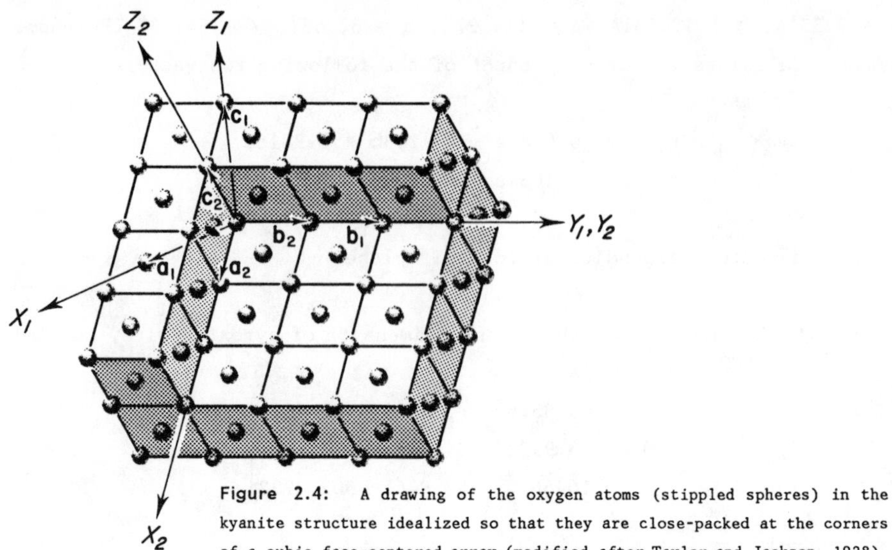

Figure 2.4: A drawing of the oxygen atoms (stippled spheres) in the kyanite structure idealized so that they are close-packed at the corners of a cubic face-centered array (modified after Taylor and Jackson, 1928). The $D_1 = \{\mathbf{a}_1,\mathbf{b}_1,\mathbf{c}_1\}$ basis vectors directed along coordinate axes X_1, Y_1, Z_1, outline a unit cell of the kyanite structure and $D_2 = \{\mathbf{a}_2,\mathbf{b}_2,\mathbf{c}_2\}$ directed along X_2, Y_2, Z_2, outlines a cell of the cubic face-centered array of oxygen atoms.

trix. This method will be illustrated with an example. Another method for finding T is illustrated in EA2.23.

(E2.17) Example - Determination of the cell dimensions of a subcell of the kyanite structure: The oxygen atoms comprising the structure of kyanite, Al_2SiO_5, can be viewed as a slightly distorted cubic close-packed face-centered structure with Si occupying 10% of the available tetrahedral voids and Al occupying 40% of the available octahedral voids (Taylor and Jackson, 1928). Before the structure of this mineral was solved, the close-packed nature of the structure was theorized because the dimensions of the unit cell of kyanite outlined by the natural basis $D_1 = \{\mathbf{a}_1,\mathbf{b}_1,\mathbf{c}_1\}$ are similar to those calculated from a set of basis vectors $D_2 = \{\mathbf{a}_2,\mathbf{b}_2\mathbf{c}_2\}$ defining a face-centered subcell of oxygen atoms.

The relationship between the D_2-basis of the subcell and the D_1-basis of the kyanite cell is displayed in Figure 2.4. Inspection of this figure shows that the D_1-basis vectors can be written in terms of the D_2-basis vectors as follows:

$$\begin{aligned} \mathbf{a}_1 &= 3/2\mathbf{a}_2 - 1/2\mathbf{b}_2 + \mathbf{c}_2 \\ \mathbf{b}_1 &= 2\mathbf{b}_2 \\ \mathbf{c}_1 &= -\mathbf{a}_2 + \mathbf{c}_2 \quad . \end{aligned}$$

A recent measurement of the cell dimensions of kyanite at 25°C (Winter and Ghose, 1979) yielded the values $a_1 = 7.126\text{A}$, $b_1 = 7.852\text{A}$, $c_1 = 5.572\text{A}$, $\alpha_1 = 89.99°$, $\beta_1 = 101.11°$ and $\gamma_1 = 106.03°$ for D_1. With this information, calculate the cell dimensions of the subcell defined by the oxygen atoms in kyanite.

Solution: Using T2.10, we can construct the change of basis matrix

$$T = \begin{bmatrix} | & | & | \\ [\mathbf{a}_1]_{D_2} & [\mathbf{b}_1]_{D_2} & [\mathbf{c}_1]_{D_2} \\ | & | & | \end{bmatrix} = \begin{bmatrix} 3/2 & 0 & -1 \\ -1/2 & 2 & 0 \\ 1 & 0 & 1 \end{bmatrix}$$

from D_1 to D_2.

Using the cell dimensions for D_1, its metrical matrix G_1 is

$$G_1 = \begin{bmatrix} 50.7799 & -15.4510 & -7.6511 \\ -15.4510 & 61.6539 & 0.0076 \\ -7.6511 & 0.0076 & 31.0472 \end{bmatrix} .$$

To compute the cell dimensions of the face-centered cell, we need to solve for the metrical matrix G_2 of D_2. According to T2.12,

$$G_2 = T^{-t} G_1 T^{-1} . \tag{2.24}$$

Inverting T (see Appendix 2), we find that

$$T^{-1} = \begin{bmatrix} 3/2 & 0 & -1 \\ -1/2 & 2 & 0 \\ 1 & 0 & 1 \end{bmatrix}^{-1} = \begin{bmatrix} 2/5 & 0 & 2/5 \\ 1/10 & 1/2 & 1/10 \\ -2/5 & 0 & 3/5 \end{bmatrix} .$$

Substituting T^{-t}, G_1 and T^{-1} into (2.24) we have

$$G_2 = \begin{bmatrix} 2/5 & 1/10 & -2/5 \\ 0 & 1/2 & 0 \\ 2/5 & 1/10 & 3/5 \end{bmatrix} \begin{bmatrix} 50.7799 & -15.4510 & -7.6511 \\ -15.4510 & 61.6539 & 0.0076 \\ -7.6511 & 0.0076 & 31.0472 \end{bmatrix} \begin{bmatrix} 2/5 & 0 & 2/5 \\ 1/10 & 1/2 & 1/10 \\ -2/5 & 0 & 3/5 \end{bmatrix}$$

$$= \begin{bmatrix} 14.9205 & -0.0090 & -0.5580 \\ -0.0090 & 15.4135 & -0.0052 \\ -0.5580 & -0.0052 & 15.0106 \end{bmatrix} = \begin{bmatrix} a_2^2 & a_2b_2\cos\gamma_2 & a_2c_2\cos\beta_2 \\ a_2b_2\cos\gamma_2 & b_2^2 & b_2c_2\cos\alpha_2 \\ a_2c_2\cos\beta_2 & b_2c_2\cos\alpha_2 & c_2^2 \end{bmatrix} .$$

Equating the entries of these two matrices and solving for the cell dimensions of the face-centered subcell, we find that $a_2 = 3.8627\text{A}$, $b_2 = 3.9260\text{A}$, $c_2 = 3.8743\text{A}$, $\alpha_2 = 90.02°$, $\beta_2 = 92.14°$ and $\gamma_2 = 90.03°$. As is evident from these results, the departure of the oxygen atoms in kyanite from an ideal cubic close packed subcell with $a_2 = b_2 = c_2$ and $\alpha_2 = \beta_2 = \gamma_2 = 90°$ is small. □

(P2.10) Problem: The crystal structure of the relatively rare mineral sapphirine, $(AlMg)_4(Al_2Si)_2O_{20}$, can also be described as a slightly distorted cubic close-packed array of oxygen atoms (Moore, 1968). The equations that define the basis vectors of the sapphirine structure $D_1 = \{\mathbf{a}_1,\mathbf{b}_1,\mathbf{c}_1\}$ in terms of the face-centered subcell $D_2 = \{\mathbf{a}_2,\mathbf{b}_2,\mathbf{c}_2\}$ are

$$\begin{aligned} \mathbf{a}_1 &= 2\mathbf{a}_2 + 2\mathbf{b}_2 \\ \mathbf{b}_1 &= -5/2\mathbf{a}_2 + 5/2\mathbf{b}_2 \\ \mathbf{c}_1 &= -\mathbf{a}_2 - \mathbf{b}_2 + 2\mathbf{c}_2 \quad . \end{aligned}$$

(1) Using the cell dimensions $a_1 = 11.286\text{A}$, $b_1 = 14.438\text{A}$, $c_1 = 9.957\text{A}$, $\beta_1 = 125.4°$, and $\alpha_1 = \gamma_1 = 90.0°$, measured for the mineral (Higgins and Ribbe, 1979), show that the metrical matrix defining the geometry of the face-centered subcell is

$$G_2 = \begin{bmatrix} 16.2991 & -0.3774 & -0.1762 \\ -0.3774 & 16.2991 & -0.1762 \\ -0.1762 & -0.1762 & 16.4722 \end{bmatrix} . \tag{2.25}$$

(2) Evaluate the entries of (2.25) and show that the cell dimensions of the subcell are $a_2 = b_2 = 4.037\text{A}$, $c_2 = 4.059$, $\alpha_2 = \beta_2 = 90.62$ and $\gamma_2 = 91.33$.

(E2.18) Example - Transforming indices of planes with change of basis: The close-packed monolayers of the kyanite structure parallel (111), $(\bar{1}11)$, $(\bar{1}\bar{1}1)$ and $(1\bar{1}1)$ of its face-centered subcell. Determine the indices of these planes in terms of the kyanite basis.

Solution: We recall that perpendicular to each stack of planes $(h_2k_2\ell_2)$ there exists a vector **s** with the coordinates such that

$$[\mathbf{s}]_{D_2^*} = \begin{bmatrix} h_2 \\ k_2 \\ \ell_2 \end{bmatrix} .$$

To find the coordinates of this vector in terms of the D_1^* basis, we use the equation (see (2.22))

$$[\mathbf{s}]_{D_1^*} = \mathrm{T}^t[\mathbf{s}]_{D_2^*} , \qquad (2.26)$$

where the components of $[\mathbf{s}]_{D_1^*}$ are the indices of the same plane but defined in terms of the kyanite basis. Transposing T and substituting this result into (2.26), we obtain

$$\begin{bmatrix} h_1 \\ k_1 \\ \ell_1 \end{bmatrix} = \begin{bmatrix} 3/2 & -1/2 & 1 \\ 0 & 2 & 0 \\ -1 & 0 & 1 \end{bmatrix} \begin{bmatrix} h_2 \\ k_2 \\ \ell_2 \end{bmatrix} = \begin{bmatrix} 3h_2/2 - k_2/2 + \ell_2 \\ 2k_2 \\ -h_2 + \ell_2 \end{bmatrix} .$$

Replacing $(h_2k_2\ell_2)$ in succession by (111), $(\bar{1}11)$, $(\bar{1}\bar{1}1)$ and $(1\bar{1}1)$, we get (220), $(\bar{1}22)$, $(0\bar{2}2)$, and $(3\bar{2}0)$. The integers in (220) and $(0\bar{2}2)$ are reduced to (110) and $(0\bar{1}1)$, respectively, because both sets contain a common factor of 2. With this reduction, we can conclude that the close-packed monolayers of oxygen atoms in kyanite are parallel to planes with indices (110), $(\bar{1}22)$, $(0\bar{1}1)$ and $(3\bar{2}0)$. □

(P2.11) Problem: The close-packed monolayers of oxygen atoms of the sapphirine structure parallel (111), $(\bar{1}11)$, $(\bar{1}\bar{1}\bar{1})$ and $(1\bar{1}1)$ of its face-centered subcell. Show that these monolayers parallel (100), (052), $(\bar{1}01)$ and $(0\bar{5}2)$ of the sapphirine structure.

(P2.12) Problem: The crystal structure of tremolite, a calcic amphibole of $Ca_2Mg_5(Si_4O_{11})_2(OH)_2$ composition, bears a close structural resemblance with that of the calcic pyroxene diopside, $CaMgSi_2O_6$, when viewed down [010] with **a**, **c**, and β being nearly the same in both minerals. However, the cell dimensions of the two minerals differ in that the *b*-cell edge of tremolite is double that of diopside. When Warren (1930) solved the structure of tremolite, he took advantage of these relationships and derived the basic structural unit (a double chain) in the amphibole structure by reflecting the diopside structure over a plane perpendicular to [010]. In a paper on this choice of basis vectors, Whittaker and Zussman (1961) have observed that although Warren's choice, $D_1 = \{\mathbf{a}_1, \mathbf{b}_1, \mathbf{c}_1\}$, illustrates the relationship to the diopside structure, it does not conform with the conventional choice of basis, $D_2 = \{\mathbf{a}_2, \mathbf{b}_2, \mathbf{c}_2\}$, defined by

$$\begin{aligned} \mathbf{a}_2 &= \mathbf{a}_1 - \mathbf{c}_1 \\ \mathbf{b}_2 &= \mathbf{b}_1 \\ \mathbf{c}_2 &= \mathbf{c}_1 \quad . \end{aligned}$$

(1) Using the cell dimensions (a_1 = 9.78A, b_1 = 17.8A, c_1 = 5.26A. β_1 = 73.97°, $\alpha_1 = \gamma_1$ = 90°) determined by Warren, show that the cell dimensions of the conventional cell are a_2 = 9.74A, b_2 = 17.8A, c_2 = 5.26A, β_2 = 105.23°, $\alpha_2 = \gamma_2$ = 90°.

(2) A Ca-rich amphibole like tremolite is commonly exsolved with Ca-poor amphibole lamallae developed along ($\bar{2}01$) of the cell adopted by Warren. Show that these lamellae parallel ($\bar{1}01$) of the conventional cell (cf. Ross *et al.*, 1969).

(3) Warren's structural analysis of tremolite showed that one of the Si atoms in the double chain has coordinates $(0.29, 0.08, 0.01)^t$. Show that the coordinates of this Si atom in the conventional cell are $(0.29, 0.08, 0.30)^t$.

(E2.19) Example - A calculation of the cell dimensions of kyanite assuming its oxygen atoms are cubic close-packed. If we assume that the face-centered subcell displayed in Figure 2.4 for kyanite has cubic geometry (i.e., the oxygen atoms in kyanite are ideally cubic close-packed so that $a_2 = b_2 = c_2$, $\alpha_2 = \beta_2 = \gamma_2 = 90°$) and the oxygen atoms are in contact along the face diagonals of the cell, then $a_2 = 2\sqrt{2}r_0$ where r_0 is the effective nonbonded radius of an oxygen atom. If we equate the volume of this cell, $v = 16\sqrt{2}r_0^3$, with that obtained by evaluating the determinant of G_2, then $16\sqrt{2}r_0^3 = 58.719A^3$. Solving for r_0, we get $r_0 = 1.374A$. Our problem is to determine the cell dimensions of kyanite assuming that its oxygen atoms are ideally cubic closest packed each with an effective radius of 1.374A. As the cell edge of such a cube is $2\sqrt{2}r_0$, then its cell edges are given by $a_2 = 2\sqrt{2} \times 1.374A = 3.886A$. The metrical matrix for this ideal structure with $a_2 = b_2 = c_2 = 3.886A$ and $\alpha_2 = \beta_2 = \gamma_2 = 90.0°$ is obtained by evaluating (2.22):

$$G_1 = T^t G_2 T$$

$$= \begin{bmatrix} 3/2 & -1/2 & 1 \\ 0 & 2 & 0 \\ -1 & 0 & 1 \end{bmatrix} \begin{bmatrix} 15.1010 & 0.0 & 0.0 \\ 0.0 & 15.1010 & 0.0 \\ 0.0 & 0.0 & 15.1010 \end{bmatrix} \begin{bmatrix} 3/2 & 0 & -1 \\ -1/2 & 2 & 0 \\ 1 & 0 & 1 \end{bmatrix}$$

$$= \begin{bmatrix} 52.8535 & -15.1010 & -7.5505 \\ -15.1010 & 60.4040 & 0.0 \\ -78.5505 & 0.0 & 30.2020 \end{bmatrix} .$$

Solving G_1 for the cell dimensions of the ideal close-packed kyanite structure, we obtain $a_1 = 7.270A$, $b_1 = 7.772A$, $c_1 = 5.496A$, $\alpha_1 = 90.0°$, $\beta_1 = 100.89°$ and $\gamma_1 = 105.50°$. These values show a close correspondence with those measured for kyanite (E2.17), indicating the centers of the oxygen atoms in the structure define rather well a face-centered cubic lattice. □

(P2.13) Problem:

(1) Evaluate the determinant of the metrical matrix G_2 for the slightly distorted subcell of sapphirine and show that its volume is $V_2 = 66.125A^3$.

(2) Equate the volume obtained in (1) with that of a cubic face-centered cell, $16\sqrt{2}r_0^3$, and show that $r_0 = 1.430A$.

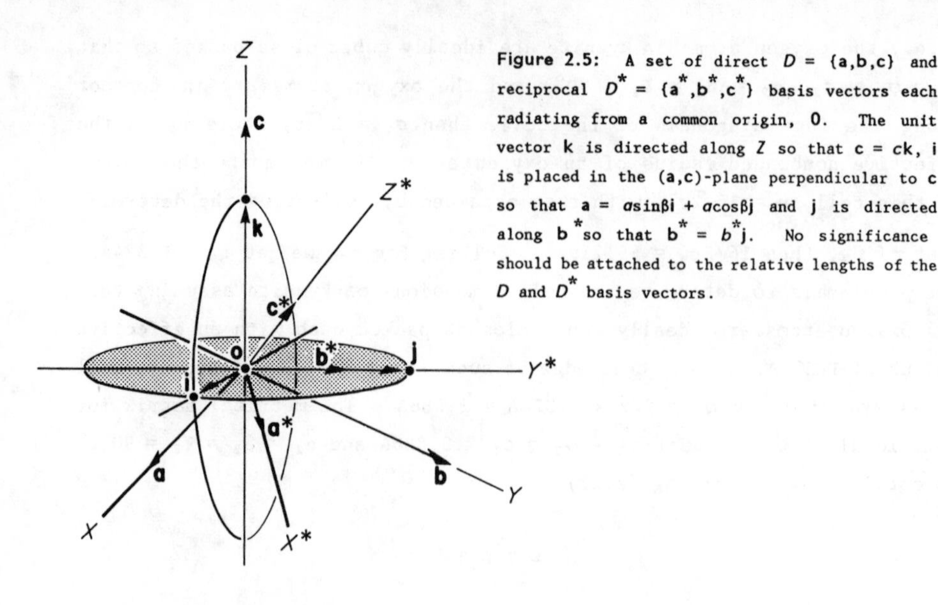

Figure 2.5: A set of direct $D = \{\mathbf{a},\mathbf{b},\mathbf{c}\}$ and reciprocal $D^* = \{\mathbf{a}^*,\mathbf{b}^*,\mathbf{c}^*\}$ basis vectors each radiating from a common origin, **0**. The unit vector $\mathbf{k}$ is directed along Z so that $\mathbf{c} = c\mathbf{k}$, $\mathbf{i}$ is placed in the $(\mathbf{a},\mathbf{c})$-plane perpendicular to $\mathbf{c}$ so that $\mathbf{a} = a\sin\beta\mathbf{i} + a\cos\beta\mathbf{j}$ and $\mathbf{j}$ is directed along $\mathbf{b}^*$ so that $\mathbf{b}^* = b^*\mathbf{j}$. No significance should be attached to the relative lengths of the D and D^* basis vectors.

(3) Assuming that the cell edge of the cubic cell is $2\sqrt{2}r_0$ and that $r_0 = 1.430$A, show that the metrical matrix for an ideal close-packed sapphirine structure is

$$G_1 = \begin{bmatrix} 130.896 & 0.0 & -65.448 \\ 0.0 & 204.525 & 0.0 \\ -65.437 & 0.0 & 98.172 \end{bmatrix} .$$

(4) Calculate the cell dimensions of the sapphirine assuming that it consists of a cubic close-packed array of oxygen atoms of radius 1.430A. Compare these calculated dimensions ($a_2 = 11.44$A, $b_2 = 14.30$A, $c_2 = 9.91$A, $\beta_2 = 125.3°$, $\alpha_2 = \gamma_2 = 90°$) with those given in P2.10.

A DESCRIPTION OF THE GEOMETRY OF A CRYSTAL IN TERMS OF A CARTESIAN BASIS

It is often useful to define a natural basis $D = \{\mathbf{a},\mathbf{b},\mathbf{c}\}$ of a crystal in terms of a cartesian basis. We shall choose the cartesian basis $C = \{\mathbf{i},\mathbf{j},\mathbf{k}\}$ such that $\mathbf{k}$ is in the direction of $\mathbf{c}$, $\mathbf{j}$ is in the direction of $\mathbf{c} \times \mathbf{a}$ and $\mathbf{i}$ is set perpendicular to the plane of $\mathbf{k}$ and $\mathbf{j}$ so as to complete a right-handed cartesian basis set (Figure 2.5). For this setup of the cartesian basis, there must exist real numbers a_{ij} such that the natural basis D for a crystal can be expressed as

$$\begin{aligned} \mathbf{a} &= a_{11}\mathbf{i} + a_{31}\mathbf{k} \\ \mathbf{b} &= a_{12}\mathbf{i} + a_{22}\mathbf{j} + a_{32}\mathbf{k} \\ \mathbf{c} &= a_{33}\mathbf{k} \quad . \end{aligned} \tag{2.27}$$

Note that by the way in which the C-basis (in particular in regard to the handedness of D and C) is defined in relation to the D-basis, a_{11}, a_{22} and a_{33} must all be positive. Thus, the change of basis matrix A from D to C is given by

$$A = \left[\begin{array}{c|c|c} [\mathbf{a}]_C & [\mathbf{b}]_C & [\mathbf{c}]_C \end{array} \right] = \begin{bmatrix} a_{11} & a_{12} & 0 \\ 0 & a_{22} & 0 \\ a_{31} & a_{32} & a_{33} \end{bmatrix} . \tag{2.28}$$

To find the entries of this matrix, we use the following circuit diagram (compare with 2.22)

$$\begin{array}{ccc} [\mathbf{r}]_D & \xrightarrow{A} & [\mathbf{r}]_C \\ G \downarrow & & \downarrow I_3 \\ [\mathbf{r}]_{D^*} & \xleftarrow{A^t} & [\mathbf{r}]_C \end{array}$$

and obtain

$$G = A^t I_3 A = A^t A \quad . \tag{2.29}$$

Expanding (2.29) we have

$$G = \begin{bmatrix} a^2 & ab\cos\gamma & ac\cos\beta \\ ab\cos\gamma & b^2 & bc\cos\alpha \\ ac\cos\beta & bc\cos\alpha & c^2 \end{bmatrix} = A^t A$$

$$= \begin{bmatrix} a_{11}^2 + a_{31}^2 & a_{11}a_{12} + a_{31}a_{32} & a_{31}a_{33} \\ a_{11}a_{12} + a_{31}a_{32} & a_{12}^2 + a_{22}^2 + a_{32}^2 & a_{32}a_{33} \\ a_{31}a_{33} & a_{33}a_{32} & a_{33}^2 \end{bmatrix} . \tag{2.30}$$

By equating corresponding elements of (2.30), expressions are found from which the a_{ij} entries of A shall be deduced. We begin by observing that $a_{33}^2 = c^2$ and so $a_{33} = c$. Next, replacing a_{33} in both $a_{32}a_{33}$ and $a_{31}a_{33}$ by c, we find that $a_{31} = a\cos\beta$ and $a_{32} = b\cos\alpha$. Replacing a_{31} in

$a_{11}^2 + a_{31}^2 = a^2$ by $a\cos\beta$, we find that $a_{11} = a\sin\beta$. When a_{11}, a_{31} and a_{32} are replaced by $a\sin\beta$, $a\cos\beta$ and $b\cos\alpha$, respectively, and $(\cos\gamma - \cos\alpha\cos\beta)/\sin\beta$ is replaced by $-\cos\gamma^*\sin\alpha$ (T2.8), we find that $a_{12} = -b\sin\alpha\cos\beta^*$. Finally, replacing a_{12} and a_{32} in $a_{12}^2 + a_{22}^2 + a_{32}^2$ by $-b\sin\alpha\cos\beta^*$ and $b\cos\alpha$, respectively, and simplifying, we find that $a_{22} = b\sin\alpha\sin\beta^*$. When these results are substituted into (2.28), the A matrix becomes

$$A = \begin{bmatrix} a\sin\beta & -b\sin\alpha\cos\gamma^* & 0 \\ 0 & b\sin\alpha\sin\gamma^* & 0 \\ a\cos\beta & b\cos\alpha & c \end{bmatrix} . \tag{2.31}$$

Thus, the vector equations in (2.27) are of the form

$$\begin{aligned} \mathbf{a} &= a\sin\beta\mathbf{i} + a\cos\beta k \\ \mathbf{b} &= -b\sin\alpha\cos\gamma^*\mathbf{i} + b\sin\alpha\sin\gamma^*\mathbf{j} + b\cos\alpha k \\ \mathbf{c} &= c\mathbf{k} \quad . \end{aligned} \tag{2.32}$$

(P2.14) Problem: The basis vectors $D^* = \{\mathbf{a}^*, \mathbf{b}^*, \mathbf{c}^*\}$ reciprocal to the D-basis are displayed in Figure 2.5. As $\mathbf{a}^*$, $\mathbf{b}^*$ and $\mathbf{c}^*$ parallel $\mathbf{b} \times \mathbf{c}$, $\mathbf{c} \times \mathbf{a}$ and $\mathbf{a} \times \mathbf{b}$, respectively, $\mathbf{b}^*$ must lie along $\mathbf{j}$ and $\mathbf{a}^*$ must lie in the $(\mathbf{i},\mathbf{j})$-plane with $\mathbf{c}^*$ adopting a general direction relative to the C-basis. With this setup of vectors, there must exist six real numbers b_{ij} such that

$$\begin{aligned} \mathbf{a}^* &= b_{11}\mathbf{i} + b_{21}\mathbf{j} \\ \mathbf{b}^* &= b_{22}\mathbf{j} \\ \mathbf{c}^* &= b_{13}\mathbf{i} + b_{23}\mathbf{j} + b_{33}\mathbf{k} \quad . \end{aligned} \tag{2.33}$$

(1) Determine the b_{ij} entries of the change of basis matrix B from D^* to C.

(2) Construct a circuit diagram comparable to the one constructed for the transformation $G = A^tA$ and show that $G^* = B^tB$.

(3) Equate the entries of the B^tB with those of G^* and show that

$$B = \begin{bmatrix} a^* \sin\gamma^* & 0 & -c^* \cos\beta \sin\alpha^* \\ a^* \cos\gamma^* & b^* & c^* \cos\alpha^* \\ 0 & 0 & c^* \sin\alpha^* \sin\beta \end{bmatrix} \tag{2.34}$$

and that the b_{ij} entries of (2.33) are of the form

$$\begin{aligned} \mathbf{a}^* &= a^* \sin\gamma^* \mathbf{i} + a^* \cos\gamma^* \mathbf{j} \\ \mathbf{b}^* &= b^* \mathbf{j} \\ \mathbf{c}^* &= -c^* \cos\beta \sin\alpha^* \mathbf{i} + c^* \cos\alpha^* \mathbf{j} + c^* \sin\alpha^* \sin\beta \mathbf{k} \quad . \end{aligned} \tag{2.35}$$

If the matrix A (2.28) is known, there is no need to construct B using equation (2.34) as it can be readily found by inverting A^t. This is because $B[\mathbf{r}]_{D^*} = [\mathbf{r}]_C$ and $A^{-t}[\mathbf{r}]_{D^*} = [\mathbf{r}]_C$ for all $\mathbf{r} \in S$, and so $B = A^{-t}$.

Calculation of angular coordinates from crystallographic data: When the D and D^* basis vectors for a crystal are both expressed as a linear combination of a C-basis, it becomes a relatively easy task to calculate the angular coordinates of the directional properties of a crystal for preparing a stereogram of the data. Not only is the method of stereographic projection useful for studying crystallographic point groups, but it has also found use in analyzing the angular relationships between fault vectors, planes, slip lines, twinned crystals, exsolved phases and data from X-ray single crystal photographs (Smith, 1982).

In the method, a crystal is placed at the center of a unit ball with the Z axis directed along the ball's south to north polar axis and with its Y^* axis directed along an west to east line (cf., Bloss, 1971, Chapter 4). The orientation of each vector $\mathbf{r}$ can then be defined in stereographic projection by the angular coordinates ϕ and θ. To find these angles for $\mathbf{r}$, we consider the unit vector $\mathbf{R} = \mathbf{r}/r$. These polar angles are then defined as in Figure 2.6 in terms of the coordinate axes of the crystal. The angle ρ is defined to be the angle between $\mathbf{R}$ and $\mathbf{k}$, i.e., $\rho = \angle(\mathbf{R}:\mathbf{k})$ and ϕ is the angle that the projection of $\mathbf{R}$ onto the $(\mathbf{i},\mathbf{j})$-plane makes with $\mathbf{j}$. The components of $[\mathbf{R}]_C$ are found by using the relationship (2.12)

$$\mathbf{R} = (\mathbf{R} \bullet \mathbf{i}^*)\mathbf{i} + (\mathbf{R} \bullet \mathbf{j}^*)\mathbf{j} + (\mathbf{R} \bullet \mathbf{k}^*)\mathbf{k} \quad .$$

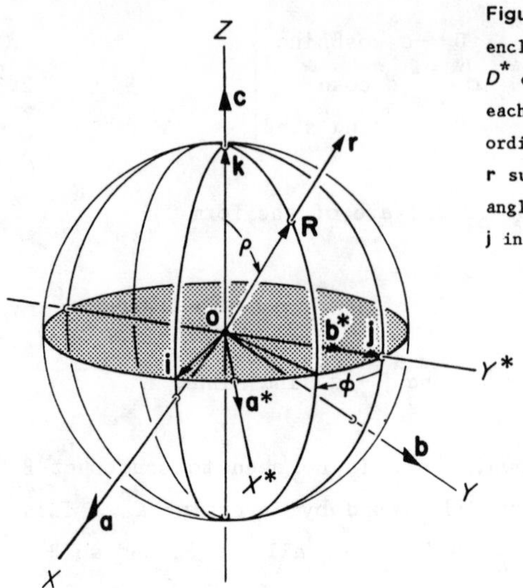

Figure 2.6: A cartesian set of basis vectors enclosed in a unit ball centered at **0** with D and D^* defined as in Figure 2.5. The direction of each vector $\mathbf{r} \in S$ is defined by the angular coordinates ϕ and ρ. The unit vector **R** lies along **r** such that $\mathbf{R} = s\phi s\rho\mathbf{i} + c\phi s\rho\mathbf{j} + c\rho\mathbf{k}$ where the angle ϕ is measured in a clockwise direction from **j** in the plane perpendicular to **k**.

Since $C^* = C$, we have the equality

$$\mathbf{R} = (\mathbf{R} \cdot \mathbf{i})\mathbf{i} + (\mathbf{R} \cdot \mathbf{j})\mathbf{j} + (\mathbf{R} \cdot \mathbf{k})\mathbf{k} \quad .$$

As **k** is a unit vector, $\mathbf{R} \cdot \mathbf{k}$ is the projection of **R** onto **k**, hence $\mathbf{R} \cdot \mathbf{k} = \cos\rho$. The projection of **R** onto the **i**,**k** plane has length $\sin\rho$ and is positioned at an angle of ϕ measured in a clockwise direction from **j**. Hence $\mathbf{R} \cdot \mathbf{j} = \sin\rho\cos\phi$ and $\mathbf{R} \cdot \mathbf{i} = \sin\rho\sin\phi$. Therefore,

$$[\mathbf{R}]_C = \begin{bmatrix} s_\phi s_\rho \\ c_\phi s_\rho \\ c_\rho \end{bmatrix} \tag{2.36}$$

where $s_\phi = \sin\phi$, $c_\phi = \cos\phi$, etc.

The strategy for finding ϕ and ρ is to calculate $[\mathbf{R}]_C$ from the description of **r** and then use the Equation (2.36) to solve for the angular coordinates, taking into account the octant of **R**.

(E2.20) Example - Calculation of angles between zones and face poles: Calculate the angular coordinates for the zone [010] and the face pole (111) of anorthite and the angle between these two directions (see Smith, 1982, Figure 4.17 for a stereogram of these data).

Solution: As a first step in the calculation of the angular coordinates of [010], we note that $\mathbf{r} = \mathbf{b}$ is along the zone [010]. To find $[\mathbf{r}]_C$ we construct the A matrix (2.31), using the cell dimensions of anorthite provided in (P1.13):

$$A = \begin{bmatrix} 7.3513 & -0.65437 & 0 \\ 0 & 12.833 & 0 \\ -3.5716 & -0.69886 & 14.165 \end{bmatrix} .$$

We know that $[\mathbf{r}]_C = A[\mathbf{r}]_D$ and we are given that

$$[\mathbf{r}]_D = \begin{bmatrix} 0 \\ 1 \\ 0 \end{bmatrix} ,$$

so that

$$[\mathbf{r}]_C = \begin{bmatrix} -0.65437 \\ 12.833 \\ -0.69886 \end{bmatrix} . \quad (2.37)$$

Thus, the vector $\mathbf{b}$ directed along the zone [010] in anorthite can be written as a linear combination of the C-basis as

$$\mathbf{b} = -0.65437\mathbf{i} + 12.833\mathbf{j} - 0.69886\mathbf{k} .$$

We next convert the components of $[\mathbf{r}]_C$ (2.37) into those of the unit vector $\mathbf{R}$, equate with (2.36) and solve for ρ and ϕ. We find $[\mathbf{R}]_C$ by dividing each component of $[\mathbf{r}]_C$ by its length which, in this case, is $b = 12.869$A:

$$[\mathbf{R}]_C = (12.869)^{-1} \begin{bmatrix} -0.65437 \\ 12.833 \\ -0.69886 \end{bmatrix} = \begin{bmatrix} -0.05085 \\ 0.99720 \\ -0.05431 \end{bmatrix} .$$

Equating $[\mathbf{R}]_C$ with (2.36), we get

$$\begin{bmatrix} s_\phi s_\rho \\ c_\phi s_\rho \\ c_\rho \end{bmatrix} = \begin{bmatrix} -0.05085 \\ 0.99720 \\ -0.05431 \end{bmatrix} ,$$

from which it follows that $c_\rho = -0.05431$ and $\rho = 93.113°$. Also, because $s_\phi s_\rho = -0.05085$ and $c_\phi s_\rho = 0.99720$, we can write the ratio

$$\frac{\sin\phi\sin\rho}{\cos\phi\sin\rho} = \tan\phi = -0.05099 \ .$$

Taking into account the octant of **R**, we can conclude that $\phi = -2.919°$. Thus, the angular coordinates of the zone [010] in the anorthite are $\rho = 93.113°$ and $\phi = -2.919°$.

In the calculation of the angular coordinates for the face pole of the (111), we observe that $\mathbf{s} = a^* + b^* + c^*$ is perpendicular to this plane. To find $[\mathbf{s}]_C$ we first compute the entries of (2.34)

$$B = \begin{bmatrix} 0.13603 & 0.0 & 0.03430 \\ 0.00694 & 0.07792 & 0.00559 \\ 0.0 & 0.0 & 0.07060 \end{bmatrix} ,$$

using the cell dimensions of anorthite. As indicated earlier, B can also be found by forming A^{-t}. Because $B[\mathbf{s}]_{D^*} = [\mathbf{s}]_C$ and because we are given that

$$[\mathbf{s}]_{D^*} = \begin{bmatrix} 1 \\ 1 \\ 1 \end{bmatrix} ,$$

we have

$$[\mathbf{s}]_C = \begin{bmatrix} 0.17033 \\ 0.09045 \\ 0.07060 \end{bmatrix} . \tag{2.38}$$

Dividing $[\mathbf{s}]_C$ by its magnitude, we convert (2.38) into the unit vector

$$[\mathbf{S}]_C = \begin{bmatrix} 0.82937 \\ 0.44042 \\ 0.34377 \end{bmatrix} = \begin{bmatrix} s_\phi s_\rho \\ c_\phi s_\rho \\ c_\rho \end{bmatrix}$$

from which it follows that the polar angles of the face pole of (111) are $\rho = 69.89°$ and $\phi = 62.03°$.

Finally, the angle between the normal to (111) and [010] is found by forming the inner product between the two unit vectors **S** and **R**. That is,

$$\mathbf{S} \bullet \mathbf{R} = \cos\angle([010]:(111)) = [\mathbf{S}]_C^t G[\mathbf{R}]_C \quad ,$$

where G is the metrical matrix for C. Since C is a cartesian basis, $G = I_3$. Hence

$$\mathbf{S} \bullet \mathbf{R} = [\mathbf{S}]_C^t[\mathbf{R}]_C$$

$$= [-0.05805 \quad 0.99720 \quad -0.05431]\begin{bmatrix} 0.82937 \\ 0.44042 \\ 0.34377 \end{bmatrix} = 0.37237$$

and $\angle([010]:(111)) = 68.14°$. □

(P2.15) Problem: Amblygonite, $LiAlFPO_4$, is a rare mineral that may be confused with feldspar but can be distinguished by its cleavage angles. Calculate the angle between the (100) and (110) cleavage planes of the mineral given its cell dimensions (a = 5.148A, b = 7.215A, c = 5.060A, α = 113.97°, β = 98.64°, γ = 67.25°) and compare your result with the feldspar cleavage angle of about 90°.

(E2.21) Example - Calculation of interaxial angles and axial ratios from goniometric data: Given the goniometric data in Table 2.1 for chalcanthite, $CuSO_4 \bullet 5H_2O$,

(1) Calculate the interaxial angles α^*, β^*, γ^* and α, β, γ and the axial ratios a^*/b^*, c^*/b^*, a/b, c/b for the reciprocal and direct lattices, respectively.

(2) Calculate a matrix that can be used to find the angular coordinates of the face pole of any plane ($hk\ell$).

(3) Calculate the angular coordinates of the face pole $(0\bar{2}1)$.

(4) Calculate the angular coordinates of the zone [112].

Solution: Using Equation (2.36), unit vectors $[\mathbf{S}(hk\ell)]_C$ perpendicular to each ($hk\ell$) in Table 2.1 are calculated:

Table 2.1 Goniometric data for chalcanthite (Fisher, 1952)

$(hk\ell)$	$\phi°$	$\rho°$	$(hk\ell)$	$\phi°$	$\rho°$
(100)	100.87	90	$(\bar{1}\bar{1}1)$	−112.63	37.73
(010)	0	90	$(0\bar{2}1)$	163.77	47.98
(001)	76.37	17.76			

$$[\mathbf{S}(100)]_C = \begin{bmatrix} 0.9821 \\ -0.1886 \\ 0.0 \end{bmatrix}, \qquad [\mathbf{S}(010)]_C = \begin{bmatrix} 0 \\ 1 \\ 0 \end{bmatrix},$$

$$[\mathbf{S}(001)]_C = \begin{bmatrix} 0.2964 \\ 0.0719 \\ 0.9523 \end{bmatrix}, \qquad [\mathbf{S}(\bar{1}\bar{1}1)]_C = \begin{bmatrix} -0.5648 \\ -0.2355 \\ 0.7909 \end{bmatrix}.$$

The interaxial angles ($\alpha^* = 85.88°$, $\beta^* = 73.89°$, $\gamma^* = 100.87°$) of the reciprocal lattice are found by evaluating the inner products

$$\mathbf{S}(010) \bullet \mathbf{S}(001), \quad \mathbf{S}(001) \bullet \mathbf{S}(100), \quad \mathbf{S}(100) \bullet \mathbf{S}(010) .$$

(Remember that $\mathbf{a}^*$, $\mathbf{b}^*$ and $\mathbf{c}^*$ are perpendicular to (100), (010) and (001), respectively.) Next, evaluating the cross products

$$\mathbf{S}(010) \times \mathbf{S}(001), \quad \mathbf{S}(001) \times \mathbf{S}(100), \quad \mathbf{S}(100) \times \mathbf{S}(010)$$

and normalizing the resulting vectors, we find the triples for unit vectors $\mathbf{R}[uvw]$ along the zone $[uvw]$. For [100], [010] and [001] of the direct lattice, we get

$$[\mathbf{R}[100]]_C = \begin{bmatrix} 0.9494 \\ 0.0 \\ -0.2972 \end{bmatrix}, \quad [\mathbf{R}[010]]_C = \begin{bmatrix} 0.1864 \\ 0.9736 \\ -0.1312 \end{bmatrix}, \quad [\mathbf{R}[001]]_C = \begin{bmatrix} 0 \\ 0 \\ 1 \end{bmatrix}.$$

The interaxial angles ($\alpha = 97.57°$, $\beta = 107.29°$, $\gamma = 77.43°$) of the direct lattice are found by evaluating the inner products

$$\mathbf{R}[010] \bullet \mathbf{R}[001],\ \mathbf{R}[001] \bullet \mathbf{R}[100],\ \mathbf{R}[100] \bullet \mathbf{R}[010] \quad .$$

To find the axial ratios a/b and c/b for the direct lattice we find an equation for the plane $(\bar{1}\bar{1}1)$ defined in terms of the cartesian basis and evaluate its intercepts along the three zones [100], [010] and [001]. Replacing t_1, t_2 and t_3 of Equation (2.4) by the components of $\mathbf{S}(\bar{1}\bar{1}1)$, the equation of a plane not passing through the origin that parallels the $(\bar{1}\bar{1}1)$ planes can be written in the form

$$-0.5648x - 0.2355y + 0.7909z = 1 \quad . \tag{2.39}$$

The intercepts of this plane along the coordinate axes of the direct lattice are proportional to $-\mathbf{a}$, $-\mathbf{b}$ and $\mathbf{c}$. Hence there exists a real number k such that $\mathbf{e} = -k\mathbf{a}$, $\mathbf{f} = -k\mathbf{b}$ and $\mathbf{g} = k\mathbf{c}$ are the vectors to the plane. Since $\mathbf{R}[\bar{1}00]$ is a unit vector in the direction of $\mathbf{e}$,

$$[\mathbf{e}]_C = e[\mathbf{R}[\bar{1}00]]_C = e \begin{bmatrix} -0.9548 \\ 0.0 \\ 0.2972 \end{bmatrix} = \begin{bmatrix} -0.9548e \\ 0.0 \\ 0.2972e \end{bmatrix} \quad .$$

When $[\mathbf{e}]_C$ is inserted into (2.39), we get

$$-0.5648(-0.9548e) - 0.2355(0.0) + 0.7909(0.2972e) = 1 \quad .$$

Solving for e, we find $e = 1.2914$. Similarly, using $f[\mathbf{R}[0\bar{1}0]]_C$ and $g[\mathbf{R}[001]]_C$ and (2.39) we find that $f = 2.2780$ and $g = 1.2644$. Hence $a{:}b{:}c = e{:}f{:}g = 1.2914{:}2.2780{:}1.2644 = 0.5669{:}1{:}0.5550$, where by $a{:}b{:}c = e{:}f{:}g$, we mean that there exists an $h \varepsilon R$ such that $a = eh$, $b = fh$ and $c = gh$. Therefore $a/b = 0.5669$ and $c/b = 0.5550$.

In order to find $a^*{:}b^*{:}c^*$, we note that if $D = \{\mathbf{a},\mathbf{b},\mathbf{c}\}$ and $D_1 = \{\mathbf{a}_1,\mathbf{b}_1,\mathbf{c}_1\}$ are such that $a{:}b{:}c = a_1{:}b_1{:}c_1$, then $a^*{:}b^*{:}c^* = a_1^*{:}b_1^*{:}c_1^*$. Let $\mathbf{a}_1 = 0.5685\mathbf{R}[100]$, $\mathbf{b}_1 = \mathbf{R}[010]$ and $\mathbf{c}_1 = 0.5542\mathbf{R}[001]$. Then $a{:}b{:}c = a_1{:}b_1{:}c_1$. Hence we will calculate

$a_1^*:b_1^*:c_1^*$ to obtain $a^*:b^*:c^*$. The metrical matrix G_1 for D_1 can now be found since we know $a_1 = 0.5669$, $b_1 = 1$, $c_1 = 0.5550$, $\alpha_1 = 97.57°$, $\beta_1 = 107.29°$ and $\gamma_1 = 77.43°$. Hence

$$G_1 = \begin{bmatrix} 0.3214 & 0.1234 & -0.0935 \\ 0.1234 & 1.0 & -0.0731 \\ -0.0935 & -0.0731 & 0.3080 \end{bmatrix} .$$

Consequently,

$$G_1^* = G_1^{-1} = \begin{bmatrix} 3.5390 & -0.3644 & 0.9879 \\ -0.3644 & 1.0552 & 0.1398 \\ 0.9879 & 0.1398 & 3.5796 \end{bmatrix}$$

from which it follows that $a_1^* = 1.8812$, $b_1^* = 1.0272$, $c_1^* = 1.8920$, $\alpha_1^* = 85.87$, $\beta_1^* = 73.89$, $\gamma_1^* = 100.87$. Hence $a^*:b^*:c^* = 1.8812:1.0272:1.8920 = 1.8314:1:1.8419$ and so $a^*/b^* = 1.8314$ and $c^*/b^* = 1.8419$.

The matrix that enables us to find the angular coordinates of the face pole of any plane $(hk\ell)$ is the change of basis matrix B from D_1^* to C. The matrix was determined in (2.34). In this case

$$B = \begin{bmatrix} 1.8474 & 0 & 0.5608 \\ -0.3548 & 1.0272 & 0.1363 \\ 0 & 0 & 1.8018 \end{bmatrix} .$$

The components of a unit vector $\mathbf{R}(0,-2,1)$ that is perpendicular to $(0\bar{2}1)$ is found by normalizing $B[0,-2,1]^t$. Hence

$$[\mathbf{R}(0,-2,1)]_C = \begin{bmatrix} 0.2084 \\ -0.7129 \\ 0.6696 \end{bmatrix} .$$

Equating this vector with (2.36) and solving for ϕ and ρ, we get $\phi = -163.7°$ and $\rho = 48.0°$.

To determine the angular coordinates of the zone [112], we calculate

$$B^{-t} = A = \begin{bmatrix} 0.5413 & 0.1870 & 0 \\ 0 & 0.9735 & 0 \\ -0.1685 & -0.1318 & 0.5550 \end{bmatrix} .$$

Multiplying $[112]^t$ by A and normalizing the resulting vector we get

$$[\mathbf{R}[112]]_C = \begin{bmatrix} 0.4986 \\ 0.6664 \\ 0.5543 \end{bmatrix} .$$

Equating this vector with (2.36), we obtain $\phi = 36.8°$ and $\rho = 56.3°$.

A measurement of the powder diffraction record for chalcanthite shows that the d-spacing of $(\bar{1}\bar{1}1)$ is 4.713A. By the discussion following (2.6), and the fact that $\mathbf{S}(\bar{1}\bar{1}1)$ is a unit vector, we know that

$$-0.5648x - 0.2355y + 0.7909z = 1$$

defines a plane 1A from the origin. Hence

$$-0.5648x - 0.2355y + 0.7909z = 4.713$$

defines a plane 4.713A from the origin. This plane intercepts the coordinate axes of the direct lattice at the termini of $-\mathbf{a}$, $-\mathbf{b}$ and $\mathbf{c}$. Replacing x, y and z in this equation by $a[\mathbf{R}(\bar{1}00)]_C$, $b[\mathbf{R}(0\bar{1}0)]_C$ and $c[\mathbf{R}(001)]_C$ in succession and solving for a, b and c, we obtain an estimate of the unit cell edges (a = 6.09A, b = 10.74A, c = 5.96A) for the mineral. □

DRAWING CRYSTAL STRUCTURES

(E2.22) Example - Atomic coordinate transformation for viewing a structure perpendicular to a plane: The ability to make an accurate drawing of a crystal structure viewed perpendicular to a plane or down a line can be a valuable aid in understanding the structure of a crystal. In this example, we shall derive a matrix for transforming the atomic coordinates of pyroxferroite so that its structure can be drawn viewed perpendicular to the plane defined by three of its atoms designated M3, M4 and M5. As discussed earlier in E2.16, pyroxferroite, is triclinic with a = 6.621A, b = 7.551A, c = 17.381A, α = 114.27°, β = 82.68°, γ = 94.58°. The atomic coordinates of selected atoms in the mineral determined by Burnham (1971) are given in Table 2.2.

Table 2.2: Atomic coordinates for selected atoms in pyroxferroite

Atom	x	y	z	Atom	x	y	z
M3	0.0663	0.4341	0.8963	0(A3)	0.1534	0.5817	0.8123
M4	0.1626	0.3098	0.6945	0(B3)	0.5460	0.7060	0.7907
M5	0.2710	0.2218	0.9892	0(C2)	0.2501	0.9367	0.8214
Si(3)	0.3251	0.7544	0.8398	0(C3)	0.3422	0.8445	0.9417

Solution: Let $\mathbf{r}_3$, $\mathbf{r}_4$ and $\mathbf{r}_5$ denote vectors that radiate from the origin to the positions of M3, M4 and M5, respectively. From Table 2.2, we can write

$$[\mathbf{r}_3]_D = \begin{bmatrix} 0.0663 \\ 0.4341 \\ 0.8963 \end{bmatrix}, \quad [\mathbf{r}_4]_D = \begin{bmatrix} 0.1626 \\ 0.3098 \\ 0.6945 \end{bmatrix}, \quad [\mathbf{r}_5]_D = \begin{bmatrix} 0.2710 \\ 0.2218 \\ 0.9892 \end{bmatrix}.$$

Our first job is to define a unit vector perpendicular to the plane of M3, M4 and M5. In this example, $\mathbf{k}$ will be taken to be this vector along which the structure is to be viewed. Since the end points of $\mathbf{r}_3$, $\mathbf{r}_4$, $\mathbf{r}_5$ are on the plane, $\mathbf{r}_{34} = \mathbf{r}_3 - \mathbf{r}_4$ and $\mathbf{r}_{54} = \mathbf{r}_5 - \mathbf{r}_4$ parallel the plane. Hence $\mathbf{r} = \mathbf{r}_{34} \times \mathbf{r}_{54}$ is perpendicular to the plane and so is in the direction of $\mathbf{k}$. Since

$$[\mathbf{r}_{34}]_D = [\mathbf{r}_3]_D - [\mathbf{r}_4]_D = \begin{bmatrix} -0.0963 \\ 0.1243 \\ 0.2018 \end{bmatrix}$$

and

$$[\mathbf{r}_{54}]_D = [\mathbf{r}_5]_D - [\mathbf{r}_4]_D = \begin{bmatrix} 0.1084 \\ -0.0880 \\ 0.2947 \end{bmatrix},$$

then

$$\mathbf{r} = \mathbf{r}_{34} \times \mathbf{r}_{54} = \det(G)^{-\frac{1}{2}} \begin{vmatrix} \mathbf{a} & & \\ \mathbf{b} & G[\mathbf{r}_{34}]_D & G[\mathbf{r}_{54}]_D \\ \mathbf{c} & & \end{vmatrix}$$

$$= 1.0219\mathbf{a} + 0.8478\mathbf{b} + 0.0888\mathbf{c}$$

(see E2.16). Hence,

$$\mathbf{k} = (1/r)\mathbf{r} = 0.1166\mathbf{a} + 0.0968\mathbf{b} + 0.0102\mathbf{c}$$

and so

$$[\mathbf{k}]_D = \begin{bmatrix} 0.1166 \\ 0.0968 \\ 0.0101 \end{bmatrix} .$$

As $\mathbf{k}$ is perpendicular to $\mathbf{r}_{34}$, we may choose $\mathbf{i} = (1/r_{34})\mathbf{r}_{34}$ which leads to

$$[\mathbf{i}]_D = \begin{bmatrix} -0.0298 \\ 0.0385 \\ 0.0625 \end{bmatrix} .$$

Then $\mathbf{j}$ is obtained by normalizing $\mathbf{k} \times \mathbf{i}$ which gives (see P2.9)

$$[\mathbf{j}]_D = \begin{bmatrix} 0.0934 \\ -0.1013 \\ -0.0034 \end{bmatrix} .$$

As before, we let A denote the change of basis matrix from D to C. That is, $A[\mathbf{r}]_D = [\mathbf{r}]_C$. Hence, $A^{-1}[\mathbf{r}]_C = [\mathbf{r}]_D$ and so

$$A^{-1} = \begin{bmatrix} | & | & | \\ [\mathbf{i}]_D & [\mathbf{j}]_D & [\mathbf{k}]_D \\ | & | & | \end{bmatrix} = \begin{bmatrix} -0.0298 & 0.0934 & 0.1166 \\ 0.0385 & -0.1013 & 0.0968 \\ 0.0625 & -0.0034 & 0.0101 \end{bmatrix} .$$

Since we have the coordinates of the atoms with respect to the D basis and want them in terms of the C basis, we need to find A. Inverting A^{-1}, we obtain

$$A = \begin{bmatrix} -0.5454 & -1.0530 & 16.3886 \\ 4.4492 & -5.9639 & 5.7952 \\ 4.8730 & 4.5082 & -0.4536 \end{bmatrix} .$$

Since $A[\mathbf{r}]_D = [\mathbf{r}]_C$, we can now transform the coordinates of all the atoms in pyroxferroite to construct a drawing of the structure defined in terms of a cartesian basis and viewed perpendicular to the plane of M3, M4, M5. For example, the atomic coordinates of O(A3) in the cartesian basis is obtained as follows

$$\begin{bmatrix} -0.5454 & -1.0530 & 16.3886 \\ 4.4492 & -5.9639 & 5.7952 \\ 4.8730 & 4.5082 & -0.4536 \end{bmatrix} \begin{bmatrix} 0.1534 \\ 0.5817 \\ 0.8123 \end{bmatrix} = \begin{bmatrix} 12.616 \\ 1.921 \\ 3.001 \end{bmatrix} .$$

Note that the new coordinates of O(A3) are also in Angstrom units and that the columns of A are the coordinates of the crystallographic basis vectors $D = \{\mathbf{a},\mathbf{b},\mathbf{c}\}$ expressed in terms of a cartesian basis, i.e.,

$$\begin{aligned} \mathbf{a} &= -0.5454\mathbf{i} + 4.4492\mathbf{j} + 4.8730\mathbf{k} \quad , \\ \mathbf{b} &= -1.0530\mathbf{i} - 5.9639\mathbf{j} + 4.5082\mathbf{k} \quad , \\ \mathbf{c} &= 16.3886\mathbf{i} + 5.7952\mathbf{j} - 0.4536\mathbf{k} \quad . \end{aligned}$$

□

(P2.16) Problem: Find the Cartesian coordinates of M3, M4 and M5 viewed down a line perpendicular to their plane.

(P2.17) Problem: The crystal structure of monoclinic pyroxene consists of chains of tetrahedra and octahedra that span the structure along the Z-axis and that interlock in layers paralleling (100). When drawings are made of the pyroxene structure, they are usually made in a view down $\mathbf{a}$. As $\mathbf{a}$ makes an angle of about 70° to the layers, such a view gives a distorted appearance to the chains. To avoid this problem, the atomic coordinates can be transformed so that the structure is viewed perpendicular to (100), i.e., down $\mathbf{a}^*$.

(1) Derive the A-matrix

$$A = \begin{bmatrix} a\sin\beta & 0 & 0 \\ 0 & b & 0 \\ a\cos\beta & 0 & c \end{bmatrix} .$$

for transforming the atomic coordinates of a monoclinic crystal ($a \neq b \neq c$, $\alpha = \gamma = 90° \neq \beta$) to give a view of the structure along $\mathbf{a}^*$.

Table 2.3: Atomic coordinates of the SiO_4 tetrahedron in jadeite.

Atom	x	y	z
Si	0.2094	0.4066	0.7723
O_1	0.3910	0.4237	0.8725
O_2	0.1392	0.2370	0.7071
O_3	0.1467	0.4930	0.9942
O_3	0.1467	0.5070	0.4942

(2) Given the atomic coordinates of a SiO_4 tetrahedron in jadeite (Table 2.3), a monoclinic pyroxene (a = 9.418A, b = 8.562A, c = 5.219A, β = 107.58°) of composition, $NaAlSi_2O_6$ use the A-matrix to transform the coordinates for a view of the tetrahedron down $\mathbf{a}^*$. Then make a drawing of the tetrahedron using the transformed coordinates. Compare your results with a drawing of the jadeite structure prepared by Prewitt and Burnham (1966).

(E2.23) Example - Atomic coordinate transformation for viewing a structure down a line: In this example, we shall derive a matrix for transforming the atomic coordinates of pyroxferroite for making a drawing of the structure viewed along its Si(3)O(A3) bond.

Solution: The atomic coordinates for Si(3) and O(A3) (Table 2.2) provide the information for writing their respective triples:

$$[\mathbf{r}_6]_D = \begin{bmatrix} 0.3251 \\ 0.7544 \\ 0.8398 \end{bmatrix}, \quad [\mathbf{r}_7]_D = \begin{bmatrix} 0.1534 \\ 0.5817 \\ 0.8123 \end{bmatrix} .$$

The vector $\mathbf{r}_{76} = \mathbf{r}_7 - \mathbf{r}_6$ parallels and has the same length as the Si(3)O(A3) bond. Since

$$[\mathbf{r}_{76}]_D = [\mathbf{r}_7]_D - [\mathbf{r}_6]_D = \begin{bmatrix} -0.1717 \\ -0.1727 \\ -0.0275 \end{bmatrix} ,$$

a unit vector **k** directed along the bond can be found by dividing each of these coordinates by 1.616A, the length of the bond to give

$$[\mathbf{k}]_D = \begin{bmatrix} -0.1063 \\ -0.1069 \\ -0.0170 \end{bmatrix} .$$

Next we will find a vector **i** perpendicular to **k**. In general, a nonzero vector **u** is perpendicular to a nonzero vector **v** if and only if their inner product is zero; that is

$$[\mathbf{u}]_D^t G[\mathbf{v}]_D = [\mathbf{0}]_D \quad . \tag{2.40}$$

Given a nonzero vector **u**, a nonzero vector **v** can be chosen perpendicular to **u** by the following procedure. Denoting $[\mathbf{u}]_D^t G$ by $[w_1 \; w_2 \; w_3]$, we shall choose **v** so that $w_1v_1 + w_2v_2 + w_3v_3 = 0$, where $[\mathbf{v}]_D^t = [v_1v_2v_3]^t$. If $w_1 \neq 0$, chose $v_1 = -w_2$, $v_2 = w_1$ and $v_3 = 0$. If $w_1 = 0$, choose $v_1 = 0$, $v_2 = w_3$ and $v_3 = -w_2$. In each case $\mathbf{v} \neq \mathbf{0}$ (since **u** is not zero) and **v** satisfies (2.40). In our case where $\mathbf{u} = \mathbf{k}$,

$$[\mathbf{k}]_D^t G = [-0.1063 \quad -0.1069 \quad -0.0170] \begin{bmatrix} 43.8376 & -3.9922 & 14.6624 \\ -3.9922 & 57.0176 & -53.9461 \\ 14.6624 & -53.9461 & 302.0992 \end{bmatrix}$$

$$= [-4.4824 \quad -4.7537 \quad -0.9275] = [w_1 \; w_2 \; w_3] \quad .$$

Since $w_1 \neq 0$, we take $\mathbf{v} = 4.7537\mathbf{a} - 4.4824\mathbf{b}$. Hence, **i** is taken to be $\mathbf{i} = \mathbf{v}/v$. That is,

$$[\mathbf{i}]_D = \begin{bmatrix} 0.0990 \\ -0.0933 \\ 0.0 \end{bmatrix} .$$

Since **j** is the unit vector in the direction of **k x i**, we have

$$[\mathbf{j}]_D = \begin{bmatrix} -0.0460 \\ 0.0314 \\ 0.0611 \end{bmatrix} .$$

From $[\mathbf{i}]_D$, $[\mathbf{j}]_D$ and $[\mathbf{k}]_D$, we form

$$A^{-1} = \begin{bmatrix} 0.0990 & -0.0460 & -0.1063 \\ -0.0933 & 0.0314 & -0.1069 \\ 0.0 & 0.0611 & -0.0170 \end{bmatrix} ,$$

where, as before, $A[\mathbf{r}]_D = [\mathbf{r}]_C$. Inverting A^{-1}, we get

$$A = \begin{bmatrix} 4.7126 & -5.7176 & 6.4863 \\ -1.2462 & -1.3224 & 16.1080 \\ -4.4791 & -4.7527 & -0.9297 \end{bmatrix} . \tag{2.41}$$

□

(P2.18) Problem: (1) Using (2.41) find the cartesian coordinates (Table 2.4) for Si(3), O(A3), O(B3), O(C2) and O(C3) in pyroxferroite.

Table 2.4: Cartesian coordinates of an SiO_4 tetrahedron in pyroxferroite.

Atom	*x*	*y*	*z*
Si(3)	2.666	12.125	-5.822
O(A3)	2.666	12.124	-4.207
O(B3)	3.665	11.123	-6.536
O(C2)	1.151	11.681	-6.336
O(C3)	2.892	13.626	-6.422

(2) With the cartesian coordinates in Table 2.4, compute the separations between Si(3) and O(A3), O(B3), O(C2) and O(C3) and compare your results with those published by Burnham (1971).

(P2.19) Problem: The basis vectors $D_1 = \{\mathbf{a}_1, \mathbf{b}_1, \mathbf{c}_1\}$ of a rhombohedral lattice in the obverse setting are defined in terms of a hexagonal set of basis vectors $D_2 = \{\mathbf{a}_2, \mathbf{b}_2, \mathbf{c}\}$ in Figure 2.7. An examination of this figure shows that

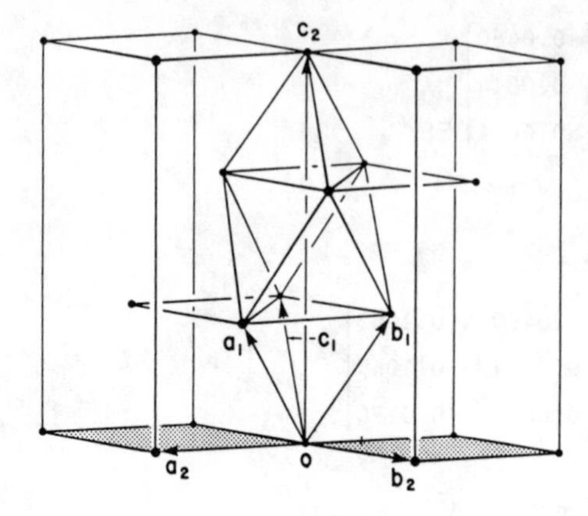

Figure 2.7: A rhombohedral lattice generated by $D_1 = \{\mathbf{a}_1, \mathbf{b}_1, \mathbf{c}_1\}$ and a hexagonal cell outlined by $D_2 = \{\mathbf{a}_2, \mathbf{b}_2, \mathbf{c}_2\}$. The lattice points of the rhombohedral lattice occur at 0, 0, 0; 2/3, 1/3, 1/3 and 1/3, 2/3, 2/3 (obverse setting) with respect to D_2. The reverse setting with lattice points at 0, 0, 0; 1/3, 2/3, 1/3 and 1/3, 2/3, 2/3) is obtained by rotating the lattice points of the observe setting 60° about $\mathbf{c}_2$.

$$\begin{aligned}\mathbf{a}_2 &= \mathbf{a}_1 - \mathbf{b}_1\\ \mathbf{b}_2 &= \mathbf{b}_1 - \mathbf{c}_1\\ \mathbf{c}_2 &= \mathbf{a}_1 + \mathbf{b}_1 + \mathbf{c}_1\ .\end{aligned}$$

(1) Show that

$$\begin{aligned}\mathbf{a}_1 &= (2/3)\mathbf{a}_2 + (1/3)\mathbf{b}_2 + (1/3)\mathbf{c}_2\\ \mathbf{b}_1 &= -(1/3)\mathbf{a}_2 + (1/3)\mathbf{b}_2 + (1/3)\mathbf{c}_2\\ \mathbf{c}_1 &= -(1/3)\mathbf{a}_2 - (2/3)\mathbf{b}_2 + (1/3)\mathbf{c}_2\ .\end{aligned}$$

(2) Given that $a_2 = b_2 = 15.951$A and $c_2 = 7.24$A, $\alpha_2 = \beta_2 = 90°$, and $\gamma_2 = 120°$, show that $a_1 = b_1 = c_1 = 9.520$A and $\alpha_1 = \beta_1 = \gamma_1 = 113.80°$.

CHAPTER 3

POINT ISOMETRIES - VEHICLES FOR DESCRIBING SYMMETRY

"In asking what operations will turn a pattern into itself, we are discovering the invisible laws that govern our space." -- J. Bronowski

INTRODUCTION

In Chapter 1 we observed that the atoms in the monosilicic acid molecule are repeated at regular intervals about a line and a point. We also observed that the atoms in α-quartz are repeated at regular intervals along straight lines. In both of these examples, the atoms in each interval are related by the class of distance-preserving mappings known as isometries. In this chapter we define the term isometry and show how it can be used to describe the symmetry of an object. Two types of point isometries, rotations and rotoinversion, will be examined, and we will see how matrices can be used to represent these isometries with respect to a given basis D of R^3. We shall also see how the external symmetry of a crystal like α-quartz can be completely described by a group of these isometries and their matrix representations.

ISOMETRIES

The mappings of S (see Appendix 1) that will enable us to describe symmetry are those that do not distort sizes and shapes of objects. Such a mapping is called an isometry and is defined as follows:

(D3.1) Definition: An *isometry* is a one-to-one and onto mapping from S onto S such that for all $\mathbf{u},\mathbf{v} \in S$, the distance between the end points of $\mathbf{u}$ and $\mathbf{v}$ equals the distance between the end points of $\alpha(\mathbf{u})$ and $\alpha(\mathbf{v})$. If, in addition, there exists a point $\mathbf{r} \in S$ such that $\alpha(\mathbf{r}) = \mathbf{r}$, then α is said to be a *point isometry*.

The first statement in this definition is described by saying "α preserves distances" and the second by saying "α is a point isometry if at least one point in S is fixed by α." As shown by Boisen and Gibbs

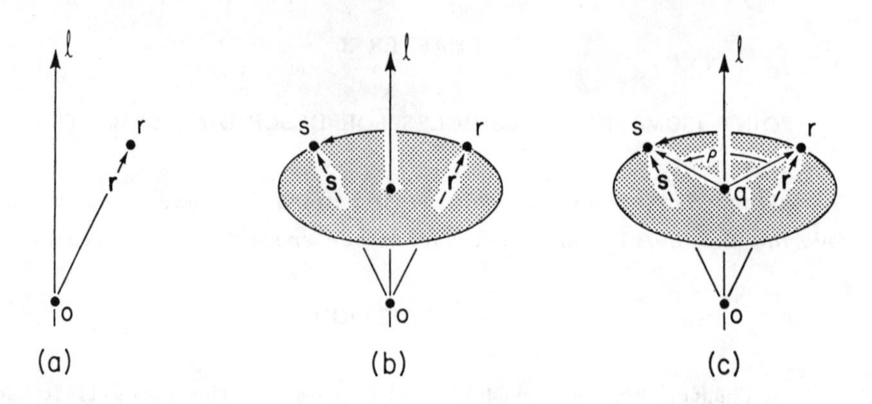

Figure 3.1: The action of rotating an arbitrary point $r \in S$ through a turn angle ρ about an axis ℓ. (a) The directed line segment ℓ passing through the origin, O, defines the rotation axis of α and $\mathbf{r}$ defines a vector emanating from O to the arbitrary point r. (b) A circle traced by rotating $\mathbf{r}$ a full turn about ℓ. The action of α rotates $\mathbf{r}$ onto $\mathbf{s}$ such that $\alpha(\mathbf{r}) = \mathbf{s}$. (c) The plane perpendicular to ℓ on which the turn angle ρ of α is measured. Note that points r and s both lie on the plane and that the point q is where the plane intersects ℓ. The turn angle of α is $\rho = \angle(pqr)$.

(1976), there are exactly two classes of point isometries referred to as rotations and rotoinversions.

Rotations: Let α denote a *rotation* about a line ℓ. Since α simply turns space about ℓ, space is neither compressed, extended, nor distorted by α. Hence, distances are preserved. Furthermore, each point on the line ℓ is left fixed by α. Therefore, α qualifies as a point isometry.

In order to describe and compare rotations, we must understand their basic properties. The action of a rotation α on S is described in Figure 3.1. In part (a) of the figure, a straight line ℓ passing through the origin is shown on which a positive direction has been defined by an arrow. An arbitrary point in space is depicted by the vector $\mathbf{r}$. When $\mathbf{r}$ is rotated about ℓ, its end point r traces a circle lying in a plane perpendicular to ℓ as shown in Figure 3.1(b). Hence α maps $\mathbf{r}$ to a vector $\mathbf{s}$ whose end point is also on the circle. We say that $\mathbf{s}$ is the *image* of $\mathbf{r}$ under α and that ℓ is the *rotation axis* of α. The angle through which space is rotated by α, measured on the plane containing the circle, is called the turn angle of α. In Figure 3.1(c), q denotes the point at which ℓ passes through this plane. The *turn angle* of α in this case is defined to be $\rho = \angle(rqs)$ so that a positive angle is measured in the counterclockwise direction when viewed from the positive end of ℓ. If the turn angle of α is ρ°, then $\rho^\circ + (360n)^\circ$ is an equivalent turn angle of α for any integer n. Hence, each rotation has an infinite number of equivalent turn angle representatives associated with it (see Appendix 1). The

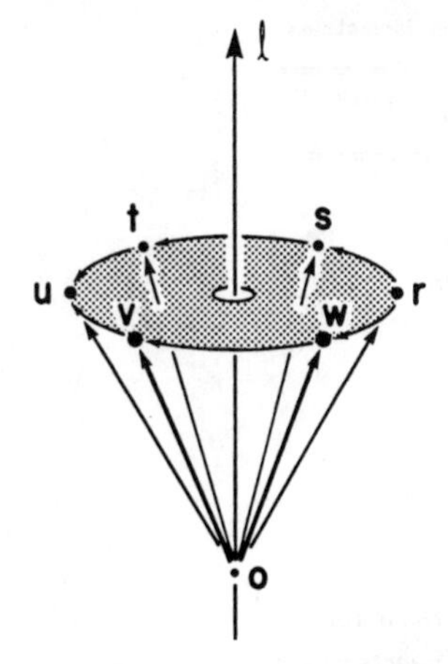

Figure 3.2: The action of each rotation of a six-fold axis on an arbitrary point **r** ε *S*. The point **r** and its images define a plane perpendicular to ℓ where **1**(**r**) = **r**, **6**(**r**) = **s**, $\mathbf{6}^{-1}$(**r**) = **w**, **3**(**r**) = **t**, $\mathbf{3}^{-1}$(**r**) = **v** and **2**(**r**) = **u**.

identity mapping **1**, where **1**(**r**) = **r** for all **r** ε *S*, is considered to be a rotation with turn angle 0° (or any multiple of 360°) and with every line in space as its rotation axis. Thus, the identity can be viewed as taking place about each line in space simultaneously.

Suppose ρ is a turn angle for a rotation α such that $\rho > 0$, and suppose that ρ is an integral divisor of 360°, i.e., 360/ρ is an integer. Let *n* denote the integer 360/ρ. If the vector **r** is rotated by α *n*-times in succession, then **r** will be mapped back onto itself. If $n > 1$, such a rotation is called an *nth*-turn. When $n = 1$, the rotation is simply the identity. An *nth*-turn is symbolized by **n** and has a turn angle of 360°/*n*. For example, **3** denotes a ***third-turn*** and represents a counterclockwise rotation of space through a turn angle of 120° about some axis. On the other hand, if the turn angle ρ of α is clockwise such that $-360 < \rho < 0$ and $n = -360/\rho$ is an integer, then α has a turn angle of $\rho = -360/n$ and is referred to as a ***negative nth-turn***. Such a rotation is symbolized by $\mathbf{n}^{-1}$ because, as we shall see later, it is the inverse of an *nth*-turn about the same axis. Thus $\mathbf{3}^{-1}$ signifies a ***negative third-turn*** and represents a clockwise rotation of 120°. In Table 3.1 several *nth* turns and negative *nth* turns are described. In order to avoid ambiguities, we shall choose as the preferred turn angle representative for any rotation α an angle ρ such that $-180 < \rho < 180$. Hence, the preferred angle for $\mathbf{3}^{-1}$ will be −120° and that for **2** will be 180°. Figure 3.2 shows an arbitrary vector **r** and its images under several rotations each taking place about a common axis ℓ. In this case, the identity maps **r** onto itself, symbolized **1**(**r**) = **r**; the sixth-turn maps **r** onto **s**, hence **6**(**r**) = **s**; the third-turn maps **r** onto **t**, hence **3**(**r**) = **t**, the half-turn maps **r** onto **u**, hence **2**(**r**) = **u**; the negative sixth-turn maps **r** onto **w** hence

Table 3.1: Symbols and names for rotation isometries.

Turn angle ρ	Symbol	Name
360°(0°)	1	*identity*
180°	2	*half-turn*
120°	3	*third-turn*
90°	4	*quarter-turn*
72°	5	*fifth-turn*
60°	6	*sixth-turn*
•	•	•
•	•	•
•	•	•
360°/*n*	n	*nth-turn*
−120°	3^{-1}	*negative third-turn*
− 90°	4^{-1}	*negative quarter-turn*
− 60°	6^{-1}	*negative sixth-turn*
•	•	•
•	•	•
•	•	•
−360°/*n*	n^{-1}	*negative nth-turn*

Table 3.2: Symbols and names for rotoinversion isometries.

Turn angle ρ	Symbol	Name
360°(0°)	*i*	*inversion*
180°	m	*reflection*
120°	$\bar{3}$	*third-turn inversion*
90°	$\bar{4}$	*quarter-turn inversion*
72°	$\bar{5}$	*fifth-turn inversion*
60°	$\bar{6}$	*sixth-turn inversion*
•	•	•
•	•	•
•	•	•
360°/*n*	$\bar{n}$	*nth-turn inversion*
−120°	$\bar{3}^{-1}$	*negative third-turn inversion*
− 90°	$\bar{4}^{-1}$	*negative quarter-turn inversion*
− 60°	$\bar{6}^{-1}$	*negative sixth-turn inversion*
•	•	•
•	•	•
•	•	•
−360°/*n*	$\bar{n}^{-1}$	*negative nth-turn inversion*

$6^{-1}(\mathbf{r}) = \mathbf{w}$, and the negative third-turn maps $\mathbf{r}$ onto $\mathbf{v}$, hence $3^{-1}(\mathbf{r}) = \mathbf{v}$. As we shall see later, these six rotations when taken as a collection form an important set of rotations, each of which take place about a common axis. This set denoted $6 = \{1,6,3,2,3^{-1},6^{-1}\}$ contains the rotations of a so-called *six-fold axis* (also called a *hexad axis*).

Orientation Symbols: The symbols for the rotations listed in Table 3.1 provide information about turn angles but lack information about the orientation of their rotation axes. When specifying a non-identity rotation using this symbolism, a description of the orientation of the rotation axis is made in terms of a natural basis, $D = \{\mathbf{a},\mathbf{b},\mathbf{c}\}$. Following Boisen and Gibbs (1976), the zone symbol $[uvw]$ will be attached as a left superscript to the rotation symbol to specify the orientation of a vector

$$\mathbf{r} = u\mathbf{a} + v\mathbf{b} + w\mathbf{c}$$

along the positive direction of the rotation axis where u, v and w are usually integers. Thus, ${}^{[uvw]}n$ symbolizes an *nth*-turn about an axis paralleling $\mathbf{r} = u\mathbf{a} + v\mathbf{b} + w\mathbf{c}$. Also, the symbols ${}^{[100]}2$, ${}^{[110]}2$ and ${}^{[010]}2$ specify half-turns about axes coincident with $\mathbf{a}$, $\mathbf{a} + \mathbf{b}$, and $\mathbf{b}$, respectively. Likewise, ${}^{[100]}4$ specifies a quarter-turn about an axis paralleling $\mathbf{a}$, whereas ${}^{[1\bar{1}1]}3^{-1}$ specifies a negative third-turn about an axis coincident with $\mathbf{a} - \mathbf{b} + \mathbf{c}$. An exception to our rule is when the rotation axis points along $\mathbf{c}$. In this case, no orientation symbol is attached to the rotation symbol; thus, symbols like 3 and 3^{-1} will be used to specify a third-turn and a negative third-turn, respectively, about an axis coincident with $\mathbf{c}$. Thus, the symbols for all rotations defining a turning of space about $\mathbf{c}$ will by convention always be written *without* an orientation symbol. Finally, as any line in space can be selected as the rotation axis of the identity, the symbol 1 will also be written without an orientation symbol.

Compositions of Isometries: Let α and β denote mappings from S onto S and let $\mathbf{r} \in S$. Then the composition of α and β, denoted $\alpha\beta$ is defined to be the mapping

$$\alpha\beta(\mathbf{r}) = \alpha(\beta(\mathbf{r})) .$$

That is, to calculate $\alpha\beta(\mathbf{r})$, we first find $\beta(\mathbf{r})$ and then apply α to it.

See Appendix 1 for an illustration of this definition.

(T3.2) Theorem: The composition of two isometries is an isometry. The composition of two point isometries leaving a common point fixed is a point isometry leaving that point fixed.

Proof: Suppose α and β are isometries. We shall show that $\alpha\beta$ is an isometry. In Appendix 1, it is shown that since α and β are both one-to-one and onto mappings, $\alpha\beta$ is a one-to-one and onto mapping. Let $\mathbf{r} \in S$. Because β is distance preserving, $\|\beta(\mathbf{r})\| = \|\mathbf{r}\|$. Also, since α is distance preserving, $\|\alpha(\beta(\mathbf{r}))\| = \|\beta(\mathbf{r})\|$ and so $\|\alpha\beta(\mathbf{r})\| = \|\mathbf{r}\|$. Therefore, the composition of two isometries is again an isometry. Suppose α and β are point isometries both leaving the origin $\mathbf{0}$ fixed. Then

$$\alpha\beta(\mathbf{0}) = \alpha(\beta(\mathbf{0})) = \alpha(\mathbf{0}) = (\mathbf{0}) \ .$$

Therefore, $\alpha\beta$ must also be a point isometry that leaves $\mathbf{0}$ fixed. □

Rotoinversions: A *rotoinversion* is a hybrid operation produced by composing a rotation with the inversion. The *inversion*, denoted by the symbol i, maps each vector $\mathbf{v} \in S$ onto its negative, $-\mathbf{v}$, i.e., $i(\mathbf{v}) = -\mathbf{v}$ for all $\mathbf{v} \in S$. Since $\|\mathbf{v}\| = \|-\mathbf{v}\|$ and $i(\mathbf{0}) = -\mathbf{0} = \mathbf{0}$, we see that i is a distance preserving mapping that leaves the origin fixed. However, unlike a rotation which maps each vector along its axis onto itself, an inversion maps exactly one vector in S, namely $\mathbf{0}$, onto itself. Thus i is not a rotation.

Because the composition of two point isometries leaving a common point fixed is a point isometry, the composition of i with any rotation whose axis passes through the origin is again a point isometry. Hence, every rotoinversion is a point isometry. If $n > 1$, then $\mathsf{n}i$ is denoted $\bar{\mathsf{n}}$. We denote $\mathsf{1}i$ simply as i. The names and symbols for some representative rotoinversions are given in Table 3.2. Following Boisen and Gibbs (1976), we assign to $\bar{\mathsf{n}}$, the turn angle and rotation axis of n. The orientation symbol for $\bar{\mathsf{n}}$ is taken to be that of n. Thus, $^{[100]}\bar{4}$ denotes a quarter-turn inversion with the axis of its associated quarter-turn directed along $\mathbf{a}$ and with a turn angle of 90°. Likewise, $\bar{3}$ denotes a third-turn inversion whose rotation axis parallels $\mathbf{c}$.

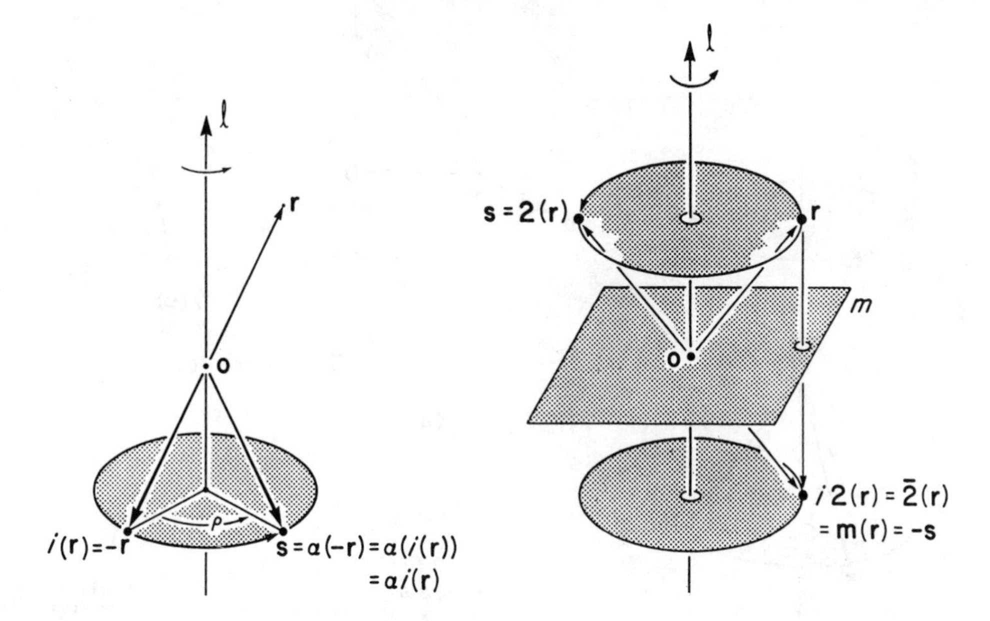

Figure 3.3: The action of a rotoinversion αi on an arbitrary point $\mathbf{r} \in S$ where **0** denotes the origin, α is a rotation about ℓ with turn angle ρ and i is the inversion isometry. The vector **r** is mapped onto $-\mathbf{r}$ by i (i.e., $i(\mathbf{r}) = -\mathbf{r}$) and α maps $(-\mathbf{r})$ onto $\mathbf{s} = \alpha(-\mathbf{r}) = \alpha(i(\mathbf{r})) = \alpha i(\mathbf{r})$. Hence, the rotoinversion isometry αi maps **r** onto **s**. The turn angle and the rotation axis of αi are inherited from α.

Figure 3.4: The action of a reflection isometry $\mathbf{m} = i2$ on an arbitrary vector $\mathbf{r} \in S$ where **0** is the origin and m is the mirror plane perpendicular to the rotation axis ℓ of a half-turn 2 over which the vector **r** is reflected. The vector **r** is mapped onto **s** by 2, **s** is mapped onto $-\mathbf{s}$ by i so that $i2(\mathbf{r}) = -\mathbf{s}$. Hence $i2$ is equivalent to a reflection **m** of the vector **r** directly over the plane m onto $-\mathbf{s}$ such that $\mathbf{m}(\mathbf{r}) = -\mathbf{s}$.

Figure 3.3 shows the effect of a rotoinversion αi on a point $\mathbf{r} \in S$ where α has a turn angle of $\rho°$. As the inversion sends **r** to $-\mathbf{r}$, we observe that $\alpha i(\mathbf{r}) = \alpha(i(\mathbf{r})) = \alpha(-\mathbf{r})$. Then α rotates $-\mathbf{r}$ through a turn angle of ρ mapping $-\mathbf{r}$ onto **s** as shown in Figure 3.3.

So far we have only considered rotoinversions of the form αi. This does not result in loss of generality since, as we shall now show, $i\alpha = \alpha i$ for any rotation α. Let **v** denote a vector in S and let α denote a rotation such that $\alpha(\mathbf{v}) = \mathbf{w}$. As $-\mathbf{v}$ lies on the same line as **v**, $\alpha(-\mathbf{v})$ must lie on the same line as **w** but point in the opposite direction such that $\alpha(-\mathbf{v}) = -\mathbf{w}$. Next, multiplying $\alpha(\mathbf{v}) = \mathbf{w}$ by -1, we observe that $-\alpha(\mathbf{v}) = \alpha(-\mathbf{v})$. Hence, $\alpha i(\mathbf{v}) = \alpha(i(\mathbf{v})) = \alpha(-\mathbf{v}) = -\alpha(\mathbf{v}) = i(\alpha(\mathbf{v})) = i\alpha(\mathbf{v})$. Therefore, αi and $i\alpha$ are the same mapping. Since $\alpha i = i\alpha$, we say that α and i commute. Hence i commutes with all rotations. As we shall see later, isometries do not in general commute and the fact that i commutes with each rotation is an important property of i.

All rotoinversions, with the exception of $\bar{2}$, leave exactly one point

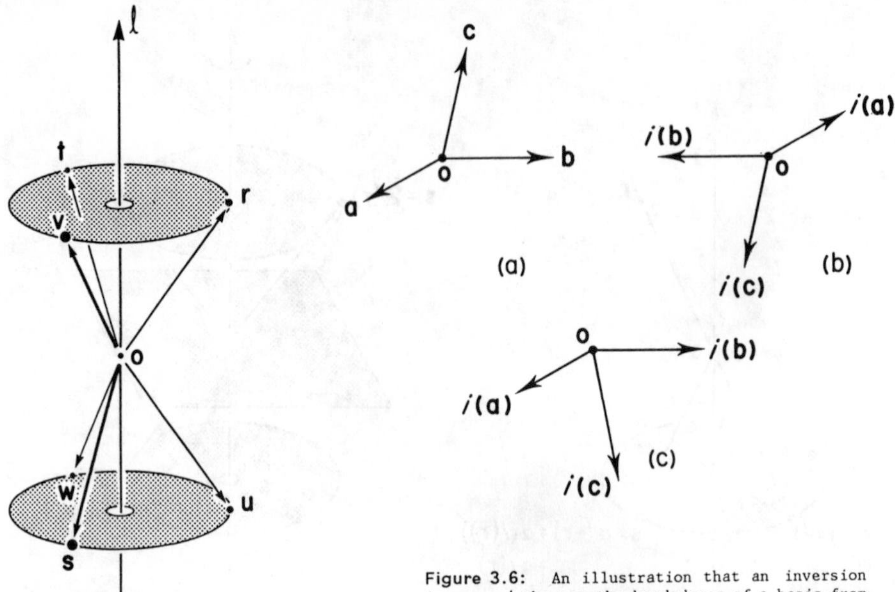

Figure 3.5: A collection of vectors that are images of an arbitrary vector **r** ε *S* under the point isometries that comprise a $\bar{6}$ inversion axis paralleling ℓ.

Figure 3.6: An illustration that an inversion isometry *i* changes the handedness of a basis from right to left. (a) A right-handed basis *D* = {**a**,**b**,**c**} emanating from an origin **0**. (b) The images of **a**, **b** and **c**, {*i*(**a**),*i*(**b**),*i*(**c**)} under *i*. (c) The figure in (b) rotated so that *i*(**a**) and *i*(**b**) are parallel to **a** and **b**, respectively, in (a). Observe that the basis vectors in *D* = {*i*(**a**),*i*(**b**),*i*(**c**)} are left-handed illustrating that an inversion isometry changes the handedness of a set of basis vectors.

in *S* fixed. The $\bar{2}$ operation leaves a plane of points fixed, as illustrated in Figure 3.4. Rather than viewing $\bar{2}$ as the composition of 2 and *i*, it is simpler to view the operation as a reflection about the plane of points left fixed by $\bar{2}$. By a reflection we mean that each point and its image lie on a line perpendicular to the plane such that the plane bisects the line segment between them. Because of this fact, $\bar{2}$ is almost always referred to as a ***reflection*** isometry and is denoted by **m**. Furthermore, the plane of fixed points is called the ***mirror plane m***. The orientation symbol assigned to **m** is inherited from that of the axis of the perpendicular half-turn part of the operation. For example, $^{[210]}\mathbf{m}$ denotes a reflection of space over a mirror plane that is perpendicular to 2**a** + **b**, whereas **m** denotes a reflection of space about the plane that is perpendicular to **c**.

(P3.1) Problem: Determine the names and symbols of the point isometries α that map the vector **r** in Figure 3.5 onto **r**, **s**, **t**, **u**, **v** and **w**, assuming that the rotation axis ℓ of α parallels **c**.

Consider the right-handed set of basis vectors $D = \{\mathbf{a},\mathbf{b},\mathbf{c}\}$ displayed in Figure 3.6(a). The images of these vectors under the inversion is the set $D = \{i(\mathbf{a}), i(\mathbf{b}), i(\mathbf{c})\}$ (Figure 3.6(b)). If we orient D with $i(\mathbf{a})$ on the left and $i(\mathbf{b})$ on the right, then $i(\mathbf{c})$ will be directed downward (Figure 3.6(c)). Consequently, the inversion transforms a right-handed basis set to a left-handed one (see Appendix 4). Since rotations do not change the handedness of a basis, the net effect of a rotoinversion αi is to change the handedness. Hence, all rotoinversions change the handedness of a basis. One important consequence of this is that no isometry is both a rotation and a rotoinversion.

(E3.3) Example - The inversion mapping is its own inverse: Show that the composition of two successive inversions about a common point is equivalent to the identity.

Solution: Since $i(\mathbf{r}) = -\mathbf{r}$ for all $\mathbf{r} \varepsilon S$, then $i(i(\mathbf{r})) = i(-\mathbf{r}) = 1(\mathbf{r})$. Hence $ii = i^2 = 1$. □

SYMMETRY ELEMENT

(D3.4) Definition: The *symmetry element* of a point isometry α is defined to be the set of all points in space left fixed by α, i.e., a point $\mathbf{p} \varepsilon S$ is on the symmetry element of α if $\alpha(\mathbf{p}) = \mathbf{p}$.

Let α denote a rotation other than the identity with a rotation axis ℓ. Since each point $\mathbf{p}$ along ℓ is left fixed by α and every other point in S is moved, ℓ is the symmetry element of α. Every point $\mathbf{p} \varepsilon S$ is fixed by the identity, and so S is the symmetry element of the identity. As a plane of points is left fixed by a reflection, this plane, referred to as the mirror plane, is the symmetry element of the reflection. Because all other rotoinversions each fix a single point, the symmetry element of these isometries is a point. It should be noted that our definition of symmetry element differs somewhat from that used in the ITFC (Hahn, 1983).

In general, the type of symmetry element possessed by a point isometry α reveals what type of point isometry α is. For example, if $\alpha(\mathbf{p}) = \mathbf{p}$ for all $\mathbf{p} \varepsilon S$, then α is the identity. If the symmetry element of α is a line ℓ, then α is a rotation. If the symmetry element of α is a plane, then α is a reflection. Finally, if the symmetry element of α is a single point, then α is a rotoinversion other than $\bar{2}$.

DEFINING SYMMETRY

The symmetry of an object is often described intuitively by such phrases as "the object is well-balanced or well-proportioned" or that "it consists of a pattern that is repeated at regular intervals around a point, line or plane or along a direction in space." However, these types of phrases are too ambiguous for our purposes. By using isometries, we can define symmetry with sufficient rigor so as to bring powerful mathematical tools to bear on the subject.

Consider an object in space occupying the subset of points B in S. Let α denote an isometry. By $\alpha(B)$ we mean the set consisting of all of the images of the points in B. That is,

$$\alpha(B) = \{\alpha(\mathbf{b}) \mid \mathbf{b} \in B\} .$$

If $\alpha(B) = B$, then the object has been moved by α such that the object after being moved coincides exactly with the object before it was moved. That is, α has mapped the object into *self-coincidence*. When this happens, we say that the object is left *invariant* by α. Note that the statement $\alpha(B) = B$ does *not* imply that $\alpha(\mathbf{b}) = \mathbf{b}$ for each $\mathbf{b} \in B$. Since, under α, each portion of B has been moved to an equivalent portion of B, α describes something of the symmetrical nature of the object.

As an example, consider as an object the block of the structure of the mineral narsarsukite, $Na_4TiO_2Si_4O_{11}$, displayed in Figure 3.7. The large spheres in the drawing represent Ti, the intermediate-sized ones represent Si and the small ones, O (Peacor and Buerger, 1962). The Na atoms were omitted for sake of simplicity. Let $\alpha = \mathbf{4}$ denote a quarter-turn about an axis perpendicular to the plane of the drawing, passing through the center of the block. Let B denote the set of points occupied by all the features of the block. Note that $\mathbf{4}(B) = B$. The fact that $\mathbf{4}(B) = B$ is a mathematical statement of the observation that each feature (atoms, bonds, etc.) of the block is repeated at 90° intervals about its central point. Note that a half-turn, a negative quarter-turn and, of course, the identity about the rotation axis also map B onto itself. Furthermore, the rotations $\mathbf{1}$, $\mathbf{4}$, $\mathbf{2}$ and $\mathbf{4}^{-1}$ all taking place about the axis perpendicular to the page form a complete list of all of the isometries that map the block into self-coincidence. Hence, the set denoted $4 = \{\mathbf{1},\mathbf{4},\mathbf{2},\mathbf{4}^{-1}\}$ completely describes the symmetry of the narsarsukite block. The rotations in 4 define a *four-fold axis* (also called a *tetrad* axis)

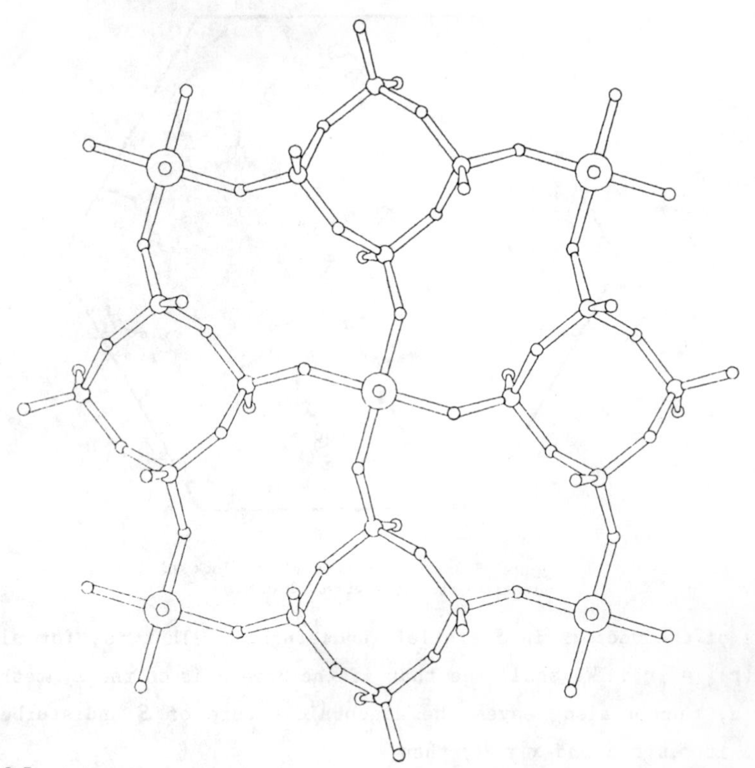

Figure 3.7: A drawing of a block of the narsarsukite structure viewed down **c**. The small spheres represent O, the intermediate-sized ones represent Si and the large ones Ti. The symmetry of this block is 4.

perpendicular to the block. We now define what we mean by the symmetry of an object.

(D3.5) Definition: Let B denote the set of points in S occupied by an object in space. The symmetry of the object is the set of all isometries α such that $\alpha(B) = B$.

(P3.2) Problem: The crystal structure of zussmanite, $KFe_{13}Si_{17}AlO_{42}(OH)_{14}$, is based on composite layers of 6- and 12-membered rings (Lopes-Vieira and Zussman, 1969). A block of the structure is displayed in Figure 3.8 in a view perpendicular to the layers. Find the set of isometries that define the symmetry of the zussmanite block, name the set and the n-fold axis that is perpendicular to the block.

LINEAR MAPPINGS

From D3.1 we see that when a point isometry α maps S to S, α does not disrupt the geometric nature of S in the sense that the lengths of

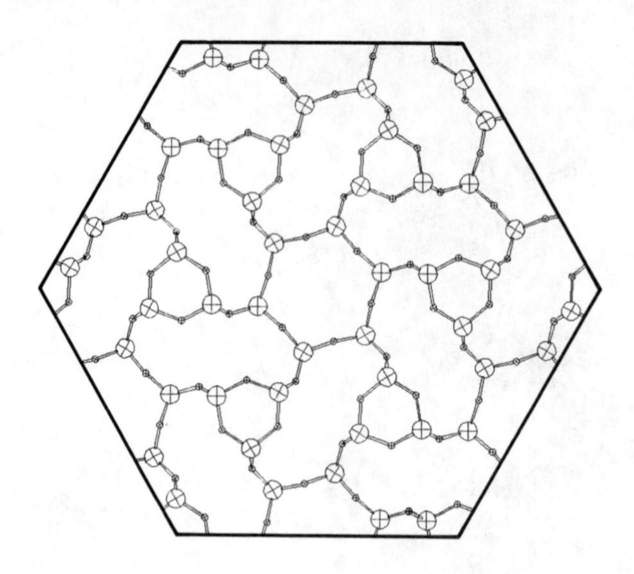

Figure 3.8: A drawing of a block of the zussmanite structure viewed down **c**.

all of the vectors in S are left undisturbed. That is, for all $\mathbf{r} \in S$, $\|\alpha(\mathbf{r})\| = \|\mathbf{r}\|$. We shall see that if the origin is on the symmetry element of α, then α also leaves the algebraic nature of S undisturbed. That is, if $\mathbf{r},\mathbf{s} \in S$ and $x \in R$, then

$$\alpha(\mathbf{r} + \mathbf{s}) = \alpha(\mathbf{r}) + \alpha(\mathbf{s})$$

and

$$\alpha(x\mathbf{r}) = x\alpha(\mathbf{r}) \; . \tag{3.1}$$

Any mapping that preserves the operations of a vector space V in this manner is called a *linear mapping* on V.

(D3.6) Definition: A mapping α from a vector space V to V is a *linear mapping* if it satisfies (3.1) for all $\mathbf{r},\mathbf{s} \in V$ and $x \in R$.

We shall now demonstrate that any rotation whose axis passes through the origin is a linear mapping. Let α denote such a rotation and, for convenience, let the Z-axis be its rotation axis. Let $\mathbf{r}$ and $\mathbf{s}$ denote vectors as shown in Figure 3.9(a). The sum of $\mathbf{r}$ and $\mathbf{s}$ is the diagonal of the parallelogram outlined by $\mathbf{r}$ and $\mathbf{s}$ as shown. Since a rotation merely turns space about an axis, all sizes and shapes are preserved. For example, if a set of points outlines a triangle, then the image of that set will outline a triangle of exactly the same size and shape, that

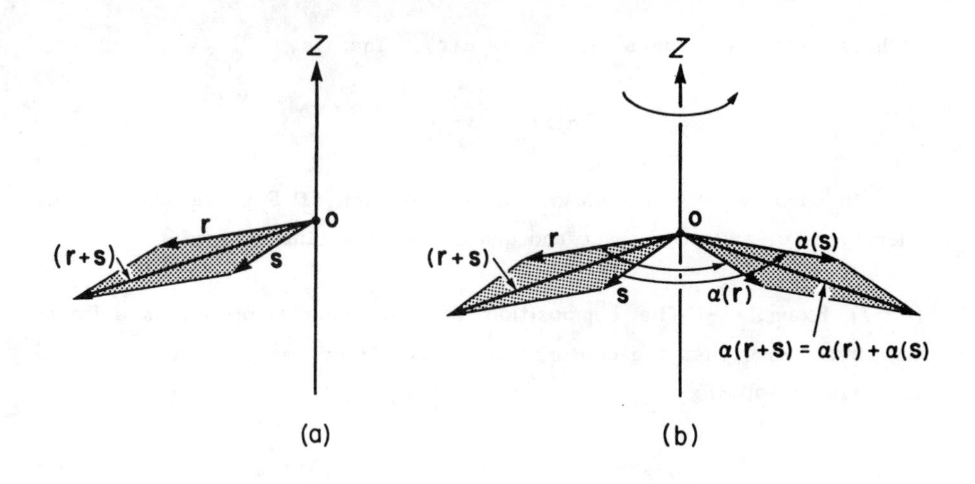

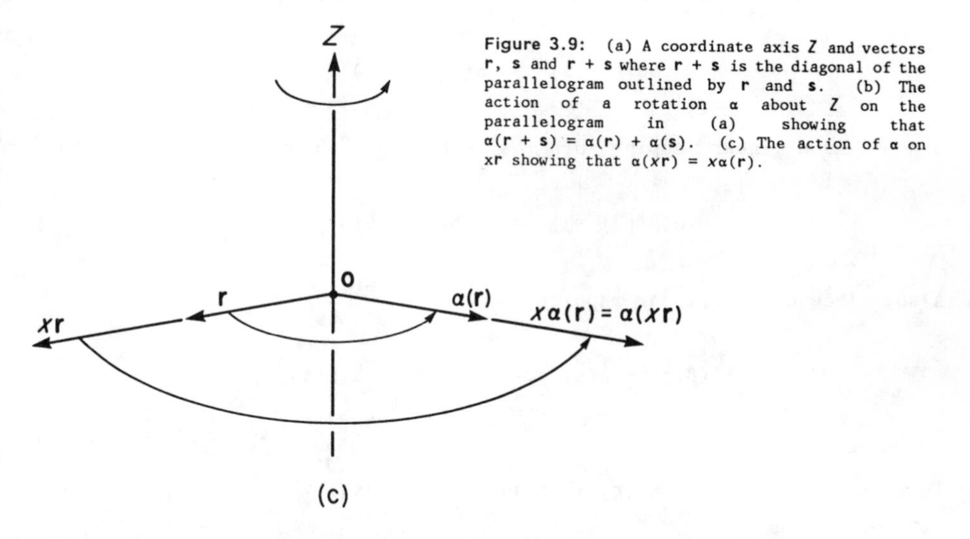

Figure 3.9: (a) A coordinate axis Z and vectors $\mathbf{r}$, $\mathbf{s}$ and $\mathbf{r} + \mathbf{s}$ where $\mathbf{r} + \mathbf{s}$ is the diagonal of the parallelogram outlined by $\mathbf{r}$ and $\mathbf{s}$. (b) The action of a rotation α about Z on the parallelogram in (a) showing that $\alpha(\mathbf{r} + \mathbf{s}) = \alpha(\mathbf{r}) + \alpha(\mathbf{s})$. (c) The action of α on $x\mathbf{r}$ showing that $\alpha(x\mathbf{r}) = x\alpha(\mathbf{r})$.

is, a congruent triangle. Similarly, a parallelogram will map to a congruent parallelogram. Consequently, the parallelogram outlined by $\mathbf{r}$ and $\mathbf{s}$ is mapped by α onto a congruent parallelogram outlined by $\alpha(\mathbf{r})$ and $\alpha(\mathbf{s})$ as shown in Figure 3.9(b). Since the diagonal of the first parallelogram, $\mathbf{r} + \mathbf{s}$, maps to the diagonal of the second, $\alpha(\mathbf{r}) + \alpha(\mathbf{s})$, we have

$$\alpha(\mathbf{r} + \mathbf{s}) = \alpha(\mathbf{r}) + \alpha(\mathbf{s}) \ .$$

In Figure 3.9(c), a vector $\mathbf{r}$ and a scalar multiple $x\mathbf{r}$ are shown. Since $\mathbf{r}$ and $x\mathbf{r}$ are in a common direction, their images $\alpha(\mathbf{r})$ and $\alpha(x\mathbf{r})$ are in a common direction. Furthermore, since $\alpha(\mathbf{r})$ has the same length as $\mathbf{r}$ and $\alpha(x\mathbf{r})$ has the same length as $x\mathbf{r}$ and since $x\mathbf{r}$ is x times as long as $\mathbf{r}$, we have $\alpha(x\mathbf{r})$ is x times as long as $\alpha(\mathbf{r})$. That is,

$$\alpha(x\mathbf{r}) = x\alpha(\mathbf{r}) .$$

In summary, we have shown that a rotation of S preserves the two operations of a vector space and qualifies as a linear mapping.

(E3.7) Example - The composition of two linear mappings is a linear mapping: Show that the composition of two linear mappings α and β on S is a linear mapping.

Solution: We need to show that $\alpha\beta(\mathbf{r} + \mathbf{s}) = \alpha\beta(\mathbf{r}) + \alpha\beta(\mathbf{s})$ and $\alpha\beta(x\mathbf{r}) = x\alpha\beta(\mathbf{r})$ for all $x \varepsilon R$ and $\mathbf{r},\mathbf{s} \varepsilon S$. By definition of composition,

$$\alpha\beta(\mathbf{r} + \mathbf{s}) = \alpha(\beta(\mathbf{r} + \mathbf{s})) .$$

Since β is a linear mapping, $\beta(\mathbf{r} + \mathbf{s}) = \beta(\mathbf{r}) + \beta(\mathbf{s})$ and so

$$\alpha(\beta(\mathbf{r} + \mathbf{s})) = \alpha(\beta(\mathbf{r}) + \beta(\mathbf{s})) .$$

Also, since α is a linear mapping,

$$\begin{aligned} \alpha(\beta(\mathbf{r}) + \beta(\mathbf{s})) &= \alpha(\beta(\mathbf{r})) + \alpha(\beta(\mathbf{s})) , \\ &= \alpha\beta(\mathbf{r}) + \alpha\beta(\mathbf{s}) . \end{aligned}$$

Therefore, $\alpha\beta(\mathbf{r} + \mathbf{s}) = \alpha\beta(\mathbf{r}) + \alpha\beta(\mathbf{s})$. Similarly,

$$\begin{aligned} \alpha\beta(x\mathbf{r}) &= \alpha(\beta(x\mathbf{r})) && \text{(definition of composition)} , \\ &= \alpha(x\beta(\mathbf{r})) && (\beta \text{ is a linear mapping}) , \\ &= x\alpha(\beta(\mathbf{r})) && (\alpha \text{ is a linear mapping}) , \\ &= x\alpha\beta(\mathbf{r}) && \text{(definition of composition)} . \end{aligned}$$

Consequently, we have shown that the composition of two linear mappings is a linear mapping. □

(E3.8) Example - The inversion operation is a linear mapping: Show that the inversion is a linear mapping.

Solution: Since $i(\mathbf{r} + \mathbf{s}) = -(\mathbf{r} + \mathbf{s}) = -\mathbf{r} - \mathbf{s} = i(\mathbf{r}) + i(\mathbf{s})$ and $i(x\mathbf{r}) = -x\mathbf{r} = x(-\mathbf{r}) = xi(\mathbf{r})$ for all $\mathbf{r},\mathbf{s} \in S$ and $x \in R$, we see that i is a linear mapping. □

(P3.3) Problem: Show that any rotoinversion that leaves the origin fixed is a linear mapping.

Matrix Representations of Linear Mappings: We are now ready to explore how a linear mapping can be represented by a matrix. See Appendix 2 for a discussion about matrices. To begin, suppose that $D = \{\mathbf{a},\mathbf{b},\mathbf{c}\}$ is a basis of S so that each vector $\mathbf{r}_1 \in S$ can be written as a linear combination of D. That is to say that corresponding to each vector $\mathbf{r}_1$, there exist three real numbers x_1, y_1 and z_1 such that $\mathbf{r}_1$ can be expressed as

$$\mathbf{r}_1 = x_1\mathbf{a} + y_1\mathbf{b} + z_1\mathbf{c} . \tag{3.2}$$

Next, consider the effect of a linear mapping (for example, any point isometry fixing the origin of space) on (3.2) such that $\alpha(\mathbf{r}_1) = \mathbf{r}_2$. As the image of $\mathbf{r}_1$ under α is a vector $\mathbf{r}_2 = \alpha(\mathbf{r}_1)$ in S, there must also exist three real numbers x_2, y_2 and z_2 such that the image vector $\mathbf{r}_2$ can be written as

$$\mathbf{r}_2 = x_2\mathbf{a} + y_2\mathbf{b} + z_2\mathbf{c} . \tag{3.3}$$

Applying the properties of a linear mapping to $\mathbf{r}_1$, we have

$$\begin{aligned} \alpha(\mathbf{r}_1) &= \alpha(x_1\mathbf{a} + y_1\mathbf{b} + z_1\mathbf{c}) \\ &= \alpha(x_1\mathbf{a}) + \alpha(y_1\mathbf{b}) + \alpha(z_1\mathbf{c}) \\ &= x_1\alpha(\mathbf{a}) + y_1\alpha(\mathbf{b}) + z_1\alpha(\mathbf{c}) . \end{aligned} \tag{3.4}$$

This tells us that if we know $\alpha(\mathbf{a})$, $\alpha(\mathbf{b})$ and $\alpha(\mathbf{c})$, then $\alpha(\mathbf{r}_1)$ is completely determined. In particular, if $\alpha(\mathbf{a})$, $\alpha(\mathbf{b})$ and $\alpha(\mathbf{c})$ are expressed as linear combinations of D, then we can use (3.4) to write $\alpha(\mathbf{r}_1)$ in that form. Thus, if we write $\alpha(\mathbf{a})$, $\alpha(\mathbf{b})$ and $\alpha(\mathbf{c})$ as linear combinations of D, we have the equations

$$\alpha(\mathbf{a}) = \ell_{11}\mathbf{a} + \ell_{21}\mathbf{b} + \ell_{31}\mathbf{c}$$
$$\alpha(\mathbf{b}) = \ell_{12}\mathbf{a} + \ell_{22}\mathbf{b} + \ell_{32}\mathbf{c}$$
$$\alpha(\mathbf{c}) = \ell_{13}\mathbf{a} + \ell_{23}\mathbf{b} + \ell_{33}\mathbf{c} \ ,$$

where, of course, the nine coefficients, ℓ_{ij}, are real numbers. Next we replace $\alpha(\mathbf{a})$, $\alpha(\mathbf{b})$ and $\alpha(\mathbf{c})$ in (3.4) by these expressions obtaining

$$\alpha(\mathbf{r}_1) = x_1(\ell_{11}\mathbf{a} + \ell_{21}\mathbf{b} + \ell_{31}\mathbf{c}) + y_1(\ell_{12}\mathbf{a} + \ell_{22}\mathbf{b} + \ell_{32}\mathbf{c}) + z_1(\ell_{13}\mathbf{a} + \ell_{23}\mathbf{b} + \ell_{33}\mathbf{c}) \ .$$

Collecting the coefficients of $\mathbf{a}$, $\mathbf{b}$ and $\mathbf{c}$, we have

$$\mathbf{r}_2 = \alpha(\mathbf{r}_1) = (\ell_{11}x_1 + \ell_{12}y_1 + \ell_{13}z_1)\mathbf{a} + (\ell_{21}x_1 + \ell_{22}y_1 + \ell_{23}z_1)\mathbf{b} + (\ell_{31}x_1 + \ell_{32}y_1 + \ell_{33}z_1)\mathbf{c} \ . \tag{3.5}$$

As this vector equals $\mathbf{r}_2$ in (3.3), we can equate the corresponding coefficients of the vectors $\mathbf{a}$, $\mathbf{b}$ and $\mathbf{c}$ for $\mathbf{r}_2$ to obtain the following set of equations:

$$\begin{aligned} x_2 &= \ell_{11}x_1 + \ell_{12}y_1 + \ell_{13}z_1 \\ y_2 &= \ell_{21}x_1 + \ell_{22}y_1 + \ell_{23}z_1 \\ z_2 &= \ell_{31}x_1 + \ell_{32}y_1 + \ell_{33}z_1 \ . \end{aligned} \tag{3.6}$$

(Recall that two vectors $\mathbf{r}_i = x_i\mathbf{a} + y_i\mathbf{b} + z_i\mathbf{c}$ and $\mathbf{r}_j = x_j\mathbf{a} + y_j\mathbf{b} + z_j\mathbf{c}$ are equal only if $x_i = x_j$, $y_i = y_j$ and $z_i = z_j$.) The set of equations in (3.6) can be written in matrix form as

$$\begin{bmatrix} x_2 \\ y_2 \\ z_2 \end{bmatrix} = \begin{bmatrix} \ell_{11} & \ell_{12} & \ell_{13} \\ \ell_{21} & \ell_{22} & \ell_{23} \\ \ell_{31} & \ell_{32} & \ell_{33} \end{bmatrix} \begin{bmatrix} x_1 \\ y_1 \\ z_1 \end{bmatrix}$$

or more briefly as

$$[\mathbf{r}_2]_D = M_D(\alpha)[\mathbf{r}_1]_D \ , \tag{3.7}$$

where, as usual,

$$[\mathbf{r}_1]_D = \begin{bmatrix} x_1 \\ y_1 \\ z_1 \end{bmatrix} \quad \text{and} \quad [\mathbf{r}_2]_D = \begin{bmatrix} x_2 \\ y_2 \\ z_2 \end{bmatrix} \ ,$$

and we define $M_D(\alpha)$ by

$$M_D(\alpha) = \begin{bmatrix} \ell_{11} & \ell_{12} & \ell_{13} \\ \ell_{21} & \ell_{22} & \ell_{23} \\ \ell_{31} & \ell_{32} & \ell_{33} \end{bmatrix} .$$

In this context, we speak of $M_D(\alpha)$ as being the *matrix representation* of α with respect to the basis D. It is important to note the $M_D(\alpha)$ can be constructed from the coordinates of $[\alpha(\mathbf{a})]_D$, $[\alpha(\mathbf{b})]_D$, and $[\alpha(\mathbf{c})]_D$ by noting that

$$M_D(\alpha) = \begin{bmatrix} | & | & | \\ [\alpha(\mathbf{a})]_D & [\alpha(\mathbf{b})]_D & [\alpha(\mathbf{c})]_D \\ | & | & | \end{bmatrix} = \begin{bmatrix} \ell_{11} & \ell_{12} & \ell_{13} \\ \ell_{21} & \ell_{22} & \ell_{23} \\ \ell_{31} & \ell_{32} & \ell_{33} \end{bmatrix} , \qquad (3.8)$$

since

$$[\alpha(\mathbf{a})]_D = \begin{bmatrix} \ell_{11} \\ \ell_{21} \\ \ell_{31} \end{bmatrix} , \quad [\alpha(\mathbf{b})]_D = \begin{bmatrix} \ell_{12} \\ \ell_{22} \\ \ell_{32} \end{bmatrix} , \quad [\alpha(\mathbf{c})]_D = \begin{bmatrix} \ell_{13} \\ \ell_{23} \\ \ell_{33} \end{bmatrix} .$$

Thus, to construct $M_D(\alpha)$ for some linear mapping α, we observe that the first column of the matrix has as its entries the coefficients of $\alpha(\mathbf{a})$ with respect to D, the second has as its entries those of $\alpha(\mathbf{b})$, and third has as its entries those of $\alpha(\mathbf{c})$.

(E3.9) Example - Matrix representation of a half-turn rotation of space about Z: Let **2** denote a half-turn rotation about Z defined with respect to the basis $D = \{\mathbf{a},\mathbf{b},\mathbf{c}\}$ such that

$$\begin{aligned} \mathbf{2}(\mathbf{a}) &= -1\mathbf{a} + 0\mathbf{b} + 1\mathbf{c} \\ \mathbf{2}(\mathbf{b}) &= 0\mathbf{a} - 1\mathbf{b} + 1\mathbf{c} \\ \mathbf{2}(\mathbf{c}) &= 0\mathbf{a} + 0\mathbf{b} + 1\mathbf{c} . \end{aligned}$$

With this information, the matrix for the half-turn can be constructed by evaluating the columns of

$$M_D(\mathbf{2}) = \begin{bmatrix} | & | & | \\ [\mathbf{2}(\mathbf{a})]_D & [\mathbf{2}(\mathbf{b})]_D & [\mathbf{2}(\mathbf{c})]_D \\ | & | & | \end{bmatrix} .$$

Since

$$[\alpha(\mathbf{a})]_D = \begin{bmatrix} -1 \\ 0 \\ 1 \end{bmatrix}, \quad [\alpha(\mathbf{b})]_D = \begin{bmatrix} 0 \\ -1 \\ 1 \end{bmatrix}, \quad [\alpha(\mathbf{c})]_D = \begin{bmatrix} 0 \\ 0 \\ 1 \end{bmatrix},$$

we conclude that

$$M_D(\mathbf{2}) = \begin{bmatrix} -1 & 0 & 0 \\ 0 & -1 & 0 \\ 1 & 1 & 1 \end{bmatrix}.$$

To find where the vector $\mathbf{r} = x\mathbf{a} + y\mathbf{b} + z\mathbf{c}$ is mapped by $\mathbf{2}$, we recall that $M_D(\mathbf{2})$ is defined so that $M_D(\mathbf{2})[\mathbf{r}]_D = [\mathbf{2}(\mathbf{r})]_D$. Hence

$$[\mathbf{2}(\mathbf{r})]_D = \begin{bmatrix} -1 & 0 & 0 \\ 0 & -1 & 0 \\ 1 & 1 & 1 \end{bmatrix} \begin{bmatrix} x \\ y \\ z \end{bmatrix} = \begin{bmatrix} -x \\ -y \\ x + y + z \end{bmatrix},$$

from which we conclude that

$$\mathbf{2}(\mathbf{r}) = \mathbf{2}(x\mathbf{a} + y\mathbf{b} + z\mathbf{c}) = -x\mathbf{a} - y\mathbf{b} + (x + y + z)\mathbf{c}\ . \qquad \square$$

The technique for constructing the change of basis matrices discussed in Chapter 2 can be used here as an alternative way of approaching the construction of $M_D(\alpha)$. In Chapter 2 we observed that the product of a matrix times the vectors $[100]^t$, $[010]^t$ and $[001]^t$ equals the columns of the matrix. Since $[\mathbf{a}]_D = [100]^t$ and $M_D(\alpha)[\mathbf{a}]_D = [\alpha(\mathbf{a})]_D$, we have that $[\alpha(\mathbf{a})]_D$ is the first column of $M_D(\alpha)$. Similarly, $[\alpha(\mathbf{b})]_D$ and $[\alpha(\mathbf{c})]_D$ are the second and third columns of the matrix.

(P3.4) Problem: Find $M_D(\alpha)$ where $D = \{\mathbf{a},\mathbf{b},\mathbf{c}\}$ and α is a quarter-turn inversion $[11\bar{1}]_{\bar{4}}$, about the vector $\mathbf{a} + \mathbf{b} - \mathbf{c}$ given that $\alpha(\mathbf{a}) = -\mathbf{a} + \mathbf{c}$, $\alpha(\mathbf{b}) = -\mathbf{a}$, and $\alpha(\mathbf{c}) = -\mathbf{a} + \mathbf{b}$.

(P3.5) Problem: Find the matrix defining a sixth-turn, $\mathbf{6}$, about the Z-axis with respect to the basis $D = \{\mathbf{a},\mathbf{b},\mathbf{c}\}$, given that $\mathbf{6}(\mathbf{a}) = \mathbf{a} + \mathbf{b}$, $\mathbf{6}(\mathbf{b}) = -\mathbf{a}$ and $\mathbf{6}(\mathbf{c}) = \mathbf{c}$. Also, find the components of the vector $\mathbf{6}(\mathbf{r})$.

Matrix Representations of Compositions of Linear Mappings. Let α and β denote linear mappings on S and let D denote a basis for S. As discussed earlier, $\alpha\beta$ is the mapping defined by $\alpha\beta(\mathbf{r}) = \alpha(\beta(\mathbf{r}))$ for all $\mathbf{r} \in S$. The corresponding statement in the context of R^3 is

$$M_D(\alpha\beta)[\mathbf{r}]_D = M_D(\alpha)(M_D(\beta)[\mathbf{r}]_D) .$$

Since matrix multiplication is associative,

$$M_D(\alpha)(M_D(\beta)[\mathbf{r}]_D) = (M_D(\alpha)M_D(\beta))[\mathbf{r}]_D .$$

Hence, $M_D(\alpha\beta)[\mathbf{r}]_D = (M_D(\alpha)M_D(\beta))[\mathbf{r}]_D$ for all $\mathbf{r} \in S$ and so

$$M_D(\alpha\beta) = M_D(\alpha)M_D(\beta) . \tag{3.9}$$

Therefore, the matrix representation of the composition of two mappings is obtained by multiplying the matrix representations of the individual mappings (in the same order). For a discussion of matrix multiplication, see Appendix 2.

(E3.10) Example - A matrix representing the composition of half-turn and quarter-turn rotations about different but intersecting axes: Let $D = \{\mathbf{a},\mathbf{b},\mathbf{c}\}$ denote a basis. Let α denote the half-turn of E3.9 and let β denote the quarter-turn inversion of P3.4. Find $M_D(\alpha\beta)$.

Solution: In E3.9, we found that

$$M_D(\alpha) = \begin{bmatrix} -1 & 0 & 0 \\ 0 & -1 & 0 \\ 1 & 1 & 1 \end{bmatrix} .$$

By a similar process, it was shown in P3.4 that

$$M_D(\beta) = \begin{bmatrix} -1 & -1 & -1 \\ 0 & 0 & 1 \\ 1 & 0 & 0 \end{bmatrix} .$$

Hence,

$$M_D(\alpha\beta) = M_D(\alpha)M_D(\beta) = \begin{bmatrix} 1 & 1 & 1 \\ 0 & 0 & -1 \\ 0 & -1 & 0 \end{bmatrix} . \tag{3.10}$$

□

(P3.6) Problem: Using $M_D(\alpha)$ and $M_D(\beta)$ given in E3.10, find $M_D(\beta\alpha)$. Note that $\alpha\beta \neq \beta\alpha$.

(P3.7) Problem: Let $D = \{\mathbf{a},\mathbf{b},\mathbf{c}\}$ denote a basis. Find $M_D(\beta\alpha)$ where α is the sixth-turn of P3.5 and β is the half-turn defined by $\beta(\mathbf{a}) = -\mathbf{a} - \mathbf{b}$, $\beta(\mathbf{b}) = \mathbf{b}$ and $\beta(\mathbf{c}) = -\mathbf{c}$. Then find $M_D(\beta\alpha)$.

In Appendix 3, methods for analyzing matrix representations of point isometries are presented. This appendix should be read before attempting to answer the following problems.

(P3.8) Problem: Determine the symbol for the point isometry represented by (3.10) in E3.10.

(P3.9) Problem: Determine the symbol for $\beta\alpha$ of P3.6.

(P3.10) Problem: Determine the symbol for $\beta\alpha$ of P3.7.

THE CONSTRUCTION OF A SET OF MATRICES DEFINING THE ROTATIONS OF *322*

In this section, we shall construct the matrix representations of the point isometries that describe the external symmetry of an α-quartz crystal, i.e., the point isometries that send such a crystal into self-coincidence. An X-ray study of the mineral shows that its rotation axes labeled ℓ_1, ℓ_2, ℓ_3 and ℓ_4 in Figure 3.10 coincide with the vectors $\mathbf{a}$, $\mathbf{a} + \mathbf{b}$, $\mathbf{b}$ and $\mathbf{c}$, respectively. An examination of this figure shows that the crystal is sent into self-coincidence by $\mathbf{3}$ and $\mathbf{3}^{-1}$ about ℓ_4 and by $[100]_2$, $[110]_2$, and $[010]_2$ about ℓ_1, ℓ_2 and ℓ_3, respectively. The crystal is also mapped into self-coincidence by the identity, $\mathbf{1}$. Furthermore, these are the only isometries that map the α-quartz crystal into self-coincidence. As the symmetry of an object is defined (D3.5) to be the set of all isometries that send it into self-coincidence, the symmetry of an α-quartz crystal is the set of rotations $\{\mathbf{1}, \mathbf{3}, \mathbf{3}^{-1}, [100]_2, [110]_2, [010]_2\}$. We will denote this set of rotations by the symbol *322* to be consistent with our derivation of the point groups. However, the conventional symbol for this set is *32*.

In the previous section we learned how to find the matrix representation of a point isometry α with respect to a basis D given $\alpha(\mathbf{a})$, $\alpha(\mathbf{b})$

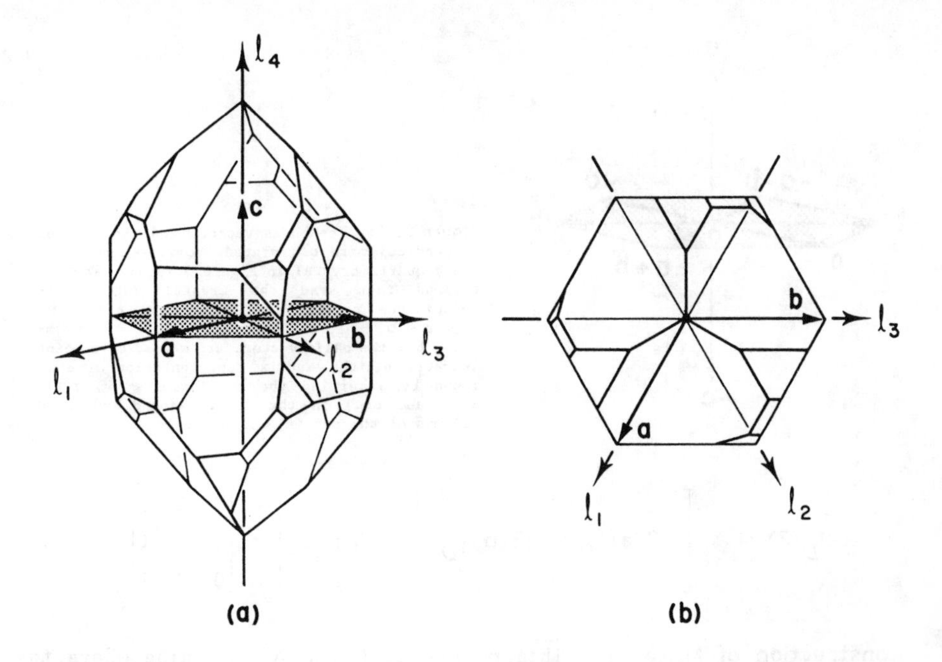

Figure 3.10: Drawing of an ideal crystal of α-quartz (a) viewed at an oblique angle to **c** and (b) viewed down **c**. The rotation axes ℓ_1, ℓ_2, ℓ_3 and ℓ_4 of $[100]_2$, $[110]_2$, $[010]_2$, and *3* are parallel to **a**, **a** + **b**, **b**, and **c**, respectively.

and α(**c**) in terms of *D*. Now we shall learn how to find the matrix representation of the point isometries that map a crystal like α-quartz into self-coincidence from the basic geometric descriptions of these mappings. To simplify our view of the geometry, we extract the vectors and the hexagon lying in the ℓ_1,ℓ_3 plane of the illustration of the α-quartz crystal displayed in Figure 3.10. Within this geometrical context, we can describe each of the point isometries of *322* and find their matrix representations. To facilitate our observations, we have labeled each of the vectors emanating to the six corners of the hexagon in Figure 3.11. We shall now derive a matrix for each rotation α ε *322* by studying the action of α on the hexagon and the basis vectors of α-quartz.

Construction of M_D(3): In the construction of M_D(**3**), we examine the action of a 120° rotation about an axis coincident with **c**. By rotating the hexagon in Figure 3.12(a) through a turn angle of 120° about **c**, the basis vectors are rotated to the positions that they occupy in the image hexagon (Figure 3.12(b)). Thus, the motion of **3** maps **a** onto **b**, **b** onto −**a** − **b** and **c** onto **c**, i.e., **3**(**a**) = **b**, **3**(**b**) = −**a** − **b** and **3**(**c**) = **c**. With these results, we see that M_D(**3**) is

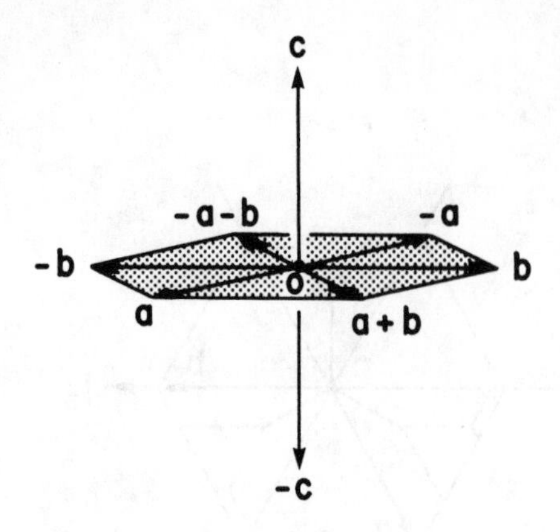

Figure 3.11: The basis vectors and the hexagon perpendicular to **c** isolated from the drawing of the α-quartz crystal in Figure 3.10(a). The rotations that send the crystal into self-coincidence permute the vectors {**a**, **a** + **b**, **b**, −**a**, −**a** − **b**, −**b**}, and permute {**c**, −**c**}. By examining how the basis vectors are mapped by a point isometry α, the matrix representation of α is found by inserting the coefficients of α(**a**), α(**b**) and α(**c**) in the first, second and third columns of the matrix, respectively.

$$M_D(\mathbf{3}) = \left[\begin{array}{c|c|c} [\mathbf{3}(\mathbf{a})]_D & [\mathbf{3}(\mathbf{b})]_D & [\mathbf{3}(\mathbf{c})]_D \end{array}\right] = \begin{bmatrix} 0 & -1 & 0 \\ 1 & -1 & 0 \\ 0 & 0 & 1 \end{bmatrix} .$$

Construction of $M_D(\mathbf{3}^{-1})$: This matrix is found by examining where the basis vectors in Figure 3.12(a) are sent by −120° rotation about an axis coincident with **c** to the configuration shown in Figure 3.12(b). In this case, **a** is mapped onto −**a** − **b**, **b** is mapped onto **a**, and **c** is mapped onto itself, i.e., $\mathbf{3}^{-1}(\mathbf{a}) = -\mathbf{a} - \mathbf{b}$, $\mathbf{3}^{-1}(\mathbf{b}) = \mathbf{a}$, and $\mathbf{3}^{-1}(\mathbf{c}) = \mathbf{c}$. Hence,

$$M_D(\mathbf{3})^{-1} = \left[\begin{array}{c|c|c} [\mathbf{3}^{-1}(\mathbf{a})]_D & [\mathbf{3}^{-1}(\mathbf{b})]_D & [\mathbf{3}^{-1}(\mathbf{c})]_D \end{array}\right] = \begin{bmatrix} -1 & 1 & 0 \\ -1 & 0 & 0 \\ 0 & 0 & 1 \end{bmatrix} .$$

We also observe that a −120° rotation of space about **c** is equivalent to two successive rotations of 120° (i.e., a positive rotation of 240° is equivalent to a negative rotation of 120°) each about the same axis. That is, $\mathbf{33} = \mathbf{3}^2 = \mathbf{3}^{-1}$. With this in mind, by (3.9), we expect that when $M_D(\mathbf{3})$ is multiplied by itself, the matrix $M_D(\mathbf{3}^{-1})$ results:

$$M_D(\mathbf{3})M_D(\mathbf{3}) = \begin{bmatrix} 0 & -1 & 0 \\ 1 & -1 & 0 \\ 0 & 0 & 1 \end{bmatrix} \begin{bmatrix} 0 & -1 & 0 \\ 1 & -1 & 0 \\ 0 & 0 & 1 \end{bmatrix}$$

$$= \begin{bmatrix} -1 & 1 & 0 \\ -1 & 0 & 0 \\ 0 & 0 & 1 \end{bmatrix}$$

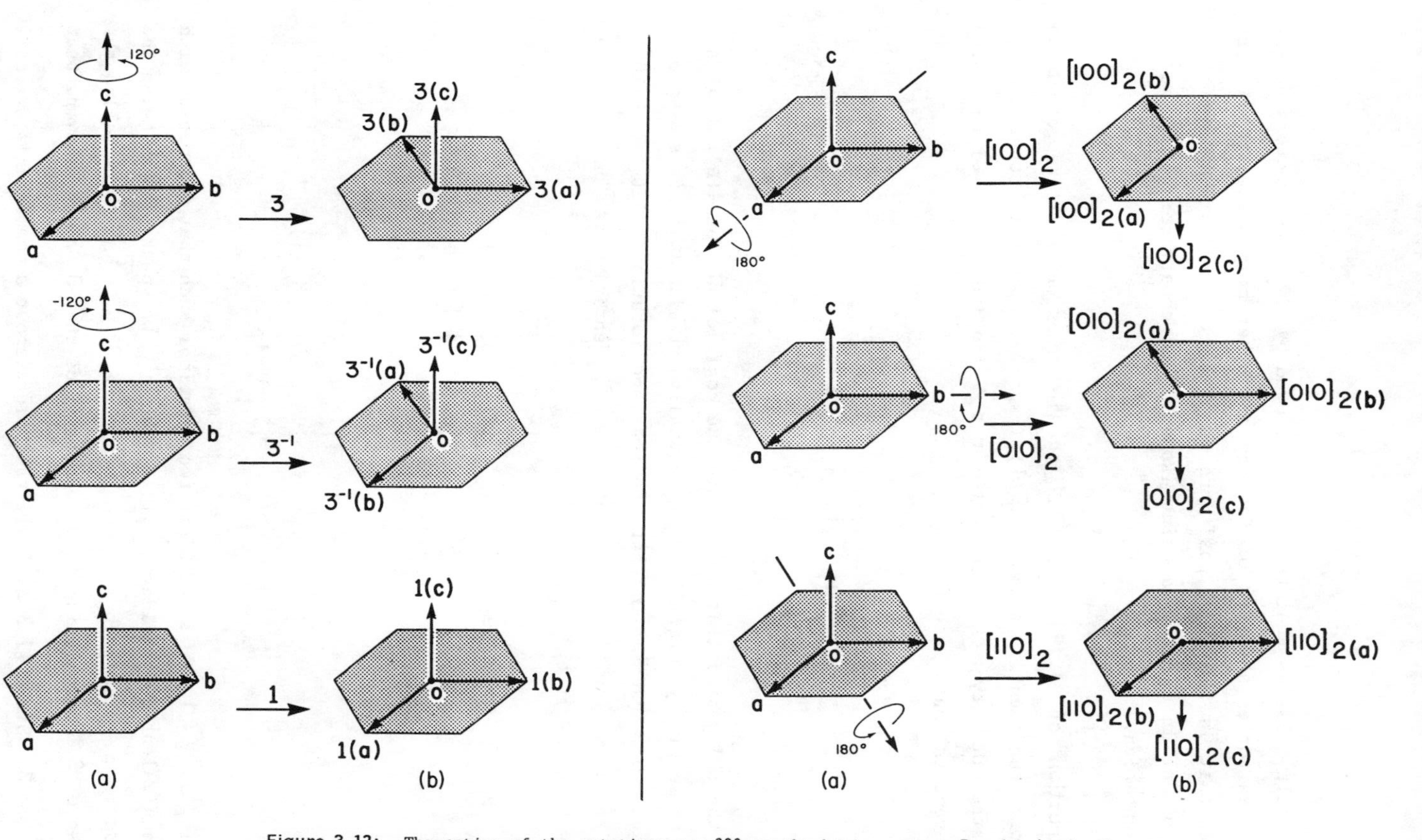

Figure 3.12: The action of the rotations $\alpha \in$ *322* on the basis vectors $D = \{\mathbf{a}, \mathbf{b}, \mathbf{c}\}$ of α-quartz. The original positions of D are given in column (a) and the positions of their images under α are given in column (b). By consulting the names of the vectors in Figure 3.11, the coefficients of $\alpha(\mathbf{a})$ $\alpha(\mathbf{b})$ and $\alpha(\mathbf{c})$ can be determined.

$$= M_D(\mathbf{3}^{-1}) .$$

This result tells us that the route followed by the basis vectors during a rotation (or as a matter of fact by any other point isometry) is irrelevant; it is the final positions to which the vectors are sent that is important in defining a point isometry and constructing its matrix representation.

Construction of $M_D(\mathbf{1})$: Of all the matrices used to represent an isometry, the identity matrix, $M_D(\mathbf{1})$, is the simplest to construct because the identity maps each vector $\mathbf{r} \in S$ (including each of the three basis vectors) onto itself. Thus,

$$M_D(\mathbf{1}) = \begin{bmatrix} 1 & 0 & 0 \\ 0 & 1 & 0 \\ 0 & 0 & 1 \end{bmatrix} ,$$

where the columns of this matrix are the coefficients of $\mathbf{1}(\mathbf{a})$, $\mathbf{1}(\mathbf{b})$ and $\mathbf{1}(\mathbf{c})$, respectively (Fig. 3.12(b)). Actually, the identity bears a special relation to $\mathbf{3}$ and $\mathbf{3}^{-1}$ in that the combined rotations $(120° + (-120°))$ about the same axis is equivalent to a $0°$ identity rotation. We can verify this result by

$$M_D(\mathbf{3}\mathbf{3}^{-1}) = M_D(\mathbf{3})M_D(\mathbf{3}^{-1}) = \begin{bmatrix} 0 & -1 & 0 \\ 1 & -1 & 0 \\ 0 & 0 & 1 \end{bmatrix} \begin{bmatrix} -1 & 1 & 0 \\ -1 & 0 & 0 \\ 0 & 0 & 1 \end{bmatrix}$$

$$= \begin{bmatrix} 1 & 0 & 0 \\ 0 & 1 & 0 \\ 0 & 0 & 1 \end{bmatrix} = M_D(\mathbf{1}) .$$

That is, $\mathbf{3}\mathbf{3}^{-1} = \mathbf{1}$. If α and β are two rotations such that α composed with β is equivalent to the identity, i.e., $\alpha\beta = \mathbf{1}$, we speak of β as being the *inverse* of α. Thus, in our example, since $\mathbf{3}\mathbf{3}^{-1} = \mathbf{1}$, $\mathbf{3}^{-1}$ is called the inverse of $\mathbf{3}$. Calculating $M_D(\mathbf{3}^{-1}\mathbf{3})$ one finds that $\mathbf{3}^{-1}\mathbf{3} = \mathbf{1}$, showing that $(\mathbf{3}^{-1})^{-1} = \mathbf{3}$. Since $\{\mathbf{1},\mathbf{3},\mathbf{3}^{-1}\}$ all take place about a common axis, we call such an axis a ***three-fold axis*** (also called a ***triad axis***). We use the symbol *3* to denote $\{\mathbf{1},\mathbf{3},\mathbf{3}^{-1}\}$ and $M_D(3)$ to denote the matrix representations of the rotations in *3*. That is,

$$M_D(3) = \left\{ \begin{bmatrix} 1 & 0 & 0 \\ 0 & 1 & 0 \\ 0 & 0 & 1 \end{bmatrix}, \begin{bmatrix} 0 & -1 & 0 \\ 1 & -1 & 0 \\ 0 & 0 & 1 \end{bmatrix}, \begin{bmatrix} -1 & 1 & 0 \\ -1 & 0 & 0 \\ 0 & 0 & 1 \end{bmatrix} \right\}.$$

While the three-fold axis and the elements of *3* are quite distinct objects, it is customary, nonetheless, to refer to this set of rotations as a three-fold axis and to symbolize it in a drawing by an equilateral triangle, ▲.

Construction of $M_D({}^{[100]}2)$: A half-turn, ${}^{[100]}2$, about an axis coincident with **a** turns the hexagon in Figure 3.12(a) over and maps **a** onto itself, **b** onto −**a** − **b** and **c** onto −**c** as shown in Figure 3.12(b). Inserting the coordinates of ${}^{[100]}2(\mathbf{a})_D$, ${}^{[100]}2(\mathbf{b})_D$ and ${}^{[100]}2(\mathbf{c})_D$ into the columns of a matrix, we have

$$M_D({}^{[100]}2) = \begin{bmatrix} 1 & -1 & 0 \\ 0 & -1 & 0 \\ 0 & 0 & -1 \end{bmatrix}.$$

If two successive half-turns are made about **a** ($\rho = 180° + 180°$), the end product would be equivalent to the identity. In terms of matrices, this means that

$$M_D({}^{[100]}2)M_D({}^{[100]}2) = M_D(1),$$

which can be checked by performing the multiplication. Thus, we see that a half-turn is its own inverse, i.e., ${}^{[100]}2 = {}^{[100]}2^{-1}$. The two rotations *1* and ${}^{[100]}2$ form a subset denoted ${}^{[100]}2 = \{1, {}^{[100]}2\}$. As both of these rotations take place about a common axis, they constitute a ***two-fold axis*** (also called a ***diad axis***). A two-fold axis is often specified with the symbol ⬮.

Construction of $M_D({}^{[110]}2)$: A half-turn, ${}^{[110]}2$, about an axis coincident with **a** + **b** turns the hexagon in Figure 3.12(a) over and maps **a** onto **b**, **b** onto **a** and **c** onto −**c** (Figure 3.12(b)). The matrix representation for this rotation is therefore

$$M_D({}^{[110]}2) = \begin{bmatrix} 0 & 1 & 0 \\ 1 & 0 & 0 \\ 0 & 0 & -1 \end{bmatrix}.$$

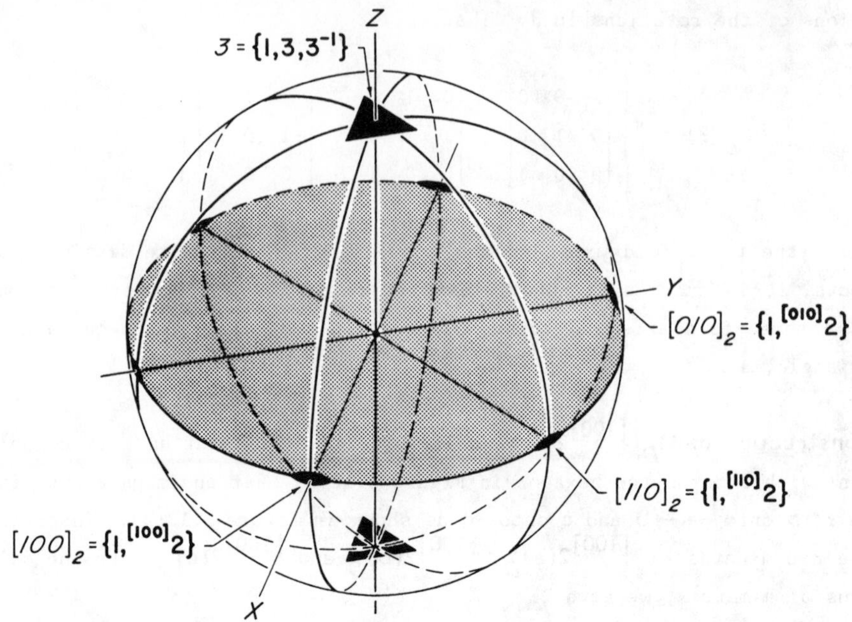

Figure 3.13: The symmetry elements of *322*. The rotation axes of *322* intersect at the center of the unit ball. The 3-fold axis is perpendicular to the plane containing each of the 2-fold axes. Adjacent 2-fold axes intersect at 60° to one another with $^{[100]}2$ paralleling X, $^{[110]}2$ making an angle of 60° with respect to X and $^{[010]}2$ paralleling Y.

Like the half-turn described in the last section, $^{[110]}2\,^{[110]}2 = 1$. As $^{[110]}2$ and **1** share a common rotation axis, they comprise the rotations of a two-fold axis denoted $^{[110]}2 = \{1, {}^{[110]}2\}$.

Construction of $M_D(^{[010]}2)$: The remaining rotation of *322*, $^{[010]}2$ turns the hexagon in Figure 3.12(a) over and maps **a** onto −**a** − **b**, **b** onto **b** and **c** onto −**c** (Figure 3.12(b)) so that

$$M_D(^{[010]}2) = \begin{bmatrix} -1 & 0 & 0 \\ -1 & 1 & 0 \\ 0 & 0 & -1 \end{bmatrix} .$$

This half-turn and the identity comprise the two rotations of the two-fold axis, $^{[010]}2$, that occur about **b**.

In our study of each of the rotations in *322* that send an α-quartz crystal into self-coincidence, we found that a three-fold rotation axis parallels **c**, and that two-fold rotation axes parallel **a**, **a** + **b** and **b**. Accordingly, the three-fold axis is perpendicular to the plane containing the two-fold axes and each two-fold makes an angle of 60° to the remaining two-folds. These axes are depicted in Figure 3.13 where they intersect at an origin located at the center of a unit ball. Such a diagram is referred to as a drawing of the symmetry elements of *322*.

Algebraic Properties of *322*: As seen earlier in this chapter, the composition of two linear mappings (such as isometries leaving the origin fixed) is mimicked with respect to R^3 by multiplying their matrix representations with respect to a basis D. In the case of *322*, we have found all of these matrix representations. Thus, the set of matrices

$$M_D(322) = \{M_D(1), M_D(3), M_D(3^{-1}), M_D({}^{[100]}2), M_D({}^{[110]}2), M_D({}^{[010]}2)\}$$

describes *322* with respect to R^3 just as well as the set of rotations

$$322 = \{1, 3, 3^{-1}, {}^{[100]}2, {}^{[110]}2, {}^{[010]}2\}$$

describes the set with respect to S. However, working with matrices avoids the difficulties encountered in visualizing the effect of a point isometry on an object. For example, it is usually difficult to visualize the combined affect of two or more successive applications of point isometries on S. On the other hand, when these point isometries are represented by matrices, their successive application can be analyzed through matrix multiplication routinely. This fact enables us to describe the algebraic properties of $M_D(322)$ with a *multiplication table* similar to the ones used to construct a "times table" in grade school. In the construction of this table for $M_D(322)$, we start by listing each matrix of the set in the first row (called the *guide row*) above the horizontal line and the first column (called the *guide column*) to the left of the vertical line (see Figure 3.14). Each matrix $M_D(\alpha)$, in the guide column is then multiplied by each matrix $M_D(\beta)$, in the guide row, with the matrix from the guide row on the right. The table is completed by entering the product $M_D(\alpha)M_D(\beta)$ at the intersection of the row headed by $M_D(\alpha)$ and the column headed by $M_D(\beta)$. For example, when $M_D({}^{[100]}2)$ of the guide column is multiplied on the right by $M_D({}^{[110]}2)$ of the guide row, we get

$$M_D({}^{[100]}2)M_D({}^{[110]}2) = \begin{bmatrix} 1 & -1 & 0 \\ 0 & -1 & 0 \\ 0 & 0 & -1 \end{bmatrix} \begin{bmatrix} 0 & 1 & 0 \\ 1 & 0 & 0 \\ 0 & 0 & -1 \end{bmatrix}$$

Figure 3.14: Group Multiplication Table for $M_D(322)$

$M_D(322)$	$M_D(1)$	$M_D(3)$	$M_D(3^{-1})$	$M_D({}^{[100]}2)$	$M_D({}^{[110]}2)$	$M_D({}^{[010]}2)$
$M_D(1)$	$M_D(1)$	$M_D(3)$	$M_D(3^{-1})$	$M_D({}^{[100]}2)$	$M_D({}^{[110]}2)$	$M_D({}^{[010]}2)$
$M_D(3)$	$M_D(3)$	$M_D(3^{-1})$	$M_D(1)$	$M_D({}^{[110]}2)$	$M_D({}^{[010]}2)$	$M_D({}^{[100]}2)$
$M_D(3^{-1})$	$M_D(3^{-1})$	$M_D(1)$	$M_D(3)$	$M_D({}^{[010]}2)$	$M_D({}^{[100]}2)$	$M_D({}^{[110]}2)$
$M_D({}^{[100]}2)$	$M_D({}^{[100]}2)$	$M_D({}^{[010]}2)$	$M_D({}^{[110]}2)$	$M_D(1)$	$M_D(3^{-1})$	$M_D(3)$
$M_D({}^{[110]}2)$	$M_D({}^{[110]}2)$	$M_D({}^{[100]}2)$	$M_D({}^{[010]}2)$	$M_D(3)$	$M_D(1)$	$M_D(3^{-1})$
$M_D({}^{[010]}2)$	$M_D({}^{[010]}2)$	$M_D({}^{[110]}2)$	$M_D({}^{[100]}2)$	$M_D(3^{-1})$	$M_D(3)$	$M_D(1)$

$$= \begin{bmatrix} -1 & 1 & 0 \\ -1 & 0 & 0 \\ 0 & 0 & 1 \end{bmatrix} .$$

The columns of this matrix, $M_D(\alpha)$, show that the rotation α produced by the combined action of ${}^{[100]}2{}^{[110]}2$ maps **a** onto $-\mathbf{a} - \mathbf{b}$, **b** onto **a** and **c** onto **c**. As **c** is left fixed by α (i.e., $\alpha(\mathbf{c}) = \mathbf{c}$), we conclude that α must take place about an axis coincident with **c**. Of the rotations in *322* that turn space about **c**, we observe that the negative third-turn maps **a** onto $-\mathbf{a} - \mathbf{b}$, **b** onto **a** and **c** onto **c**. Thus, $M_D({}^{[100]}2)M_D({}^{[110]}2) = M_D(3^{-1})$. Hence, $M_D(3^{-1})$ must occur at the intersection of the row headed by $M_D({}^{[100]}2)$ and the column headed by $M_D({}^{[110]}2)$ in Figure 3.14. It is left to the reader to verify the remaining entries of the table.

Because matrix multiplication mimics the composition of point isometries, we can immediately obtain the multiplication table for the point isometries of *322* where the multiplication table is constructed by simply dropping the matrix notation $M_D(\ )$ from the entires of the table in Figure 3.14. For example, when the matrix notation is dropped from the product $M_D({}^{[100]}2)M_D({}^{[110]}2) = M_D(3^{-1})$, we obtain the composition ${}^{[100]}2{}^{[110]}2 = 3^{-1}$. When this is done for each matrix, we obtain the table displayed in Figure 3.15. The relationship between *322* and $M_D(322)$ that we have used in this development is described by saying that

Figure 3.15: Group Multiplication Table for *322*.

322	1	3	3^{-1}	$[100]_2$	$[110]_2$	$[010]_2$
1	1	3	3^{-1}	$[100]_2$	$[110]_2$	$[010]_2$
3	3	3^{-1}	1	$[110]_2$	$[010]_2$	$[100]_2$
3^{-1}	3^{-1}	1	3	$[010]_2$	$[100]_2$	$[110]_2$
$[100]_2$	$[100]_2$	$[010]_2$	$[110]_2$	1	3^{-1}	3
$[110]_2$	$[110]_2$	$[100]_2$	$[010]_2$	3	1	3^{-1}
$[010]_2$	$[010]_2$	$[110]_2$	$[100]_2$	3^{-1}	3	1

322 and $M_D(322)$ are isomorphic where the isomorphism is the mapping that takes each $\alpha \varepsilon$ *322* to $M_D(\alpha)$ in $M_D(322)$. More information on isomorphisms can be found in Appendix 8.

The multiplication table for *322* (above) illustrates several general and fundamental properties of the set under composition. An inventory of five of these properties is given below:

(1) If $\alpha,\beta \varepsilon$ *322*, then $\alpha\beta \varepsilon$ *322*. In other words, the compositions in Figure 3.15 are all elements of *322*. As this property holds for all elements of *322* with respect to composition, we say that *322* is *closed* under the binary operation of composition.

(2) There exists one row and one column in the table that is an exact copy of the guide row and the guide column, respectively. This property tells us that *322* contains an identity element (1 in this case) with the property that $1\alpha = \alpha 1 = \alpha$ for all $\alpha \varepsilon$ *322*.

(3) Each column and row of the table contains the identity element 1 once and only once. This means that for each element $\alpha \varepsilon$ *322*, there exists a corresponding element β such that $\alpha\beta = \beta\alpha = 1$. We call β the inverse of α, i.e., $\beta = \alpha^{-1}$. Hence, we can conclude that each point isometry $\alpha \varepsilon$ *322* has a unique inverse denoted by α^{-1}.

(4) The elements of *322* obey the associative law because they are mappings (see Appendix 1). That is, $\alpha(\beta\gamma) = (\alpha\beta)\gamma$ for all $\alpha,\beta,\gamma \varepsilon$ *322*.

(5) The elements of *322* obey the left and right cancellation laws. That is, if $\alpha, \beta, \gamma \in 322$ such that $\alpha\beta = \alpha\gamma$, then $\beta = \gamma$ (the ***left cancellation law***) or if $\alpha\gamma = \beta\gamma$, then $\alpha = \beta$ (the ***right cancellation law***). The left cancellation law can be verified by observing that if $\alpha\beta = \alpha\gamma$, then in the α row of the table the same entry would appear in the β column and the γ column. Since in each row of the multiplication table no entry is repeated, β must equal γ.

(P3.11) Problem: Verify that the right cancellation law holds for *322*.

When a set of elements like *322* is closed under some associative binary operation, such as composition, with respect to which an identity element exists and each element has a unique inverse, the set qualifies as an important algebraic structure known as a ***group***. In addition to being a group, *322* belongs to a special class of groups known as the ***point groups*** because each of its elements is a point isometry that leaves the origin fixed. The definitions and theorems that provide the underpinnings of the group theory used to derive all of the point groups will be discussed in the following chapters.

(P3.12) Problem: The symmetry elements of *322* consist of a three-fold rotation axis along **c**, two-fold rotation axes along **a**, **a + b** and **c** and the identity (see Figure 3.13). The rotations of the three-fold axis are $3 = \{1, 3, 3^{-1}\}$, those of the two-folds are $^{[100]}2 = \{1, {}^{[100]}2\}$, $^{[110]}2 = \{1, {}^{[110]}2\}$ and $^{[010]}2 = \{1, {}^{[010]}2\}$ and that of the identity is $1 = \{1\}$. Using the multiplication table in Figure 3.15, create a multiplication table for each of these sets.

(P3.13) Problem: Examine each of the multiplication tables completed in P3.12 and note that each set is closed under composition, that each contains an identity element, **1**, and that each element has a unique inverse. Also, because composition is associative, we conclude that the sets of rotations comprising the symmetry elements of *322* are also point groups. It is not uncommon for groups to contain smaller groups imbedded in their structures. As we shall learn in the next chapter, such groups (called subgroups) play an indispensable role in our development of the symmetry groups of crystals.

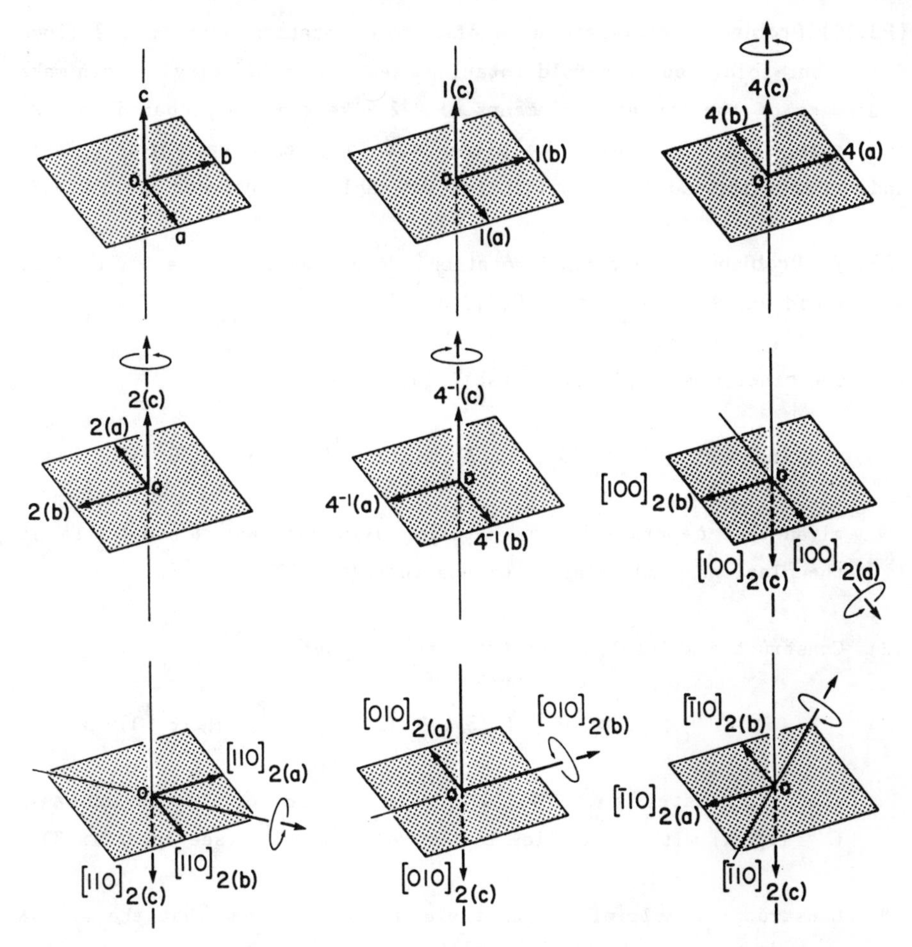

Figure 3.16: The action of each rotation $\alpha \in 422$ on a tetragonal set of basis vectors $D = \{\mathbf{a},\mathbf{b},\mathbf{c}\}$ where $a = b \neq c$ and $\alpha = \beta = \gamma = 90°$. The square in the upper left corner represents the original positions of D with **a** and **b** in the plane of the square and **c** perpendicular thereto. The remaining eight squares are the image squares under the rotations of *422*.

(P3.14) Problem: Consider the set of eight rotations denoted

$$422 = \{1,4,2,4^{-1},[100]_2,[010]_2,[110]_2,[\bar{1}10]_2\}$$

whose symmetry elements leave the origin, **O**, fixed. The basis vectors D for this problem are such that $a = b \neq c$ and that $\alpha = \beta = \gamma = 90°$. Find the matrix representation, $M_D(\alpha)$ for each point isometry $\alpha \in 422$ by inspecting where the basis vectors are mapped by α (see Figure 3.16).

(P3.15) Problem: Create a multiplication table for the matrices of *422*, $M_D(422)$, and then for *422* itself.

(P3.16) Problem: Enumerate each of n-fold rotation axes in 422 (look for a four-fold, four two-fold rotation axes and an identity). Then make a drawing of the symmetry elements of 422 like the one prepared for 322 in Figure 3.13. Enumerate the elements belonging to each n-fold axis and the identity and form a multiplication table for each set as in P3.12.

(P3.17) Problem: The set of rotations described in Figure 3.2 defines a six-fold axis denoted $6 = \{1,6,3,2,3^{-1},6^{-1}\}$.

(1) Construct a multiplication table for the set

$$M_D(6) = \{M_D(1),\ M_D(6),\ M_D(3),\ M_D(2),\ M_D(3^{-1}),\ M_D(6^{-1})\}$$

given that **c** parallels the rotation axis and that **a** and **b** lie in the plane perpendicular to **c** such that $\gamma = 120°$.

(2) Construct a multiplication table for the set

$$M_C(6) = \{M_C(1),\ M_C(6),\ M_C(3),\ M_C(2),\ M_C(3^{-1}),\ M_C(6^{-1})\}$$

where each matrix in $M_C(6)$ is written in terms of a cartesian basis $C = \{\mathbf{i},\mathbf{j},\mathbf{k}\}$ with $\mathbf{k}$ parallel to the rotation axis (see Appendix 3).

(3) Construct a multiplication table for 6 and show that the set is closed under composition, that the associative property holds, that an identity elements exists and that each element has an inverse and that 6 is a group.

(P3.18) Problem: The set of point isometries described in P3.1 defines $\bar{6}$ inversion axis denoted $\bar{6} = \{1,\bar{6},3,m,3^{-1},\bar{6}^{-1}\}$. Construct a multiplication table for $M_D(\bar{6})$.

CHAPTER FOUR

THE MONAXIAL CRYSTALLOGRAPHIC POINT GROUPS

"There are only certain kinds of symmetries which our space can support, not only in man made problems, but in regularties which nature herself imposes on her fundamental atomic structures." -- J. Bronowski

INTRODUCTION

In Chapter 3 we discussed two kinds of point isometries referred to as rotations and rotoinversions, their matrix representations and how they can be assembled into an algebraic system called a group. In this chapter we shall give the definition of a group and then deduce some elementary properties of groups that will be used in the derivation of the crystallographic point groups. It will be shown that the set of all point isometries that leaves some lattice invariant forms a finite group. In addition, it will be shown that the turn angle representatives of these isometries must be one of the following possibilities: 0°; ±60°; ±90°; ±120°; 180°. These sets of isometries are called the crystallographic point groups. A derivation will be undertaken of the five proper monaxial crystallographic point groups and these will in turn be used to derive the eight improper monaxial crystallographic point groups.

ALGEBRAIC CONCEPTS

We have already seen several examples of groups and binary operations. In Chapter 3 we observed that *322* and *422* are groups under the binary operation of composition. In this section, we shall present a formal definition for each of these important concepts.

Binary Operations: As the definition of a group depends on the notion of a binary operation, we begin with its definition.

(D4.1) Definition: Let T denote a nonempty set (that is, a set containing at least one element). A *binary operation* * on T is a rule that assigns to each ordered pair of elements **a**, **b** $\in T$, a uniquely determined element in T which we denote by **a** * **b** (**a** * **b** is read "**a** star **b**").

The addition, subtraction and multiplication of integers are well known binary operations defined on the set of integers, Z. If, for example, one integer is multiplied by another, an integer results. In other words, multiplication is a rule that assigns to two integers a third integer in Z that is their product. Also, the addition or subtraction of two integers results in an integer. Besides these familiar binary operations on the integers, we have already encountered several other binary operations of interest to our study of crystallography. In Chapter 1, we defined geometric addition of vectors on S and a component-wise addition of triples on R^3. Since both of these operations combine any two elements from a given set to yield a uniquely determined element in that set, they qualify as binary operations. According to D1.6, a vector space requires vector addition as a binary operation that assigns to each pair of vectors a third vector that is their vector sum. In Chapter 3, we defined the composition of isometries which is a binary operation on the set of all isometries such that if α and β are isometries, then their composition $\beta\alpha$ is an isometry. We also observe that composition is a binary operation on *322* by observing that each of the elements within its multiplication table (Fig. 3.15) comes from the composition of elements taken from the guide row and guide column. Furthermore, each such composition yields a unique result. Note that composition is not a binary operation on the subset $T = \{1, [100]_2, [110]_2\}$ of *322* since $[100]_2[110]_2 = 3^{-1}$ is not in T.

(P4.1) Problem: Determine which of the following rules qualify as a binary operation on a given set:

(1) The set E of all even integers under the rule of multiplication.

(2) The set O of odd integers under the rule of multiplication.

(3) The set E under the rule of ordinary addition.

(4) The subset $222 = \{1, 2, [100]_2, [010]_2\}$ of *422* under composition.

(5) The set $Z^3 = \{[u,v,w]^t \mid u,v,w \in Z\}$ under component-wise addition.

(6) The set of all 3×3 matrices over the reals under matrix multiplication (The phrase "over the reals" means that the entries of these matrices are real numbers.)

(7) The set Z of all integers under division.

(8) The set $R^4 = \{[x,y,z,w]^t \mid x,y,z,w \in R\}$ under component-wise addition.

Groups: A group is an algebraic system consisting of a set of elements, a single binary operation and a set of rules that the operation must obey. These rules are stated in the following definition.

(D4.2) Definition: Let G denote a nonempty set and let $*$ denote a binary operation defined on G. Then $(G,*)$ forms a ***group*** if it obeys the following rules:

(1) $\mathbf{a} * \mathbf{b} \in G$ for all $\mathbf{a}$, $\mathbf{b} \in G$ (Closure rule)

(2) $\mathbf{a} * (\mathbf{b} * \mathbf{c}) = (\mathbf{a} * \mathbf{b}) * \mathbf{c}$ for all $\mathbf{a}$, $\mathbf{b}$, $\mathbf{c} \in G$ (Associative property)

(3) There exists an element $\mathbf{e} \in G$ such that $\mathbf{a} * \mathbf{e} = \mathbf{e} * \mathbf{a} = \mathbf{a}$ for all $\mathbf{a} \in G$ (The existence of an identity element, $\mathbf{e}$)

(4) For each element $\mathbf{a} \in G$, there exists a corresponding element $\mathbf{b}$ such that $\mathbf{a} * \mathbf{b} = \mathbf{b} * \mathbf{a} = \mathbf{e}$ (The existence of a unique inverse for each element).

Rule (1) is actually redundant since the definition of a binary operation includes closure. It has been restated here for emphasis. We also note that a group $(G,*)$ is not just a set G, but rather it consists of three important ingredients: a set of elements G, a binary operation $*$ defined on G and a set of rules that must be obeyed. We shall use a wide variety of groups to simplify our description and understanding of molecules, crystals and crystal structures. Some of the binary operations we have already defined together with their respective sets form groups, as will be demonstrated in the next few examples.

(E4.3) Example - $(Z,+)$ is a group: The integers under addition $(Z,+)$ form a group under the binary operation of addition. Addition is a binary operation on Z because given any pair of integers, addition is a rule that assigns to that pair a uniquely determined integer in Z (namely their sum). The identity element is 0 and the inverse of a given element a is $(-1)a$. For example, the inverse of 4 is $(-1)4 = -4$ while the inverse of -6 is $(-1)(-6) = 6$. Also, the addition of integers is associative; for example $(-1) + (6 + 4) = (-1 + 6) + 4 = 8$.

(E4.4) Example - *322* is a group under composition: The set $322 = \{1, 3, 3^{-1}, [100]_2, [110]_2, [010]_2\}$ forms a group under composition, as was shown in Chapter 3.

(E4.5) Example: The set P of all integers greater than or equal to zero under addition is not a group. Note that addition is a binary operation on the set, that the operation is associative and that 0 is the identity element in the set. However, there are elements in the set that do not have inverses. For example, 4 does not have an additive inverse in P. In fact, the only element in P with an inverse is 0.

(E4.6) Example - $(Z,-)$ is not a group: The set of integers Z under subtraction is not a group. Subtraction is a binary operation and there appears to be an identity since $a - 0 = a$ for all $a \in Z$. However, the definition of an identity element requires that $0 - a = a$ as well. But $0 - 4 = -4 \neq 4$ shows that this is not the case. Furthermore, $(Z,-)$ is not associative since, for example, $8 - (3 - 2) = 7$ while $(8 - 3) - 2 = 3$.

(E4.7) Example - $GL(3,R)$ is a group under matrix multiplication: Show that the set of all 3×3 invertible matrices, denoted $GL(3,R)$, under matrix multiplication is a group.

Solution: Since matrix multiplication mimics composition of linear mappings and, as shown in Appendix 1, composition of mappings is associative, it is reasonable to expect that matrix multiplication is also associative. This is true but its proof is tedious and so will be omitted. Let M and N denote invertible matrices. That is, M^{-1} and N^{-1} exist. Then the inverse of the product MN is $N^{-1}M^{-1}$ since

$$\begin{aligned}(MN)(N^{-1}M^{-1}) &= M(NN^{-1})M^{-1} \\ &= M(I_3)M^{-1} \\ &= MM^{-1} = I_3.\end{aligned}$$

Similarly, $(N^{-1}M^{-1})(MN) = I_3$. Therefore, $(MN)^{-1}$ exists. Consequently, $GL(3,R)$ is closed under matrix multiplication. Since $I_3^{-1} = I_3$, the identity matrix is in $GL(3,R)$ and is its identity element. Let $M \in GL(3,R)$, then M^{-1} exists. Note that since

$$M^{-1}M = MM^{-1} = I_3 ,$$

M is the inverse of M^{-1}. Hence, M^{-1} is invertible. Furthermore, the inverse of a matrix with real entries is a matrix with real entries (see Appendix 2). We have now shown that all of the rules for a group hold for $GL(3,R)$ under matrix multiplication. □

The group $GL(3,R)$ under matrix multiplication is called the *general linear group of 3×3 matrices over the reals*. The word "linear" refers to the fact that these matrices mimic linear mappings.

(P4.2) Problem: Decide which of the following systems are groups:

(1) The set R^3 under component-wise addition;

(2) The set Z^3 under component-wise addition;

(3) The set of all 3×3 matrices over the reals under matrix multiplication, assuming that matrix multiplication is associative;

(4) The set $6 = \{1,6,3,2,3^{-1},6^{-1}\}$ under matrix multiplication;

(5) The set $M_C(6) = \{M_C(1),M_C(6),M_C(3),M_C(2),M_C(3^{-1}),M_C(6^{-1})\}$ under matrix multiplication.

(6) The set $M_D({}^{[1\bar{1}1]}4)$

$$= \left\{ \begin{bmatrix} 1 & 0 & 0 \\ 0 & 1 & 0 \\ 0 & 0 & 1 \end{bmatrix}, \begin{bmatrix} 0 & -1 & 0 \\ 0 & 0 & -1 \\ 1 & 1 & 1 \end{bmatrix}, \begin{bmatrix} 0 & 0 & 1 \\ -1 & -1 & -1 \\ 1 & 0 & 0 \end{bmatrix}, \begin{bmatrix} 1 & 1 & 1 \\ -1 & 0 & 0 \\ 0 & -1 & 0 \end{bmatrix} \right\}$$

under matrix multiplication.

(7) The set $M_D({}^{[110]}4)$

$$= \left\{ \begin{bmatrix} 1 & 0 & 0 \\ 0 & 1 & 0 \\ 0 & 0 & 1 \end{bmatrix}, \begin{bmatrix} 0 & 1 & 0 \\ 0 & 1 & -1 \\ -1 & 1 & 0 \end{bmatrix}, \begin{bmatrix} 0 & 1 & -1 \\ 1 & 0 & -1 \\ 0 & 0 & -1 \end{bmatrix}, \begin{bmatrix} 1 & 0 & -1 \\ 1 & 0 & 0 \\ 1 & -1 & 0 \end{bmatrix} \right\}$$

under matrix multiplication.

(8) The set R^4 under component-wise addition.

Symmetry groups: A set of mappings describing the symmetry of an object was defined in D3.5. In the next example, we show that this important set of mappings forms a group.

(E4.8) Example - The set of all isometries that leave an object invariant forms a group under composition: Let I denote the set of all isometries that map an object B into self-coincidence (By D3.5, these isometries define the symmetry of B). Show that the set I under composition is a group.

Solution: Let $\alpha, \beta \in I$. Then $\alpha\beta$ is an isometry by T3.2. Furthermore,

$$\begin{aligned} \alpha\beta(B) &= \alpha(\beta(B)) && \text{(Definition of composition)} \\ &= \alpha(B) && \text{(Since } \beta(B) = B\text{)} \\ &= B \ . \end{aligned}$$

Hence, $\alpha\beta \in I$ and I is closed under composition. The associative law holds because the elements of I are mappings (see Appendix 1). The identity mapping $\mathbf{1}$ is such that $\mathbf{1}(\mathbf{b}) = \mathbf{b}$ for all $\mathbf{b} \in B$ and so $\mathbf{1}(B) = B$. Therefore, $\mathbf{1} \in I$ and since $\mathbf{1}\alpha = \alpha\mathbf{1} = \alpha$ for all $\alpha \in I$, $\mathbf{1}$ is the identity element of I. Let $\alpha \in I$. Then α^{-1} is a mapping since α is one-to-one and onto. Furthermore

$$\alpha^{-1}(B) = \alpha^{-1}(\alpha(B)) = (\alpha^{-1}\alpha)(B) = \mathbf{1}(B) = B \ .$$

Hence, $\alpha^{-1} \in I$. Therefore, I is a group under composition. □

(P4.3) Problem: Show that the set of all point isometries leaving an origin fixed that map an object B into self-coincidence forms a group under composition.

The relationship between group theory and study of symmetry is sufficiently important to warrant a special vocabulary of its own. We now introduce the vocabulary through a series of definitions.

(D4.9) Definition: The group G of all point isometries leaving a common point fixed and mapping an object B into self-coincidence is called the *point symmetry group of B*. In this case we say that G is the *point symmetry of B*.

(D4.10) Definition: The group G of all isometries mapping an object B into self-coincidence is called the *symmetry group of B*. In this case, we say that G is the *symmetry of B*.

One type of object whose symmetry is of particular interest in our study of crystallography is the space lattice L associated with a crystal. We shall now focus our attention on those isometries that map such a lattice onto itself.

(D4.11) Definition: An isometry α is called a *crystallographic isometry* if there exists a space lattice L that is mapped into self-coincidence by α.

(D4.12) Definition: A *crystallographic group* is a group of isometries that map some space lattice L into self-coincidence.

(D4.13) Definition: A *crystallographic point group* is a *crystallographic group* G such that each element of G leaves a common point fixed.

CRYSTALLOGRAPHIC RESTRICTIONS

Because a crystallographic point group G maps a space lattice

$$L_{D_1} = \{u\mathbf{a}_1 + v\mathbf{b}_1 + w\mathbf{c}_1 \mid u,v,w \in Z\}$$

generated by a basis D_1 into self-coincidence, severe restrictions are imposed on the turn angles of its isometries and the number of distinct elements in G. It is helpful, when analyzing the relationship between a point isometry α and the lattice L_{D_1} left invariant by α, if the origin of L_{D_1} is a point fixed by α. However, as we see in Figure 4.1(a), it is sometimes the case that the origin of L_{D_1} is not a fixed point of α. When this happens, we shall define a new lattice L_{D_2} whose origin is any chosen point p fixed by α and whose geometry is identical with that of L_{D_1} (Figure 4.1(b)). In fact, the basis $\mathbf{D}_1$ for L_{D_1} with origin $\mathbf{0}$ will have the same metrical matrix as the basis D_2 for L_{D_2} with origin $\mathbf{p}$. In order to correctly describe this change of origin, we shall need to distinguish between the points in space and the vectors emanating to them from an origin since the vector associated with a point is dependent upon which origin is used.

(T4.14) Theorem: If α is a crystallographic isometry that maps the lattice L_{D_1} into self-coincidence and if p is a fixed point of α, then there exists a lattice L_{D_2} whose origin is p such that α maps L_{D_2} into self-coincidence. Furthermore, with respect to the origin of L_{D_2}, the

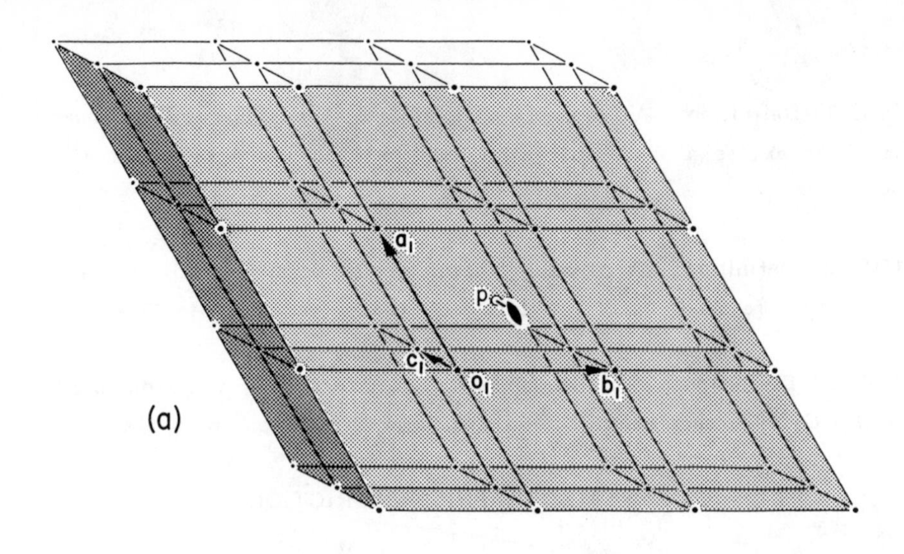

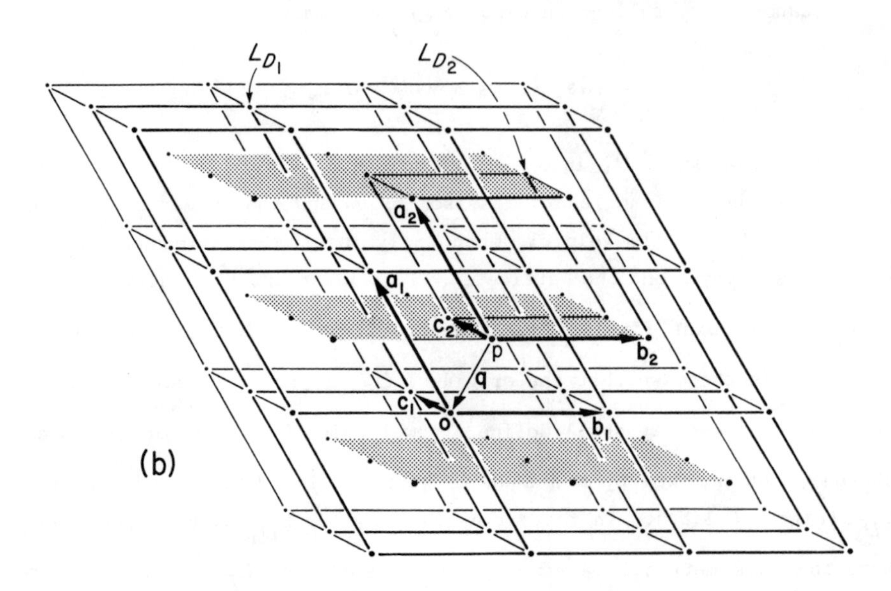

Figure 4.1: (a) A lattice L_{D_1} with basis vectors $D_1 = \{\mathbf{a}_1, \mathbf{b}_1, \mathbf{c}_1\}$ radiating from an origin $\mathbf{o}_1$. This lattice is mapped into self-coincidence by a half-turn rotation α about an axis that parallels $\mathbf{c}_1$ and passes through the point p at $[\frac{1}{2},\frac{1}{2},0]^t$. Note that each lattice point in L_{D_1} including the origin is moved by α. (b): Two interpenetrating lattices L_{D_1} and L_{D_2} where L_{D_2} is a parallel copy of L_{D_1} displaced to the point p fixed by α.

set of points L_{D_1} is the set of end points of the vectors

$$\{\mathbf{q} + \ell_2 \mid \ell_2 \in L_{D_2}\} ,$$

where $\mathbf{q}$ is the vector emanating from p to the origin of L_{D_1}.

Proof: Let $D_1 = \{\mathbf{a}_1, \mathbf{b}_1, \mathbf{c}_1\}$ denote the basis that generates L_{D_1}. Note that the vectors in D_1 emanate from the origin $\mathbf{0}$ of L_{D_1}. Define $D_2 = \{\mathbf{a}_2, \mathbf{b}_2, \mathbf{c}_2\}$ to be the set of vectors emanating from p such that $\mathbf{a}_2$ is parallel to $\mathbf{a}_1$ and $a_2 = a_1$, $\mathbf{b}_2$ is parallel to $\mathbf{b}_1$ and $b_2 = b_1$ and $\mathbf{c}_2$ is parallel to $\mathbf{c}_1$ and $c_2 = c_1$. Then the set of end points of the vectors in

$$L_{D_1} = \{u\mathbf{a}_1 + v\mathbf{b}_1 + w\mathbf{c}_1 \mid u,v,w \in Z\}$$

is described in terms of D_2 as

$$\{\mathbf{q} + u\mathbf{a}_2 + v\mathbf{b}_2 + w\mathbf{c}_2 \mid u,v,w \in Z\} ,$$

where $\mathbf{q}$ is the vector emanating from p with terminus $\mathbf{0}$. If we let $L_{D_2} = \{u\mathbf{a}_2 + v\mathbf{b}_2 + w\mathbf{c}_2 \mid u,v,w \in Z\}$, then the set of points in L_{D_1} equals

$$\{\mathbf{q} + \ell_2 \mid \ell_2 \in L_{D_2}\} .$$

To show that α maps our new lattice L_{D_2} into self-coincidence, we shall show that if $\ell_2 \in L_{D_2}$, then $\alpha(\ell_2) \in L_{D_2}$. By adding and subtracting $\mathbf{q}$, we have

$$\alpha(\ell_2) = \alpha(\mathbf{q} + \ell_2 - \mathbf{q}) .$$

Since the origin p is fixed by α, α is a linear mapping with respect to the vectors emanating from p, and so

$$\alpha(\mathbf{q} + \ell_2 - \mathbf{q}) = \alpha(\mathbf{q} + \ell_2) - \alpha(\mathbf{q}) .$$

Since end points $\mathbf{q} + \ell_2$ and $\mathbf{q}$ are both in L_{D_1}, $\alpha(\mathbf{q} + \ell_2)$ and $\alpha(\mathbf{q})$ have end points in L_{D_1}. Therefore,

$$\alpha(\mathbf{q} + \ell_2) = \mathbf{q} + \ell_2' \text{ and } \alpha(\mathbf{q}) = \mathbf{q} + \ell_2'' ,$$

where $\ell_2', \ell_2'' \in L_{D_2}$. Hence,

$$\begin{aligned}\alpha(\ell_2) &= \alpha(\mathbf{q} + \ell_2) - \alpha(\mathbf{q}) \\ &= [\mathbf{q} + \ell_2'] - [\mathbf{q} - \ell_2''] \\ &= \ell_2' - \ell_2'' \in L_{D_2} \quad .\end{aligned}$$

Since α is an isometry, the fact that $\alpha(L_{D_2})$ is a subset of L_{D_2} implies that $\alpha(L_{D_2}) = L_{D_2}$ and so L_{D_2} is mapped into self-coincidence by α. Thus, if a lattice is mapped into self-coincidence by α, then the origin of the lattice can always be chosen as a fixed point of α. □

(T4.15) Theorem: If G is a crystallographic point group, then G has a finite number of elements.

Proof: Let L_D denote the lattice that is mapped into self-coincidence by each of the isometries in G. By T4.14, without loss of generality, the origin $\mathbf{0}$ of L_D may be taken to be a common point left fixed by each of the isometries in G. Consider a ball centered at $\mathbf{0}$ with large enough radius so that it contains all three basis vectors $\mathbf{a}$, $\mathbf{b}$, $\mathbf{c} \in L_D$ and let α be an element of G. Since α does not stretch vectors, and since $\alpha(L_D) = L_D$, $\alpha(\mathbf{a})$, $\alpha(\mathbf{b})$ and $\alpha(\mathbf{c})$ must also be vectors in L_D contained within the ball. Moreover, inasmuch as the radius of the ball is finite, there are only a finite number of lattice points of L_D in the ball (Newman, 1972, p.90). Consequently, there are only a finite number of choices for $\alpha(\mathbf{a})$, $\alpha(\mathbf{b})$ and $\alpha(\mathbf{c})$. Since α is completely determined by $\{\alpha(\mathbf{a}), \alpha(\mathbf{b}), \alpha(\mathbf{c})\}$, we conclude that there is only a finite number of distinct point isometries in G that leave L_D invariant. □

(D4.16) Definition: When a group G is finite, the order of G, denoted $\#(G)$, is the number of distinct elements in G.

For example, since *322* consists of 6 elements, we write $\#(322) = 6$.

(P4.4) Problem: Determine $\#(422)$.

(T4.17) Theorem: Let α denote a crystallographic point isometry. Then the turn angle ρ associated with α is a multiple of either $60°$ or $90°$.

Proof: Since α is a crystallographic point isometry, there exists a lattice L_D such that α maps L_D into self-coincidence. Let $D = \{\mathbf{a},\mathbf{b},\mathbf{c}\}$ denote a basis for L_D. Since $\alpha(L_D) = L_D$, $\alpha(\mathbf{a})$, $\alpha(\mathbf{b})$, and $\alpha(\mathbf{c})$ are in L_D and so $[\alpha(\mathbf{a})]_D$, $[\alpha(\mathbf{b})]_D$ and $[\alpha(\mathbf{c})]_D$ have integer coordinates. Since these triples are the columns of $M_D(\alpha)$, all of the entries of $M_D(\alpha)$ are integers. Consequently, the trace of $M_D(\alpha)$ is an integer. By TA3.6,

$$tr(M_D(\alpha)) = \pm(1 + 2\cos\rho) \quad , \tag{4.1}$$

where the "+" sign is used when α is a rotation and the "−" sign when α is a rotoinversion. The solutions to (4.1) given that

$$tr(M_D(\alpha)) = N$$

for some integer N are enumerated in Table 4.1. Note that since $|\cos\rho| \leq 1$, we have the inequalities

$$-1 \leq 1 + 2\cos\rho \leq 3 \quad \text{and} \quad -3 \leq -(1 + 2\cos\rho) \leq 1 \quad .$$

An examination of Table 4.1, where those values of N that satisfy these inequalities are listed and the corresponding ρ values are found, reveals that ρ is always either a multiple of 60° or 90°. The point isometries associated with each ρ value is also shown in Table 4.1. □

Table 4.1: Solutions sets of the turn angles ρ for any crystallographic point isometry α.

(1 + 2cρ) Solutions				−(1 + 2cρ) Solutions			
N	$c\rho = (N-1)/2$	$\rho°$	α	N	$c\rho = -(N-1)/2$	$\rho°$	α
−1	−1	180°	**2**	−3	1	0°	$\bar{1}$
0	$-\frac{1}{2}$	±120°	$3,3^{-1}$	−2	$\frac{1}{2}$	±60°	$\bar{6},\bar{6}^{-1}$
1	0	±90°	$4,4^{-1}$	−1	0	±90°	$\bar{4},\bar{4}^{-1}$
2	$\frac{1}{2}$	±60°	$6,6^{-1}$	0	$-\frac{1}{2}$	±120°	$\bar{3},\bar{3}^{-1}$
3	1	0°	**1**	1	−1	180°	**m**

The 16 isometries found in Table 4.1 will serve as the elements for all of the crystallographic point groups. However, before we can conclude that such a group G is a *bona fida* crystallographic point group, we must show that there is some lattice that every element of G maps into self-coincidence. In this chapter and the next we will find all of the finite groups that can be constructed from these isometries. In Chapter 6, we will find a lattice that is left invariant for each of these groups.

MONAXIAL ROTATION GROUPS

The structure of a group G can be better understood by studying certain special subsets of G called subgroups.

(D4.18) Definition: A subset H of a group G is a ***subgroup*** of G if, under the binary operation of G, H is a group.

For example, in Chapter 3, we observed that *322* has several subgroups including $3 = \{1, 3, 3^{-1}\}$, $^{[100]}2 = \{1, {}^{[100]}2\}$, $^{[110]}2 = \{1, {}^{[110]}2\}$ and $^{[010]}2 = \{1, {}^{[010]}2\}$. We confirmed that these were subgroups by creating their multiplication tables. However, some of the rules defining a group are difficult to see directly from such a table. The rule of associativity is not readily verifiable from a simple inspection of the table. The determination of which element is the identity and which element is the inverse of a given element can be seen from a table, but it is somewhat tedious. Furthermore, if a subset of a group is very large, then creating its multiplication table can be impractical. The following theorem eliminates most of these difficulties in the case of finite groups.

(T4.19) Theorem: If H is a finite nonempty subset of a group G, then H is a subgroup of G if and only if H is closed under the binary operation of G.

We will not prove T4.19 here but its proof can be found in any standard modern alegbra text (e.g., Herstein, 1964).

(E4.20) Example - *222* is a subgroup of *422*: Consider the subset

$$222 = \{1, 2, {}^{[100]}2, {}^{[010]}2\}$$

Figure 4.2: The multiplication table for the group *222* under composition.

222	1	2	$[100]_2$	$[010]_2$
1	1	2	$[100]_2$	$[010]_2$
2	2	1	$[010]_2$	$[100]_2$
$[100]_2$	$[100]_2$	$[010]_2$	1	2
$[010]_2$	$[010]_2$	$[100]_2$	2	1

and show that it is a subgroup of *422* (see P3.15).

Solution To determine whether *222* is a subgroup of *422*, we construct the multiplication table shown in Figure 4.2. Since we have a nonempty list of elements in *222*, we know that *222* is a nonempty set. Also, since each entry in the table belongs to the set *222*, *222* is closed under the binary operation (composition) of *422* and so is a subgroup. In other words, if H is a nonempty subset of a group $(G,*)$, then $(H,*)$ is a subgroup of G if each entry in the multiplication table of H is an element of H. □

(E4.21) Example: Let K denote a crystallographic point group and let D denote a basis. Then the set of matrices

$$G = M_D(K) = \{M_D(\alpha) \mid \alpha \in K\}$$

forms a finite group under matrix multiplication. Since K is a finite set (T4.15), then $M_D(K)$ must be a finite group. Let H denote the subset of G consisting of all matrices M in G such that $\det(M) = 1$, i.e., M represents a rotation isometry. Show that H is closed and therefore a group under the binary operation of G.

Solution: Since $\mathbf{1} \in K$ (it is the identity element of K), then $M_D(\mathbf{1}) = I_3 \in G$. But $\det(I_3) = 1$ and so $I_3 \in H$. Therefore, H is nonempty. Let $M_1, M_2 \in H$. Then $M_1M_2 \in G$ since G is a group and $\det(M_1M_2) = \det(M_1)\det(M_2) = 1 \cdot 1 = 1$. Hence, $M_1M_2 \in H$ and so H is closed under the binary operations of G. This result also shows that the composition of two proper crystallographic operations is a proper crystallographic operation. □

(P4.5) Problem: Let α and β be elements of a crystallographic point group K (E4.21).

(1) Show that $\alpha\beta$ is a proper operation when α and β are both improper operations.

(2) Show that $\alpha\beta$ is an improper operation when α is a proper and β is an improper operation.

(D4.22) Definition: Let $\mathbf{g}$ denote an element of the group G under the binary operation $*$. The *nth* power of $\mathbf{g}$, denoted $\mathbf{g}^n$ is defined to be

$$\mathbf{g}^n = \mathbf{g} * \mathbf{g} * \ldots * \mathbf{g} \qquad (n \ \mathbf{g}\text{'s})$$

where n is a positive integer. When $n = 0$, $\mathbf{g}^0$ is defined to be the identity element of G and when $n < 0$, $\mathbf{g}^n$ is defined to be $(\mathbf{g}^{-1})^{-n}$; that is

$$\mathbf{g}^n = (\mathbf{g}^{-1} * \mathbf{g}^{-1} * \ldots * \mathbf{g}^{-1}) \qquad (-n \ \mathbf{g}\text{'s})$$

when $n < 0$. The usual rules of exponents hold true for the powers of an element $\mathbf{g} \varepsilon G$ including the following:

$$(1) \quad \mathbf{g}^n * \mathbf{g}^m = \mathbf{g}^{n+m} \text{ for all } n, m \varepsilon Z ,$$

$$(2) \quad (\mathbf{g}^n)^m = \mathbf{g}^{nm} \text{ for all } n, m \varepsilon Z .$$

(E4.23) Example: Let G denote a crystallographic point group. Let $\mathbf{g} \varepsilon G$ and define

$$H = \{\mathbf{g}^n \mid n \varepsilon Z\} ,$$

then show that H is a subgroup of G.

Solution: According to T4.19, we need only show that H is closed under composition. Let $\mathbf{x},\mathbf{y} \varepsilon H$. By the definition of H, there exists integers n and m such that $\mathbf{x} = \mathbf{g}^n$ and $\mathbf{y} = \mathbf{g}^m$. Hence,

$$\mathbf{xy} = \mathbf{g}^n\mathbf{g}^m = \mathbf{g}^{n+m} .$$

Since $n + m$ is an integer, $\mathbf{xy} \varepsilon H$. Hence, H is a subgroup of G. □

(D4.24) Definition: Let $(G,*)$ denote a group and let $\mathbf{g} \varepsilon G$. The *cyclic subgroup* of G generated by $\mathbf{g}$, denoted $<\mathbf{g}>$, is defined to be

$$\langle \mathbf{g} \rangle = \{\mathbf{g}^n \mid n \in Z\} .$$

We will now justify calling $\langle \mathbf{g} \rangle$ a subgroup. In (E4.23) we showed that when G is a crystallographic point group, then $\langle \mathbf{g} \rangle$ is a subgroup. In the next problem, you shall be asked to prove this fact in general.

(P4.6) Problem: Show that if $\mathbf{g} \in G$ where G is a group, then $\langle \mathbf{g} \rangle$ is a subgroup of G (do not assume that G is finite).

(D4.25) Definition: Let G denote a group and let $\mathbf{g} \in G$. If $\mathbf{g}^i = \mathbf{e}$ for some positive integer i where $\mathbf{e}$ is the identity of G, then the smallest such positive integer is called the ***order of*** $\mathbf{g}$ and is denoted $o(\mathbf{g})$.

Let $\mathbf{g} \in G$ where G is a group such that there is a positive integer i such that $\mathbf{g}^i = \mathbf{e}$, the identity element of G. Let $k = o(\mathbf{g})$. Then

$$\langle \mathbf{g} \rangle = \{\mathbf{g}, \mathbf{g}^2, \mathbf{g}^3, \ldots, \mathbf{g}^k = \mathbf{e}\} .$$

This means that any element of $\{\mathbf{g}^n \mid n \in Z\}$ can be written as $\mathbf{g}^j$ where $1 \le j \le k$. For example, $\mathbf{g}^{-1} = \mathbf{g}^{k-1}$ since

$$\mathbf{g}\mathbf{g}^{k-1} = \mathbf{g}^k = \mathbf{e} ,$$

and $\mathbf{g}^{k+1} = \mathbf{g}$ since

$$\mathbf{g}^{k+1} = \mathbf{g}^k\mathbf{g} = \mathbf{e}\mathbf{g} = \mathbf{g} .$$

Furthermore, it is straightforward to show that $\mathbf{g}, \mathbf{g}^2, \ldots, \mathbf{g}^k$ are all distinct. That is, if $\mathbf{g}^i = \mathbf{e}$ for some positive integer i, then $\#(\langle \mathbf{g} \rangle) = o(\mathbf{g})$. A cyclic subgroup of a crystallographic point group is a set of isometries having a common axis. These form such symmetry elements as an n-fold rotation axis or rotoinversion axis.

(E4.26) Example - The elements of $\langle 4 \rangle = 4$: Let G denote the group of all point isometries leaving the origin fixed and consider a quarter-turn $\mathbf{4} \in G$. Since $\mathbf{4}^4 = \mathbf{1}$ and $\mathbf{1}$ is the identity element of G,

$$\langle 4 \rangle = \{4, 4^2, 4^3, 4^4 = 1\} .$$

We give this cyclic group the symbol 4 and note that the elements of this group define a 4-fold axis. Since all of these proper rotations fix the same axis, 4 is an example of a proper monaxial group. By convention, 4^3 is usually written in the equivalent form 4^{-1} and 4^2 is written as 2 to conform with the convention adopted with respect to the turn angle in Chapter 3. Hence,

$$4 = \langle 4 \rangle = \{1, 4, 2, 4^{-1}\} .$$

(D4.27) Definition: A group consisting of rotations that fix a common axis is called a *proper monaxial group*.

A finite proper monaxial group G is cyclic and is generated by the rotation in G with the least non-negative turn angle. Thus 4 is the generator for 4. If G is a proper monaxial crystallographic point group, then by T4.17, a generator for G can be found with a turn angle 0°, 60°, 90°, 120° or 180°. This can be shown by observing that rotations with turn angles greater than 180° generate groups that include rotations with turn angles less than 180°. For example, if α is a rotation of 270°, then α^3 is a rotation of 90°. A complete list of all of the possible proper monaxial crystallographic point groups is given below:

$$\begin{aligned}
1 &= \langle 1 \rangle = \{1^\nu \mid \nu \in Z\} = \{1\} \\
2 &= \langle 2 \rangle = \{2^\nu \mid \nu \in Z\} = \{1, 2\} \\
3 &= \langle 3 \rangle = \{3^\nu \mid \nu \in Z\} = \{1, 3, 3^{-1}\} \\
4 &= \langle 4 \rangle = \{4^\nu \mid \nu \in Z\} = \{1, 4, 2, 4^{-1}\} \\
6 &= \langle 6 \rangle = \{6^\nu \mid \nu \in Z\} = \{1, 6, 3, 2, 3^{-1}, 6^{-1}\} .
\end{aligned}$$

We observe that each of these proper monaxial groups define an n-fold axis where n = 1, 2, 3, 4 and 6.

Now that we have all of the possible proper monaxial crystallographic point groups, we can use the following theorem to obtain all of the possible improper monaxial crystallographic point groups.

(T4.28) Theorem: (*The improper point group generating theorem.*) If I is an improper crystallographic point group, then there exists a proper crystallographic point group G such that either

(1) $I = G \cup Gi$, where i is the inversion (here $Gi = \{gi \mid g \in G\}$), or

(2) $I = H \cup (G \setminus H)i$ where H is a subgroup of G such that

$$\#(G)/\#(H) = 2 \text{ (here } (G \setminus H)i = \{gi \mid g \in G \text{ and } g \notin H\}).$$

Furthermore, all of the sets constructed from a proper crystallographic group G as in (1) or (2) are groups and hence improper crystallographic point groups.

By the symbol "U" used in $I = G \cup Gi$, we mean the union of the two sets G and Gi where the union of two sets A and B, $A \cup B$ is defined to be the set consisting of all of the elements of A together with all of the elements of B. By $(G \setminus H)$ we mean the set of elements in G but not in H. The set $(G \setminus H)$ is obtained by deleting each element in H from G. A subgroup H of G such that $\#(G)/\#(H) = 2$ is called a *halving group* of G (since it has half of the elements of G). The proof of T4.28 is given by Boisen and Gibbs (1976).

(E4.29) Example: Use T4.28 to find the improper crystallographic point groups for the case where $G = 4$.

Solution: Applying part (1) of T4.28, we obtain

$$G \cup Gi = 4 \cup 4i = \{1,4,2,4^{-1}\} \cup \{1,4,2,4^{-1}\}i$$

$$= \{1,4,2,4^{-1},1i,4i,2i,4^{-1}i\}$$

$$= \{1,4,2,4^{-1},i,\bar{4},m,\bar{4}^{-1}\} .$$

This group contains a 4-fold rotation axis, $4 = \{1,4,2,4^{-1}\}$, perpendicular to a mirror plane, $m = \{1,m\}$, and is consequently designated $4/m$ read "4 upon m". Applying part (2) of T4.28, we note that $H = 2 = \{1,2\}$ is the only halving group of 4, since both 4 and 4^{-1} generate all of G. Then

$$H \cup (G \setminus H)i = 2 \cup (4 \setminus 2)i = \{1,2\} \cup (\{1,4,2,4^{-1}\} \setminus \{1,2\})i$$

$$= \{1,2\} \cup \{4,4^{-1}\}i$$

$$= \{1,2,4i,4^{-1}i\}$$

$$= \{1,2,\bar{4},\bar{4}^{-1}\} .$$

This crystallographic point group is designated $\bar{4}$ because it is the cyclic group generated by $\bar{4}$. We also observe that T4.28 states the groups constructed using (1) and (2) are improper crystallographic point groups and so there is no need to examine whether these groups obey the rules of a group. □

(E4.30) Example - Derivation of the possible improper point groups based on *3*: Use T4.28 to find the improper crystallographic point groups for the case where $G = 3$.

Solution: Applying part (1) of the theorem, we obtain

$$3 \cup 3i = \{1,3,3^{-1}\} \cup \{i,\bar{3},\bar{3}^{-1}\}$$

$$= \{1,3,3^{-1},i,\bar{3},\bar{3}^{-1}\} .$$

Since this is a cyclic group generated by $\bar{3}$, the group is denoted by $\bar{3}$ read "three bar". Since 3 has an odd number of elements, it has no halving groups and therefore (2) does not apply. □

Note that if G is a proper crystallographic point group, then $\#(G \cup Gi) = 2\#(G)$ and if H is a halving group of G, $\#(H \cup (G\backslash H)i) = \#(G)$. Furthermore, the nature of $G \cup G_i$ differs from that of $H \cup (G\backslash H)i$ and G in the sense that $G \cup Gi$ contains the inversion. The reason $H \cup (G\backslash H)i$ does not contain i is that $\mathbf{1}$ is not an element of $G\backslash H$. A point group containing **i** is called a *centrosymmetric group*.

Note that when G is monaxial, then $G \cup Gi$ and $H \cup (G \backslash H)i$ are also monaxial with the same axis. In the case of $G \cup Gi$, we observe that if $\alpha \in G \cup Gi$, then either $\alpha = \mathbf{g}$ for some $\mathbf{g} \in G$ or $\alpha = \mathbf{g}i$ for some $\mathbf{g} \in G$. In either case α and $\mathbf{g}$ have the same axis. In the case of $H \cup (G \backslash H)i$, if $\alpha \in H \cup (G \backslash H)i$, either $\alpha = \mathbf{g}$ for some $\mathbf{g} \in H$, implying that α and $\mathbf{g}$ have the same axis, or $\alpha = \mathbf{g}i$ where $\mathbf{g} \in G \backslash H$, again implying that α and $\mathbf{g}$ have the same axis. Therefore, we see that the set of axes in any improper crystallographic point group is the same as that of the proper group from which it is derived.

(P4.7) Problem: Use T4.28 to find the improper crystallographic point groups for the case where $G = 2$. Show that

$$2 \cup 2i = \{1,2,i,\mathrm{m}\}$$

($2 \cup 2i$ is denoted by $2/m$). Show that $H = \{1\}$ is a halving group of $2 = \{1,2\}$ and that

$$1 \cup (2 \setminus 1)i = \{1,\mathrm{m}\} .$$

This group is called m because it is a cyclic group generated by **m**.

(P4.8) Problem: Use T4.28 to find the improper crystallographic point groups for the case $G = 1$. Show that

$$1 \cup 1i = \{1,i\}$$

($1 \cup 1i$ is denoted $\bar{1}$). Explain why 1 has no halving group.

(P4.9) Problem: Use T4.28 to find the improper crystallographic point group for the case $G = 6$. Show that

$$6 \cup 6i = \{1,6,3,2,3^{-1},6^{-1},i,\bar{6},\bar{3},\mathrm{m},\bar{3}^{-1},\bar{6}^{-1}\}$$

($6 \cup 6i$ is called $6/m$ because it has a 6-fold axis perpendicular to a reflection plane). Show that 3 is a halving group of 6 and that

$$3 \cup (6 \setminus 3)i = \{1,3,3^{-1},\bar{6},\mathrm{m},\bar{6}^{-1}\} \quad .$$

This group is denoted $\bar{6}$ since it is $\langle\bar{6}\rangle$).

We have now derived all 13 of the possible monaxial crystallographic point groups. These are listed and their derivations summarized in Table 4.2.

Matrix representations and basis vectors: In order to obtain a matrix representation for each of the point isometries for a given group, we need to select a basis for each group.

To accomplish this, we will define, for each crystallographic point group G, a set of bases such that the matrix representations of the isometries is the same for any basis in the set and such that the matrices are of a simple form. In fact, each entry of the resulting matrices will be either 1, 0 or −1. In chapter 6 we will show that every lattice left invariant by a point group G will contain a sublattice that has a basis

Table 4.2: The 13 monaxial crystallographic point groups and their orders as derived from the proper monaxial crystallographic point groups.

Proper Monaxial Point Groups		Halving Groups*	Improper Monaxial Point Groups			
			Containing i (centrosymmetric)		Not Containing i	
G	$\#(G)$	H	$G \cup Gi$	$\#(G \cup Gi)$	$H \cup (G \setminus H)i$	$\#(H \cup (G \setminus H)i)$
1	1	none	$\bar{1}$	2	none	—
2	2	1	$2/m$	4	m	2
3	3	none	$\bar{3}$	6	none	—
4	4	2	$4/m$	8	$\bar{4}$	4
6	6	3	$6/m$	12	$\bar{6}$	6

of the type we define for G in this chapter and the next. Since such a sublattice will be called a primitive lattice, we will use the letter P to denote these bases.

We begin by considering the nth-turns of the monaxial crystallographic point groups in Table 4.2. In all cases we will assume that the bases described are right-handed.

1 (identity): Since **1** is represented by the identity matrix for every basis, any basis can be chosen.

2 (half-turn): For $n > 1$, we shall choose **c** to be any nonzero vector in the positive direction of the rotation axis. The **a** and **b** vectors are chosen to be any non-collinear, nonzero vectors in the plane perpendicular to **c**. Since any vector perpendicular to the rotation axis of **2** will be mapped to its negative,

$$M_P(\mathbf{2}) = \begin{bmatrix} -1 & 0 & 0 \\ 0 & -1 & 0 \\ 0 & 0 & 1 \end{bmatrix} .$$

3 (third-turn): As in the case of a half-turn, **c** is chosen to be a nonzero vector in the positive direction of the rotation axis. We then choose **a** to be any nonzero vector perpendicular to **c**. As **3**(**a**) is not collinear with **a**, we choose **b** = **3**(**a**). With this choice of basis, we recall from our discussion of *322* in Chapter 3 that **3**(**b**) = −**a** − **b** and so

$$M_p(\mathbf{3}) = \begin{bmatrix} 0 & -1 & 0 \\ 1 & -1 & 0 \\ 0 & 0 & 1 \end{bmatrix} . \tag{4.1}$$

4 (quarter-turn): As before we choose **c** to be a nonzero vector in the positive direction of the rotation axis. Also, we let **a** be any nonzero vector perpendicular to **c** and let **b** = **4**(**a**). For this choice of basis, we obtain

$$M_p(\mathbf{4}) = \begin{bmatrix} 0 & -1 & 0 \\ 1 & 0 & 0 \\ 0 & 0 & 1 \end{bmatrix} .$$

6 (sixth-turn): To keep the number of basis types small, we choose as the basis for **6** that which is used for **3**, where **c** is chosen as a nonzero vector in the positive direction of the rotation axis of **6**. Then **6**(**a**) = **a** + **b** and **6**(**b**) = −**a**. Hence,

$$M_p(\mathbf{6}) = \begin{bmatrix} 1 & -1 & 0 \\ 1 & 0 & 0 \\ 0 & 0 & 1 \end{bmatrix} .$$

(E4.31) Example: Find $M_p(\mathit{3}) = \{M_p(\mathbf{3}^k) \mid k \in Z\}$ by finding the powers of $M_p(\mathbf{3})$ and recognizing that $M_p(\mathbf{3}^k) = (M_p(\mathbf{3}))^k$. After finding $M_p(\mathit{3})$, find

$$M_p(\bar{\mathit{3}}) = M_p(\mathit{3}) \cup M_p(\mathit{3})M_p(\mathit{i}).$$

Solution: By (4.1),

$$M_p(\mathbf{3}^2) = \begin{bmatrix} 0 & -1 & 0 \\ 1 & -1 & 0 \\ 0 & 0 & 1 \end{bmatrix}^2 = \begin{bmatrix} -1 & 1 & 0 \\ -1 & 0 & 0 \\ 0 & 0 & 1 \end{bmatrix} = M_p(\mathbf{3}^{-1}) ;$$

$$M_p(3^3) = \begin{bmatrix} 0 & -1 & 0 \\ 1 & -1 & 0 \\ 0 & 0 & 1 \end{bmatrix}^3 = \begin{bmatrix} 1 & 0 & 0 \\ 0 & 1 & 0 \\ 0 & 0 & 1 \end{bmatrix} = M_p(1) .$$

Hence,

$$M_p(3) = \left\{ \begin{bmatrix} 1 & 0 & 0 \\ 0 & 1 & 0 \\ 0 & 0 & 1 \end{bmatrix}, \begin{bmatrix} 0 & -1 & 0 \\ 1 & -1 & 0 \\ 0 & 0 & 1 \end{bmatrix}, \begin{bmatrix} -1 & 1 & 0 \\ -1 & 0 & 0 \\ 0 & 0 & 1 \end{bmatrix} \right\} . \qquad (4.2)$$

Since $M_p(\alpha)M_p(i) = -M_p(\alpha)$ for any point isometry α,

$$M_p(3)M_p(i) = \{-M_p(1), -M_p(3), -M_p(3^{-1})\} .$$

Hence

$$M_p(3)M_p(i) = \left\{ \begin{bmatrix} -1 & 0 & 0 \\ 0 & -1 & 0 \\ 0 & 0 & -1 \end{bmatrix}, \begin{bmatrix} 0 & 1 & 0 \\ -1 & 1 & 0 \\ 0 & 0 & -1 \end{bmatrix}, \begin{bmatrix} 1 & -1 & 0 \\ 1 & 0 & 0 \\ 0 & 0 & -1 \end{bmatrix} \right\} . \qquad (4.3)$$

Then $M_p(\bar{3})$ consists of the matrices in (4.2) together with those in (4.3). □

(P4.10) Problem: Find $M_p(n) = \{M_p(\mathbf{n}^k) \mid k \in Z\}$ for $n = 1, 2, 4, 6$. Then find $M_p(n) \cup M_p(n)M_p(i)$. For each $G = n$ and each halving group H of G that exists, find

$$M_p(H \cup (G \setminus H)i) = M_p(H) \cup (M_p(G) \setminus M_p(H))M_p(i).$$

Equivalent Points and Planes: We defined the symmetry group G of an object to be the set of all isometries that map the object into self-coincidence. Utilizing G we can impose an "equivalence" on the points in S.

(D4.32) Definition: Let G denote a group of isometries. The points $\mathbf{x}$ and $\mathbf{y}$ in S are said to be *G-equivalent* if there exists $\alpha \in G$ such that $\alpha(\mathbf{x}) = \mathbf{y}$.

In the case where G is the symmetry group of an object B, then the intuitive interpretation of G-equivalent points is that if $\mathbf{x}$ and $\mathbf{y}$ are G-quivalent, then the object B appears the same in every respect to a person viewing the object from either point $\mathbf{x}$ or point $\mathbf{y}$.

The notion of G-equivalent points is a generalization of the notion of equality. That is, while $\mathbf{x}$ and $\mathbf{y}$ may be distinct points, the statement that they are G-equivalent expresses the idea that in some specific sense they are equal. This is similar to our usual handling of fractions in arithmetic where we think of the fractions 2/3 and 4/6 as being "equal" even though they are clearly not identical. If G-equivalence is to mimic equality, we would hope that the basic properties of equality hold.

(T4.33) Theorem: Let G denote a group of isometries and let $\mathbf{x} \sim \mathbf{y}$ denote "$\mathbf{x}$ is G-equivalent to $\mathbf{y}$". Then

(1) $\mathbf{x} \sim \mathbf{x}$ for all $\mathbf{x} \in S$ (reflexive property) ;

(2) For all $\mathbf{x}, \mathbf{y} \in S$, if $\mathbf{x} \sim \mathbf{y}$ then $\mathbf{y} \sim \mathbf{x}$ (symmetric property) ;

(3) For all $\mathbf{x}, \mathbf{y}, \mathbf{z} \in S$, if $\mathbf{x} \sim \mathbf{y}$ and $\mathbf{y} \sim \mathbf{z}$, then $\mathbf{x} \sim \mathbf{z}$ (transitive property) .

Proof: Since G is a group of isometries, $\mathbf{1} \in G$. Since $\mathbf{1}(\mathbf{x}) = \mathbf{x}$ for all $\mathbf{x} \in S$, we have $\mathbf{x} \sim \mathbf{x}$ for all $\mathbf{x} \in S$. Let $\mathbf{x}$ and $\mathbf{y}$ denote elements in S such that $\mathbf{x} \sim \mathbf{y}$. Then there exists $\alpha \in G$ such that $\alpha(\mathbf{x}) = \mathbf{y}$. Since G is a group, $\alpha^{-1} \in G$. Then

$$\begin{aligned} \alpha^{-1}(\alpha(\mathbf{x})) &= \alpha^{-1}(\mathbf{y}) \\ (\alpha^{-1}\alpha)(\mathbf{x}) &= \alpha^{-1}(\mathbf{y}) \\ 1(\mathbf{x}) &= \alpha^{-1}(\mathbf{y}) \\ \mathbf{x} &= \alpha^{-1}(\mathbf{y}) . \end{aligned}$$

Hence $\mathbf{y} \sim \mathbf{x}$. Let $\mathbf{x},\mathbf{y},\mathbf{z} \in S$ such that $\mathbf{x} \sim \mathbf{y}$ and $\mathbf{y} \sim \mathbf{z}$. Then there exists $\alpha, \beta \in G$ such that $\alpha(\mathbf{x}) = \mathbf{y}$ and $\beta(\mathbf{y}) = \mathbf{z}$. Then

$$\beta\alpha(\mathbf{x}) = \beta(\alpha(\mathbf{x})) = \beta(\mathbf{y}) = \mathbf{z}.$$

Since G is a group, it is closed and so $\beta\alpha \in G$. Therefore $\mathbf{x} \sim \mathbf{z}$. □

Any relation defined on the elements of a set satisfying the three parts of T4.33 is called an *equivalence relation*. Since equivalence relations are fundamental to any true understanding of the mathematical principles of crystallography, we have devoted much of Appendix 7 to this concept. Those who are unfamiliar with this concept should read this appendix carefully.

Equivalence relations organize the set on which they are defined into subsets of related elements called equivalence classes. In the case of G-equivalent points, if $\mathbf{x} \in S$, then the *equivalence class* of $\mathbf{x}$, denoted $[\mathbf{x}]$ is

$$\begin{aligned}[\mathbf{x}] &= \{\mathbf{y} \in S \mid \mathbf{x} \sim \mathbf{y}\} \\ &= \{\mathbf{y} \in S \mid \text{there exists } \mathbf{g} \in G \text{ such that } \mathbf{g}(\mathbf{x}) = \mathbf{y}\} \\ &= \{\mathbf{g}(\mathbf{x}) \mid \mathbf{g} \in G\}.\end{aligned}$$

Since, in the case of G-quivalence, $[\mathbf{x}]$ is the set of images of $\mathbf{x}$ under G, $[\mathbf{x}]$ is called the *orbit* of $\mathbf{x}$ under G and is designated $\mathrm{orb}_G(\mathbf{x})$. When the number of points in $\mathrm{orb}_G(\mathbf{x})$ equals $\#(G)$, then $\mathbf{x}$ is called a *point of general position*. Otherwise, $\mathbf{x}$ is a *point of special position*. When $\mathbf{x}$ is a point of special position, there exists one or more nonidentity elements $\mathbf{g} \in G$ such that $\mathbf{g}(\mathbf{x}) = \mathbf{x}$. Let D denote a basis of S, $\mathbf{x}$ be an element of S and G be a point group. Then we define

$$\mathrm{orb}_{G,D}([\mathbf{x}]_D) = \{[\mathbf{g}(\mathbf{x})]_D \mid \mathbf{g} \in G\} .$$

(E4.34) Example - The symmetry of H_4SiO_4 is $\bar{4}$: The atomic coordinates of H_4SiO_4, are given in Table 1.1. Show that the coordinates with respect to C of the oxygen atoms designated O_1, O_2, O_3 and O_4 are mapped into self-coincidence by the matrix representations of the monaxial group $\bar{4}$ where

$$M_C(\bar{4}) = \{M_C(1), M_C(\bar{4}), M_C(2), M_C(\bar{4}^{-1})\} .$$

That is, the four oxygen atoms of the molecule are $\bar{4}$-equivalent.

Solution: This will be done by showing that O_1 is $\bar{4}$ equivalent to each of O_2, O_3, and O_4. Since the coordinates of the vector $\mathbf{r}$ emanating from the origin to O_1 given in Table 1.1 are

$$[\mathbf{r}]_C = [1.281,\ 0.466,\ 0.877]^t \quad ,$$

and since

$$M_C(1)[\mathbf{r}]_C = \begin{bmatrix} 1 & 0 & 0 \\ 0 & 1 & 0 \\ 0 & 0 & 1 \end{bmatrix} \begin{bmatrix} 1.281 \\ 0.466 \\ 0.877 \end{bmatrix} = \begin{bmatrix} 1.281 \\ 0.466 \\ 0.877 \end{bmatrix} ,$$

$$M_C(\bar{4})[\mathbf{r}]_C = \begin{bmatrix} 0 & 1 & 0 \\ -1 & 0 & 0 \\ 0 & 0 & -1 \end{bmatrix} \begin{bmatrix} 1.281 \\ 0.466 \\ 0.877 \end{bmatrix} = \begin{bmatrix} 0.466 \\ -1.281 \\ -0.877 \end{bmatrix},$$

$$M_C(2)[\mathbf{r}]_C = \begin{bmatrix} -1 & 0 & 0 \\ 0 & -1 & 0 \\ 0 & 0 & 1 \end{bmatrix} \begin{bmatrix} 1.281 \\ 0.466 \\ 0.877 \end{bmatrix} = \begin{bmatrix} -1.281 \\ -0.466 \\ 0.877 \end{bmatrix},$$

$$M_C(\bar{4}^{-1})[\mathbf{r}]_C = \begin{bmatrix} 0 & -1 & 0 \\ 1 & 0 & 0 \\ 0 & 0 & -1 \end{bmatrix} \begin{bmatrix} 1.281 \\ 0.466 \\ 0.877 \end{bmatrix} = \begin{bmatrix} -0.466 \\ 1.281 \\ -0.877 \end{bmatrix},$$

we see that the orbit of $\mathbf{r}$ is

$$\begin{aligned} \mathrm{orb}_{\bar{4},C}(\mathbf{r}) &= \{M_C(\alpha)[\mathbf{r}]_C \mid \alpha \in \bar{4}\} \\ &= \{M_C(1)[\mathbf{r}]_C, M_C(\bar{4})[\mathbf{r}]_C, M_C(2)[\mathbf{r}]_C, M_C(\bar{4}^{-1})[\mathbf{r}]_C\} \\ &= \left\{ \begin{bmatrix} 1.281 \\ 0.466 \\ 0.877 \end{bmatrix}, \begin{bmatrix} 0.466 \\ -1.281 \\ -0.877 \end{bmatrix}, \begin{bmatrix} -1.281 \\ -0.466 \\ 0.877 \end{bmatrix}, \begin{bmatrix} -0.466 \\ 1.281 \\ -0.877 \end{bmatrix} \right\}. \end{aligned}$$

But these are the coordinates of O_1, O_2, O_3 and O_4. Hence O_1 is $\bar{4}$-equivalent to each of the other oxygen atoms. The reader should repeat this process for each of the remaining oxygens in the molecule and observe that a similar result is obtained. □

(P4.11) Problem: Show that the set of hydrogen atoms designated H_1, H_2, H_3 and H_4 in H_4SiO_4 is mapped into self-conincidence by the elements of $M_C(\bar{4})$. Do this by forming the orbit of $[\mathbf{r}]_C$ where $[\mathbf{r}]_C$ is triple of the coordinates of one of the hydrogen atoms in Table 1.1.

(P4.12) Problem: Show that the silicon atom in H_4SiO_4 is on a point of special position and that its orbit consists of a single point.

(E4.35) Example: Show that the faces on the crystal in Figure 4.3 are

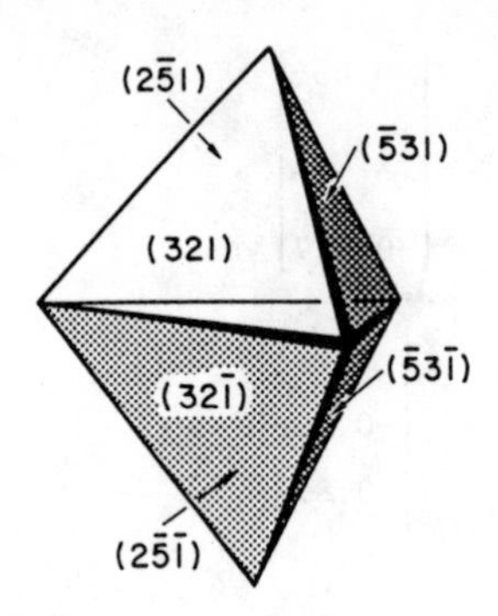

Figure 4.3: A set of faces comprising a trigonal dipyramid that are $\bar{6}$-equivalent to (321).

mapped into self-coincidence by the isometries of the monaxial group $\bar{6}$. The natural basis for the crystal is the basis $P = \{\mathbf{a},\mathbf{b},\mathbf{c}\}$ described earlier for 6. Recall that the face poles $\mathbf{s} = h\mathbf{a}^* + k\mathbf{b}^* + \ell\mathbf{c}^*$ of a crystal are defined in terms of its basis P^*. In Appendix 3 it is shown that

$$M_{P^*}(\alpha) = M_P(\alpha)^{-t}.$$

Solution: This problem will be solved by showing that the image under $\bar{6}$ of each face pole of the crystal is another of its face poles. That is, if $[\mathbf{s}]_{P^*}$ represents a face pole, then we must show that $M_{P^*}(\alpha)[\mathbf{s}]_{P^*}$ also represents a face pole for each $\alpha \in \bar{6}$. For example, consider the face pole $[\mathbf{s}]_{P^*} = (321)^t$ and note that

$$M_{P^*}(1)[\mathbf{s}]_{P^*} = \begin{bmatrix} 1 & 0 & 0 \\ 0 & 1 & 0 \\ 0 & 0 & 1 \end{bmatrix}\begin{bmatrix} 3 \\ 2 \\ 1 \end{bmatrix} = \begin{bmatrix} 3 \\ 2 \\ 1 \end{bmatrix},$$

$$M_{P^*}(\bar{6})[\mathbf{s}]_{P^*} = \begin{bmatrix} 0 & 1 & 0 \\ -1 & -1 & 0 \\ 0 & 0 & -1 \end{bmatrix}\begin{bmatrix} 3 \\ 2 \\ 1 \end{bmatrix} = \begin{bmatrix} 2 \\ -5 \\ -1 \end{bmatrix},$$

$$M_{P^*}(3)[\mathbf{s}]_{P^*} = \begin{bmatrix} -1 & -1 & 0 \\ 1 & 0 & 0 \\ 0 & 0 & 1 \end{bmatrix}\begin{bmatrix} 3 \\ 2 \\ 1 \end{bmatrix} = \begin{bmatrix} -5 \\ 3 \\ 1 \end{bmatrix},$$

$$M_{P^*}(m)[\mathbf{s}]_{P^*} = \begin{bmatrix} 1 & 0 & 0 \\ 0 & 1 & 0 \\ 0 & 0 & -1 \end{bmatrix}\begin{bmatrix} 3 \\ 2 \\ 1 \end{bmatrix} = \begin{bmatrix} 3 \\ 2 \\ -1 \end{bmatrix},$$

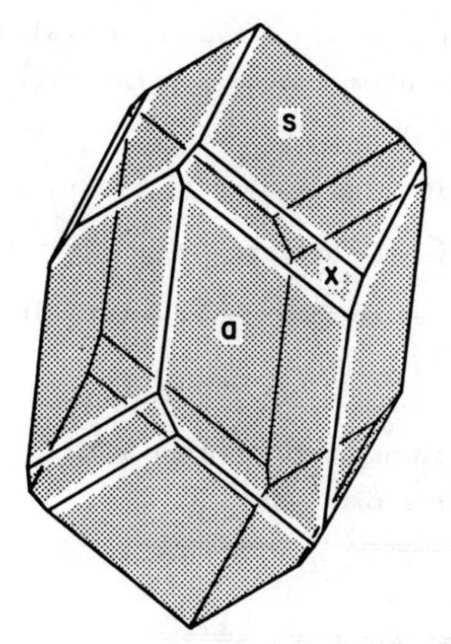

Figure 4.4: A dioptase crystal with a (110) face of a hexagonal prism labelled (*a*), and (021) and (131) faces of rhombohedra labelled *s* and *x*, respectively.

$$M_{P^*}(3^{-1})[s]_{P^*} = \begin{bmatrix} 0 & 1 & 0 \\ -1 & -1 & 0 \\ 0 & 0 & 1 \end{bmatrix} \begin{bmatrix} 3 \\ 2 \\ 1 \end{bmatrix} = \begin{bmatrix} 2 \\ -5 \\ 1 \end{bmatrix} ,$$

$$M_{P^*}(\bar{6}^{-1})[s]_{P^*} = \begin{bmatrix} -1 & -1 & 0 \\ 1 & 0 & 0 \\ 0 & 0 & -1 \end{bmatrix} \begin{bmatrix} 3 \\ 2 \\ 1 \end{bmatrix} = \begin{bmatrix} -5 \\ 3 \\ -1 \end{bmatrix} .$$

When these triples are collected together into a set, we have the orbit of face poles that are $\bar{6}$-equivalent to *(321)*:

$$\text{orb}_{\bar{6},P^*}\left(\begin{bmatrix} 3 \\ 2 \\ 1 \end{bmatrix}\right) = \{M_{P^*}(\alpha) \begin{bmatrix} 3 \\ 2 \\ 1 \end{bmatrix} \mid \alpha \in \bar{6} \}$$

$$= \left\{ \begin{bmatrix} 3 \\ 2 \\ 1 \end{bmatrix}, \begin{bmatrix} 2 \\ -5 \\ -1 \end{bmatrix}, \begin{bmatrix} -5 \\ 3 \\ 1 \end{bmatrix}, \begin{bmatrix} 3 \\ 2 \\ -1 \end{bmatrix}, \begin{bmatrix} 2 \\ -5 \\ 1 \end{bmatrix}, \begin{bmatrix} -5 \\ 3 \\ -1 \end{bmatrix} \right\} .$$

This collection of equivalent faces is called a ***crystal face form*** and is designated by placing the indices (***hkℓ***) of the representative plane of the orbit between braces, {***hkℓ***}. Thus, *{321}* denotes the equivalence class of faces on the crystal in Figure 4.3 that are $\bar{6}$-equivalent to *(321)*.

□

(P4.13) Problem: Figure 4.4 is a drawing of an idealized crystal of dioptase, $CuSiO_3 \cdot H_2O$, showing a (110) face of an hexagonal prism labelled *a* and (021) and (131) faces of rhombohedra labelled *s* and *x*, respectively. Assuming that the point symmetry of the dioptase crystal is $\bar{3}$, find the indices of each of the faces on the crystal that are $\bar{3}$-equivalent to (*110*), (*021*) and (*131*). Then assign indices to each of the $\bar{3}$-equivalent faces on the crystal.

Table 4.3: Coordinates of the atoms comprising an SiO_4 group in narsarsukite (Peacor and Buerger, 1962).

Atom	*x*	*y*	*z*
Si	0.0118	0.3085	0.1921
O_3	−0.0400	0.3024	0
O_4	0.0488	0.1754	0.2684
O_5	0.1324	0.4023	0.1934
O_6	−0.0977	0.3676	0.3062

(P4.14) Problem: The symmetry of the atoms about the origin chosen for narsarsukite ($a = b = 10.727$A, $c = 7.948$A, $\alpha = \beta = \gamma = 90°$) is $4/m$.

(1) Determine the elements of $M_D(4/m)$ and the coordinates of the atoms in narsarsukite that are $4/m$-equivalent to those in Table 4.3.
(2) With atomic coordinates obtained in (1), prepare a drawing of the atoms in narsarsukite that are $4/m$-equivalent to those in Table 4.3 viewed down the *Z*-axis.
(3) As the elements of $4/m$ are point isometries, the bond lengths and angles between the $4/m$-equivalent atoms in the narsarsukite structure must be identical. Calculate these bond lengths and angles and observe that this is indeed the case.

(P4.15) Problem: Stereoscopic drawings of *G*-equivalent ellipsoids are presented on the next four pages for each of the 13 monaxial crystallographic point groups *G*. Examine these drawings in Figure 4.5 and confirm their point symmetries.

Figure 4.5 (on this and the following pages):
Stereoscopic pair plots of G-equivalent ellipsoids for each of the 13 monaxial crystallographic point groups G.

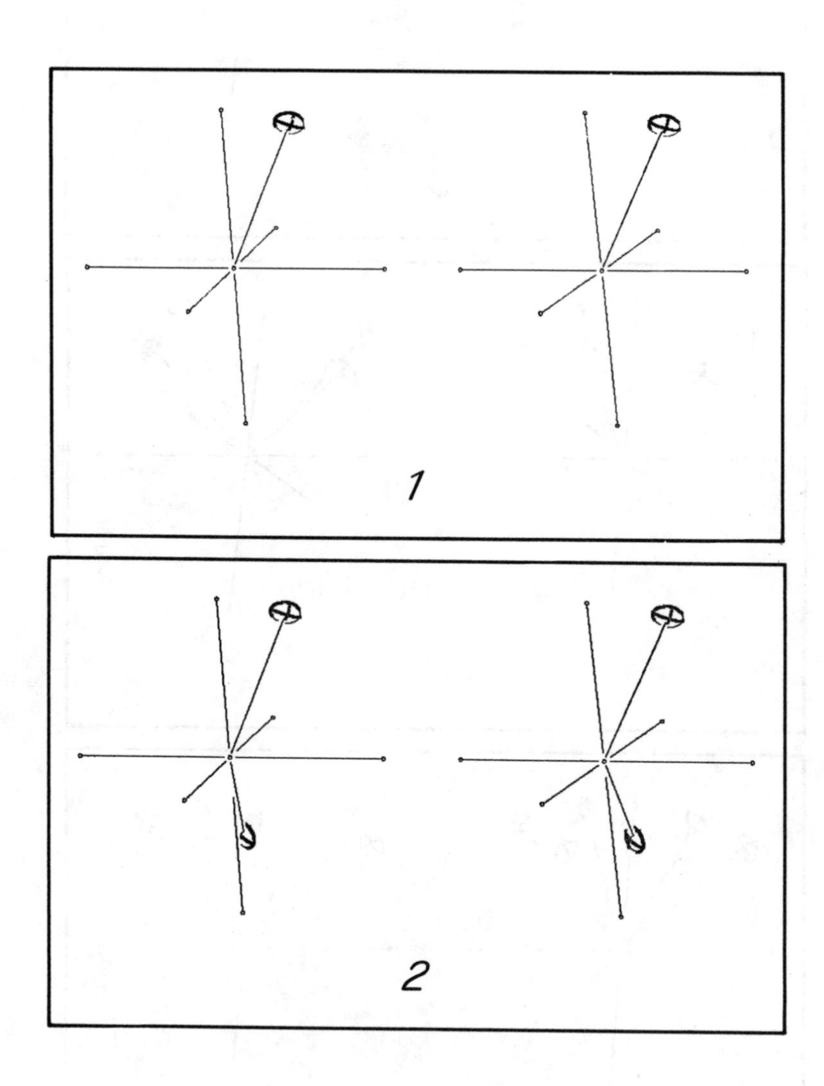

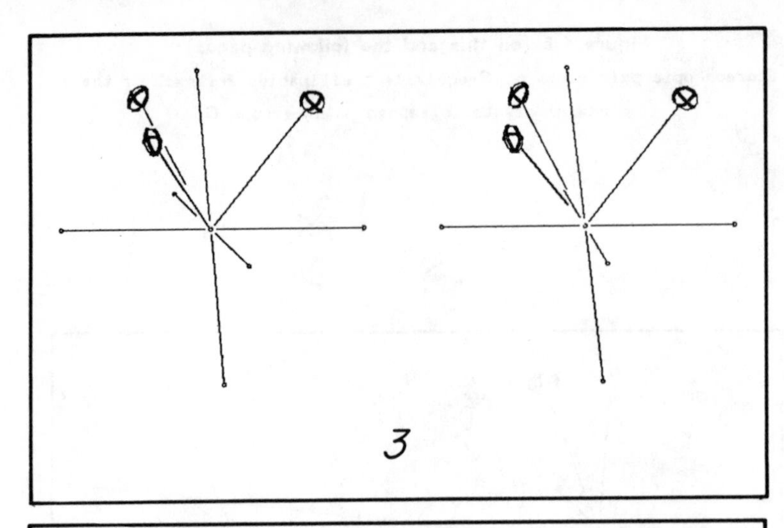
3

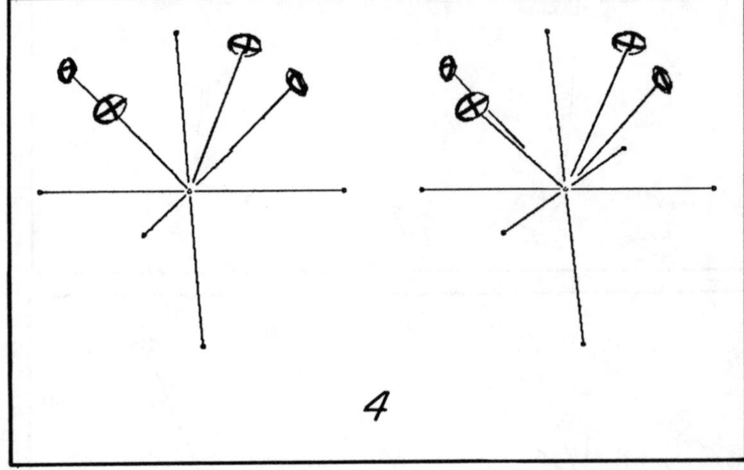
4

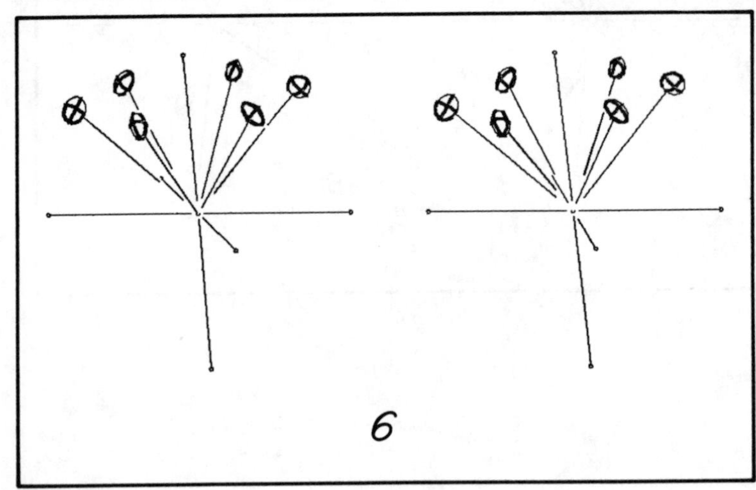
6

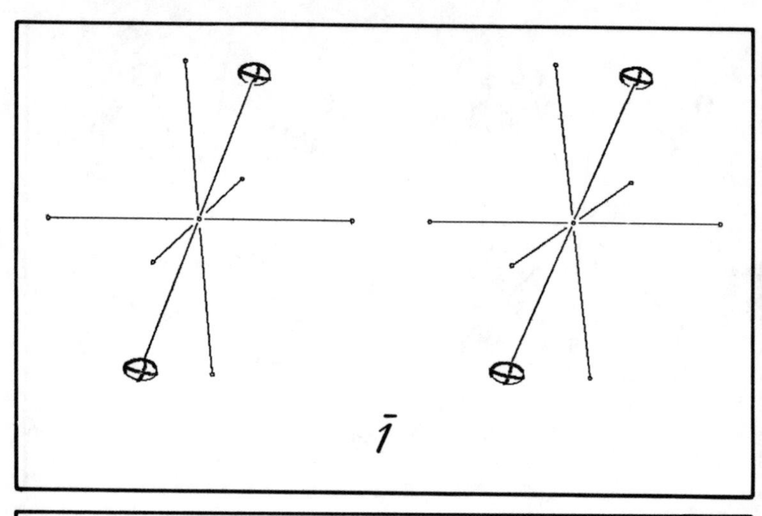
$\bar{1}$

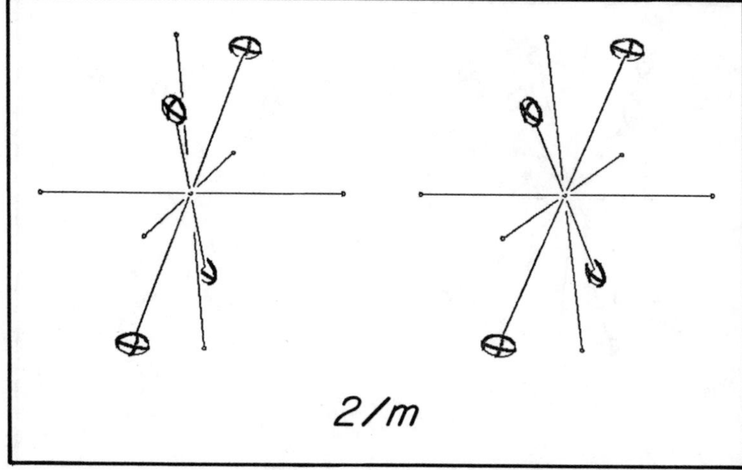
2/m

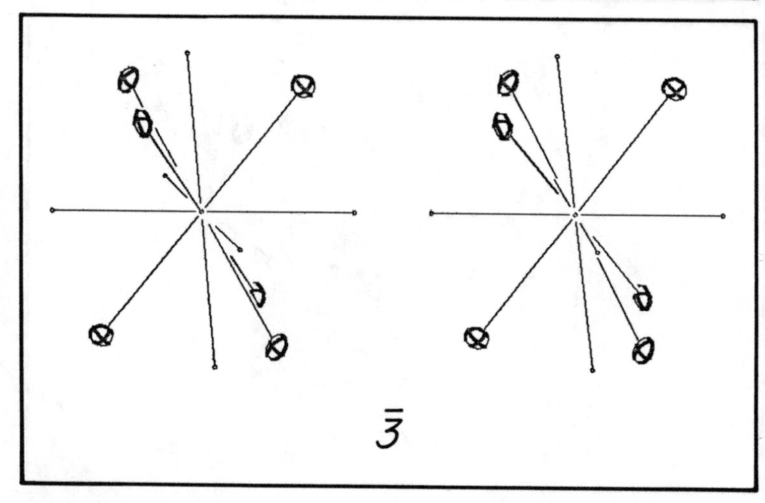
$\bar{3}$

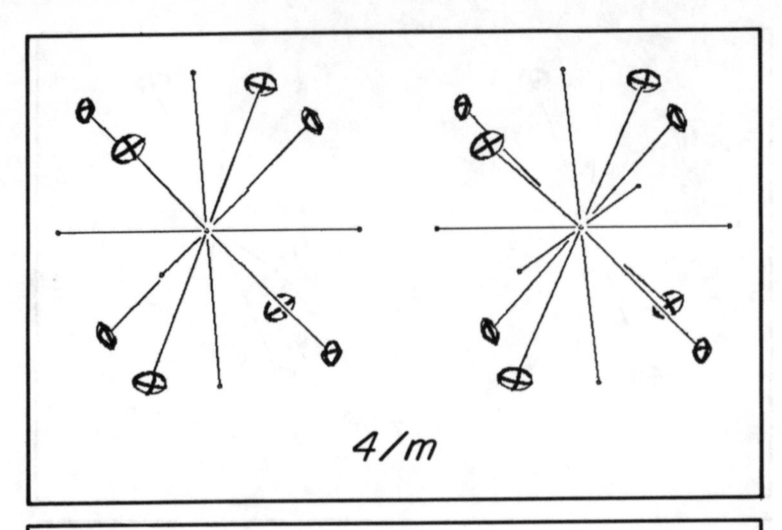
4/m

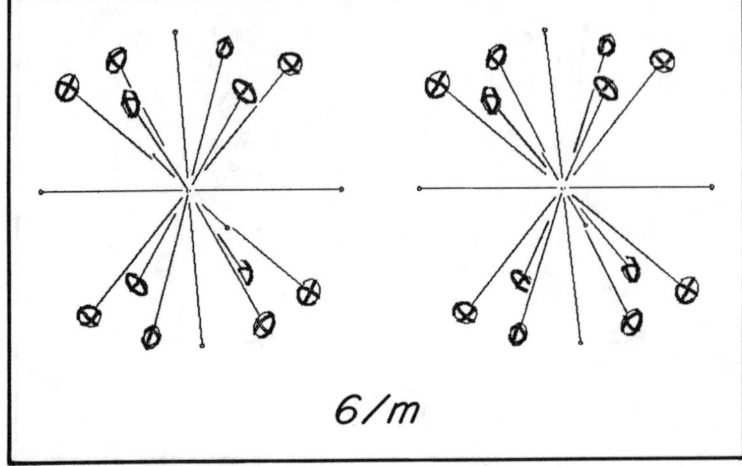
6/m

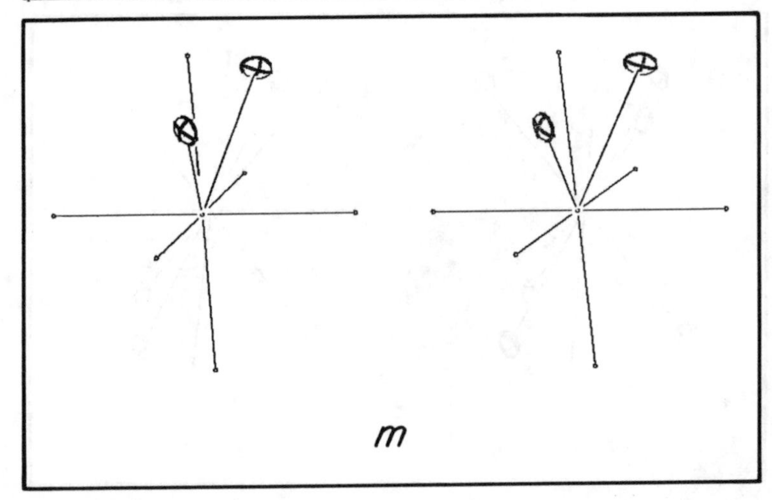
m

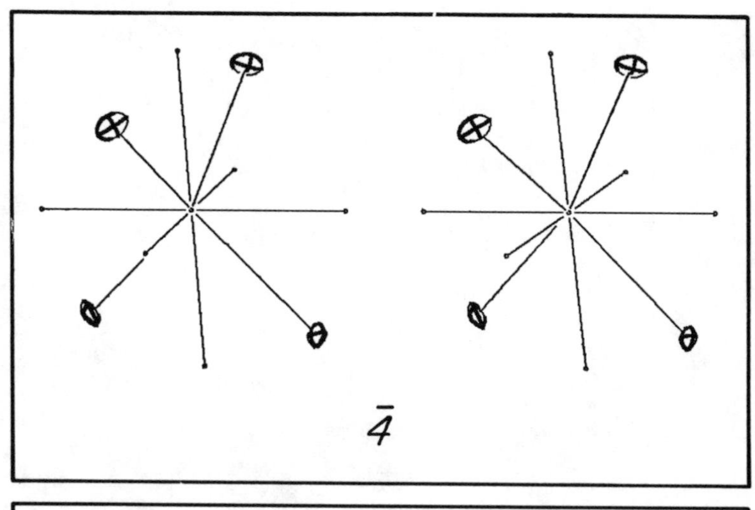
$\bar{4}$

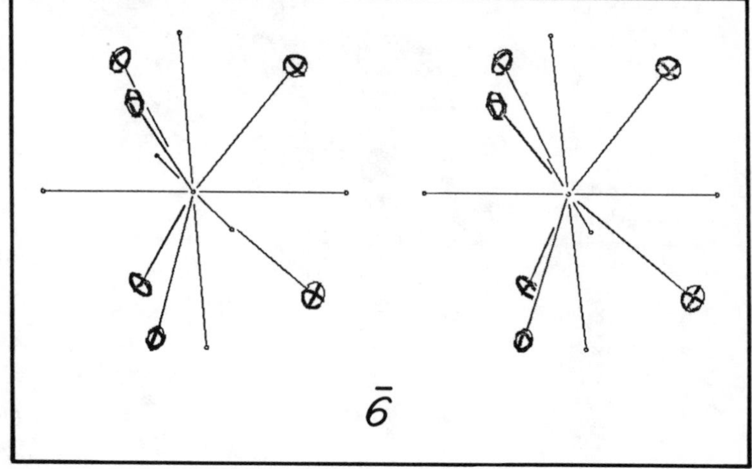
$\bar{6}$

CHAPTER 5

THE POLYAXIAL CRYSTALLOGRAPHIC POINT GROUPS

"Much of the importance of groups comes from their connection with symmetry. Just as numbers can be used to measure size (once a unit of measurement has been chosen), groups can be used to measure symmetry. With each figure we associate a group, and this group characterizes the symmetry of the figure." -- J. R. Durbin

INTRODUCTION

In Chapter 4 we derived all of the possible monaxial crystallographic point groups. In this chapter we shall learn how all of the possible proper polyaxial point groups are constructed from the proper monaxial point groups. Then, using the Improper Point Group Generating Theorem (T4.28), we shall derive all of the possible improper polyaxial crystallographic point groups. The interaxial angles between the rotation axes of the rotations participating in any one of these groups are then determined. As in Chapter 4, a special basis P will be chosen for each point group.

PROPER POLYAXIAL POINT GROUPS

(D5.1) Definition: Let G denote a proper point group. If G has more than one axis associated with its nonidentity rotations, then we call G *a proper polyaxial group*.

In our investigation of the polyaxial groups, we will be examining combinations of monaxial groups. The task of finding the possible proper polyaxial point groups will be considerably more difficult than that of finding the monaxial point groups. Our first goal is to establish the inequality which will state that three monaxial groups associated with nonequivalent pole points with orders ν_1, ν_2 and ν_3, respectively, can be used to form a proper polyaxial point group only if

$$1/\nu_1 + 1/\nu_2 + 1/\nu_3 > 1 \quad .$$

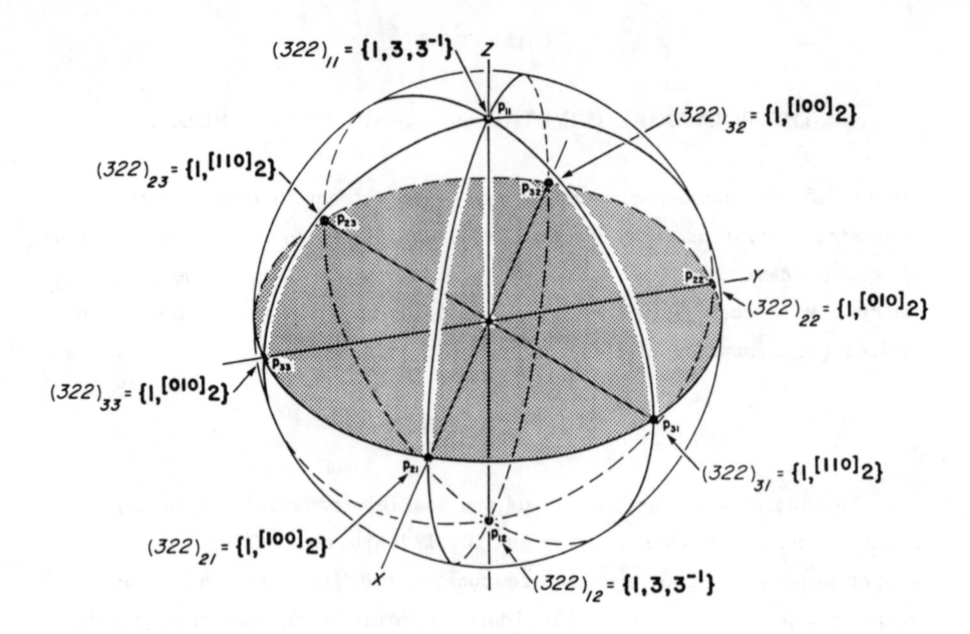

Figure 5.1: The set of all pole points belonging to the nonidentity rotations of group *322* (see Figure 3.13 for a drawing of rotation axes of the group). The basis vectors **a**, **b** and **c** coincide with the coordinate axes X, Y and Z, respectively, where $\alpha = \beta = 90°$ and $\gamma = 120°$. The nonidentity rotations in *3* leave the antipodal points $\mathbf{p}_{11}$ and $\mathbf{p}_{12}$ fixed, those in ${}^{[100]}2$ leave $\mathbf{p}_{21}$ and $\mathbf{p}_{32}$ fixed, those in ${}^{[110]}2$ leave $\mathbf{p}_{31}$ and $\mathbf{p}_{23}$ fixed and those in ${}^{[010]}2$ leave $\mathbf{p}_{22}$ and $\mathbf{p}_{33}$ fixed.

To facilitate our proof that will establish this inequality, we shall consider the surface B of a unit ball centered at **0**. Any point isometry α acting on B maps B onto itself. In fact, α is completely determined by its action on B because the effect of α on B determines the images $\{\alpha(\mathbf{i}),\alpha(\mathbf{j}),\alpha(\mathbf{k})\}$, where $C = \{\mathbf{i},\mathbf{j},\mathbf{k}\}$ is a cartesian basis, which in turn completely determines α. A nonidentity proper rotation h about the rotation axis ℓ leaves exactly two antipodal points on B unmoved. These points are precisely the points, on opposite sides of the ball, at which ℓ and B intersect and are called *the pole points* belonging to h. If these two points are labelled $\mathbf{p}$ and $\mathbf{q}$, then they are the only points $\mathbf{x}$ on B that satisfy the equality $h(\mathbf{x}) = \mathbf{x}$. The set of all pole points belonging to the nonidentity rotations in G will be denoted by $P(G)$ and will be called the set of pole points belonging to G. Note that G-equivalence is an equivalence relation on $P(G)$. Applying G-equivalence to $P(G)$, we partition $P(G)$ into its set of equivalence classes. For example, the pole points of *322*

$$P(322) = \{\mathbf{p}_{11},\mathbf{p}_{12},\mathbf{p}_{21},\mathbf{p}_{22},\mathbf{p}_{23},\mathbf{p}_{31},\mathbf{p}_{32},\mathbf{p}_{33}\}$$

are denoted in Figure 5.1 by $\mathbf{p}_{ij}$ where i indicates to which equivalence class $\mathbf{p}_{ij}$ belongs. We denote the ith equivalence class of pole points belonging to G by $C_i(G)$.

(E5.2) Example - The *322*-equivalence class of $\mathbf{p}_{11}$: Find the equivalence class under *322*-equivalence of $\mathbf{p}_{11}$ shown in Figure 5.1. That is, find $C_1(322)$.

Solution: By definition of an equivalence class,

$$[\mathbf{p}_{11}] = \{\mathbf{q} \in P(G) \mid \mathbf{p}_{11} \sim \mathbf{q}\} \quad .$$

By definition of *322*-equivalence,

$$\begin{aligned}
[\mathbf{p}_{11}] &= \{\mathbf{q} \in P(322) \mid \text{there exists } \mathbf{g} \in 322 \text{ such that } \mathbf{g}(\mathbf{p}_{11}) = \mathbf{q}\} \\
&= \{\mathbf{g}(\mathbf{p}_{11}) \mid \mathbf{g} \in 322\} \\
&= \{1(\mathbf{p}_{11}), 3(\mathbf{p}_{11}), 3^{-1}(\mathbf{p}_{11}), {}^{[100]}2(\mathbf{p}_{11}), {}^{[010]}2(\mathbf{p}_{11}), {}^{[110]}2(\mathbf{p}_{11})\} \\
&= \{\mathbf{p}_{11}, \mathbf{p}_{12}\} \quad .
\end{aligned}$$

Hence $C_1(322) = \{\mathbf{p}_{11}, \mathbf{p}_{12}\}$. □

(P5.1) Problem: Verify that $C_2(322) = \{\mathbf{p}_{21}, \mathbf{p}_{22}, \mathbf{p}_{23}\}$ and $C_3(322) = \{\mathbf{p}_{31}, \mathbf{p}_{32}, \mathbf{p}_{33}\}$ as shown in Figure 5.1.

Note that $\{C_1(322), C_2(322), C_3(322)\}$ partitions $P(322)$. That is

$$P(322) = C_1(322) \cup C_2(322) \cup C_3(322)$$

and

$$C_i(322) \cap C_j(322) = \emptyset \text{ when } i \neq j \quad ,$$

where $\emptyset$ is the empty set. Note that $\#(C_1(322)) = 2$, $\#(C_2(322)) = 3$ and $\#(C_3(322)) = 3$.

(P5.2) Problem: Draw a diagram showing the rotation axes of *422*. Let $\mathbf{p}_{11}$ denote a pole point of *4*, $\mathbf{p}_{21}$ a pole point of ${}^{[100]}2$ and $\mathbf{p}_{31}$ a pole point of ${}^{[110]}2$. Find $C_1(422)$, $C_2(422)$ and $C_3(422)$. Note that every pole point of 422 is in one of these equivalence classes.

Now we shall turn our attention to polyaxial rotation groups G and the subgroups of G associated with its pole points.

(D5.3) Definition: Let $\mathbf{p}$ denote a pole point belonging to G. Then the collection of all rotations of G that have $\mathbf{p}$ as a pole point is called the *stabilizer* of $\mathbf{p}$ in G and is denoted by $G_{\mathbf{p}}$. That is

$$G_{\mathbf{p}} = \{\mathbf{g} \in G \mid \mathbf{g}(\mathbf{p}) = \mathbf{p}\} \quad .$$

The stabilizer of a pole point $\mathbf{p}_{ij}$ is denoted by G_{ij}.

(T5.4) Theorem: Let $\mathbf{p}$ denote a pole point of G. Then $G_{\mathbf{p}}$ is a subgroup of G.

Proof: Since G is finite, we need only show that $G_{\mathbf{p}}$ is closed. Let $\mathbf{g}_1, \mathbf{g}_2 \in G_{\mathbf{p}}$. Then

$$\mathbf{g}_1\mathbf{g}_2(\mathbf{p}) = \mathbf{g}_1(\mathbf{g}_2(\mathbf{p})) = \mathbf{g}_1(\mathbf{p}) = \mathbf{p} \quad .$$

Since $\mathbf{g}_1\mathbf{g}_2(\mathbf{p}) = \mathbf{p}$ and $\mathbf{g}_1\mathbf{g}_2 \in G$ (G is closed) $\mathbf{g}_1\mathbf{g}_2 \in G_{\mathbf{p}}$. Since $G_{\mathbf{p}}$ is nonempty, $G_{\mathbf{p}}$ is a subgroup. □

Since the rotations of $G_{\mathbf{p}}$ leave $\mathbf{p}$ fixed, they all have the line ℓ containing $\mathbf{0}$ and $\mathbf{p}$ as their rotation axis. Consequently, $G_{\mathbf{p}}$ is a proper monaxial group and therefore isomorphic to one of the cyclic groups listed in Table 4.2.

In the case of 322,

$$(322)_{21} = (322)_{32} = \{{}^{[100]}2, 1\} \quad ,$$

$$(322)_{31} = (322)_{23} = \{{}^{[110]}2, 1\} \quad ,$$

$$(322)_{22} = (322)_{33} = \{{}^{[010]}2, 1\} \quad ,$$

and

$$(322)_{11} = (322)_{12} = \{3, 3^{-1}, 1\} \quad .$$

We observe in this example that two pole points are associated with each monaxial group (Figure 5.1). It is also evident that the pole points $\mathbf{p}_{21}$, $\mathbf{p}_{22}$, $\mathbf{p}_{23}$, $\mathbf{p}_{31}$, $\mathbf{p}_{32}$, and $\mathbf{p}_{33}$ are each associated with the group *2* where the rotation axes have different orientations. The remaining pole points $\mathbf{p}_{11}$ and $\mathbf{p}_{12}$ are similarly associated with group *3*.

(T5.5) Theorem: Let G denote a proper crystallographic point group. Let $\mathbf{p}$ denote a pole point and let $\mathbf{q}$ be G-equivalent to $\mathbf{p}$. Then $G_{\mathbf{p}}$ is isomorphic to $G_{\mathbf{q}}$ and hence they are both isomorphic to the same proper monaxial point group of Table 4.2.

Proof: Since $\mathbf{q}$ is G-equivalent to $\mathbf{p}$, there exists a rotation $\mathbf{g} \varepsilon G$ such that $\mathbf{g}(\mathbf{p}) = \mathbf{q}$. Consider $\theta: G_{\mathbf{p}} \to G_{\mathbf{q}}$ defined by $\theta(\mathbf{h}) = \mathbf{ghg}^{-1}$. Note that $\mathbf{g}^{-1}(\mathbf{q}) = \mathbf{p}$ and so $\mathbf{ghg}^{-1}(\mathbf{q}) = \mathbf{gh}(\mathbf{p}) = \mathbf{g}(\mathbf{p}) = \mathbf{q}$. Hence $\theta(\mathbf{h}) \varepsilon G_{\mathbf{q}}$ for all $\mathbf{h} = G_{\mathbf{p}}$. By EA8.5, θ is an isomorphism. Since isomorphism defines an equivalence relation, $G_{\mathbf{p}}$ and $G_{\mathbf{q}}$ are isomorphic to the same proper monaxial point group of Table 4.2. □

(E5.6) Example: Show that $(322)_{22} = \{\mathbf{3h3}^{-1} \mid \mathbf{h} \varepsilon (322)_{21}\}$.

Solution: As shown above

$$(322)_{21} = {}^{[100]}2 = \{\mathbf{1}, {}^{[100]}\mathbf{2}\}$$

and

$$(322)_{22} = {}^{[010]}2 = \{\mathbf{1}, {}^{[010}\mathbf{2}\} \quad .$$

Then

$$\{\mathbf{3h3}^{-1} \mid \mathbf{h} \varepsilon (322)_{21}\} = \{\mathbf{313}^{-1}, \mathbf{3}{}^{[100]}\mathbf{23}^{-1}\}$$

$$= \{\mathbf{1}, {}^{[010]}\mathbf{2}\}$$

$$= (322)_{22} \quad .$$

□

(P5.3) Problem: Confirm the results in E5.6 with matrices showing that

$$M_D((322)_{22}) = \{M_D(\mathbf{3})M_D(\mathbf{h})M_D(\mathbf{3}^{-1}) \mid \mathbf{h} \varepsilon (322)_{21}\} \quad .$$

Note that since $M_D(\mathbf{3}^{-1}) = M_D(\mathbf{3})^{-1}$, the matrices described in P5.3 for $M_D((322)_{21})$ are similar (see Appendix 7) to the matrices in

$M_D((322)_{22})$. Hence while $\mathbf{p}_{21}$ and $\mathbf{p}_{22}$ are *322*-equivalent under the equivalence relation of D4.32, the matrices in $M_D((322)_{21})$ and $M_D((322)_{22})$ are equivalent under the equivalence relation of DA3.3 (see also EA7.4). A relation corresponding to similarity can be defined on the subgroups of G. This relation is called *conjugation*.

(T5.7) Theorem: Let G denote a group. The relation defined on the set of subgroups of G defined by

$$H_1 \sim H_2 \iff \text{there exists a } \mathbf{g} \in G \text{ such that } \mathbf{g}H_1\mathbf{g}^{-1} = H_2 \quad ,$$

where H_1 and H_2 are subgroups of G, is an equivalence relation. The relation $\sim$ is called *conjugation* and if $H_1 \sim H_2$, H_1 is said to be *conjugate* to H_2.

The proof of T5.7 is essentially the same as that of EA7.4 where we demonstrated that similarity of matrices is an equivalence relation. Conjugation and G-equivalence are very closely related concepts. Consider

$$T = \{G_{\mathbf{p}} \mid \mathbf{p} \in P(G)\} \quad .$$

Conjugation is an equivalence relation on T while G-equivalence is an equivalence relation on $P(G)$. Furthermore, we have

$$\mathbf{p} \sim \mathbf{q} \leftrightarrow G_{\mathbf{p}} \sim G_{\mathbf{q}} \quad ,$$

where G-equivalence is used on the left and conjugation on the right. Consequently, if $\mathbf{p}$ is a pole point, then an isomorphic image of $G_{\mathbf{p}}$ occurs about any axis having $\mathbf{g}(\mathbf{p})$ as a pole point for each $\mathbf{g} \in G$. Suppose we have located the position of the axis of one of the rotations of G. Then we can find others by mapping the axis under the operations of G. Furthermore, the axes found in this manner will be associated with cyclic groups of the same order as the original axis. This observation will be extremely important in our construction of the rotation groups.

(L5.8) Lemma: Let G denote a proper point group such that $\#(G) > 1$. Then

$$2(N - 1) = \sum_{i=1}^{t} n_i(\nu_i - 1)$$

where $N = \#(G)$, t is the number of equivalence classes of pole points, $n_i = \#(C_i(G))$ and $\nu_i = \#(G_{ij})$.

Proof: The basic strategy for establishing this theorem will be to find two distinct ways of counting the nonidentity rotations of G. The result will be two expressions each equaling twice the number of nonidentity rotations of G. This will establish the result since these two equal expressions will be precisely those appearing in the equation. We begin by taking each pole point $\mathbf{p}_{ij}$ in $P(G)$ one at a time and counting the number of nonidentity rotations of G leaving $\mathbf{p}_{ij}$ fixed. The sum of these numbers taken over all the pole points in $P(G)$ will equal twice the number of nonidentity rotations in G because each of these rotations leaves exactly two pole points fixed and hence is counted twice. The number of nonidentity rotations leaving $\mathbf{p}_{ij}$ fixed is $\#(G_{ij}) - 1 = \nu_i - 1$. Thus, for the pole points in $C_i(G)$ we have

$$\left.\begin{array}{l} \#(\text{nonidentity rotations leaving } \mathbf{p}_{i1} \text{ fixed}) = \nu_i - 1 \\ \#(\text{nonidentity rotations leaving } \mathbf{p}_{i2} \text{ fixed}) = \nu_i - 1 \\ \quad\vdots \qquad \vdots \qquad \vdots \\ \#(\text{nonidentity rotations leaving } \mathbf{p}_{in_i} \text{ fixed}) = \nu_i - 1 \end{array}\right\} n_i \text{ equations}$$

Summing up these numbers we find that the contribution from $C_i(G)$ is $n_i(\nu_i - 1)$ for each $1 \le i \le t$. Adding the contribution from each of the t equivalence classes, we find that the sum is $\sum_{i=1}^{t} n_i(\nu_i - 1)$. Since $N - 1$ is the number of nonidentity rotations in G, it follows that twice the number of nonidentity rotations in G is $2(N - 1)$ and so we have established that

$$2(N - 1) = \sum_{i=1}^{t} n_i(\nu_i - 1) \quad . \qquad \square$$

From E5.2 and P5.1, we see that $n_1 = \#(C_1(322)) = 2$, $n_2 = \#(C_2(322)) = 3$ and $n_3 = \#(C_3(322)) = 3$. Also $\nu_1 = \#(3) = 3$, $\nu_2 = \#({}^{[100]}2) = 2$ and $\nu_3 = \#({}^{[110]}2) = 2$. Since $N = \#(322) = 6$, we can verify L5.8 in this case by observing that

$$2(6 - 1) = 2(3 - 1) + 3(2 - 1) + 3(2 - 1) \quad .$$

(P5.4) Problem: Using the information you have developed about *422* (including the solution to P5.2) verify the equation in L5.8 for *422*.

(T5.9) Theorem: Let $\mathbf{p}$ and $\mathbf{q}$ denote G-equivalent pole points and let T denote the set of all elements of G that map $\mathbf{p}$ to $\mathbf{q}$. Then T is a left coset of $G_{\mathbf{p}}$.

Proof: A discussion of cosets and related topics can be found in Appendix 7. Since $\mathbf{p}$ is G-equivalent to $\mathbf{q}$, there exists $\mathbf{g} \varepsilon G$ such that $\mathbf{g}(\mathbf{p}) = \mathbf{q}$. We shall show that $T = \mathbf{g}G_{\mathbf{p}}$. Let $\mathbf{t} \varepsilon T$. By definition of T, $\mathbf{t}(\mathbf{p}) = \mathbf{q}$. Since $\mathbf{g}(\mathbf{p}) = \mathbf{q}$, we have

$$\begin{aligned} \mathbf{g}^{-1}\mathbf{t}(\mathbf{p}) &= \mathbf{g}^{-1}(\mathbf{q}) \\ &= \mathbf{p} \ . \end{aligned}$$

Hence $\mathbf{g}^{-1}\mathbf{t} \varepsilon G_{\mathbf{p}}$. Therefore $\mathbf{t} \varepsilon \mathbf{g}G_{\mathbf{p}}$. Conversely, suppose $\mathbf{h} \varepsilon \mathbf{g}G_{\mathbf{p}}$. Then $\mathbf{h} = \mathbf{gk}$ where $\mathbf{k} \varepsilon G_{\mathbf{p}}$. Hence

$$\begin{aligned} \mathbf{h}(\mathbf{p}) &= \mathbf{gk}(\mathbf{p}) \\ &= \mathbf{g}(\mathbf{p}) \\ &= \mathbf{q} \ . \end{aligned}$$

Hence $\mathbf{h} \varepsilon T$. Consequently $T = \mathbf{g}G_{\mathbf{p}}$. □

One consequence of T5.9 is that if G is a finite proper point group such that $\#(G) > 1$ and $\mathbf{p}$ is a pole point of G, then there is a one-to-one correspondence between the cosets of $G_{\mathbf{p}}$ and the pole points that are G-equivalent to $\mathbf{p}$. Hence, if $\mathbf{p} \varepsilon C_i(G)$, then n_i, which is defined to be the number of pole points in $C_i(G)$, equals the number of cosets of $G_{\mathbf{p}}$ in G. Recall (see the proof of TA7.13) that each coset of $G_{\mathbf{p}}$ has the same number of elements as does $G_{\mathbf{p}}$, that is ν_i elements. Since the cosets of $G_{\mathbf{p}}$ partition G into n_i cosets each having ν_i elements we have established that

$$\#(G) = N = n_i\nu_i \text{ for each } 1 \leq i \leq t \ . \tag{5.1}$$

In the case of *322*, we observe that since $n_1 = 2$, $n_2 = n_3 = 3$ and $\nu_1 = 3$, $\nu_2 = \nu_3 = 2$ and that $N = 6 = n_i\nu_i$ in all three cases.

(P5.5) Problem: Show that $n_i\nu_i = N$ for each $1 \leq i \leq 3$ in the case of *422*.

(T5.10) Theorem: Let t denote the number of equivalence classes of pole points of a finite proper point group G where $N = \#(G) > 1$. Then $t = 2$ or 3 and

(1) if $t = 2$, G is a monaxial group and

(2) if $t = 3$, G is a polyaxial group such that

$$1/\nu_1 + 1/\nu_2 + 1/\nu_3 > 1$$

and

$$\#(G) = 2/(1/\nu_1 + 1/\nu_2 + 1/\nu_3 - 1) \quad .$$

Proof: By L5.8, we have

$$2(N - 1) = \sum_{i=1}^{t} n_i(\nu_i - 1) \quad . \tag{5.2}$$

and by (5.1), we have $N = n_i\nu_i$. Dividing the left side of (5.2) by N and the right side of (5.2) by $n_i\nu_i$, we obtain

$$2 - 2/N = \sum_{i=1}^{t} (1 - 1/\nu_i) \quad . \tag{5.3}$$

Since ν_i is the order of the stabilizer of a pole point, $\nu_i \geq 2$. Hence

$$\sum_{i=1}^{t} (1 - (1/\nu_i)) \geq \sum_{i=1}^{t} (1 - \tfrac{1}{2}) = \sum_{i=1}^{t} (\tfrac{1}{2}) = t/2 \quad .$$

Also, since $N \geq 2$,

$$2 > 2 - 2/N \quad .$$

Hence, from (5.3)

$$2 > t/2 \quad .$$

Hence $4 > t$. Therefore, t can only equal 1, 2 or 3. We can thus conclude that there are no rotation groups having more than 3 equivalence classes of pole points. We now examine each of these three cases for the value of t.

Case where $t = 1$: In this case, Equation (5.3) becomes

$$2 - 2/N = \sum_{i=1}^{t} (1 - 1/\nu_i) = \sum_{i=1}^{1} (1 - 1/\nu_1)$$

$$\text{or } 1 - 2/N = -1/\nu_1 \quad .$$

The left member of this equation is always nonnegative because $N \geq 2$, but the right member is always negative because $\nu_1 \geq 2$, which is a contradiction. Therefore, t cannot equal 1, from which we conclude that $P(G)$ must contain more than one equivalence class of pole points.

Case where $t = 2$: In this case Equation (5.3) becomes

$$2 - 2/N = \sum_{i=1}^{2} (1 - 1/\nu_i) = (1 - 1/\nu_1) + (1 - 1/\nu_2) \quad .$$

By a little algebraic manipultion we find that

$$2 = N/\nu_1 + N/\nu_2 \quad .$$

From Equation (5.1) we have that $N/\nu_i = n_i$, and so the above expression simplifies to

$$2 = n_1 + n_2 \quad .$$

Because n_1 and n_2 are positive integers, we conclude that $n_1 = n_2 = 1$ is the only possible solution. Hence, for a rotation group with $t = 2$, we have two equivalence classes consisting of one pole point each. Altogether G has a total of two pole points, which defines one and only one rotation axis. Therefore, those groups with two equivalence classes of pole points must be the proper monaxial groups given in Table 4.2. The number of elements in each of these possible monaxial groups is equal to the order of the rotation axis, $\#(G) = \nu_1 = \nu_2$.

Case where $t = 3$: In this case Equation (5.3) expands to

$$2 - 2/N = (1 - 1/\nu_1) + (1 - 1/\nu_2) + (1 - 1/\nu_3) \quad .$$

Rewriting this result we see that

$$1 + 2/N = 1/\nu_1 + 1/\nu_2 + 1/\nu_3 \quad . \tag{5.4}$$

Since $N \geq 2$, it follows that $1 + 2/N > 1$ and so

Table 5.1: Possible finite proper polyaxial point groups.

Symbol for $G = \nu_1\nu_2\nu_3$	$\#(G) = N$	$\#(C_1(G)) = N/\nu_1$	$\#(C_2(G)) = N/\nu_2$	$\#(C_3(G)) = N/\nu_3$	Group Name
222	4	2	2	2	
322	6	2	3	3	
422	8	2	4	4	
*522**	10	2	5	5	
622	12	2	6	6	*Dihedral*
.	.	.	.	.	
.	.	.	.	.	
.	.	.	.	.	
*n22***	2n	2	n	n	
332	12	4	4	6	*Tetrahedral*
432	24	6	8	12	*Octahedral*
*532**	60	12	20	30	*Icosahedral*

* non-crystallographic

** non-crystallographic when $n > 6$.

The group *332* is usually designated by *23* and the group *532* is usually designated by *235*.

$$1/\nu_1 + 1/\nu_2 + 1/\nu_3 > 1 \quad . \tag{5.5}$$

Solving for N in (5.4) we obtain

$$\#(G) = N = 2/(1/\nu_1 + 1/\nu_2 + 1/\nu_3 - 1) \quad . \qquad \square$$

Using part (2) of T5.10 where, for convenience, we assume that $\nu_1 \geq \nu_2 \geq \nu_3$, we shall construct all of the possible finite proper polyaxial point groups. Note that if $\nu_3 > 2$, then each of the fractions $1/\nu_i$ would be less than or equal to 1/3 for each i and so (5.5) would not be satisfied. Hence $\nu_3 = 2$. If $\nu_2 > 3$, then $1/\nu_1 + 1/\nu_2$ is less than or equal to $\frac{1}{2}$ and so (5.5) would again not be satisfied. Therefore, $\nu_2 = 2$ or 3. Suppose $\nu_2 = \nu_3 = 2$. Then any value of $\nu_1 > 1$ would satisfy (5.5). These groups, denoted *n22* when $n = \nu_1$, are called the ***dihedral groups*** of which there are an infinite number. Using (T5.10) we see that

$$\#(n22) = 2n \quad .$$

If $\nu_2 = 3$ and $\nu_3 = 2$, then if $\nu_1 > 5$, (5.5) would not be satisfied. Hence the only groups of this type are ***332***, ***432*** and ***532***. All of these possible finite proper point groups are recorded Table 5.1. Note that we have

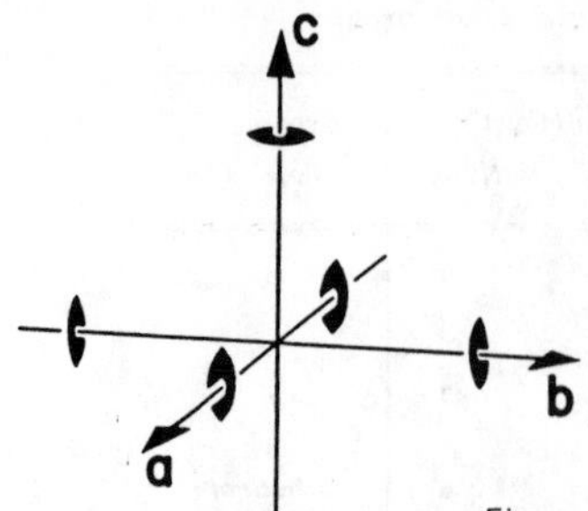

Figure 5.2: The orientation of the three mutually perpendicular 2-fold axes in *222*. These axes define a natural coordinate system with *2* lying along **c**, $^{[100]}2$ along **a** and $^{[010]}2$ along **b**. Each 2-fold axis is represented by a *diad* symbol.

Figure 5.3: Multiplication table for $M_P(222)$.

$M_P(222)$	$M_P(1)$	$M_P(2)$	$M_P(^{[100]}2)$	$M_P(^{[010]}2)$
$M_P(1)$	$M_P(1)$	$M_P(2)$	$M_P(^{[101]}2)$	$M_P(^{[010]}2)$
$M_P(2)$	$M_P(2)$	$M_P(1)$	$M_P(^{[010]}2)$	$M_P(^{[100]}2)$
$M_P(^{[100]}2)$	$M_P(^{[100]}2)$	$M_P(^{[010]}2)$	$M_P(1)$	$M_P(2)$
$M_P(^{[010]}2)$	$M_P(^{[010]}2)$	$M_P(^{[100]}2)$	$M_P(2)$	$M_P(1)$

yet to show that each of these possibilities actually occurs as a point group.

CONSTRUCTION OF THE DIHEDRAL GROUPS

In Appendix 6 we proved, for *n22*, that the *n*-fold axis is perpendicular to each 2-fold symmetry axis and adjacent 2-fold axes in this group must intersect at an angle of $180/n$ (TA6.1). To confirm that each *n22* is actually a group, we shall define a basis of *S* for each, write the elements of *n22* as described in TA6.1 with respect to this basis and then form the multiplication table to check closure.

The construction of ***222***: Since the 2-fold axes are mutually perpendicular, we define $P = \{\mathbf{a},\mathbf{b},\mathbf{c}\}$ to be a basis where **a**, **b** and **c** are also mutually perpendicular such that each lies along an axis (see Figure 5.2). Hence the metrical matrix G for *P* is

$$G = \begin{bmatrix} g_{11} & 0 & 0 \\ 0 & g_{22} & 0 \\ 0 & 0 & g_{33} \end{bmatrix} .$$

With respect to this choice of basis, the half-turns are denoted ${}^{[100]}2$, ${}^{[010]}2$, **2** along **a**, **b**, **c**, respectively. By Table 5.1, *222* can only have 4 elements and so we conjecture that

$$222 = \{1, 2, {}^{[100]}2, {}^{[010]}2\}$$

is a group. The matrix representation of each of these can be found employing the approach used for *322* in Chapter 3. Hence

$$M_P(222) = \{M_P(1), M_P(2), M_P({}^{[100]}2), M_P({}^{[010]}2)\} .$$

These matrices are described in Table 5.2. To confirm that *222* is a group, we form the multiplication table of $M_P(222)$ shown in Figure 5.3. Since no new entries resulted in the formation of the table, $M_P(222)$ is closed under matrix multiplication and, since it is finite, it is a group. Since the mapping from $M_P(222)$ to *222* that maps $M_P(\alpha)$ to α for each α preserves the operation, the multiplication table of *222* under composition can be obtained by deleting the $M_P(\ )$ for each element (Figure 4.2). Hence, *222* is a group.

The construction of ***322***: As in Chapter 3 (see Figure 3.10), we choose $P = \{\mathbf{a}, \mathbf{b}, \mathbf{c}\}$ where **c** coincides with the 3-fold axis, **a** coincides with one of the two-fold axes and $\mathbf{b} = \mathbf{3}(\mathbf{a})$ (Figure 5.4). Hence the metrical matrix G of *P* is

$$G = \begin{bmatrix} g_{11} & -g_{11}/2 & 0 \\ -g_{11}/2 & g_{11} & 0 \\ 0 & 0 & g_{33} \end{bmatrix} .$$

Since 60 = 180/3, there are two-fold axes at 60° intervals starting with the one coinciding with **a**. Hence, a two-fold axis lies along **b**. Recall that this is the same basis that was used for the monaxial groups *3*, $\bar{3}$, *6*, $\bar{6}$ and *6/m*. With respect to *P*,

Table 5.2: The nonzero entries of the matrix representations $M_P(\alpha)$ for rotation isometries α groups 1, 2, 4, 222, 422, 23 and 432 for $P = \{a,b,c\}$.

$M_P(1) : \ell_{11} = \ell_{22} = \ell_{33} = 1$	$M_P(^{[\bar{1}11]}3) : \ell_{23} = 1;\ \ell_{31} = \ell_{12} = -1$
$M_P(2) : \ell_{11} = \ell_{22} = -1;\ \ell_{33} = 1$	$M_P(^{[\bar{1}11]}3^{-1}) : \ell_{32} = 1;\ \ell_{21} = \ell_{13} = -1$
$M_P(^{[100]}2) : \ell_{11} = 1;\ \ell_{22} = \ell_{33} = -1$	$M_P(^{[\bar{1}\bar{1}1]}3) : \ell_{21} = 1;\ \ell_{32} = \ell_{13} = -1$
$M_P(^{[010]}2) : \ell_{22} = 1;\ \ell_{11} = \ell_{33} = -1$	$M_P(^{[\bar{1}\bar{1}1]}3^{-1}) : \ell_{12} = 1;\ \ell_{31} = \ell_{23} = -1$
$M_P(^{[101]}2) : \ell_{31} = \ell_{13} = 1;\ \ell_{22} = 1$	$M_P(^{[1\bar{1}1]}3) : \ell_{31} = -1;\ \ell_{12} = \ell_{23} = -1$
$M_P(^{[011]}2) : \ell_{32} = \ell_{23} = 1;\ \ell_{11} = -1$	$M_P(^{[1\bar{1}1]}3^{-1}) : \ell_{13} = 1;\ \ell_{21} = \ell_{32} = -1$
$M_P(^{[\bar{1}01]}2) : \ell_{31} = \ell_{22} = \ell_{13} = -1$	$M_P(4) : \ell_{21} = \ell_{33} = 1;\ \ell_{12} = -1$
$M_P(^{[0\bar{1}1]}2) : \ell_{11} = \ell_{32} = \ell_{23} = -1$	$M_P(4^{-1}) : \ell_{12} = \ell_{33} = 1;\ \ell_{21} = -1$
$M_P(^{[110]}2) : \ell_{21} = \ell_{12} = 1;\ \ell_{33} = -1$	$M_P(^{[100]}4) : \ell_{11} = \ell_{32} = 1;\ \ell_{23} = -1$
$M_P(^{[\bar{1}10]}2) : \ell_{21} = \ell_{12} = \ell_{33} = -1$	$M_P(^{[100]}4^{-1}) : \ell_{11} = \ell_{23} = 1;\ \ell_{32} = -1$
$M_P(^{[111]}3) : \ell_{21} = \ell_{32} = \ell_{13} = 1$	$M_P(^{[010]}4) : \ell_{22} = \ell_{13} = 1;\ \ell_{31} = -1$
$M_P(^{[111]}3^{-1}) : \ell_{31} = \ell_{12} = \ell_{23} = 1$	$M_P(^{[010]}4^{-1}) : \ell_{31} = \ell_{22} = 1;\ \ell_{13} = -1$

$$322 = \{1, 3, 3^{-1}, {}^{[100]}2, {}^{[110]}2, {}^{[010]}2\} .$$

In Figures 3.14 and 3.15, the multiplication tables for $M_D(322)$ and *322*, respectively, are displayed. From these tables, we observed that *322* is a group.

The construction of *422*: As in Chapter 4, we choose $P = \{\mathbf{a},\mathbf{b},\mathbf{c}\}$ where **c** coincides with the 4-fold axis, **a** is along a 2-fold axis and $\mathbf{b} = 4(\mathbf{a})$. The metrical matrix G for P is

$$G = \begin{bmatrix} g_{11} & 0 & 0 \\ 0 & g_{11} & 0 \\ 0 & 0 & g_{33} \end{bmatrix} .$$

Since 45 = 180/4, there are two-fold axes at 45° intervals starting with the one coinciding with **a**. Hence, there is a two-fold axis along **b**. The elements of *422* with respect to P, given in P3.14 are (see Figure 5.5)

Table 5.3: Non-zero entries of the matrix representations $M_P(\alpha)$ and $M_{P^*}(\alpha)$ for the rotation isometries α of point groups 3, 6, 322, and 622 for $P = \{a,b,c\}$ and $P^* = \{a^*,b^*,c^*\}$.

$M_P(\alpha)$	
$M_P(1) : \ell_{11} = \ell_{22} = \ell_{33} = 1$	$M_P({}^{[010]}2) : \ell_{22} = 1;\ \ell_{11} = \ell_{21} = \ell_{33} = -1$
$M_P(2) : \ell_{11} = \ell_{22} = -1;\ \ell_{33} = 1$	$M_P({}^{[\bar{1}10]}2) : \ell_{21} = \ell_{12} = \ell_{33} = -1$
$M_P({}^{[100]}2) : \ell_{11} = 1;\ \ell_{12} = \ell_{22} = \ell_{33} = -1$	$M_P(3) : \ell_{21} = \ell_{33} = 1;\ \ell_{12} = \ell_{22} = -1$
$M_P({}^{[210]}2) : \ell_{11} = \ell_{21} = 1;\ \ell_{22} = \ell_{33} = -1$	$M_P(3^{-1}) : \ell_{12} = \ell_{33} = 1;\ \ell_{11} = \ell_{21} = -1$
$M_P({}^{[110]}2) : \ell_{21} = \ell_{12} = 1;\ \ell_{33} = -1$	$M_P(6) : \ell_{11} = \ell_{21} = \ell_{33} = 1;\ \ell_{12} = -1$
$M_P({}^{[120]}2) : \ell_{12} = \ell_{22} = 1;\ \ell_{11} = \ell_{33} = -1$	$M_P(6^{-1}) : \ell_{12} = \ell_{22} = \ell_{33} = 1;\ \ell_{21} = -1$
$M_{P^*}(\alpha)$	
$M_{P^*}(1) : \ell_{11} = \ell_{22} = \ell_{33} = 1$	$M_{P^*}({}^{[010]}2) : \ell_{22} = 1;\ \ell_{11} = \ell_{12} = \ell_{33} = -1$
$M_{P^*}(2) : \ell_{11} = \ell_{22} = -1;\ \ell_{33} = 1$	$M_{P^*}({}^{[\bar{1}10]}2) : \ell_{21} = \ell_{12} = \ell_{33} = -1$
$M_{P^*}({}^{[100]}2) : \ell_{11} = 1;\ \ell_{21} = \ell_{22} = \ell_{33} = -1$	$M_{P^*}(3) : \ell_{21} = \ell_{33} = 1;\ \ell_{11} = \ell_{12} = -1$
$M_{P^*}({}^{[210]}2) : \ell_{11} = \ell_{12} = 1;\ \ell_{22} = \ell_{33} = -1$	$M_{P^*}(3^{-1}) : \ell_{12} = \ell_{33} = 1;\ \ell_{21} = \ell_{22} = -1$
$M_{P^*}({}^{[110]}2) : \ell_{21} = \ell_{12} = 1;\ \ell_{33} = -1$	$M_{P^*}(6) : \ell_{21} = \ell_{22} = \ell_{33} = 1;\ \ell_{12} = -1$
$M_{P^*}({}^{[120]}2) : \ell_{21} = \ell_{22} = 1;\ \ell_{11} = \ell_{33} = -1$	$M_{P^*}(6^{-1}) : \ell_{11} = \ell_{12} = \ell_{33} = 1;\ \ell_{21} = -1$

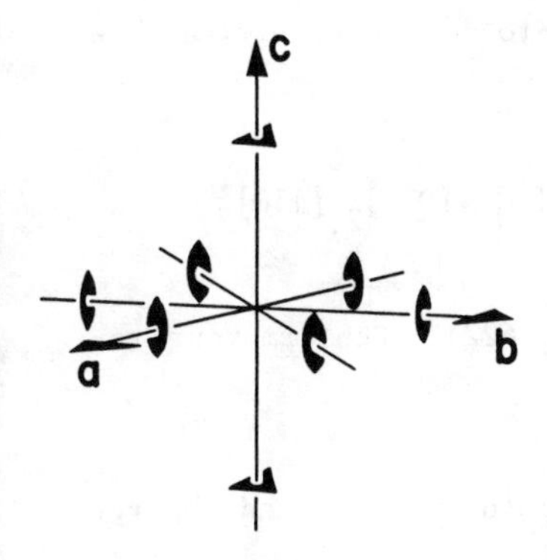

Figure 5.4: The orientation of the rotation axes in *322*. The 3-fold axis of the group is perpendicular to a plane of three 2-fold axes with adjacent 2-fold axes in the plane intersecting at an angle of 60°. The basis vector **c** is defined to lie along the 3-fold axis, **a** is defined to lie along one of the 2-fold axes and **b** is defined to lie along another at 120° to **a** so that **b** = 3(**a**). Thus, *3* parallels **c**, ${}^{[100]}2$ parallels **a**, ${}^{[110]}2$ parallels **a** + **b** and ${}^{[010]}2$ parallels **b**. The 3-fold axis is represented by a triad and each 2-fold by a diad symbol.

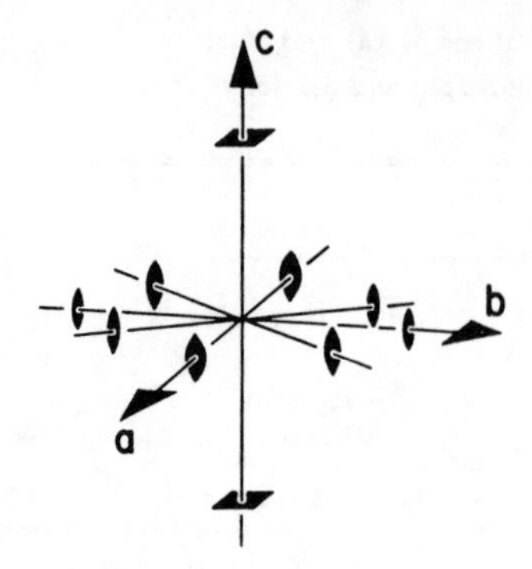

Figure 5.5 The orientation of the rotation axes in *422*. The 4-fold axis of the group is perpendicular to a plane of four 2-fold axes with adjacent 2-fold axes in the plane intersecting at 45°. The basis vector **a** is defined to lie along one of the 2-folds and **b** is defined to lie along another at 90° to **a** so that **b** = 4(**a**). Thus, *4* parallels **c**, $[100]_2$ and $[010]_2$ are disposed as in *222*, $[110]_2$ parallels **a** + **b** and $[\bar{1}10]_2$ parallels −**a** + **b**. The 4-fold axis is represented by a tetrad and each 2-fold by a diad symbol.

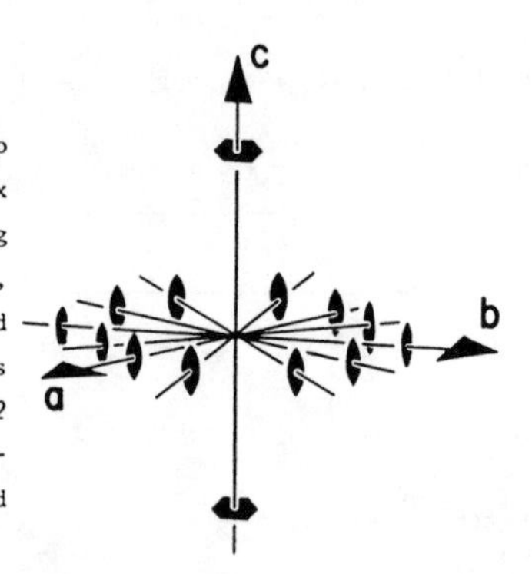

Figure 5.6: The orientation of the rotation axes of group 622. The 6-fold axis is perpendicular to a plane of six 2-fold axes with adjacent 2-folds in the plane intersecting at 30°. The vector **c** is defined to lie along the 6-fold axes, **a** is defined to lie along one of the 2-folds and **b** is defined as in *322* at 120° to **a** so that **b** = 3(**a**). Hence, *6* parallels **c**, $[100]_2$, $[110]_2$ and $[010]_2$ are disposed as in *322* and $[210]_2$ parallels 2**a** + **b**, $[120]_2$ parallels **a** + 2**b** and $[\bar{1}10]_2$ parallels −**a** + **b**. The 6-fold axis is represented by a hexad and the 2-folds by diad symbols.

$$422 = \{1, 4, 2, 4^{-1}, [100]_2, [110]_2, [010]_2, [\bar{1}10]_2\} \quad .$$

The multiplication table for $M_p(422)$ and (422) were found in P3.15. An examination of the multiplication table of *422* shows that *422* is closed under composition and hence is a group.

The construction of *622*: We choose the basis used for *322* for *622*. Since 30 = 180/6, there are two-fold axes at intervals of 30° starting with the one coinciding with **a**. Thus, there is a two-fold axis along **b**. The elements of *622* (see Figure 5.6) are

$$622 = \{1, 6, 3, 2, 3^{-1}, 6^{-1}, [100]_2, [210]_2, [110]_2, [120]_2, [010]_2, [\bar{1}10]_2\} \quad .$$

(P5.6) Problem: Find all of the matrices in $M_p(622)$. Confirm your results with those given in Table 5.3.

(P5.7) Problem: Prepare a multiplication table for $M_p(622)$ and observe that $M_p(622)$ is closed under multiplication, demonstrating that *622* is a group.

(P5.8) Problem: For *522* no basis exists that gives all of the pole points and matrices with integer entries. Therefore, a cartesian basis is used where **k** is along the five-fold axis and **i** is along one of the two-fold axes. Find the matrices in $M_C(522)$ and form the multiplication table showing that $M_C(522)$ is closed and hence *522* is a group.

Note that *522* is not a crystallographic group and hence does not map a lattice into self-coincidence. This is why no basis can be found such that the representation of the pole points and the matrices of the mappings consist entirely of integers. This fact is true for all non-crystallographic point groups. In particular, *n22* groups with $n > 6$ are all of this type. However, using a cartesian basis as in P5.8, each of these can be shown to be groups.

CONSTRUCTION OF THE CUBIC AXIAL GROUPS

In the construction of these groups we shall need to map pole points using the matrix representation of the constituent rotations. Since the pole points are constrained to be on a unit ball, the triple representing a given pole point can be somewhat complicated. For example, consider the pole point $[\sqrt{3}/3,\sqrt{3}/3,\sqrt{3}/3]^t$. As this pole point lies on the zone [111], we shall use $[111]^t$ to represent the pole point to simplify the computations. We shall see that each pole can be easily represented in this manner.

(E5.11) Example: Using the zone symbols to represent the pole points of *222*, form the equivalence classes of pole points with respect to *222*-equivalence.

Solution: Since *222* has three 2-fold rotation axes, it has six pole points. Recall that by the way the basis $P = \{\mathbf{a},\mathbf{b},\mathbf{c}\}$ for *222* was defined, the triples for these pole points on the unit ball B are

$$\{\mathbf{a}/a,-\mathbf{a}/a,\mathbf{b}/b,-\mathbf{b}/b,\mathbf{c}/c,-\mathbf{c}/c\} \quad .$$

The representatives formed from the zone symbol associated with these pole points are

$$\{[\mathbf{a}]_p,[-\mathbf{a}]_p,[\mathbf{b}]_p,[-\mathbf{b}]_p,[\mathbf{c}]_p,[-\mathbf{c}]_p]\} =$$

$$\{[100]^t,[\bar{1}00]^t,[010]^t,[0\bar{1}0]^t,[001]^t,[00\bar{1}]^t\} \quad .$$

The equivalence class of a given pole point $\mathbf{p}$ is the set $\{\mathbf{g}(\mathbf{p}) \mid \mathbf{g} \in 222\}$. Starting with $[\mathbf{a}]_p$, the equivalence class $[\mathbf{a}]$ of $\mathbf{a}$ is

$$[\mathbf{a}] = \{M_p(\mathbf{g})[\mathbf{a}]_p \mid \mathbf{g} \in 222\} = \{[\mathbf{a}]_p,[-\mathbf{a}]_p\} \quad .$$

Similarly,

$$[\mathbf{b}] = \{[\mathbf{b}]_p,[-\mathbf{b}]_p\}$$

and

$$[\mathbf{c}] = \{[\mathbf{c}]_p,[-\mathbf{c}]_p\} \quad .$$

This is consistent with the information given in Table 5.1 where we observe that the pole points of *222* are partitioned into 3 equivalence classes with two pole points each. □

(P5.9) Problem: Using the zone symbols to represent the pole points of *322*, form the equivalence classes of pole points with respect to *322*-equivalence. Confirm your answer by referring to Figure 5.1.

Construction of *332*(≡*23*): We shall discover what conditions must be satisfied if *332* is to be a group. Once these conditions have been established, we shall use them to determine the interaxial angles that must occur between the generating rotations and then construct *332* and show it is a group. We begin by noting that if there does not exist two third-turns whose composition is a half-turn, then the set of all third turns (recall that the inverse third-turn is a third-turn about the other end of the axis) together with the identity would be a closed set under composition and hence we would have a polyaxial group of the form *333* which would violate part (2) of T5.10. Hence, there exist two third-turns in *332* whose composition is a half-turn. In Appendix 6, we showed (PA6.2 and PA6.5) that when $\mathbf{33} = \mathbf{2}$, then

$$\angle(\mathbf{3}:\mathbf{3}) = \cos^{-1}(1/3) \simeq 70.53°, \text{ and}$$
$$\angle(\mathbf{3}:\mathbf{2}) = \cos^{-1}(\sqrt{3}/3) \simeq 54.74° \quad .$$

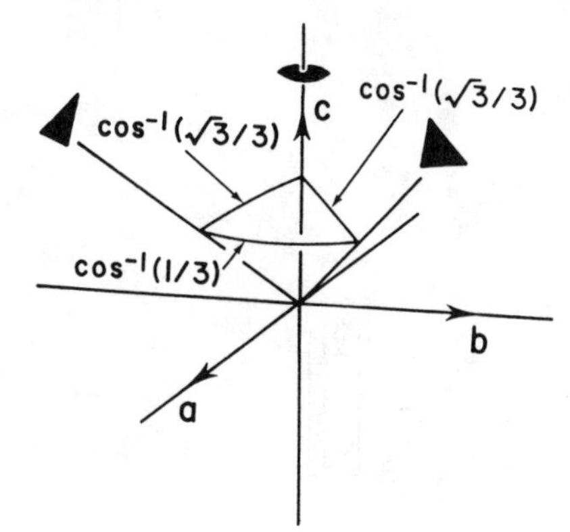

Figure 5.7: The placement of the generating rotations for *332≡23* given that $<(3:3) = \cos^{-1}(1/3) \simeq 70.53°$ and $<(3:2) = \cos^{-1}(\sqrt{3}/3) \simeq 54.74°$. The basis vectors chosen for this group are mutually perpendicular with $\alpha = \beta = \gamma = 90°$ and $a = b = c$. By convention, **c** is defined to lie along the 2-fold axis, **a** is oriented at angle of $\cos^{-1}(\sqrt{3}/3)$ with respect to each of the 3-fold axes and **b** is oriented at an angle of $\cos^{-1}(\sqrt{3}/3)$ with one of the 3-folds and an angle of $\cos^{-1}(-\sqrt{3}/3) \simeq 125.3°$ with respect to the other. For this choice of basis, the 2-fold along **c** is designated *2*, the 3-fold along **a** + **b** + **c** is designated $[111]_3$ and that along **a** − **b** + **c** is designated $[1\bar{1}1]_3$.

A convenient basis P = {**a**,**b**,**c**} for describing the orientation of these axes is the basis whose metrical matrix G is

$$G = \begin{bmatrix} g_{11} & 0 & 0 \\ 0 & g_{11} & 0 \\ 0 & 0 & g_{11} \end{bmatrix} .$$

where we orient the two-fold axis along the zone [001], and the three-fold axes along the zones [111] and $[1\bar{1}1]$. The fact that this can be done is shown in the next problem. This placement of axes is illustrated in Figure 5.7.

(P5.10) Problem: Show that

$$<([111]:[001]) = <([1\bar{1}1]:[001]) = \cos^{-1}(\sqrt{3}/3) \simeq 54.74° \quad ,$$

and that

$$<([111]:[1\bar{1}1]) = \cos^{-1}(1/3) \simeq 70.53°.$$

To find the remaining axes belonging to the rotations of *332*, we shall search for the remaining pole points. Each time we obtain a new pole point, the stabilizer of that pole point is determined according to T5.5 and we add these to our list of elements of *332*. By Table 5.1, we know that there is a total of 14 pole points, 8 of which are associated with 3-fold axes and 6 with 2-fold axes.

To begin with we have the following list representing pole points associated with 3-fold axes in Figure 5.7: $\{[111],[1\bar{1}1]\}$. The matrix representation of the half-turn about [001] is

$$M_p(2) = \begin{bmatrix} -1 & 0 & 0 \\ 0 & -1 & 0 \\ 0 & 0 & 1 \end{bmatrix} .$$

Hence

$$M_p(2) \begin{bmatrix} 1 \\ 1 \\ 1 \end{bmatrix} = \begin{bmatrix} -1 \\ -1 \\ 1 \end{bmatrix} \text{ and } M_p(2) \begin{bmatrix} 1 \\ -1 \\ 1 \end{bmatrix} = \begin{bmatrix} -1 \\ 1 \\ 1 \end{bmatrix}$$

also represent pole points. By T5.5, these pole points are associated with 3-fold axes. It is helpful to note that if $[uvw]$ represents a pole point associated with an n-fold axis, then $[\bar{u}\bar{v}\bar{w}]$ also represents a pole point associated with the n-fold axis. Hence we have found all of the 8 pole points. Their zone representatives are

$$\{[111],[\bar{1}11],[\bar{1}\bar{1}1],[1\bar{1}1],[\bar{1}\bar{1}\bar{1}],[1\bar{1}\bar{1}],[11\bar{1}],[\bar{1}1\bar{1}]\} .$$

While [111] and $[\bar{1}\bar{1}\bar{1}]$ both designate the same axis about which a 3-fold takes place, we denote the monaxial group along this direction as $[111]_3$. In general, when possible, we shall choose the zone symbol with a positive third component. Hence, we have found the following 3-fold axes:

$$[111]_3 = \{1, {}^{[111]}3, {}^{[111]}3^{-1}\}$$

$$[\bar{1}11]_3 = \{1, {}^{[\bar{1}11]}3, {}^{[\bar{1}11]}3^{-1}\}$$

$$[\bar{1}\bar{1}1]_3 = \{1, {}^{[\bar{1}\bar{1}1]}3, {}^{[\bar{1}\bar{1}1]}3^{-1}\}$$

$$[1\bar{1}1]_3 = \{1, {}^{[1\bar{1}1]}3, {}^{[1\bar{1}1]}3^{-1}\} .$$

To find the 6 pole points associated with the half-turns, we write the matrix for ${}^{[111]}3$ and apply it to the pole point represented by [001].

$$M_p({}^{[111]}3) \begin{bmatrix} 0 \\ 0 \\ 1 \end{bmatrix} = \begin{bmatrix} 0 & 0 & 1 \\ 1 & 0 & 0 \\ 0 & 1 & 0 \end{bmatrix} \begin{bmatrix} 0 \\ 0 \\ 1 \end{bmatrix} = \begin{bmatrix} 1 \\ 0 \\ 0 \end{bmatrix} .$$

Applying $M_p({}^{[111]}3)$ to this new pole point we obtain [010]. Including

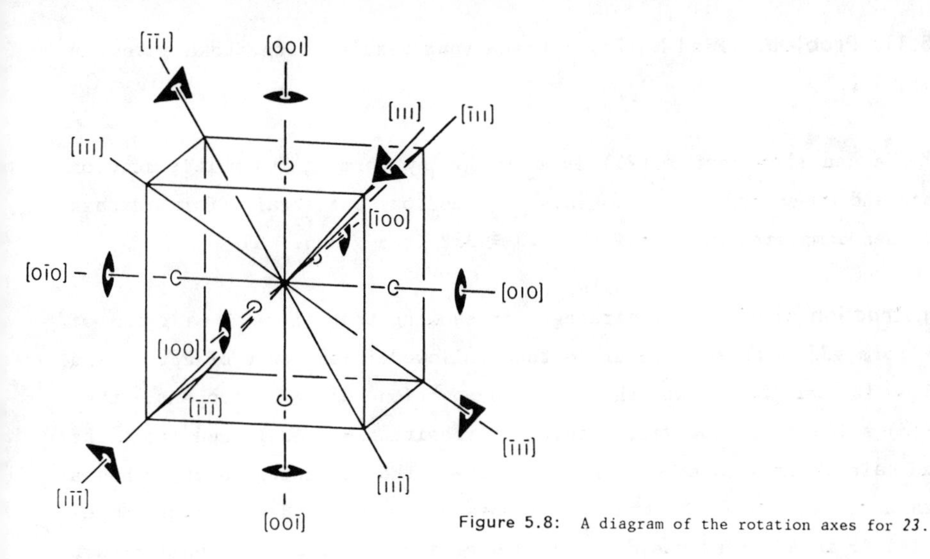

Figure 5.8: A diagram of the rotation axes for *23*.

the negatives of these, we have the following 6 pole points associated with the two-fold axes:

$$\{[100],[010],[001],[\bar{1}00],[0\bar{1}0],[00\bar{1}]\} \quad .$$

Hence we have the following two fold axes in *332*,

$$[100]_2 = \{1,[100]_2\}$$

$$[010]_2 = \{1,[010]_2\}$$

$$2 = \{1,2\} \quad .$$

As predicted in Table 5.1, we have found a total 12 rotations, 3 half-turns, 8 third-turns or negative third-turns and an identity. Hence

$$23 \equiv 332 = \{1,2,[100]_2,[010]_2,[111]_3,[111]_3{-1} \quad ,$$
$$[\bar{1}11]_3,[\bar{1}11]_3{-1},[\bar{1}\bar{1}1]_3,[\bar{1}\bar{1}1]_3{-1},[1\bar{1}1]_3,[1\bar{1}1]_3{-1}\} \quad .$$

Figure 5.8 shows the placement of the rotation axes of *23* in terms of the basis vector *P*.

(P5.11) Problem: Find $M_p(23)$. Check your results with those given in Table 5.2.

We can show that $M_p(23)$ is a group, by forming its multiplication table and observing closure. This is a tedious but straightforward task and when completed will show that *23* ≡ *332* is a group.

Construction of *432*: The strategy for showing that there is a group of the form *432* will be similar to that followed in the case of *23*. Using TA6.4 it can be shown that if composition **43** = **3** happens, then $<(\mathbf{4}:\mathbf{3}) \simeq 103.84°$. However, when the composition $\mathbf{4}^2\mathbf{3}$ is formed about the same pair of axes intersecting at ~103.84°, the resulting rotation has a turn angle of 156.094° which is impossible in *432*. Hence, each composition **43** in *432* must yield either a **4** or a **2**. Fixing one quarter-turn **4**, since *432* has 8 third-turns, we obtain 8 rotations of the form **43** using the fixed quarter-turn. By the cancellation law these must all be distinct rotations. Since there are only 6 quarter-turns, some of the **43** compositions must be half-turns. Let **4** and **3** be such that the composition **43** = **2**. From Appendix 6 we see that

$$<(\mathbf{4}:\mathbf{3}) = \cos^{-1}(\sqrt{3}/3) \simeq 54.74° \ ,$$
$$<(\mathbf{4}:\mathbf{2}) = \cos^{-1}(\sqrt{2}/2) = 45° \ , \quad \text{and}$$
$$<(\mathbf{3}:\mathbf{2}) = \cos^{-1}(\sqrt{6}/3) \simeq 35.26° \ .$$

There cannot be a quarter-turn along the 2-fold axis since two quarter-turns whose axes intersect at a 45° compose to yield a rotation of 163.158° which does not exist in *432*. Using the same basis as in *23*, we place the 4-fold axis along [001], the 3-fold along [111] and the 2-fold along [101].

(P5.12) Problem: Show that

$$<([001]:[101]) = \cos^{-1}(\sqrt{2}/2) = 45° \text{ and}$$
$$<([101]:[111]) = \cos^{-1}(\sqrt{6}/3) \simeq 35.26° \ .$$

(P5.13) Problem: Find the 8 pole points belonging to the 3-fold axes. (The answer is the same as found for *23* except they all occur in the same *432*-equivalence class).

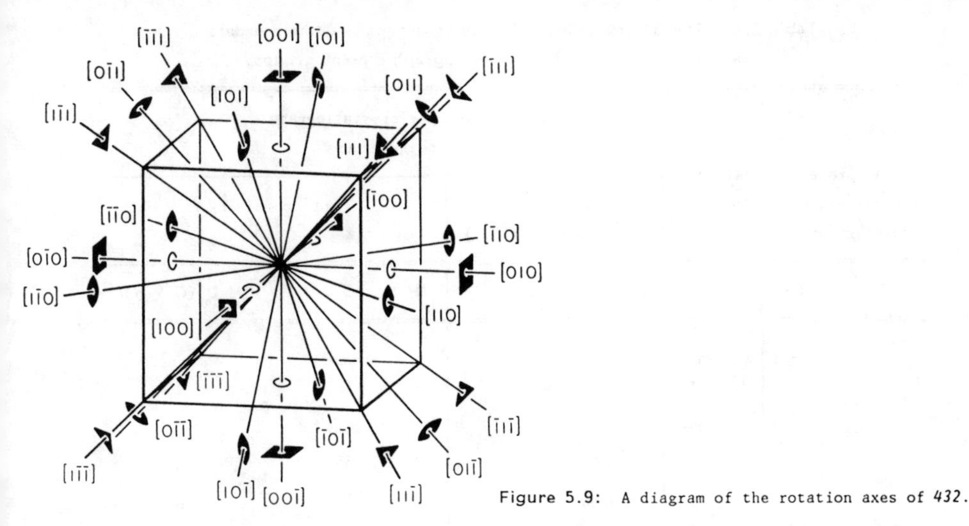

Figure 5.9: A diagram of the rotation axes of *432*.

(P5.14) Problem: Show that the zone representations of the pole points associated with the 2-fold axes are

$$\{[101],[011],[\bar{1}01],[0\bar{1}1],[110],[\bar{1}10],[\bar{1}0\bar{1}],[0\bar{1}\bar{1}],[10\bar{1}],[01\bar{1}],[\bar{1}\bar{1}0],[1\bar{1}0]\}$$

(P5.15) Problem: Show that the zone representations of the pole points associated with the 4-fold axes are

$$\{[001],[100],[010],[00\bar{1}],[\bar{1}00],[0\bar{1}0]\} \quad .$$

(P5.16) Problem: Enumerate the rotations in each of the 3-fold axes $[111]_3$, $[\bar{1}11]_3$, $[\bar{1}\bar{1}1]_3$, $[1\bar{1}1]_3$ in *432*.

(P5.17) Problem: Enumerate the rotations in each of the 4-fold axes *4*, $[100]_4$ and $[010]_4$ in *432*.

(P5.18) Problem: Enumerate the rotations in each of the 2-fold axes, $[101]_2$, $[011]_2$, $[\bar{1}01]_2$, $[0\bar{1}1]_2$, $[110]_2$, $[\bar{1}10]_2$ in *432*.

(P5.19) Problem: Enumerate the rotations in *432* by collecting together the distinct rotations found in P5.16, P5.17, and P5.18. As in the case of *23*, one need only form a multiplication table for $M_D(432)$ and observe closure to conclude that *432* is a group. Figure 5.9 shows the rotation axes of *432* defined in terms of its basis vectors.

Table 5.4: The 32 crystallographic point groups and their orders as derived from the proper crystallographic point groups.

The 11 proper crystallographic point groups		Halving Groups	The 21 improper crystallographic point groups			
			containing i (centrosymmetrical)		Not containing i	
G	$\#(G)$	H	$G \cup Gi$	$\#(G \cup Gi)$	$(H \cup (G \setminus H)i)$	$\#(H \cup (G \setminus H)i)$
1	1	none	$\bar{1}$	2	none	—
2	2	*1*	*2/m*	4	*m*	2
3	3	none	$\bar{3}$	6	none	—
4	4	*2*	*4/m*	8	$\bar{4}$	4
6	6	*3*	*6/m*	12	$\bar{6}$	6
222	4	*2*	*mmm*	8	*mm2*	4
322	6	*3*	$\bar{3}2/m$	12	*3mm*	6
422	8	*4*	*4/mmm*	16	*4mm*	8
		222			$\bar{4}2m$	8
622	12	*6*	*6/mmm*	24	*6mm*	12
		322			$\bar{6}2m$	12
322 = 23	12	none	$2/m\bar{3}$	24	none	—
432	24	*23*	$4/m\bar{3}2/m$	48	$\bar{4}3m$	24

The noncrystallographic group *532* ≡ *235* will be discussed at the end of the chapter.

CONSTRUCTION OF THE IMPROPER CRYSTALLOGRAPHIC POINT GROUPS

The construction of these groups will be accomplished by applying the Improper Point Group Generating Theorem T4.28. In column one of Table 5.4, each of the proper crystallographic point groups are listed. Part (1) of T4.28 yields one centrosymmetric point group $G \cup Gi$, listed in column 4, from each proper point group G from column 1. The application of part (2) of T4.28 requires a list of the halving groups of each proper crystallographic point group. Since any subgroup of a crystallographic point group is again a crystallographic point group, these halving groups can be found in column one. For example, *322* has 6 elements. Examining column two, we see that the group *3* has 3 elements and hence is a candidate to be a halving group. Since *3* is a subset of *322*, it is a halving group and hence is listed opposite *322* in column 3. Since *23* has 12 elements, the candidates for the halving groups of *23* are *322* and *6*, both of order 6. However, neither of these are subsets of *23* and hence *23* has no halving groups. A similar analysis of the remaining groups from column one gives the results shown in column 3. Applying part (2) of T4.28 to each *G* with halving groups, we obtain the results shown in column 6.

The name given to each of the improper crystallographic point groups, is derived from the names of the resulting monaxial groups that occur along the axes of the generators of the possible proper crystallographic groups.

(E5.12) Example: Consider $G = 322$. Since $G \cup Gi = 322 \cup 322i$

$$\{1,3,3^{-1},{}^{[100]}2,{}^{[110]}2,{}^{[010]}2\} \cup \{i,\bar{3},\bar{3}^{-1},m,{}^{[110]}m,{}^{[010]}m\} \quad .$$

The generators of *322* are 3, ${}^{[100]}2$, ${}^{[110]}2$. In $322 \cup 322i$, the symmetry elements lie along the three zones:

$$[001] : (\{1,\bar{3},3^{-1},i,3,\bar{3}^{-1}\} \text{ which is the monaxial group } \bar{3}) \quad ,$$

$$[100] : (\{1,i,{}^{[100]}2,{}^{[100]}m\} \text{ which is the monaxial group } 2/m) \quad ,$$

$$[110] : (\{1,i,{}^{[110]}2,{}^{[110]}m\} \text{ which is the monaxial group } 2/m) \quad .$$

Hence the full symbol for $322 \cup 322i$ is $\bar{3}(2/m)(2/m)$. The Hermann-Mauguin symbol for this group is $\bar{3}m$. This symbol is reasonable since the elements of $\bar{3}$ and any of the mirrors generate the remaining elements of the group.

As indicated in Column 3, $H = 3$ is a halving group of $G = 322$. Hence, we construct

$$H \cup (G \setminus H)i = 3 \cup (322 \setminus 3)i$$

$$= \{1,3,3^{-1}\} \cup \{{}^{[100]}2,{}^{[110]}2,{}^{[010]}2\}i$$

$$= \{1,3,3^{-1},{}^{[100]}m,{}^{[110]}m,{}^{[010]}m\} \quad .$$

Along [001], we have *3*, perpendicular to [100] and [110] we have planes associated with mirror operations *m*. Hence, the full symbol is *3mm*. The Hermann-Mauguin symbol is *3m*. Note that **3** and either **m** generates the group. □

(P5.20) Problem: Find *422* U *422i* and show that its full Hermann-Mauguin symbol is *(4/m)(2/m)(2/m)*.

(E5.13) Example: Consider *G* = *422*. The halving groups of *G* are *4* and *222*. Find *222* U (*422* \ *222*)*i* and derive its full Hermann-Mauguin symbol.

Solution: We begin by generating the elements of the set

$$222 \cup (422 \setminus 222)i = \{1,2,{}^{[100]}2,{}^{[010]}2\} \cup \{4,4^{-1},{}^{[110]}2,{}^{[\bar{1}10]}2\}i$$

$$= \{1,2,{}^{[100]}2,{}^{[010]}2,\bar{4},\bar{4}^{-1},{}^{[110]}m,{}^{[\bar{1}10]}m\} \quad .$$

As a 4-rotoinversion axis, $\{1,\bar{4},2,\bar{4}^{-1}\}$, parallels [001], a 2-fold axis, $\{1,{}^{[100]}2\}$, parallels [100] and a mirror plane is perpendicular to [110], the Hermann-Mauguin symbol of this group is $\bar{4}2m$. □

(P5.21) Problem: Construct and determine the full Hermann-Mauguin symbol for *4* U (*422*) \ *4*)*i*.

(P5.22) Problem: Given that *G* = *622*, show that

$$622 \cup 622i = 6/mmm$$

$$= \{1,6,3,2,3^{-1},6^{-1},{}^{[100]}2,{}^{[210]}2,{}^{[110]}2,{}^{[120]}2,{}^{[010]}2,{}^{[\bar{1}10]}2,$$

$$i,\bar{6},\bar{3},m,\bar{3}^{-1},\bar{6}^{-1},{}^{[100]}m,{}^{[210]}m,{}^{[110]}m,{}^{[120]}m,{}^{[010]}m,{}^{[\bar{1}10]}m\} \quad .$$

(P5.23) Problem: Given that *G* = *432*, show that

$$432 \cup 432i = (4/m)\bar{3}(2/m)$$

$$= m\bar{3}m = \{1,4,2,4^{-1},{}^{[100]}4,{}^{[100]}2,{}^{[100]}4^{-1},{}^{[010]}4,{}^{[010]}2,$$

$[010]_4{}^{-1}, [111]_3, [111]_3{}^{-1}, [\bar{1}11]_3, [\bar{1}11]_3{}^{-1}, [\bar{1}\bar{1}1]_3, [\bar{1}\bar{1}1]_3{}^{-1},$

$[1\bar{1}1]_3, [1\bar{1}1]_3{}^{-1}, [101]_2, [011]_2, [\bar{1}01]_2, [0\bar{1}1]_2, [110]_2, [\bar{1}10]_2,$

$i, \bar{4}, m, \bar{4}^{-1}, [100]_{\bar{4}}, [100]_m, [100]_{\bar{4}}{}^{-1}, [010]_{\bar{4}}, [010]_m, [010]_{\bar{4}}{}^{-1},$

$[111]_{\bar{3}}, [111]_{\bar{3}}{}^{-1}, [\bar{1}11]_3, [\bar{1}11]_3{}^{-1}, [\bar{1}\bar{1}1]_{\bar{3}}, [\bar{1}\bar{1}1]_{\bar{3}}{}^{-1}, [1\bar{1}1]_{\bar{3}},$

$[1\bar{1}1]_{\bar{3}}{}^{-1}, [101]_m, [011]_m, [\bar{1}01]_m, [0\bar{1}1]_m, [110]_m, [\bar{1}10]_m\}$.

Using the procedures followed in the preceeding examples and problems, one can construct all of the improper point groups shown in Table 5.4. According to T4.28, these are all of the possible improper point groups. Hence, in summary, we have found 11 proper crystallographic point groups, 11 centrosymmetric crystallographic point groups and 10 non-centrosymmetric improper crystallographic point groups for a total of 32 possible crystallographic point groups.

(P5.24) Problem: Determine the point symmetry G of the tricyclosiloxane molecule displayed in Figure 1.2. Find the elements of the set $M_C(G)$ and show that the coordinates of the atoms in Table 1.2 are permuted (interchanged) by the elements of $M_C(G)$. Ascertain the atoms in the molecule that are G-equivalent to O_1, H_1, and Si_1. Determine which of the atoms are on special positions.

(P5.25) Problem: Stereoscopic pair drawings of G-equivalent ellipsoids are presented in Figure 5.10 for each of the 19 polyaxial point groups G. Study each of these drawings and confirm the point symmetry of each.

THE CRYSTAL SYSTEMS

It is customary to organize the 32 crystallographic point groups into classes according to geometrical considerations. The geometry involved in each point group G is used to determine a natural basis with respect to which the matrix representation of the elements of G are simply written (so that all of the entries of the matrices are either 0, 1 or −1). *In toto*, we use only 6 different bases. In Table 5.5 the metrical matrix for each of these bases is listed together with a list of all of the point groups using the basis. This gives rise to the 6 crystal systems whose names are given in column 3. The metrical matrix given in Table 5.5 for the monoclinic system is that for the so called first setting where **c** is chosen along the axis of order 2. In the second setting, which is more commonly used, **b** is chosen along the axis of order 2, resulting in the metrical matrix

Figure 5.10: Stereoscopic pair drawings of G-equivalent ellipsoids for each of the 19 polyaxial point groups G.

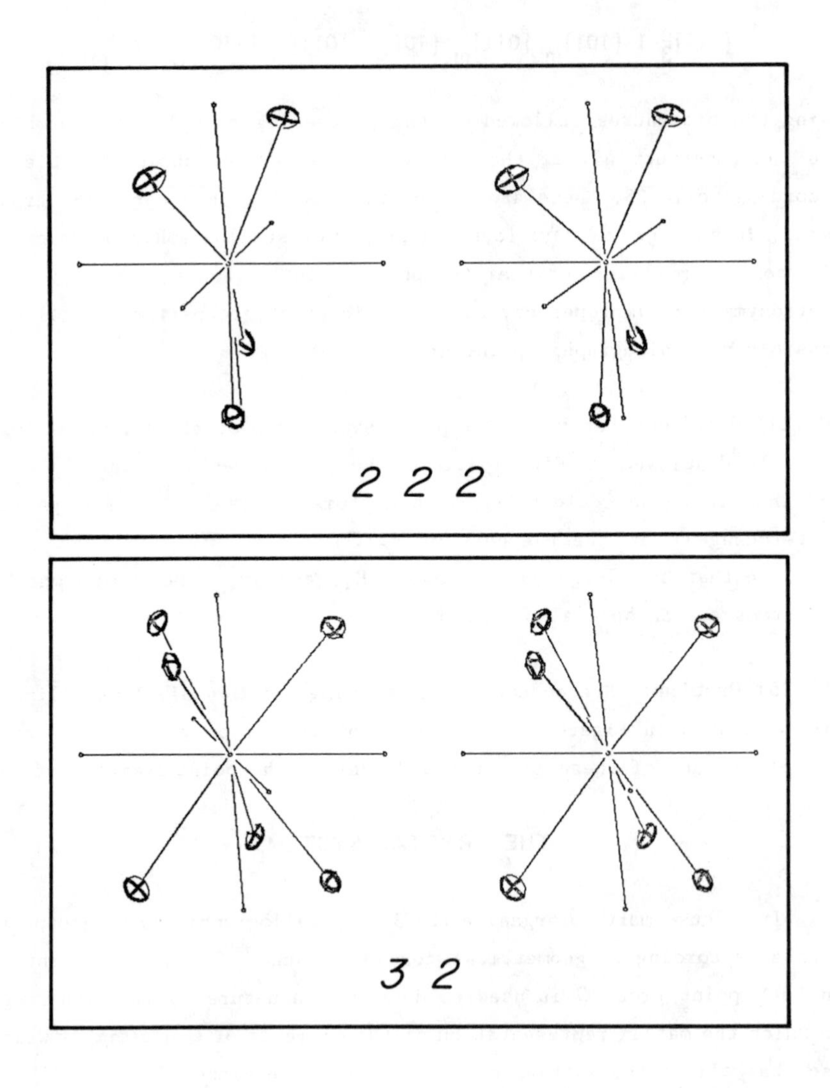

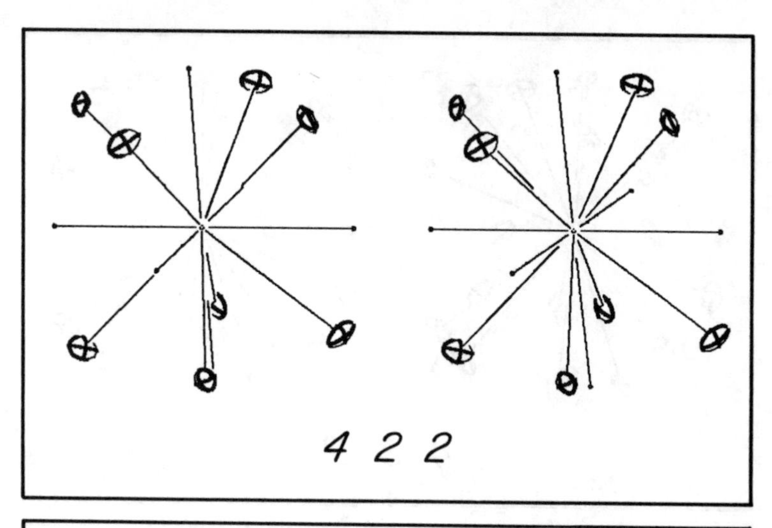
4 2 2

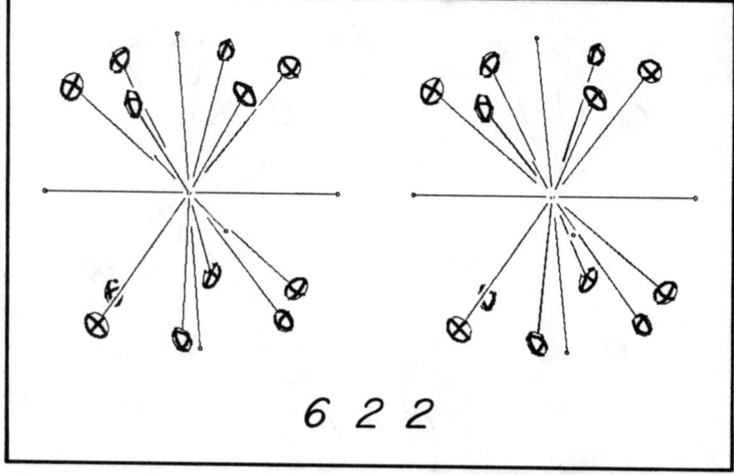
6 2 2

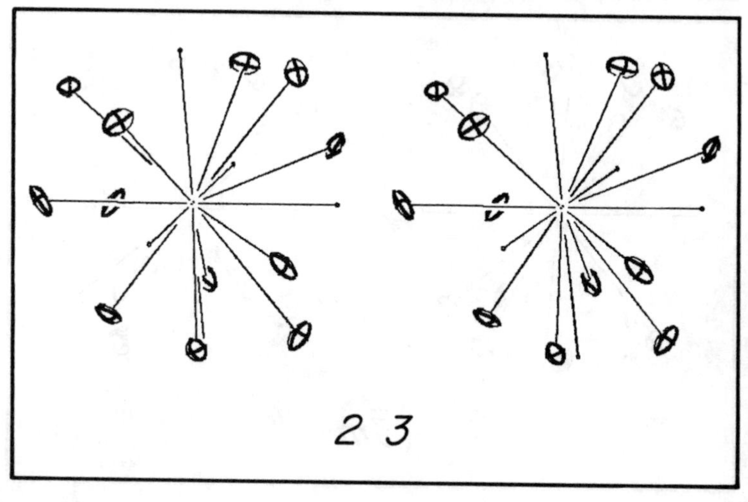
2 3

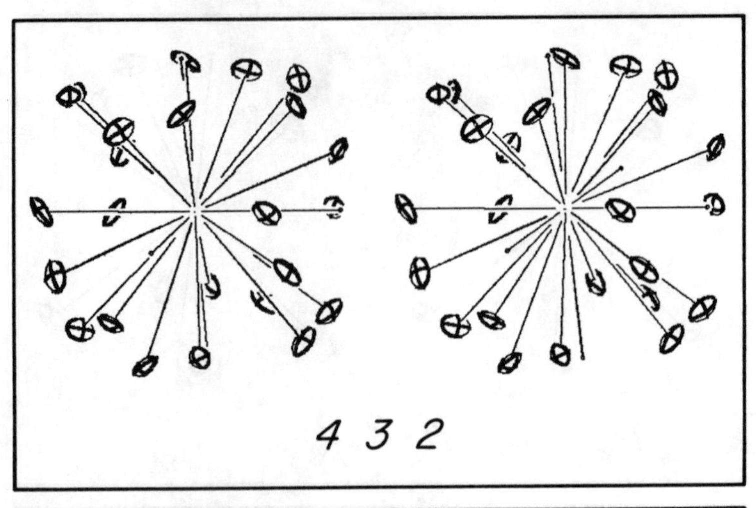
4 3 2

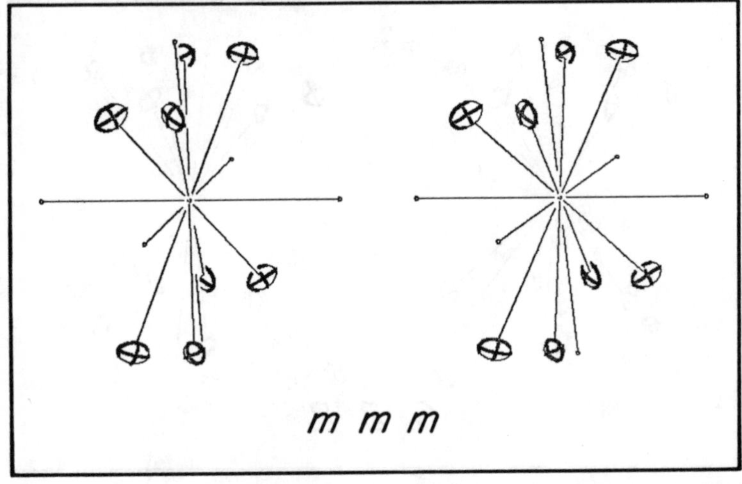
m m m

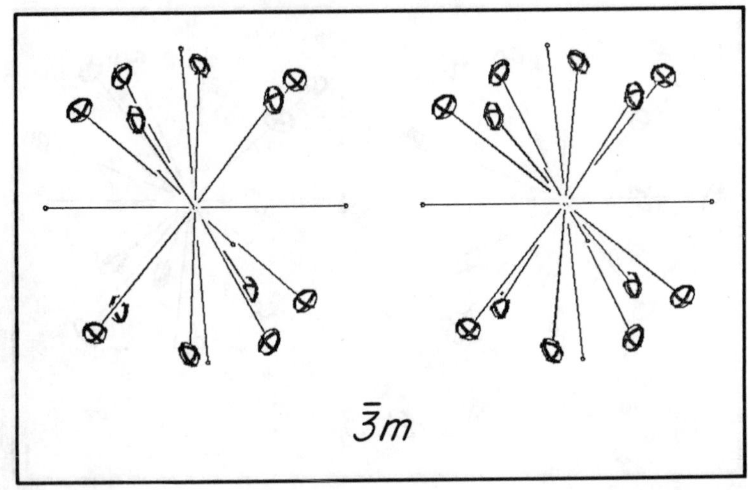
$\bar{3}m$

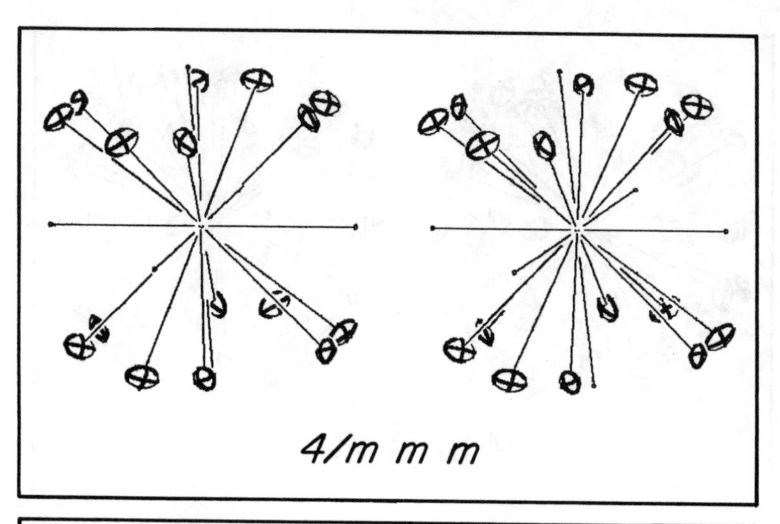
4/m m m

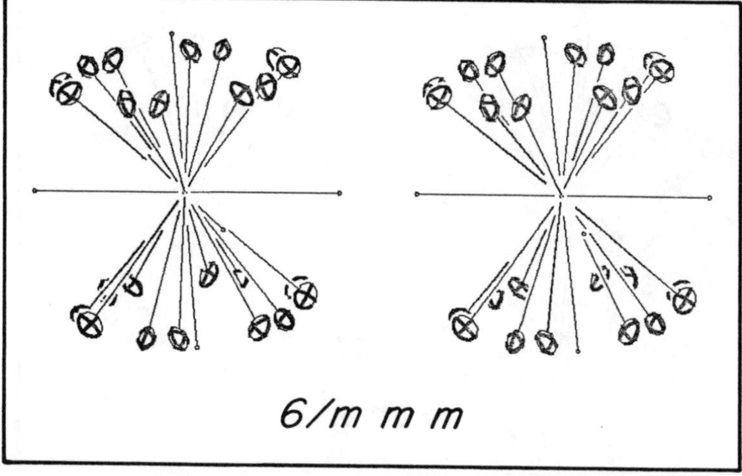
6/m m m

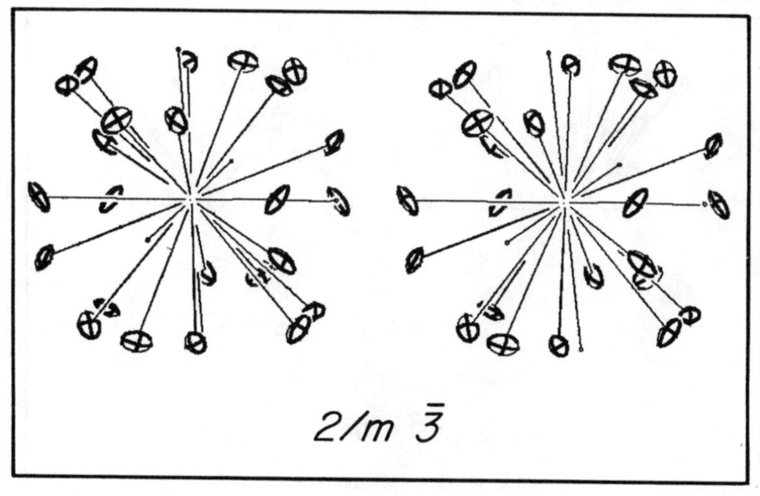
2/m $\bar{3}$

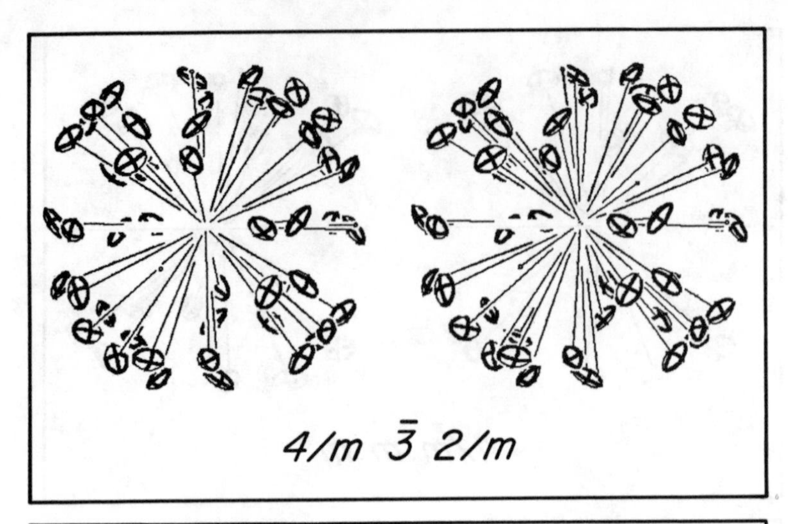
4/m $\bar{3}$ 2/m

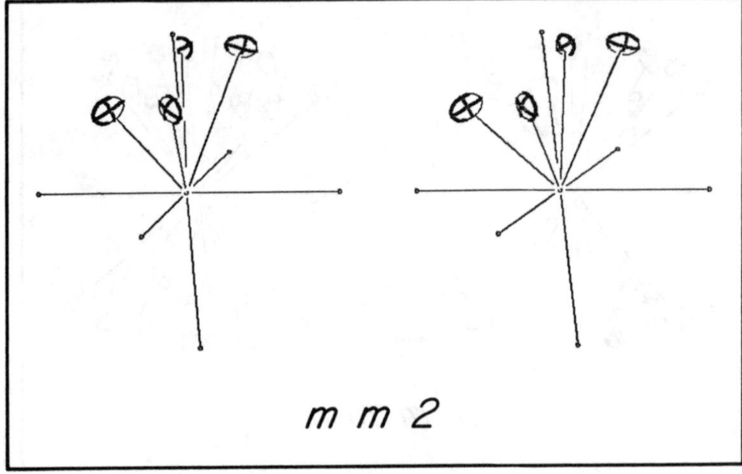
m m 2

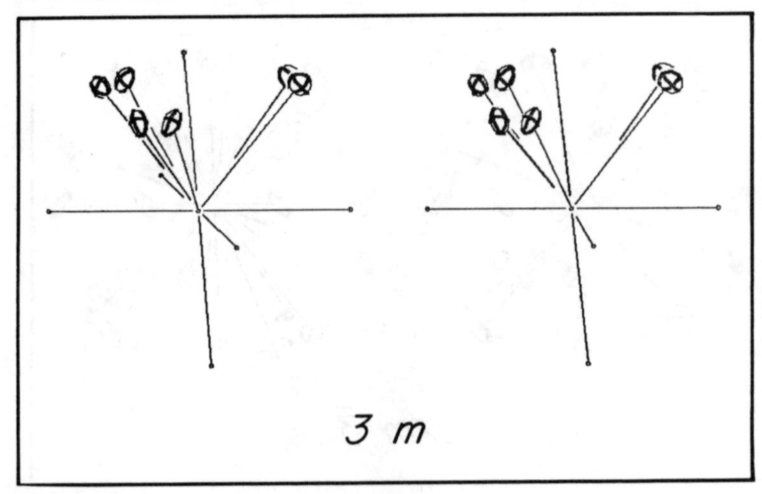
3 m

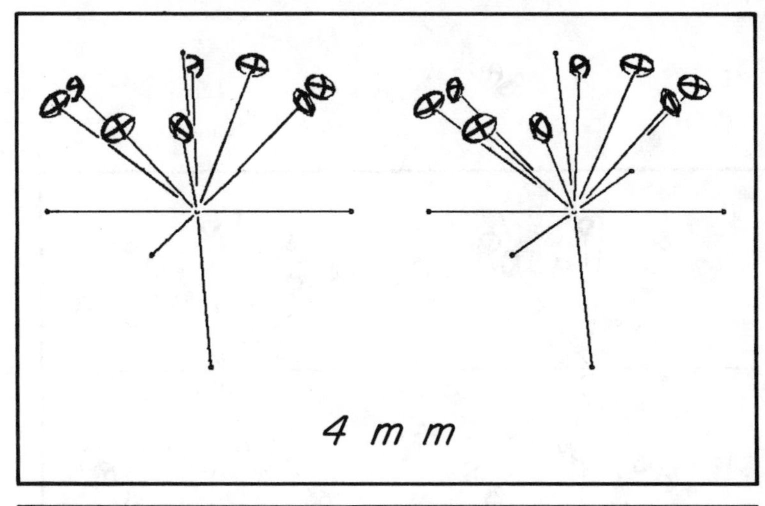
4 m m

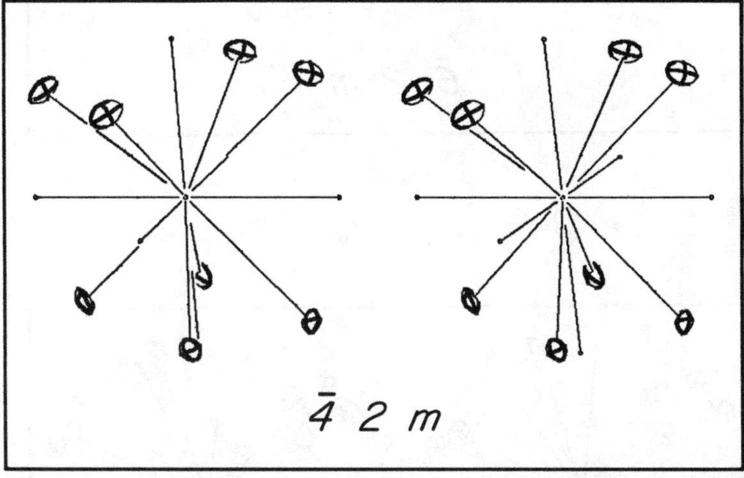
$\bar{4}$ 2 m

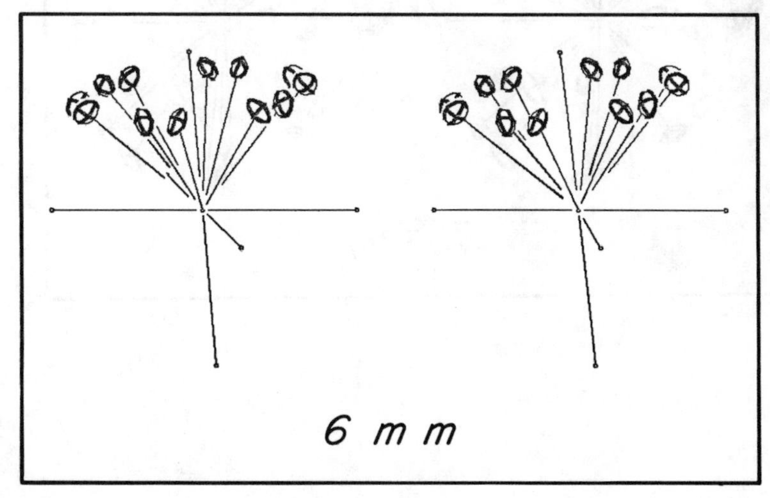
6 m m

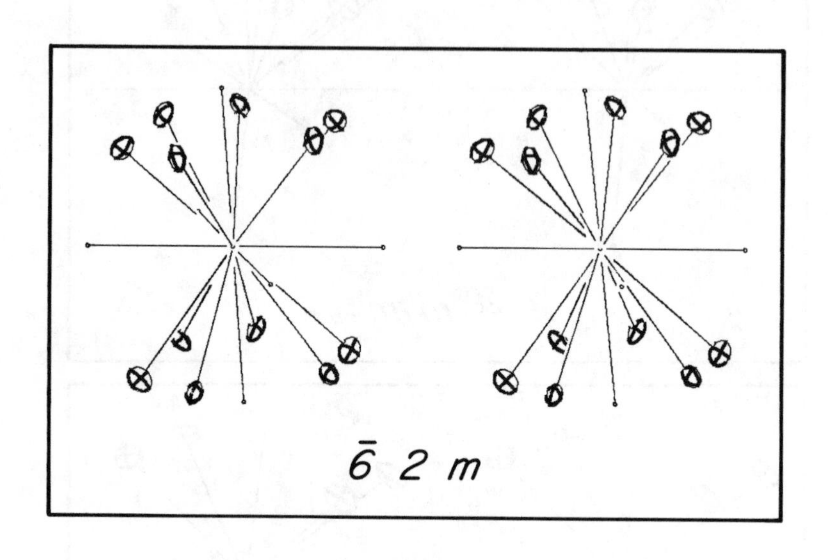

$\bar{6}\ 2\ m$

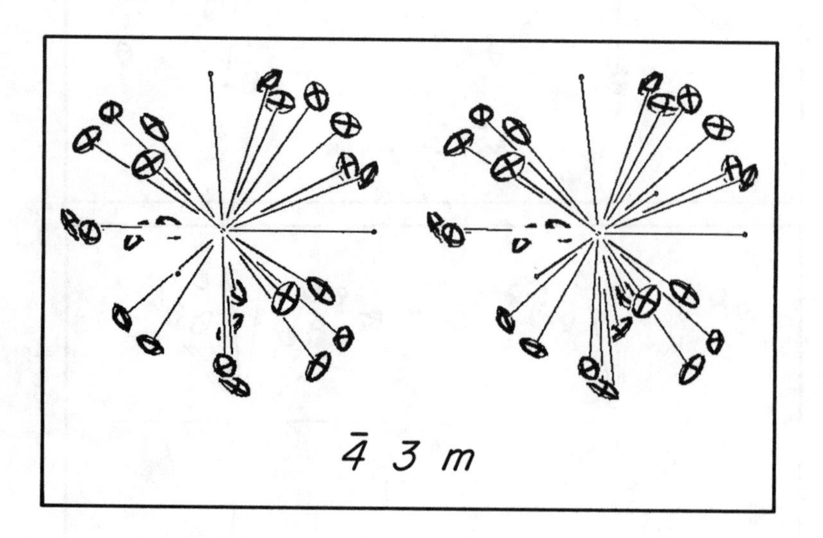

$\bar{4}\ 3\ m$

Table 5.5: Metrical Matrices for the crystallographic point groups.

METRICAL MATRICES	*POINT GROUPS*	*CRYSTAL SYSTEM*
$\begin{bmatrix} g_{11} & g_{12} & g_{13} \\ g_{12} & g_{22} & g_{13} \\ g_{13} & g_{23} & g_{33} \end{bmatrix}$	*1,1*	*Triclinic*
$\begin{bmatrix} g_{11} & g_{12} & 0 \\ g_{12} & g_{22} & 0 \\ 0 & 0 & g_{33} \end{bmatrix}$	*2,m,2/m*	*Monoclinic*
$\begin{bmatrix} g_{11} & 0 & 0 \\ 0 & g_{22} & 0 \\ 0 & 0 & g_{33} \end{bmatrix}$	*222,mm2,mmm*	*Orthorhombic*
$\begin{bmatrix} g_{11} & 0 & 0 \\ 0 & g_{11} & 0 \\ 0 & 0 & g_{33} \end{bmatrix}$	*4,4,4/m* *422,4mm,42m* *4/mmm*	*Tetragonal*
$\begin{bmatrix} g_{11} & -g_{11}/2 & 0 \\ -g_{11}/2 & g_{11} & 0 \\ 0 & 0 & g_{22} \end{bmatrix}$	*3,3,32,3m,3m* *6,6,6/m,622,6mm* *62m,6/mmm*	*Hexagonal*
$\begin{bmatrix} g_{11} & 0 & 0 \\ 0 & g_{11} & 0 \\ 0 & 0 & g_{11} \end{bmatrix}$	*23,m3,432* *43m,m3m*	*Cubic*

$$\begin{bmatrix} g_{11} & 0 & g_{13} \\ 0 & g_{22} & 0 \\ g_{13} & 0 & g_{33} \end{bmatrix} .$$

Schoenflies Symbols: Besides the Hermann-Mauguin symbolism for the point groups used here, there is another important symbolism, called the *Schoenflies symbolism*, which is widely used by chemists. The equivalence between the Schoenflies symbols and the Hermann-Mauguin symbols are given in Table 5.6. The proper cyclic groups $1, 2, \ldots, n$ are denoted $C_1, C_2, \ldots, C_n$ in the Schoenflies symbolism whereas the improper cyclic groups $\bar{1}$, $\bar{3}$, m, $\bar{4}$, $\bar{6}$ are denoted C_i, C_{3i}, C_s, S_4, C_{3h}, respectively. The subscript i signifies that the group contains the inversion isometry and h signifies that it contains a horizontal mirror plane perpendicular to a rotation axis. The subscript s stands for the German word *Spiegelung* for reflection and signifies that $C_s \equiv m$. The improper centrosymmetric monaxial groups $2/m, 3/m, \ldots, n/m$ are symbolized as $C_{2h}, C_{3h}, \ldots, C_{nh}$.

Table 5.6: Schoenflies (S) and Hermann-Mauguin (HM) point group symbols.

HM	S	HM	S	HM	S
1	C_1	$\bar{1}$	C_i		
2	C_2	$2/m$	C_{2h}	m	C_s
3	C_3	$\bar{3}$	C_{3i}		
4	C_4	$4/m$	C_{4h}	$\bar{4}$	S_4
6	C_6	$6/m$	C_{6h}	$\bar{6}$	C_{3h}
222	D_2	mmm	D_{2h}	$mm2$	C_{2v}
32	D_3	$\bar{3}m$	D_{3d}	$3m$	C_{3v}
422	D_4	$4/mmm$	D_{4h}	$\bar{4}2m$	D_{2d}
				$4mm$	C_{4v}
622	D_6	$6/mmm$	D_{6h}	$\bar{6}2m$	D_{3h}
23	T	$m\bar{3}$	T_h	$6mm$	C_{6v}
432	O	$m\bar{3}m$	O_h	$\bar{4}3m$	T_d
235	I	$m\bar{3}\bar{5}$	I_h		

The improper noncentrosymmetric monaxial groups $2mm(\equiv mm2)$, $3mm,\ldots,nmm$ with vertical mirror planes are symbolized C_{2v}, $C_{3v},\ldots,C_{nv}$. The proper dihedral groups 222, $322(\equiv 32),\ldots,n22$ are denoted $D_2,D_3,\ldots,D_n$ whereas the improper dihedral groups mmm, $\bar{3}m$, $\bar{4}2m$, $4/mmm$, $6/mmm$ are denoted D_{2h}, D_{3d}, D_{2d}, D_{4h}, D_{6h}. The subscript d denotes the presence of diagonal reflection planes bisecting adjacent 2-fold axes. The tetrahedral cubic groups 23, $m\bar{3}$, $\bar{4}3m$ are denoted T, T_h, T_d and the octahedral cubic groups 432 and $m\bar{3}m$ are denoted O and O_h, respectively. Finally, the icosahedral groups 235 and $m\bar{3}\bar{5}$ are symbolized as I and I_h, respectively.

The icosahedral point groups: These point groups occur when $\nu_1 = 5$, $\nu_2 = 3$ and $\nu_3 = 2$. Regular icoshedral arrangements are not found as crystals because they require a five-fold axis of symmetry, and bęcause a set of regular icosahedra cannot be packed together so as to fill space. However, the combination of an icosahedral unit as the As_{12} unit in skutterudite, Co_4As_{12}, with interstitial Co atoms can fill space to form a cubic crystal with $2/m\bar{3}$ point symmetry. Also, an icosahedral unit dominates in the structure types (allotropes) of crystalline boron where B_{12} icosahedra are packed together inefficiently leaving regularly spaced voids. Although the B_{12} groups in boron can be viewed as being close-packed, only 37% of space is occupied. On the other hand, evidence for icosahedral geometry and cation-anion mixed layer packings has been used to explain the "super" dense packing in the glaserite, $K_3Na(SO_4)_2$ structure types (Moore, 1976). The icosahedron also occurs in the world of viruses where virus particles pack together with icosahedral point symmetry (PA6.4). The recent discovery by Schectmann et al. (1984) that the alloy Al_6Mn yields an electron diffraction record that conforms with icosahedral symmetry has added new importance to this type of symmetry. Of course, the presence of a five-fold axis of symmetry violates T4.17 and so the periodicity of the atoms in the alloy cannot be repres- ented by a lattice. Until now solids have been classified as either crystalline or amorphous. However, the work on Al_6Mn suggests a new state of matter called *quasi-crystals*. In such substances the atoms are believed to be arranged in rows as in a crystal but the spacing within and between these rows is believed to exhibit a more complicated rapport. The result is a clustering of atoms that is repeated over and over again but at irregular intervals. However, as the directions of the bonds in each cluster is maintained as in a crystal, quasi-crystals are believed to exhibit orientational rather than translational symmetry. Physicists, material scientists and mathematicians are actively studying this new form of matter. Some argue that the atoms in such matter are quasiperiodic whereas others argue that they exhibit incommensurate ordering. Whatever the outcome, it is clearly important that we devote some time to the icosahedral groups so that we may gain a better appreciation of substances like Al_6Mn. Finally, the fact that skutterudite crystallizes as a pyritohedron, which closely resembles the dual of an icosahedron, and that its point symmetry is a subgroup of the icosahedral group, suggests that it may have passed through a quasi-crystalline state prior to its final crystallization, leaving a remnant As_{12} iscosahedral unit in the structure.

(P5.26) Problem: Use TA6.2 to show that the composition of two fifth-turns cannot be a half-turn.

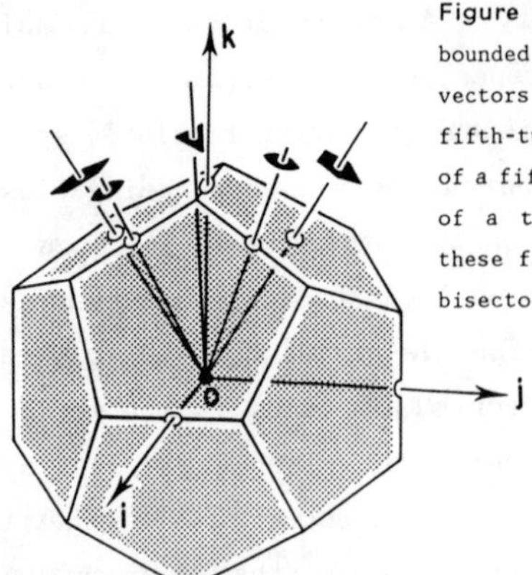

Figure 5.11: A regular dodecahedron(dual of an icosahedron) bounded by 12 pentagonal faces with a cartesian set of basis vectors $C = \{\mathbf{i},\mathbf{j},\mathbf{k}\}$. The generating rotations of *235* are a fifth-turn, a third-turn and an half-turn. The rotation axis of a fifth turn is perpendicular to each pentagonal face, that of a third-turn passes through each corner where three of these faces meet and that of a half-turn passes through the bisector of each edge.

(P5.27) Problem: Use TA6.4 to show that if the composition of two fifth-turns is a third turn, **55** = **3**, then the angle between the fifth-turn axes is approximately 63.435°.

(P5.28) Problem: Use TA6.4 to show that if the composition of two fifth-turns is a fifth turn, **55** = **5**, then the angle between the fifth-turn axes is approximately 116.565°.

Note that since 63.435° and 116.565° are supplementary angles, the results given in P5.27 and P5.28 are compatible. Hence the *532* groups must be such that the angle between two of the five-fold axes is 63.435°. Let C denote a cartesian coordinate system. We place two five-fold axes in the **j**, **k** plane such that the angle between them is bisected by **k** (Figure 5.11). Hence the angle between **k** and each of these is 31.718°. Hence the unit vectors in the directions of these axes are

$$(0,\ \sin(31.718),\ \cos(31.718))^t = (0,\ .52574,\ .85065)^t$$

and

$$(0,\ \sin(-31.718),\ \cos(-31.718))^t = (0,\ -.52574,\ .85065)^t\ .$$

As will become apparent as we generate the pole point representatives for *532*, certain numbers occur again and again. In order to take advantage of these repeating numbers, we define τ to be

$$\tau = (1 + \sqrt{5})/2 \approx 1.618034 \approx \text{ctn}(31.718)$$

the "golden mean". Using τ, representatives of the pole points of these five-fold axes becomes $(0,1,\tau)^t$ and $(0,-1,\tau)^t$ which are at a distance of

$\sqrt{1 + \tau^2} \simeq 1.902113$ units from the origin. One fact that will aid us in our calculations is that $\tau^2 = \tau + 1$.

(P5.29) Problem: Confirm that

$$M_C([01\tau]_5) = \frac{1}{2}\begin{bmatrix} \tau - 1 & -\tau & 1 \\ \tau & 1 & \tau - 1 \\ -1 & \tau - 1 & \tau \end{bmatrix}$$

$$= \begin{bmatrix} .309017 & -.809017 & .500000 \\ .809017 & .500000 & .309017 \\ -.500000 & .309017 & .809017 \end{bmatrix}$$

by using the unit vector $(0, .52574, .85065)^t$ and $\rho = 72°$ in (A3.1). Also show that

$$M_C([0\ -1\ \tau]_5) = \frac{1}{2}\begin{bmatrix} \tau - 1 & -\tau & -1 \\ \tau & 1 & 1 - \tau \\ 1 & 1 - \tau & \tau \end{bmatrix}$$

$$= \begin{bmatrix} .309017 & -.809017 & -.500000 \\ .809017 & .500000 & -.309017 \\ .500000 & -.309017 & .809017 \end{bmatrix}$$

by using the unit vector $(0, -.52574, .85065)^t$ and $\rho = 72°$ in (A3.1).

(P5.30) Problem: Show that

$$M_C([01\tau]_5)M_C([0\bar{1}\tau]_5)$$

$$= \frac{1}{2}\begin{bmatrix} 1 - \tau & -\tau & 1 \\ \tau & -1 & 1 - \tau \\ 1 & \tau - 1 & \tau \end{bmatrix}$$

$$= \begin{bmatrix} -.309017 & -.809017 & .500000 \\ .809017 & -.500000 & -.309017 \\ .500000 & .309017 & .809017 \end{bmatrix} .$$

For example the (1,2) entry is found as follows:

$$\begin{aligned}\ell_{12} &= \tfrac{1}{4}[(\tau - 1)(-\tau) + (-\tau)(1) + (1 - \tau)] \\ &= \tfrac{1}{4}[-\tau^2 - \tau + 1] \\ &= \tfrac{1}{4}[-(\tau + 1) - \tau + 1] \\ &= \tfrac{1}{4}[-2\tau] = -\tau/2 \quad .\end{aligned}$$

Analyze this matrix and show that

$$M_{\mathcal{C}}([01\tau]_5)M_{\mathcal{C}}([0\bar{1}\tau]_5) = M_{\mathcal{C}}([1, 0, \tau + 1]_3) \quad .$$

(P5.31) Problem: Show that $M_{\mathcal{C}}([01\tau]_5)\ M_{\mathcal{C}}([\tau\tau\tau]_3)$

$$= \frac{1}{2}\begin{bmatrix} -1 & \tau - 1 & \tau \\ \tau - 1 & -\tau & 1 \\ \tau & 1 & \tau - 1 \end{bmatrix} = M_{\mathcal{C}}([\tau, 1, \tau + 1]_2)$$

$$\simeq \begin{bmatrix} -.500000 & .309017 & .809017 \\ .309017 & -.809017 & .500000 \\ .809017 & .500000 & .309017 \end{bmatrix} \quad .$$

(P5.32) Problem: Show that

$$M_{\mathcal{C}}([01\tau]_5)M_{\mathcal{C}}([1, 0, \tau + 1]_3)M_{\mathcal{C}}([01\tau]_5)^{-1}$$

$$= \begin{bmatrix} 0 & 0 & 1 \\ 1 & 0 & 0 \\ 0 & 1 & 0 \end{bmatrix} = M_{\mathcal{C}}([\tau\tau\tau]_3) \quad .$$

Note that $M_{\mathcal{C}}([01\tau]_5)^{-1} = M_{\mathcal{C}}([01\tau]_5)^t$. □

Setting $A = M_{\mathcal{C}}([01\tau]_5)$ and $B = M_{\mathcal{C}}([\tau\tau\tau]_3)$ and recalling that we have already found the pole point representatives $[01\tau]^t$ for the 5-folds, $[1, 0, \tau + 1]^t$ and $[\tau\tau\tau]^t$ for the 3-folds and $[\tau,1,\tau + 1]^t$ for the 2-folds, we can find the remaining pole point representatives in the first octant as follows:

(1) 5-fold pole points:

$$\{[01\tau]^t, B[01\tau]^t, B^{-1}[01\tau]^t\} = \{[01\tau]^t, [\tau 01]^t, [1,\tau,0]^t\} \ ;$$

(2) 3-fold pole points:

$$\{[0,1,\tau + 1]^t,[\tau\tau\tau]^t,A[\tau\tau\tau]^t,B[1,0,\tau + 1]^t\}$$
$$= \{[0,1,\tau + 1]^t,[\tau\tau\tau]^t,[0,\tau + 1,1]^t,[\tau + 1,1,0]^t\} \quad ;$$

(3) 2-fold pole points:

$$\{[\tau,1,\tau + 1]^t,A[\tau,1,\tau + 1]^t,A^{-1}[\tau,1,\tau + 1]^t,$$
$$B[\tau,1,\tau + 1]^t,B(A^{-1}[\tau,1,\tau + 1]^t),B^{-1}(A^{-1}[\tau,1,\tau + 1]^t)\}$$
$$= \{[\tau,1,\tau + 1]^t,[0,0,2\tau]^t,[1,\tau + 1,\tau]^t,$$
$$[\tau + 1,\tau,1]^t,[2\tau,0,0]^t,[0,2\tau,0]^t\} \quad .$$

By applying $[2\tau,0,0]_2$, $[0,2\tau,0]_2$ and $[0,0,2\tau]_2$ to each of the pole point representatives in the first octant and then including the negatives of the resulting representatives, representatives of all of the pole points are obtained. One can now find all 60 matrix representatives and show that $M_C(532)$, and hence *532*, is a group. Note that since we have no proper polyaxial subgroup of *532* of order 30 (note that the dihedral group *n22* with $n = 15$ is not a subset of *532*). Hence the only improper group created from *532* is

$$(532) \cup (532)i = \bar{5}\bar{3}(2/m) \quad .$$

The ITFC (Hahn, 1983) denotes *532* by *235* and $\bar{5}\bar{3}2/m$ by $m\bar{3}\bar{5}$.

CHAPTER 6

THE BRAVAIS LATTICE TYPES

For some minutes Alice stood without speaking, looking out in all directions over the country - and a most curious country it was. There were a number of tiny brooks running straight across it from side to side, and the ground between was divided up into squares by a number of green hedges that reached from brook to brook. "I declare, it's marked out just like a large chess-board," Alice said at last. "... all over the world - if this is the world at all." -- Lewis Carroll

INTRODUCTION

In Chapters 4 and 5, all of the possible point groups were found that contain isometries with turn angles of multiples of either 60° or 90°. To confirm that all of these are crystallographic point groups, a lattice must be found for each that is left invariant under the group. In this chapter, we not only show that such a lattice exists for each of these groups, but also we find a description that includes all such lattices for each group. We shall see that the bases chosen in Chapter 5 are fundamental to the discovery of these lattices. *In toto* we shall find 14 different lattice types.

Frankenheim (1842), the first to study lattices, concluded in his derivation that there are 15 distinct lattice types. Several years later, Auguste Bravais (1849) undertook a more rigorous derivation and showed that there are only 14 types. These lattices were named the 14 Bravais lattice types in his honor. The American Crystallographic Association has published an English translation of Bravais' derivation (Bravais, 1945) that makes interesting reading. However, the strategy followed in our derivation of the lattice types is more closely akin to that used by Zachariasen (1945) in his beautiful book entitled *Theory of X-ray Diffraction in Crystals*.

LATTICES

We recall from Chapter 1 that any basis $D = \{\mathbf{a},\mathbf{b},\mathbf{c}\}$ generates a lattice L_D where

$$L_D = \{u\mathbf{a} + v\mathbf{b} + w\mathbf{c} \mid u,v,w \in Z\} \quad .$$

(T6.1) Theorem: Let $D_1 = \{\mathbf{a}_1,\mathbf{b}_1,\mathbf{c}_1\}$ and $D_2 = \{\mathbf{a}_2,\mathbf{b}_2,\mathbf{c}_2\}$ denote bases of S. Then D_1 and D_2 generate the same lattice if and only if each vector in D_1 is an integral combination of D_2 and each vector in D_2 is an integral combination of D_1. That is, $L_{D_1} = L_{D_2}$ if and only if $[\mathbf{a}_1]_{D_2}$, $[\mathbf{b}_1]_{D_2}$, $[\mathbf{c}_1]_{D_2}$ and $[\mathbf{a}_2]_{D_1}$, $[\mathbf{b}_2]_{D_1}$, $[\mathbf{c}_2]_{D_1}$ are all in Z^3.

Proof: Suppose $L_{D_1} = L_{D_2}$. Consider $\mathbf{a}_1$. Since $\mathbf{a}_1 \in L_{D_1}$, $\mathbf{a}_1 \in L_{D_2}$. Hence $\mathbf{a}_1$ can be written as an integral combination of the vectors in D_2. Likewise, $\mathbf{b}_1$ and $\mathbf{c}_1$ can be written as integral combinations of vectors in D_2. Similarly $\mathbf{a}_2$, $\mathbf{b}_2$ and $\mathbf{c}_2$ can be written as integral combinations of D_1. Now suppose each vector in D_1 can be written as an integral combination of vectors in D_2 and vice versa. Let T denote the change of basis matrix from D_1 to D_2. Then

$$T = \begin{bmatrix} | & | & | \\ [\mathbf{a}_1]_{D_2} & [\mathbf{b}_1]_{D_2} & [\mathbf{c}_1]_{D_2} \\ | & | & | \end{bmatrix} \quad .$$

Note that, by assumption, the entries of T are all integers (that is, T is an integral matrix). Let $\mathbf{v}$ denote a vector in L_{D_1}. Then $[\mathbf{v}]_{D_1} \in Z^3$ by definition of L_{D_1}. Hence $[\mathbf{v}]_{D_2} = T[\mathbf{v}]_{D_1} \in Z^3$ and so $\mathbf{v} \in L_{D_2}$. Therefore L_{D_1} is a subset of L_{D_2}. Similarly, since T^{-1} is constructed from $[\mathbf{a}_2]_{D_1}$, $[\mathbf{b}_2]_{D_1}$ and $[\mathbf{c}_2]_{D_1}$, it is an integral matrix and so $T^{-1}[\mathbf{v}]_{D_2} = [\mathbf{v}]_{D_1} \in Z^3$, L_{D_2} is a subset of L_{D_1} and so $L_{D_1} = L_{D_2}$. □

(D6.2) Definition: A matrix T all of whose entries are integers is called an *integral matrix*. If, in addition det(T) = ±1, T is said to be *unimodular over the integers* or, for our purposes, simply *unimodular*. If T is integral and det(T) = 1, then T is said to be a *proper unimodular matrix*.

(P6.1) Problem: Show that the product of two proper unimodular matrices is again a proper unimodular matrix.

(T6.3) Theorem: Let D_1 and D_2 be bases of S. Then $L_{D_1} = L_{D_2}$ if and only if the change of basis matrix T from D_1 to D_2 is a unimodular matrix. Furthermore, if $L_{D_1} = L_{D_2}$ then D_1 and D_2 are of the same handedness if and only if T is a proper unimodular matrix.

Proof: In the proof of T6.1, we showed that $L_{D_1} = L_{D_2}$ if and only if the change of basis matrix from D_1 to D_2 is integral and the change of basis matrix from D_2 to D_1 is integral. Hence if $L_{D_1} = L_{D_2}$, the change of basis matrix T from D_1 to D_2 and the change of basis matrix from D_2 to D_1, T^{-1}, are both integral. Hence det(T) and det(T^{-1}) are integers. By TA2.24, det(T^{-1}) = 1/det(T). Therefore, det(T) = ±1 and so T is unimodular. Now suppose the change of basis matrix T from D_1 to D_2 is unimodular, then T^{-1} is also an integral matrix (TA2.25). Suppose $L_{D_1} = L_{D_2}$ and M is the change of basis matrix from D_2 to a right-handed cartesian coordinate basis C. Then TM is the change of basis matrix from D_1 to C. Since det(TM) = det(T)det(M), the signs of det(M) and det(TM) agree if det(T) > 0 and disagree if det(T) < 0. Since D_1 and D_2 are of the same handedness (see the cross product section of Chapter 1) if and only if det(TM) and det(T) agree in sign, we have established the theorem. □

(E6.4) Example - When two bases generate the same lattice: Let D = {**a**,**b**,**c**} denote a basis of S. Consider

$$D_1 = \{\mathbf{b},\ (1/3)\mathbf{a} + (2/3)\mathbf{b} + (2/3)\mathbf{c},\ (2/3)\mathbf{a} + (1/3)\mathbf{b} + (1/3)\mathbf{c}\}$$

and

$$D_2 = \{(2/3)\mathbf{a} + (1/3)\mathbf{b} + (1/3)\mathbf{c},\ -(1/3)\mathbf{a} + (1/3)\mathbf{b} + (1/3)\mathbf{c},\ -(1/3)\mathbf{a} - (2/3)\mathbf{b} + (1/3)\mathbf{c}\}\ .$$

Show that D_1 and D_2 generate the same lattice and are of the same handedness. □

Solution: Since D_1 and D_2 are both expressed in terms of the basis D, we can find the change of basis matrix from D_1 to D_2 using the following circuit diagram.

$$\begin{array}{ccc} [\mathrm{v}]_{D_1} & \xrightarrow{T} & [\mathrm{v}]_{D_2} \\ T_1 \downarrow & & \uparrow T_2 \\ [\mathrm{v}]_D & \xrightarrow{I_3} & [\mathrm{v}]_D \end{array}$$

From the diagram, we see that $T = T_2T_1$. From the definition of D_1 we have

$$T_1 = \begin{bmatrix} 0 & 1/3 & 2/3 \\ 1 & 2/3 & 1/3 \\ 0 & 2/3 & 1/3 \end{bmatrix} .$$

From P2.19, when D was denoted by D_1 and D_2 by D_2, we see that

$$T_2 = \begin{bmatrix} 1 & 0 & 1 \\ -1 & 1 & 1 \\ 0 & -1 & 1 \end{bmatrix} .$$

Hence

$$T = T_2T_1 = \begin{bmatrix} 0 & 1 & 1 \\ 1 & 1 & 0 \\ -1 & 0 & 0 \end{bmatrix} .$$

Since $\det(T) = 1$ and T is an integral matrix, T is a proper unimodular matrix. Hence D_1 and D_2 generate the same lattice and are of the same handedness. □

We shall now consider generators of lattices of one-, two- or three-dimensions. Let $\mathbf{a}$ denote a nonzero vector. Then $D = \{\mathbf{a}\}$ generates a one-dimensional lattice L_D,

$$L_D = \{u\mathbf{a} \mid u \in Z\} .$$

Next, let $\mathbf{a}$ and $\mathbf{b}$ denote non-collinear vectors, then $D = \{\mathbf{a},\mathbf{b}\}$ generates a two-dimensional lattice L_D,

$$L_D = \{u\mathbf{a} + v\mathbf{b} \mid u,v \in Z\} \quad .$$

Note that the term "lattice" will be used by us to denote a three-dimensional lattice unless otherwise indicated. In any lattice (of dimension 1, 2 or 3) there exists shortest nonzero lattice vectors (Newman, 1972). That is, there exists a lattice vector $\mathbf{v}$ whose length is less than or equal to the length of every other nonzero vector in the lattice. The next theorem uses the existence of these vectors to find a basis for the lattice.

(T6.5) Theorem: Let $\mathbf{a}$ denote the shortest vector in a one-dimensional lattice L_D. Then $\{\mathbf{a}\}$ generates L_D. Let $\mathbf{a}$ and $\mathbf{b}$ be the shortest non-collinear vectors in a two-dimensional lattice L_D. Then $\{\mathbf{a},\mathbf{b}\}$ generates L_D. Let $\mathbf{a}$, $\mathbf{b}$ and $\mathbf{c}$ be the shortest non-coplanar vectors in a three dimensional lattice L_D, then $\{\mathbf{a},\mathbf{b},\mathbf{c}\}$ generates L_D.

The proof of this theorem would require a geometric digression that we do not have space for in this book.

We will now consider the situation where one lattice is contained in another.

(D6.6) Definition: Let L denote a lattice. By a sublattice L' of L, we mean a subset of L that is a lattice in its own right.

We will show that every lattice left invariant under a point group G has a sublattice of the form L_P where P is a basis of the type defined for G in Chapters 4 and 5. Hence we will be searching for lattices containing L_P that are also left invariant under G. Suppose L_D is a lattice such that L_P is contained in L_D. As groups, L_P is a normal subgroup of L_D since lattices are abelian groups. Hence L_D/L_P is a group. We will study the relationship between L_P and L_D by considering the elements of L_D/L_P. Note that if $\mathbf{v}$ is a vector in L_D and not in L_P, then $[\mathbf{v}]_P$ must contain fractional coordinates. We can restrict the type of fractional coordinates used if we note that L_D/L_P is a subgroup of S/L_P and use the equivalence relation associated with this factor group. We call this equivalence relation L_P-equivalence and explicitly state the relation in the following definition.

(D6.7) Definition: Let L_P denote a lattice and let $\mathbf{v}$ and $\mathbf{w}$ denote vectors in S. We define the relation ~ on S by

$$\mathbf{v} \sim \mathbf{w} \iff \mathbf{w} - \mathbf{v} \in L_P \quad .$$

If $\mathbf{v} \sim \mathbf{w}$, we say that $\mathbf{v}$ is L_P-equivalent to $\mathbf{w}$.

See Appendix 7 for a discussion of factor groups and their related equivalence relations. Let $\mathbf{v} \in S$ and suppose $[\mathbf{v}]_P = [v_1, v_2, v_3]^t$. Let u_i be the largest integer such that $u_i \leq v_i$. Consider

$$\mathbf{u} = u_1\mathbf{a} + u_2\mathbf{b} + u_3\mathbf{c} \text{ and } \mathbf{w} = (v_1 - u_1)\mathbf{a} + (v_2 - u_2)\mathbf{b} + (v_3 - u_3)\mathbf{c} \quad ,$$

where $P = \{\mathbf{a},\mathbf{b},\mathbf{c}\}$. Then $\mathbf{u} \in L_P$ and $\mathbf{v} - \mathbf{w} = \mathbf{u}$. Consequently $\mathbf{w} \sim \mathbf{v}$. By the way the u_i's were defined, $[\mathbf{w}]_P = [w_1, w_2, w_3]^t$ where $0 \leq w_i < 1$ for each i. Hence each vector in S is equivalent to a vector whose coordinates with respect to P are greater than or equal to zero and less than 1.

(D6.8) Definition: Let L_D denote a lattice where $D = \{\mathbf{a},\mathbf{b},\mathbf{c}\}$. Then the unit cell U_D of L_D is

$$U_D = \{\mathbf{v} \in S \mid [\mathbf{v}]_D = [v_1, v_2, v_3]^t \text{ such that } 0 \leq v_i < 1 \text{ or } i = 1,2,3\}$$

$$= \{x\mathbf{a} + y\mathbf{b} + z\mathbf{c} \mid 0 \leq x < 1,\ 0 \leq y < 1,\ 0 \leq z < 1\} \quad .$$

Consequently, relative to a lattice L_D, each vector in S has an L_D-equivalent vector in U_D.

(P6.2) Problem: Show that if $[\mathbf{v}]_D = [-13.7, 12.3, 6]$, then $\mathbf{v} \sim \mathbf{w}$ where $[\mathbf{w}]_D = [0.3, 0.3, 0]^t \in U_D$.

(T6.9) Theorem: Let L denote a lattice and let L_D denote a sublattice of L. Each element of the factor group L/L_D has a representative in the unit cell U_D of L_D.

Proof: The elements of the factor group L/L_D are the right cosets of L_D in L. By EA7.8 we see that the right cosets of L_D in L are the equivalence classes of the vectors in L with respect to L_D-equivalence (when applying EA7.7 recall that the statement "ba^{-1}" in multiplicative notion is "b − a" in additive notation). By the discussion preceding T6.9, each vector in L is L_D-equivalent to some vector in U_D. Hence each right coset has a representative in U_D. □

T6.9 enables us to describe a lattice L in terms of a sublattice L_D by listing relatively few vectors with fractional components.

(E6.10) Example - The two cosets of a body-centered lattice: The body-centered lattice L constructed from a lattice L_D, $D = \{\mathbf{a},\mathbf{b},\mathbf{c}\}$, is the lattice consisting of two right cosets:

$$L_D + \mathbf{0} \text{ and } L_D + (\tfrac{1}{2}\mathbf{a} + \tfrac{1}{2}\mathbf{b} + \tfrac{1}{2}\mathbf{c}) \quad .$$

Hence the representatives with respect to D of these right cosets are $[000]^t$ and $[\frac{1}{2},\frac{1}{2},\frac{1}{2}]^t$. Hence

$$L = \{n_1\mathbf{a} + n_2\mathbf{b} + n_3\mathbf{c} + n_4(\tfrac{1}{2}\mathbf{a} + \tfrac{1}{2}\mathbf{b} + \tfrac{1}{2}\mathbf{c}) \mid n_1,n_2,n_3,n_4 \ \varepsilon \ Z\} \quad .$$

(T6.11) Theorem: A lattice L is left invariant under a point group of the form $G \cup Gi$ or $H \cup (G \setminus H)i$ (see T4.28) if and only if L is left invariant under G.

Proof: Let L denote a lattice left invariant by $G \cup Gi$. Since G is a subset of $G \cup Gi$, L is invariant under G. Suppose that L is left invariant by $H \cup (G \setminus H)i$. Since L is invariant under i, and $(G \setminus H)i$, it is invariant under each operation in $[(G \setminus H)i]i = G \setminus H$. Hence L is invariant under $G = H \cup (G \setminus H)$. Now suppose that L is invariant under G. Then L is invariant under Gi and hence under $G \cup Gi$. Since $(G \setminus H)i$ is a subset of Gi and H is a subset of G, L is invariant under $(G \setminus H)i$ and H. Hence L is invariant under $H \cup (G \setminus H)i$. □

(T6.12) Theorem: Let G denote a point group. Then a lattice L is left invariant under G if and only if L is left invariant under each of the generators of G.

Proof: Let $\{\mathbf{g}_1,\mathbf{g}_2,\ldots,\mathbf{g}_n\}$ denote the generators of G. If L is invariant under G then it is invariant under each of $\{\mathbf{g}_1,\mathbf{g}_2,\ldots,\mathbf{g}_n\}$. Now suppose is invariant under each of $\{\mathbf{g}_1,\mathbf{g}_2,\ldots,\mathbf{g}_n\}$. Let $\mathbf{g} \ \varepsilon \ G$.

Then $\mathbf{g}$ is a finite product (composition) of $\{\mathbf{g}_1,\ldots,\mathbf{g}_n\}$, say $\mathbf{g} = \mathbf{h}_1\mathbf{h}_2\ldots\mathbf{h}_t$ where $\mathbf{h}_i \in \{\mathbf{g}_1,\mathbf{g}_2,\ldots,\mathbf{g}_n\}$. Then

$$\begin{aligned} \mathbf{g}(L) &= \mathbf{h}_1\mathbf{h}_2\ldots\mathbf{h}_t(L) \\ &= \mathbf{h}_1\mathbf{h}_2\ldots\mathbf{h}_{t-1}(L) \\ &\;\;\vdots \\ &= \mathbf{h}_1(L) \\ &= L\ . \end{aligned}$$

□

Given a point isometry α and a lattice L_D we need a strategy for determining whether L_D is left invariant under α or not. Note that $\alpha(L_D)$ is the lattice generated by $\alpha(D) = \{\alpha(\mathbf{a}),\alpha(\mathbf{b}),\alpha(\mathbf{c})\}$ since

$$\alpha(u\mathbf{a} + v\mathbf{b} + w\mathbf{c}) = u\alpha(\mathbf{a}) + v\alpha(\mathbf{b}) + w\alpha(\mathbf{c}) \quad .$$

Hence $L_D = \alpha(L_D)$ if and only if D and $\alpha(D)$ generate the same lattice. Using T6.1, to show that L_D is left invariant under α, we need only show that each vector in D can be expressed as an integral combination of $\alpha(D)$ and vice versa.

(T6.13) Theorem: Let G denote a point group and let $D = \{\mathbf{a},\mathbf{b},\mathbf{c}\}$ denote a basis of S. Then L_D is invariant under G if and only if $M_D(\alpha)$ is an integral matrix for each generator α of G.

Proof: Recall that

$$M_D(\alpha) = \left[\begin{array}{c|c|c} [\alpha(\mathbf{a})]_D & [\alpha(\mathbf{b})]_D & [\alpha(\mathbf{c})]_D \end{array}\right]$$

which is the change of basis matrix from $\alpha(D)$ to D. Since we know (see CA3.8) that $\det(M_D(\alpha)) = \pm 1$, T6.3 yields the result that $L_D = L_{\alpha(D)}$ if and only if $M_D(\alpha)$ is integral.

□

(E6.14) Example - A lattice left invariant under a point group G: Suppose α and β are generators for some point group G, $D = \{\mathbf{a},\mathbf{b},\mathbf{c}\}$ denotes a basis for S and $\alpha(\mathbf{a}) = \mathbf{b}$, $\alpha(\mathbf{b}) = -\mathbf{a} + \mathbf{c}$, $\alpha(\mathbf{c}) = \mathbf{c}$, $\beta(\mathbf{a}) = -\mathbf{b}$, $\beta(\mathbf{b}) = -\mathbf{a}$, and $\beta(\mathbf{c}) = -\mathbf{c}$. Show that L_D is invariant under G.

Solution: To show that the lattice L_D is invariant under G, we must show that $M_D(\alpha)$ and $M_D(\beta)$ are both unimodular. Since

$$M_D(\alpha) = \begin{bmatrix} 0 & -1 & 0 \\ 1 & 0 & 0 \\ 0 & 1 & 1 \end{bmatrix} \quad \text{and} \quad M_D(\beta) = \begin{bmatrix} 0 & -1 & 0 \\ -1 & 0 & 0 \\ 0 & 0 & -1 \end{bmatrix}$$

are both integral and $\det(M_D(\alpha)) = \det(M_D(\beta)) = +1$, we may conclude that L_D is mapped into self coincidence by each point isometry in G. □

(P6.3) Problem: Use T6.13 to show that L_P (for the appropriate choice of P) is left invariant by each of the point groups whose generators are represented by the matrices in Tables 5.2 and 5.3.

We have established some results about properties of G that would insure that a given lattice is left invariant under G. We now shift the emphasis and ask what properties must hold in a lattice L in order for it to be left invariant under a given point group G.

(T6.15) Theorem: Suppose that α is either a half-turn or a third-turn whose rotation axis ℓ passes through the origin. Let L denote a lattice that is left invariant under α. Then there is a nonzero lattice vector along ℓ and a two-dimensional lattice plane perpendicular to ℓ passing through the origin.

The proof of T6.15 is given in Appendix 4.

A DERIVATION OF THE 14 BRAVAIS LATTICE TYPES

In this section, we shall consider the proper crystallographic point groups. For each such group G, we shall determine the structure of a lattice L_P left invariant under G such that any lattice L left invariant under G will have a sublattice with the same structure as L_P. For convenience we will denote L_P by P. The fact that both the lattice and its basis are denoted by P will not cause confusion since the context will always make the meaning of P clear. If L contains P and $L \neq P$, we call L a *centered lattice with respect to P*.

We will now determine the possible lattice types for each of the point groups.

Lattices invariant under *1*: All lattices are left invariant under *1* and so no conditions beyond those of being a lattice are needed. By convention the basis $P = \{\mathbf{a},\mathbf{b},\mathbf{c}\}$ for the lattice P is chosen to be such that **a** is the shortest nonzero vector in the lattice, **b** is the shortest lattice vector not collinear with **a** and **c** is the shortest lattice vector not coplanar with **a** and **b** chosen so that the resulting coordinate system is right-handed. Since all lattices satisfy the condition for P, no further lattices need be sought. Hence there is only one lattice type for *1* which is denoted P and is called the *primitive lattice type* (see Figure 6.1).

Lattices invariant under *2*: As in Chapter 5, for *2* we take the basis $D = \{\mathbf{a},\mathbf{b},\mathbf{c}\}$ such that **c** is along the 2-fold axis and **a** and **b** are perpendicular to **c** so that P is right-handed. Then the lattice P is left invariant under *2* because $M_P(2)$ consists only of integral matrices (see T6.13 and Table 5.2).

Let L denote any lattice left invariant under *2*. By T6.15, there exists a nonzero lattice vector along the 2-fold axis and there is a lattice plane passing through the origin perpendicular to the 2-fold axis. Let $\{\mathbf{a},\mathbf{b}\}$ denote a basis for the lattice plane perpendicular to the 2-fold axis. Since no geometric constraints have been placed on this lattice plane, no preference will be given to one basis over another. Let **c** denote the shortest nonzero vector along the two-fold axis so that $\{\mathbf{a},\mathbf{b},\mathbf{c}\}$ is a right-handed system. The lattice generated by $\{\mathbf{a},\mathbf{b},\mathbf{c}\}$ is of the same type as P described in the previous paragraph. Hence we denote both $\{\mathbf{a},\mathbf{b},\mathbf{c}\}$ and the lattice generated by it by P. If $P \neq L$, then P is a proper subset of L and so there are vectors in L whose coordinates with respect to P are fractional. In order to facilitate the discussion, we shall use P-equivalence to discuss the vectors in L. Hence, by T6.9 each vector in L has a representative whose coordinates f_1, f_2 and f_3 with respect to P are such that $0 \le f_i < 1$. Suppose $f \in L$ such that $[\mathbf{f}]_P = [f_1,f_2,f_3]^t$ where $0 \le f_i < 1$. Then

$$[\mathbf{2(f)}]_P = M_P(2)[\mathbf{f}]_P = \begin{bmatrix} -1 & 0 & 0 \\ 0 & -1 & 0 \\ 0 & 0 & 1 \end{bmatrix} \begin{bmatrix} f_1 \\ f_2 \\ f_3 \end{bmatrix} = \begin{bmatrix} -f_1 \\ -f_2 \\ f_3 \end{bmatrix}$$

If $\mathbf{e}$ is defined to be $\mathbf{e} = \mathbf{2(f)} + \mathbf{f}$, then $\mathbf{e} \in L$ and

$$[\mathbf{e}]_P = M_P(2)[\mathbf{f}]_P + [\mathbf{f}]_P = \begin{bmatrix} 0 \\ 0 \\ 2f_3 \end{bmatrix} .$$

Since $\mathbf{c}$ is the shortest vector along the two-fold axis in L, $\mathbf{e}$ must be a multiple of $\mathbf{c}$ and so $2f_3 \in Z$. Hence, $f_3 = 0$ or $\frac{1}{2}$. Similarly, if we define $\mathbf{e}$ to be $\mathbf{e} = \mathbf{f} - \mathbf{2(f)}$, then $\mathbf{e} \in L$ and $[\mathbf{e}]_P = [2f_1, 2f_2, 0]^t$. Since $\{\mathbf{a},\mathbf{b}\}$ is a basis for the lattice plane perpendicular to $\mathbf{c}$, we have $2f_1$, $2f_2 \in Z$. Therefore, $f_1 = 0$ or $\frac{1}{2}$ and $f_2 = 0$ or $\frac{1}{2}$. We tabulate the various combinations of these fractional coordinates in Table 6.1. Those combinations that violate the choice of the basis P are noted as contradictions. By the discussion following D6.6, L/P is a group. Hence, the only combination of the four possible fractional vectors that are suitable are those that form groups modulo P. Each of these vectors generates a cyclic group modulo P as shown below:

$$\begin{array}{c|c} P/P & \begin{bmatrix} 0 \\ 0 \\ 0 \end{bmatrix} \\ \hline\hline \begin{bmatrix} 0 \\ 0 \\ 0 \end{bmatrix} & \begin{bmatrix} 0 \\ 0 \\ 0 \end{bmatrix} \end{array} \qquad \begin{array}{c|cc} A/P & \begin{bmatrix} 0 \\ 0 \\ 0 \end{bmatrix} & \begin{bmatrix} 0 \\ \frac{1}{2} \\ \frac{1}{2} \end{bmatrix} \\ \hline\hline \begin{bmatrix} 0 \\ 0 \\ 0 \end{bmatrix} & \begin{bmatrix} 0 \\ 0 \\ 0 \end{bmatrix} & \begin{bmatrix} 0 \\ \frac{1}{2} \\ \frac{1}{2} \end{bmatrix} \\ \begin{bmatrix} 0 \\ \frac{1}{2} \\ \frac{1}{2} \end{bmatrix} & \begin{bmatrix} 0 \\ \frac{1}{2} \\ \frac{1}{2} \end{bmatrix} & \begin{bmatrix} 0 \\ 0 \\ 0 \end{bmatrix} \end{array} \qquad \begin{array}{c|cc} B/P & \begin{bmatrix} 0 \\ 0 \\ 0 \end{bmatrix} & \begin{bmatrix} \frac{1}{2} \\ 0 \\ \frac{1}{2} \end{bmatrix} \\ \hline\hline \begin{bmatrix} 0 \\ 0 \\ 0 \end{bmatrix} & \begin{bmatrix} 0 \\ 0 \\ 0 \end{bmatrix} & \begin{bmatrix} \frac{1}{2} \\ 0 \\ \frac{1}{2} \end{bmatrix} \\ \begin{bmatrix} \frac{1}{2} \\ 0 \\ \frac{1}{2} \end{bmatrix} & \begin{bmatrix} \frac{1}{2} \\ 0 \\ \frac{1}{2} \end{bmatrix} & \begin{bmatrix} 0 \\ 0 \\ 0 \end{bmatrix} \end{array} \qquad \begin{array}{c|cc} I/P & \begin{bmatrix} 0 \\ 0 \\ 0 \end{bmatrix} & \begin{bmatrix} \frac{1}{2} \\ \frac{1}{2} \\ \frac{1}{2} \end{bmatrix} \\ \hline\hline \begin{bmatrix} 0 \\ 0 \\ 0 \end{bmatrix} & \begin{bmatrix} 0 \\ 0 \\ 0 \end{bmatrix} & \begin{bmatrix} \frac{1}{2} \\ \frac{1}{2} \\ \frac{1}{2} \end{bmatrix} \\ \begin{bmatrix} \frac{1}{2} \\ \frac{1}{2} \\ \frac{1}{2} \end{bmatrix} & \begin{bmatrix} \frac{1}{2} \\ \frac{1}{2} \\ \frac{1}{2} \end{bmatrix} & \begin{bmatrix} 0 \\ 0 \\ 0 \end{bmatrix} \end{array}$$

No other groups can be formed from the four possible fractional vectors. This can be seen by observing that combining any two of the nonzero fractional vectors yields an impossible vector. For example,

$$[\tfrac{1}{2},0,\tfrac{1}{2}]^t + [0,\tfrac{1}{2},\tfrac{1}{2}] = [\tfrac{1}{2},\tfrac{1}{2},0]^t \qquad \text{(modulo } P\text{)}$$

which, as noted in Table 6.1, violates the choice of {**a**,**b**}.

Table 6.1: Combinations of fractional coordinates for *2*.

Possible Coordinates			Impossible Coordinates			Contradictions
f_1	f_2	f_3	f_1	f_2	f_3	
0	0	0	$\frac{1}{2}$	0	0	*choice of* **a**
0	$\frac{1}{2}$	$\frac{1}{2}$	0	$\frac{1}{2}$	0	*choice of* **b**
$\frac{1}{2}$	0	$\frac{1}{2}$	0	0	$\frac{1}{2}$	*choice of* **c**
$\frac{1}{2}$	$\frac{1}{2}$	$\frac{1}{2}$	$\frac{1}{2}$	$\frac{1}{2}$	0	*choice of* {**a**,**b**}

(P6.4) Problem: Show that $[\frac{1}{2},0,\frac{1}{2}]^t$ and $[\frac{1}{2},\frac{1}{2},\frac{1}{2}]^t$ cannot both be in L by forming their sum modulo P. Show that $[0,\frac{1}{2},\frac{1}{2}]^t$ and $[\frac{1}{2},\frac{1}{2},\frac{1}{2}]^t$ cannot both be in L.

Consider the lattice $A = (P + \mathbf{0}) \cup (P + [0,\frac{1}{2},\frac{1}{2}]^t)$. If we change basis from $P = \{\mathbf{a},\mathbf{b},\mathbf{c}\}$ to $P_1 = \{\mathbf{a}_1,\mathbf{b}_1,\mathbf{c}_1\}$ where $\mathbf{a}_1 = \mathbf{b}$, $\mathbf{b}_1 = \mathbf{a}$ and $\mathbf{c}_1 = -\mathbf{c}$, then the change of basis matrix from P to P_1 is

$$T = \begin{bmatrix} 0 & 1 & 0 \\ 1 & 0 & 0 \\ 0 & 0 & -1 \end{bmatrix}$$

which is a proper unimodular matrix. Hence P_1 is another basis for the lattice P and, furthermore, P_1 satisfies the criteria to qualify as a P basis. That is, $\mathbf{a}_1$ and $\mathbf{b}_1$ forms a basis for the lattice plane perpendicular to the 2-fold axis and $\mathbf{c}_1$ is the shortest lattice vector along the axis. Since

$$\begin{bmatrix} 0 & 1 & 0 \\ 1 & 0 & 0 \\ 0 & 0 & -1 \end{bmatrix} \begin{bmatrix} 0 \\ \frac{1}{2} \\ \frac{1}{2} \end{bmatrix} = \begin{bmatrix} \frac{1}{2} \\ 0 \\ -\frac{1}{2} \end{bmatrix} = \begin{bmatrix} \frac{1}{2} \\ 0 \\ \frac{1}{2} \end{bmatrix} \quad (\text{modulo } P_1) \quad ,$$

the A lattice of P is the same set of points as the B lattice of P_1. Similarly, it can be shown (see P6.5) that the I lattice also occurs as a B lattice with respect to a different allowable basis. Hence if we take the A lattice with respect to every allowable basis, we will obtain all of B and I lattice types as well.

(P6.5) Problem: Show that the A lattice with respect to the P basis is the I lattice with respect to the $P_2 = \{\mathbf{a}_2, \mathbf{b}_2, \mathbf{c}_2\}$ where $\mathbf{a}_2 = \mathbf{a}$, $\mathbf{b}_2 = \mathbf{a} + \mathbf{b}$, $\mathbf{c}_2 = \mathbf{c}$. (Hint: Show that P_2 is a basis for the lattice P and that P_2 satisfies the conditions that $\{\mathbf{a}_2, \mathbf{b}_2\}$ is a basis for the lattice plane perpendicular to the 2-fold axis and that $\mathbf{c}_2$ is the shortest vector along the axis. Then show that P_2 is a right-handed basis and that $[0,\frac{1}{2},\frac{1}{2}]^t$ with respect to P becomes $[\frac{1}{2},\frac{1}{2},\frac{1}{2}]^t$ with respect to P_2).

Hence, we need only use one centering for each basis to obtain all possible centered lattices. When $\mathbf{c}$ is along the rotation axis, as in our case, the ITFC (Hahn, 1983) gives as the first choice $[0,\frac{1}{2},\frac{1}{2}]^t$. Thus we obtain the lattice type

$$A = (P + \mathbf{0}) \cup (P + \tfrac{1}{2}\mathbf{b} + \tfrac{1}{2}\mathbf{c})$$

$$= \{m_1\mathbf{a} + m_2\mathbf{b} + m_3\mathbf{c} \mid m_1, m_2, m_3 \in Z\}$$

$$\cup \{m_1\mathbf{a} + m_2\mathbf{b} + m_3\mathbf{c} + (\tfrac{1}{2}\mathbf{b} + \tfrac{1}{2}\mathbf{c}) \mid m_1, m_2, m_3 \in Z\}$$

$$= \{u\mathbf{a} + v(\tfrac{1}{2}\mathbf{b} + \tfrac{1}{2}\mathbf{c}) + w\mathbf{c} \mid u, v, w \in Z\} \quad .$$

Hence a basis for A is $D = \{\mathbf{a}, \frac{1}{2}\mathbf{b} + \frac{1}{2}\mathbf{c}, \mathbf{c}\}$. If $[\mathbf{f}]_P = [f_1, f_2, f_3]^t$, the coset $P + \mathbf{f}$ is called a *colattice* and is denoted by $C_P(f_1, f_2, f_3)$. Note that if $\mathbf{f}$ is not an element of P, then $C_P(f_1, f_2, f_3)$ is not a lattice since $\mathbf{0}$ is not an element $C_P(f_1, f_2, f_3)$. However, $C_P(f_1, f_2, f_3)$ bears a strong geometric resemblance to P since it is an image of P displaced by the vector $\mathbf{f}$. In the case of A, $A = P \cup C_P(0,\frac{1}{2},\frac{1}{2})$. Note that the points of $C_P(0,\frac{1}{2},\frac{1}{2})$ lie in planes perpendicular to the 2-fold axis, located halfway between the planes containing the lattice points of the sublattice P. To show that A is invariant under 2, we need to show that $M_D(2)$ is an integral matrix. Since

$$M_D(2) = \begin{bmatrix} -1 & 0 & 0 \\ 0 & -1 & 0 \\ 0 & 1 & 1 \end{bmatrix} \quad ,$$

A is invariant under 2. Thus, there are two types of lattices left invariant by 2. The first type is called a primitive lattice and is denoted by P. Every lattice plane of P perpendicular to the 2-fold axis has a lattice point on the axis. The second lattice type denoted by A, is called an *A-centered lattice*. The lattice planes of A perpendicular to the 2-fold axis alternate, one having a point on the axis (i.e., in P) and the next not on the axis (i.e., in $C_P(0,\frac{1}{2},\frac{1}{2})$).

(P6.6) Problem:

(1) Given a basis $D = \{\mathbf{a},\mathbf{b},\mathbf{c}\}$ with $\mathbf{b}$ taken along the 2-fold axis and $\mathbf{a}$ and $\mathbf{c}$ perpendicular to $\mathbf{b}$ forming a right-handed system, show that the lattice L_D generated by D is left invariant under $[010]_2$.

(2) Let L denote any lattice type left invariant by $[010]_2$. Show that the coordinates of the fractional vectors in L with respect to D left invariant by $[010]_2$ are $[0,0,0]^t$, $[\frac{1}{2},\frac{1}{2},0]^t$, $[0,\frac{1}{2},\frac{1}{2}]^t$, $[\frac{1}{2},\frac{1}{2},\frac{1}{2}]^t$.

(3) Using the four fractional vectors obtained in the part (2) as generators, construct multiplication tables for the L/L_D groups for each of the following choices of L:

$$L_D/L_D = \{[0,0,0]^t\}$$
$$C/L_D = \{[0,0,0]^t,[\tfrac{1}{2},\tfrac{1}{2},0]^t\}$$
$$A/L_D = \{[0,0,0]^t,[0,\tfrac{1}{2},\tfrac{1}{2}]^t\}$$
$$I/L_D = \{[0,0,0]^t,[\tfrac{1}{2},\tfrac{1}{2},\tfrac{1}{2}]^t\} \quad .$$

(4) Show that no L/L_D groups contain $[\frac{1}{2},\frac{1}{2},0]^t$ and $[0,\frac{1}{2},\frac{1}{2}]^t$.

(5) Show that the A-centered and I-centered lattices with respect to D can be realized as C-centered lattices with respect to other allowable bases for L_D. When $\mathbf{b}$ is chosen to lie along the 2-fold axis, the ITFC (Hahn, 1983) gives as the first choice $[\frac{1}{2},\frac{1}{2},0]^t$. This results in the lattice type

$$\begin{aligned} C &= (L_D + \mathbf{0}) \cup (L_D + (\tfrac{1}{2}\mathbf{a} + \tfrac{1}{2}\mathbf{b})) \\ &= \{U(\tfrac{1}{2}\mathbf{a} + \tfrac{1}{2}\mathbf{b}) + v\mathbf{b} + w\mathbf{c} \mid u,v,w \in Z\} \end{aligned}$$

with the basis $D_C = \{\frac{1}{2}\mathbf{a} + \frac{1}{2}\mathbf{b},\ \mathbf{b},\ \mathbf{c}\}$. Show that C is left invariant by ${}^{[010]}2$. With the completion of this problem, we may conclude that there are two types of lattices left invariant by ${}^{[010]}2$: P and C. The lattice type denoted C is called a *C-centered lattice type* (see Figure 6.1). This setting for the one-face centered lattices in the monoclinic systems is the choice used in the ITFC (Hahn, 1983).

Lattices invariant under *3*: As in Chapter 5, we choose the basis $P = \{\mathbf{a},\mathbf{b},\mathbf{c}\}$ for 3 such that $\mathbf{c}$ is in the positive direction of the 3-fold axis, $\mathbf{a}$ and $\mathbf{b}$ are perpendicular to the axis such that $\sphericalangle(\mathbf{a}:\mathbf{b}) = 120°$ and P forms a right-handed system. Since $M_P(3)$ consists of integral matrices, L_P is left invariant under 3. Let L denote a lattice left invariant under 3. By T6.15 all lattices left invariant under 3 have a nonzero lattice vector along the three-fold axis and a lattice plane perpendicular to the axis passing through the origin. Let $\mathbf{a}$ be a lattice vector in the lattice plane perpendicular to $\mathbf{c}$ of shortest length. Then $\mathbf{b} = \mathbf{3}(\mathbf{a})$ is in the lattice plane and its length is also the shortest. Hence $\{\mathbf{a},\mathbf{b}\}$ is a basis for the lattice plane (T6.5). Let $\mathbf{c}$ be the shortest nonzero lattice vector along the positive direction of the 3-fold axis. Then $P = \{\mathbf{a},\mathbf{b},\mathbf{c}\}$ is a right-handed basis. Hence P is a lattice of the same form as the lattice P discussed in the previous paragraph. As in the case of 2, we consider the vectors in L under P-equivalence. Hence we seek vectors $\mathbf{f}$ in L of the form $[\mathbf{f}]_P = [f_1,f_2,f_3]^t$ where $0 \le f_i < 1$. Since L is invariant under 3, $\mathbf{f} + \mathbf{3}(\mathbf{f}) + \mathbf{3}^{-1}(\mathbf{f})$ is in the lattice, and, (see Table 5.3)

$$[\mathbf{f} + \mathbf{3}(\mathbf{f}) + \mathbf{3}^{-1}(\mathbf{f})]_P = [\mathbf{f}]_P + M_P(\mathbf{3})[\mathbf{f}]_P + M_P(\mathbf{3}^{-1})[\mathbf{f}]_P$$

$$= \begin{bmatrix} 0 \\ 0 \\ 3f_3 \end{bmatrix} .$$

Since $\mathbf{c}$ is the shortest vector in L along the 3-fold axis, $3f_3$ must be an integer. Hence $f_3 = 0$, 1/3 or 2/3. Also $\mathbf{f} - \mathbf{3}(\mathbf{f})$ is in L and

$$[\mathbf{f} - \mathbf{3}(\mathbf{f})]_P = \begin{bmatrix} f_1 + f_2 \\ 2f_2 - f_1 \\ 0 \end{bmatrix} .$$

Since {**a**,**b**} is a basis for the two-dimensional lattice perpendicular to the 3-fold axis, $f_1 + f_2$ and $2f_2 - f_1$ are integers. Hence there exist integers u and v such that

$$f_1 + f_2 = u$$
$$-f_1 + 2f_2 = v \quad .$$

Using the technique of Appendix 2, we solve this system by row reducing the augmented matrix

$$\left[\begin{array}{cc|c} 1 & 1 & u \\ -1 & 2 & v \end{array}\right]$$

obtaining

$$\left[\begin{array}{cc|c} 1 & 0 & u - (u + v)/3 \\ 0 & 1 & (u + v)/3 \end{array}\right] \quad .$$

Hence $f_1 = u - t/3$ and $f_2 = t/3$ where $t = u + v$. Since u and v are integers and $0 \le f_i < 1$, we obtain the three solutions $f_1 = 0$, $f_2 = 0$; $f_1 = 1/3$, $f_2 = 2/3$; and $f_1 = 2/3$, $f_2 = 1/3$. All combinations of f_1, f_2 and f_3 are considered in Table 6.2 where the possible combinations appear on the left and impossible combinations together with their contradictions appear on the right.

Table 6.2: Combinations of fractional coordinates for *3*.

Possible Coordinates			*Impossible Coordinates*			*Contradictions*
f_1	f_2	f_3	f_1	f_2	f_3	
0	0	0	1/3	2/3	0	*choice of* {**a**,**b**}
1/3	2/3	1/3	2/3	1/3	0	*choice of* {**a**,**b**}
2/3	1/3	1/3	0	0	1/3	*choice of* **c**
1/3	2/3	2/3	0	0	2/3	*choice of* **c**
2/3	1/3	2/3				

We shall now determine the possible factor groups L/P. There are three cyclic groups that occur:

$$P/P = \left\{ \begin{bmatrix} 0 \\ 0 \\ 0 \end{bmatrix} \right\} , \quad R(rev)/P = \left\{ \begin{bmatrix} 0 \\ 0 \\ 0 \end{bmatrix} , \begin{bmatrix} 1/3 \\ 2/3 \\ 1/3 \end{bmatrix} , \begin{bmatrix} 2/3 \\ 1/3 \\ 2/3 \end{bmatrix} \right\} .$$

$$R(obv)/P = \left\{ \begin{bmatrix} 0 \\ 0 \\ 0 \end{bmatrix} , \begin{bmatrix} 2/3 \\ 1/3 \\ 1/3 \end{bmatrix} , \begin{bmatrix} 1/3 \\ 2/3 \\ 2/3 \end{bmatrix} \right\} .$$

By $R(rev)$ and $R(obv)$ we mean the ***reverse*** and ***obverse*** settings of the ***rhombohedral lattice*** (see ITFC (Hahn, 1983)).

(P6.7) Problem: Prepare multiplication tables for each of the sets L mod P for P, $R(rev)$ and $R(obv)$ and show that each is closed and therefore a factor group. For $R(obv)$ we have

$$\begin{aligned} R(obv) &= P \cup C_P(2/3,1/3,1/3) \cup C_P(1/3,2/3,2/3) \\ &= \{n_1\mathbf{a} + n_2\mathbf{b} + n_3\mathbf{c} \mid n_1,n_2,n_3 \in Z\} \\ &\quad \cup \{m_1\mathbf{a} + m_2\mathbf{b} + m_3\mathbf{c} + (2/3)\mathbf{a} + (1/3)\mathbf{b} + (1/3)\mathbf{c} \mid m_1,m_2,m_3 \in Z\} \\ &\quad \cup \{k_1\mathbf{a} + k_2\mathbf{b} + k_3\mathbf{c} + (1/3)\mathbf{a} + (2/3)\mathbf{b} + (2/3)\mathbf{c} \mid k_1,k_2,k_3 \in Z\} . \end{aligned}$$

We shall select a basis for $R(obv)$ by including the two vectors that define the colattices and one of the basis vectors for P. We choose $D_R = \{(1/3)\mathbf{a} + (2/3)\mathbf{b} + (2/3)\mathbf{c},\ \mathbf{b},\ (2/3)\mathbf{a} + (1/3)\mathbf{b} + (1/3)\mathbf{c}\}$ and confirm in P6.8 that it is a basis.

(P6.8) Problem: Show that each vector in D_R is in $R(obv)$ and that each vector in $R(obv)$ can be expressed as an integral combination of D_R. That is show that D_R is a basis for the lattice $R(obv)$.

(P6.9) Problem: Show that

$$\{\mathbf{a},\ (2/3)\mathbf{a} + (1/3)\mathbf{b} + (1/3)\mathbf{c},\ (1/3)\mathbf{a} + (2/3)\mathbf{b} + (2/3)\mathbf{c}\}$$

is not a basis for $R(obv)$ by demonstrating that the vectors are not linearly independent.

The conventional basis for $R(obv)$ is

$$D_{R(obv)} = \{(2/3)\mathbf{a} + (1/3)\mathbf{b} + (1/3)\mathbf{c},\ -(1/3)\mathbf{a} + (1/3)\mathbf{b} + (1/3)\mathbf{c},\ -(1/3)\mathbf{a} - (2/3)\mathbf{b} + (1/3)\mathbf{c}\} \quad . \tag{6.1}$$

By E6.4, $D_{R(obv)}$ is a lattice equivalent to D_R. Hence $D_{R(obv)}$ is a basis for $R(obv)$. Hence,

$$\mathbf{3}(D_{R(obv)}) = \{\mathbf{3}((2/3)\mathbf{a} + (1/3)\mathbf{b} + (1/3)\mathbf{c}),\ \mathbf{3}(-(1/3)\mathbf{a} + (1/3)\mathbf{b} + (1/3)\mathbf{c},\ \mathbf{3}(-(1/3)\mathbf{a} - (2/3)\mathbf{b} + (1/3)\mathbf{c}\}$$

$$= \{-(1/3)\mathbf{a} + (1/3)\mathbf{b} + (1/3)\mathbf{c},\ -(1/3)\mathbf{a} - (2/3)\mathbf{b} + (1/3)\mathbf{c},\ (2/3)\mathbf{a} + (1/3)\mathbf{b} + (1/3)\mathbf{c}\} \quad .$$

Since $\mathbf{3}(D_{R(obv)}) = D_{R(obv)}$), $R(obv)$ is left invariant under $\mathbf{3}$. Thus, $R(obv)$ is left invariant under 3.

Note that $P_1 = \{\mathbf{a} + \mathbf{b}, -\mathbf{a}, \mathbf{c}\}$ is a basis for P, $\mathbf{3}(\mathbf{a} + \mathbf{b}) = -\mathbf{a}$ and $\mathbf{c}$ is perpendicular to $\mathbf{a} + \mathbf{b}$ and $-\mathbf{a}$. Hence P_1 satisfies all of the geometric criteria imposed on P. Furthermore with respect to P_1 the vectors in $R(rev)$ are found, using the change of basis matrix from P to P_1

$$\begin{bmatrix} 0 & 1 & 0 \\ -1 & 1 & 0 \\ 0 & 0 & 1 \end{bmatrix} ,$$

to be

$$\left\{ \begin{bmatrix} 0 \\ 0 \\ 0 \end{bmatrix}, \begin{bmatrix} 2/3 \\ 1/3 \\ 1/3 \end{bmatrix}, \begin{bmatrix} 1/3 \\ -1/3 \\ 2/3 \end{bmatrix} \right\} .$$

In the factor group an entry of $-1/3$ is the same as $2/3$ modulo P and so $R(rev)$ in this new basis is identical with $R(obv)$ in the old. Hence the two factor groups yield lattices that are not considered to be different. By convention the obverse setting is used and is called a ***rhombohedral lattice type***. Hence there are two lattice types left invariant under $\mathbf{3}$. The lattice type P is called the ***primitive hexagonal lattice type***. The centered rhombohedral lattice type is denoted R (see Figure 6.1).

Lattices invariant under *4*: As in Chapter 5, we choose the basis $P = \{\mathbf{a},\mathbf{b},\mathbf{c}\}$ for *4* such that **c** is along the positive direction along the 4-fold axis, **a** and **b** are perpendicular to the axis such that $\angle(\mathbf{a}:\mathbf{b}) = 90°$ and P forms a right-handed coordinate system. Since $M_P(4)$ consists of integral matrices (Table 5.2), L_P is left invariant under *4*.

Let L denote a lattice left invariant under *4*. By T6.15, since $2 \in 4$, all lattices left invariant under *4* have a nonzero lattice vector along the 4-fold axis and a lattice plane perpendicular to the axis passing through **0**. Let **a** be a lattice vector in this lattice plane of shortest length. Then $\mathbf{b} = 4(\mathbf{a})$ is in the lattice plane and it is also the shortest. Hence $\{\mathbf{a},\mathbf{b}\}$ is a basis for the lattice plane (T6.5). Let **c** be the shortest nonzero lattice vector along the positive direction of the 4-fold axis. Then $\{\mathbf{a},\mathbf{b},\mathbf{c}\}$ is a right-handed basis. Note that **2** is an element of *4*. Hence the possible lattice types left invariant under *4* include P, A, B and I as described for *2*. However, the A-centered lattice fails to be invariant under **4** since $4(\frac{1}{2}\mathbf{b} + \frac{1}{2}\mathbf{c}) = -\frac{1}{2}\mathbf{a} + \frac{1}{2}\mathbf{c}$ would also be in A implying that

$$(\tfrac{1}{2}\mathbf{b} + \tfrac{1}{2}\mathbf{c}) - (-\tfrac{1}{2}\mathbf{a} + \tfrac{1}{2}\mathbf{c}) = \tfrac{1}{2}\mathbf{a} + \tfrac{1}{2}\mathbf{b}$$

would also be in A. But $\frac{1}{2}\mathbf{a} + \frac{1}{2}\mathbf{b}$ in A would contradict the fact that $\{\mathbf{a},\mathbf{b}\}$ is a basis for the lattice plane. Similarly, B is rejected. Hence the only possible lattice types are P and I where

$$\begin{aligned} I &= (P + \mathbf{0}) \cup (P + (\tfrac{1}{2}\mathbf{a} + \tfrac{1}{2}\mathbf{b} + \tfrac{1}{2}\mathbf{c})) \\ &= \{u\mathbf{a} + v\mathbf{b} + w(\tfrac{1}{2}\mathbf{a} + \tfrac{1}{2}\mathbf{b} + \tfrac{1}{2}\mathbf{c}) \mid u,v,w \in Z\} \quad . \end{aligned}$$

(P6.10) Problem: Confirm that $D_I = \{\mathbf{a},\ \mathbf{b},\ \frac{1}{2}\mathbf{a} + \frac{1}{2}\mathbf{b} + \frac{1}{2}\mathbf{c}\}$ is a basis for I.

Note that

$$4(D_I) = \{\mathbf{b},\ -\mathbf{a},\ -\tfrac{1}{2}\mathbf{a} + \tfrac{1}{2}\mathbf{b} + \tfrac{1}{2}\mathbf{c}\} \quad .$$

(P6.11) Problem: Prove that $4(D_I)$ and D_I generate the same lattice.

Since the lattice points in the colattice $C_P(\frac{1}{2},\frac{1}{2},\frac{1}{2})$ appear at the center of each of the parallelepipeds outlined by the lattice points of P, I is called a *body-centered lattice type* (see Figure 6.1).

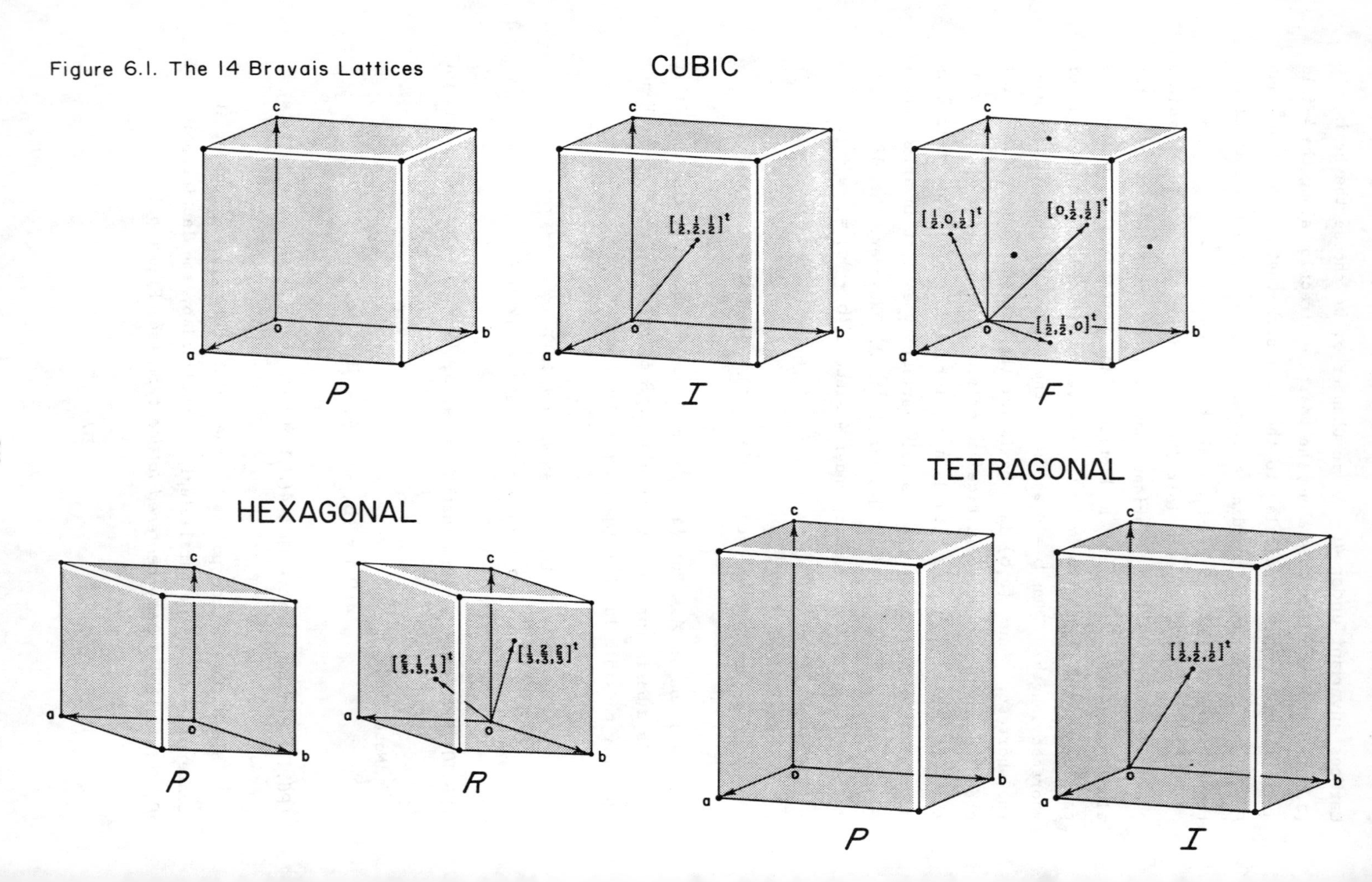

Figure 6.1. The 14 Bravais Lattices

Figure 6.1 (continued).

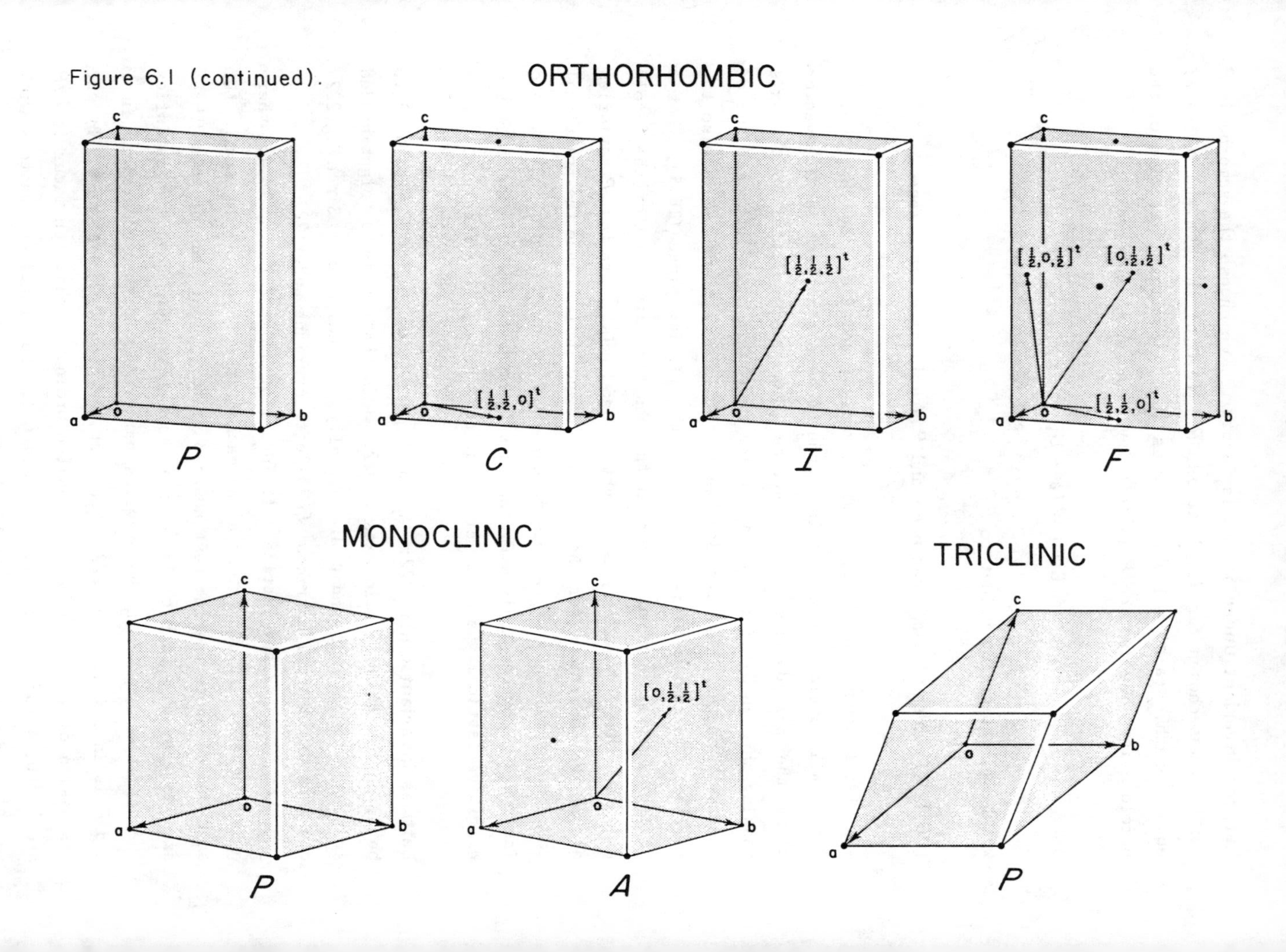

Lattices invariant under *6*: As in Chapter 5, we choose the same basis *P* for *6* as was chosen for *3*. That is, *P* = {**a**,**b**,**c**} where **a** and **b** are of equal length, <(**a**:**b**) = 120°, and **c** is in the positive direction of the 6-fold axis. The lattice *P* is left invariant under *6* because

$$6(P) = \{\mathbf{a} + \mathbf{b},\ -\mathbf{a},\ \mathbf{c}\}$$

which is lattice equivalent to *P*.

(P6.12) Problem: Show that **6**(*P*) and *P* are lattice equivalent.

Let *L* denote any lattice left invariant under *6*. By T6.15, since 2 ε *6*, all lattices left invariant under *6* have a nonzero lattice vector along the 6-fold axis and a lattice plane perpendicular to the axis passing through **0**. Let **c** be a lattice vector in this lattice plane of shortest length. Then **b** = **3**(**a**) is in the lattice plane and is also the shortest. Hence {**a**,**b**} is a basis for the lattice plane (T6.5). Let **c** be the shortest nonzero vector in the positive direction of the *6*-fold axis such that *P* = {**a**,**b**,**c**} is a right-handed basis. Note that *P* satisfies all of the conditions for the *P* used in *2* and in *3*. Hence the fractional coordinates allowed for vectors in *L*/*P* must appear in both Table 6.1 and Table 6.2. However, only 0,0,0 appears in both tables. Hence only the primitive lattice type *P* is left invariant under *6* (see Figure 6.1).

Lattices invariant under *222*: As in Chapter 5, we define *P* = {**a**,**b**,**c**} to be along the three 2-fold axes of *222* such that *P* is a right-handed system. Hence **a**, **b** and **c** are mutually perpendicular. Since $M_P(222)$ consists of integral matrices, *P* is left invariant under *222*.

Let *L* denote any lattice left invariant under *222*. By T6.15, there exist nonzero lattice vectors along each of the three 2-fold axes. Let **a**, **b** and **c** denote the shortest nonzero vectors along each of these three axes so that *P* = {**a**,**b**,**c**} is a right-handed system. Hence *P* satisfies the conditions for *P* in the previous paragraph and *P* is contained in *L* and *P* is invariant under *222*. Consider the 2-fold axis along **c**. By our discussion of *2*, the only fractional coordinates that can occur in *L*/*P* are f_i = 0, ½ for each *i*. As in Table 6.1, the fractional vectors with two 0's and one ½ are impossible by the choice of **a**, **b** and **c** as these are the shortest vectors in their directions. However, **a** and **b** do not necessarily form a basis for the lattice plane perpendicular to **c** and so the choice $[\frac{1}{2},\frac{1}{2},0]^t$ is possible for *222*. Hence we obtain the same *L/P* groups as in *2* together with

C/P	$\begin{bmatrix}0\\0\\0\end{bmatrix}$	$\begin{bmatrix}\frac{1}{2}\\\frac{1}{2}\\0\end{bmatrix}$
$\begin{bmatrix}0\\0\\0\end{bmatrix}$	$\begin{bmatrix}0\\0\\0\end{bmatrix}$	$\begin{bmatrix}\frac{1}{2}\\\frac{1}{2}\\0\end{bmatrix}$
$\begin{bmatrix}\frac{1}{2}\\\frac{1}{2}\\0\end{bmatrix}$	$\begin{bmatrix}\frac{1}{2}\\\frac{1}{2}\\0\end{bmatrix}$	$\begin{bmatrix}0\\0\\0\end{bmatrix}$

and

F/P	$\begin{bmatrix}0\\0\\0\end{bmatrix}$	$\begin{bmatrix}0\\\frac{1}{2}\\\frac{1}{2}\end{bmatrix}$	$\begin{bmatrix}\frac{1}{2}\\0\\\frac{1}{2}\end{bmatrix}$	$\begin{bmatrix}\frac{1}{2}\\\frac{1}{2}\\0\end{bmatrix}$
$\begin{bmatrix}0\\0\\0\end{bmatrix}$	$\begin{bmatrix}0\\0\\0\end{bmatrix}$	$\begin{bmatrix}0\\\frac{1}{2}\\\frac{1}{2}\end{bmatrix}$	$\begin{bmatrix}\frac{1}{2}\\0\\\frac{1}{2}\end{bmatrix}$	$\begin{bmatrix}\frac{1}{2}\\\frac{1}{2}\\0\end{bmatrix}$
$\begin{bmatrix}0\\\frac{1}{2}\\\frac{1}{2}\end{bmatrix}$	$\begin{bmatrix}0\\\frac{1}{2}\\\frac{1}{2}\end{bmatrix}$	$\begin{bmatrix}0\\0\\0\end{bmatrix}$	$\begin{bmatrix}\frac{1}{2}\\\frac{1}{2}\\0\end{bmatrix}$	$\begin{bmatrix}\frac{1}{2}\\0\\\frac{1}{2}\end{bmatrix}$
$\begin{bmatrix}\frac{1}{2}\\0\\\frac{1}{2}\end{bmatrix}$	$\begin{bmatrix}\frac{1}{2}\\0\\\frac{1}{2}\end{bmatrix}$	$\begin{bmatrix}\frac{1}{2}\\\frac{1}{2}\\0\end{bmatrix}$	$\begin{bmatrix}0\\0\\0\end{bmatrix}$	$\begin{bmatrix}0\\\frac{1}{2}\\\frac{1}{2}\end{bmatrix}$
$\begin{bmatrix}\frac{1}{2}\\\frac{1}{2}\\0\end{bmatrix}$	$\begin{bmatrix}\frac{1}{2}\\\frac{1}{2}\\0\end{bmatrix}$	$\begin{bmatrix}\frac{1}{2}\\0\\\frac{1}{2}\end{bmatrix}$	$\begin{bmatrix}0\\\frac{1}{2}\\\frac{1}{2}\end{bmatrix}$	$\begin{bmatrix}0\\0\\0\end{bmatrix}$

.

(P6.13) Problem: Find a basis D_C for $C = (P + \mathbf{0}) \cup (P + (\frac{1}{2}\mathbf{a} + \frac{1}{2}b))$ and show that C is invariant under *222*.

In the case of F,

$$F = (P + \mathbf{0}) \cup (P + (\tfrac{1}{2}\mathbf{b} + \tfrac{1}{2}\mathbf{c})) \cup (P + (\tfrac{1}{2}\mathbf{a} + \tfrac{1}{2}\mathbf{c})) \cup (P + (\tfrac{1}{2}\mathbf{a} + \tfrac{1}{2}\mathbf{b}))$$
$$= \{u(\tfrac{1}{2}\mathbf{b} + \tfrac{1}{2}\mathbf{c}) + v(\tfrac{1}{2}\mathbf{a} + \tfrac{1}{2}\mathbf{c}) + w(\tfrac{1}{2}\mathbf{a} + \tfrac{1}{2}\mathbf{b}) \mid u,v,w \in Z\} \quad .$$

(P6.14) Problem: Confirm that $D_F = \{\frac{1}{2}\mathbf{b} + \frac{1}{2}\mathbf{c}, \frac{1}{2}\mathbf{a} + \frac{1}{2}\mathbf{c}, \frac{1}{2}\mathbf{a} + \frac{1}{2}\mathbf{b}\}$ is a basis for F.

(P6.15) Problem: Show that F is left invariant under *222*.

(P6.16) Problem: Show that by a change of basis the *A* and *B* lattices can be expressed as *C* lattices

In summary, there are four lattice types left invariant under *222*, the primitive lattice type *P*, the body-centered lattice type *I*, the all-face centered lattice type *F*, and the one-faced centered lattice type *C* (see Figure 6.1).

The Lattices invariant under *322*: In our discussion of *3* we found that there are two lattice types left invariant under *3*. If we orient **c** along the positive direction of the 3-fold axis and **a** along one of the 2-fold axes, then **3**(**a**) is along another 2-fold axis. With this orientation of *P* = {**a**,**b**,**c**}, *P* is left invariant under *322* since as was shown in Chapter 5, $M_P(322)$ consists of integral matrices.

(P6.17) Problem: With the orientation of **a**, **b** and **c** as described above, show that *R*(*obv*) (see (6.1)) is invariant under *322*. (Hint: We already showed that *R*(*obv*) is invariant under **3**. Since *322* is generated by **3** and $^{[100]}\mathbf{2}$, you need only show that *R*(*obv*) is invariant under $^{[100]}\mathbf{2}$).

Hence the types of lattices left invariant under *322* are the same as those left invariant under *3*.

THE LATTICES INVARIANT UNDER *422* AND UNDER *622*

(P6.18) Problem: Show that the types of lattices left invariant under *422* are the same as those left invariant under *4*.

(P6.19) Problem: Show that the types of lattices left invariant under *622* are the same as those left invariant under *6*.

The lattices invariant under *23* **and under** *432*: Since *222* is a subgroup group of *23*, the lattices left invariant under *23* must be found amongst those left invariant under *222*. Since *23* is generated by the elements of *222* together with $^{[111]}\mathbf{3}$, we need only decide which of the lattices invariant under *222* are invariant under $^{[111]}\mathbf{3}$. Let *P* = {**a**,**b**,**c**} be defined as for *222*. Since $^{[111]}\mathbf{3}(\mathbf{a}) = \mathbf{b}$ and $^{[111]}\mathbf{3}(\mathbf{b}) = \mathbf{c}$, we have $a = b = c$. Hence the basis *P* for **23** will be such that **a**, **b** and **c** lie along perpendicular 2-fold axes and $a = b = c$.

(P6.20) Problem: Show that the lattices *P*, *F* and *I* based on the *P* for *23* are left invariant under *23*.

(P6.21) Problem: Show that the lattice *C* based on the *P* for *23* is not left invariant under *23*.

In view of P6.20 and P6.21, the types of lattices left invariant by *23* are *P*, *F* and *I*.

Note that *23* is a subgroup of *432* and that *432* is generated by the elements of *23* together with **4**. Hence the types of lattices left invariant under *432* are those left invariant under *23* and **4**.

(P6.22) Problem: Show that the lattices *P*, *F* and *I* based on the *P* for *23* are left invariant under *432*.

Diagrams of the three lattice types *P*, *F* and *I* left invariant under these cubic groups are given in Figure 6.1.

THE 14 BRAVAIS LATTICE TYPES

We have seen that corresponding to each of the eleven proper point groups from Table 5.4, a set of lattice types that include every lattice left invariant under the group can be found. The set of lattice types that are associated with a group *G* forms a crystal system. This classification scheme leads to the formation of seven crystal systems. With the exception of the primitive hexagonal lattice type, no type appears in more than one system. Hence of the 15 lattice types listed in column 2 of Table 6.3, 14 are different. These are called the 14 Bravais lattices (Figure 6.1). In column 3 of the table, the metrical matrix for the basis *P* of the primitive lattice *P* is given. Because the primitive lattices of the trigonal and hexagonal systems are the same there are only 6 different metrical matrices. This is why only 6 different crystal systems were considered in Chapter 5 where only the bases were evident. Recall that the 21 improper point groups of the form $G \cup Gi$ and $H \cup (G \setminus H)i$ leave the same lattices invariant as their underlying proper point group *G*. Hence we have shown that all 32 of the point groups listed in Table 5.4 are indeed crystallographic point groups.

Table 6.3: The 14 Bravais lattice types and the metrical matrices for the primitive bases.

Proper Point Group G	*Bravais Lattice Invariant under G*	*Metrical Matrix for P*	*Crystal System*
1	P	$g_{11}\ g_{12}\ g_{13}$ $g_{22}\ g_{23}$ g_{33}	*Triclinic*
2	P C	$g_{11}\ 0\ g_{13}$ $g_{22}\ 0$ g_{33}	*Monoclinic (Y unique axis)*
	P B	$g_{11}\ g_{12}\ 0$ $g_{22}\ 0$ g_{33}	*Monoclinic (Z unique axis)*
222	P C I F	$g_{11}\ 0\ 0$ $g_{22}\ 0$ g_{33}	*Orthorhombic*
4,422	P I	$g_{11}\ 0\ 0$ $g_{11}\ 0$ g_{33}	*Tetragonal*
3,32	P* R	$g_{11}\ -\frac{1}{2}g_{11}\ 0$ $g_{11}\ 0$ g_{33}	*Trigonal*
6,622	P*	$g_{11}\ -\frac{1}{2}g_{11}\ 0$ $g_{11}\ 0$ g_{33}	*Hexagonal*
23,432	P I F	$g_{11}\ 0\ 0$ $g_{11}\ 0$ g_{11}	*Cubic*

* These two *P* lattice types have the same metrical matrix and therefore are considered to be identical.

In determining when two lattice types should be considered the same we have given geometrical reasons where possible. Other times we have made the decision on conventional grounds. Firm definitions can be made based on the form of $M_P(G)$ for the bases P of the various lattices (Brown *et al.* 1978) that lead to the same decisions we have presented here.

MATRIX GROUPS REPRESENTING THE CRYSTALLOGRAPHIC POINT GROUPS

Thus far, we have found all of the lattice types and all of the point groups. In the next chapter we will find that the crystallographic space groups can be created by combining a crystallographic point group H (with some modifications) with each of the lattice types (viewed as translation groups) associated with H. To do this we will need to have the matrix representations of each of the point groups with respect to lattice bases. In most cases, we will use the basis for the primitive lattice P associated with H even when centered lattices are being considered. In Tables 5.2 and 5.3, we have presented the matrix representations of the individual point isometries that compose the point groups with respect to the basis P of the primitive lattice. Consequently, for most of the groups, H, we can immediately construct $M_P(H)$ without further work. There are, however, groups for which some choices must be made. That is, groups in which the placement of the symmetry elements of H with respect to the lattice can be made in more than one way. For example, in the group *322*, the 3-fold axis must be along **c**, but the 2-fold axes can be taken such that either $\{[100]_2, [110]_2, [010]_2\}$ represents the half-turns in *322* or such that $\{[210]_2, [120]_2, [\bar{1}10]_2\}$ represents the half-turns. These are considered to be different representations of *322* because, as shown below, the matrix representations of one cannot be transformed into those of the other by a change of basis of the lattice. Consequently, when constructing the space groups based on P and *322*, we have two distinct cases to consider. We denote this fact by calling *322* in the first case *321* and calling *322* in the second case *312*. We can show that there are no other ways to place the symmetry elements and still map the lattice into self-coincidence by observing that (see Figure 3.11) the image of **a** under any of the half-turns must map **a** into **a**, **a** + **b**, **b**, −**a**, −**a** − **b** or −**b** and noting that each of these cases occur in either *321* or *312*. Since the third-turn and any one of the half-turns generate *322*, we have exhausted all of the possibilities. The difficulty of orienting the point group with respect to the lattice in more than one way only occurs in the dihedral groups. In Table 6.4, we have listed each dihedral group with each of the possible orientations relative to the lattice P. The placement of each is determined by the orientation symbols used to describe the generators. Consequently there are a total of 37 matrix groups representing the 32 crystallographic point groups with respect to the appropriate primitive bases.

If H is a point group, we consider $M_1(H)$ and $M_2(H)$ to be different representations if there is no proper unimodular matrix T (the change of basis matrix) such that $M_1(H) = \{TMT^{-1} \mid M \in M_2(H)\}$. We will not give a detailed discussion on how to prove that such a matrix does not exist. However, we will illustrate a brute force method for doing this.

(E6.16) Example - $M_P(321)$ and $M_P(312)$ are different matrix representations for 322: Show that there does not exist a proper unimodular matrix T such that $M_P({}^{[100]}2) = TM_P({}^{[210]}2)T^{-1}$.

Solution: Suppose T is such a matrix. Then

$$M_P({}^{[100]}2)T = TM_P({}^{[210]}2)$$

$$\begin{bmatrix} t_{11} - t_{21} & t_{12} - t_{22} & t_{13} - t_{23} \\ -t_{21} & -t_{22} & -t_{23} \\ -t_{31} & -t_{32} & -t_{33} \end{bmatrix} = \begin{bmatrix} t_{11} + t_{12} & -t_{12} & -t_{13} \\ t_{21} + t_{22} & -t_{22} & -t_{23} \\ t_{31} + t_{32} & -t_{32} & -t_{33} \end{bmatrix}$$

which implies that $t_{21} = t_{22} = t_{12} = 0$ and that $2t_{13} = t_{23}$. Since $t_{21} = t_{22} = t_{12} = 0$, $\det(T) = -t_{23}t_{11}t_{32}$. Since $\det(T) = \pm 1$ and t_{23}, t_{11}, t_{32} are all integers, t_{23} equals either ± 1. But then $t_{13} = \pm\frac{1}{2}$ which contradicts the fact that $t_{13} \in Z$. □

Note that E6.16 can be used to show that neither $M_P({}^{[120]}2)$ nor $M_P([\bar{1}10]_2)$ can be transformed into $M_P({}^{[100]}2)$. This is because $3^{[210]}2 3^{-1} = [\bar{1}10]_2$ and $3^{-1}\,{}^{[210]}2\,3 = {}^{[120]}2$. Hence, if a proper unimodular matrix T existed such that

$$TM_P([\bar{1}10]_2)T^{-1} = M_P({}^{[100]}2) \quad ,$$

then,

$$(TM_P(3))M_P({}^{[210]}2)(M_P(3)^{-1}T^{-1}) = M_P({}^{[100]}2)$$

and so

$$(TM_P(3))M_P({}^{[210]}2)(TM_P(3))^{-1} = M_P({}^{[100]}2) \quad .$$

Since $M_P(3)$ is a proper unimodular matrix, $TM_P(3)$ is also one. But it was shown in E6.16 that no such matrix exists. Similarly, no such matrix exists from $M_P({}^{[120]}2)$ to $M_P({}^{[100]}2)$. Hence we have shown that the two matrix representations of *322*, $M_P(321)$ and $M_P(312)$, are not equivalent.

Table 6.4: The placements of the generating rotations α and β, of the dihedral groups relative to *P* giving rise to non-equivalent matrix groups.

H	α	β	*H*	α	β
222	2	$^{[100]}2$	*422*	4	$^{[100]}2$
mm2	2	$^{[100]}m$	*4mm*	4	$^{[100]}m$
321	3	$^{[100]}2$	$\bar{6}2m$	$\bar{6}$	$^{[100]}2$
312	3	$^{[210]}2$	$\bar{6}m2$	$\bar{6}$	$^{[100]}m$
3m1	3	$^{[100]}m$	*6mm*	6	$^{[100]}m$
31m	3	$^{[210]}2$	*622*	6	$^{[100]}2$
$\bar{3}m1$	$\bar{3}$	$^{[100]}2$	$\bar{4}m2$	$\bar{4}$	$^{[100]}m$
$\bar{3}1m$	$\bar{3}$	$^{[210]}2$	$\bar{4}2m$	$\bar{4}$	$^{[100]}2$

(P6.23) Problem: Consider the two following matrix representations that can be taken for point group $\bar{4}2m$:

$$M_P(\bar{4}2m) = \{M_P(1), M_P(\bar{4}), M_P(2), M_P(\bar{4}^{-1}), M_P({}^{[100]}2), M_P({}^{[110]}m), M_P({}^{[010]}2), M_P({}^{[1\bar{1}0]}m)\};$$

$$M_P(\bar{4}m2) = \{M_P(1), M_P(\bar{4}), M_P(2), M_P(\bar{4}^{-1}), M_P({}^{[100]}m), M_P({}^{[110]}2), M_P({}^{[010]}m), M_P({}^{[1\bar{1}0]}2)\} \quad .$$

Show that no change of basis matrix T exists that will transform the matrices of $M_P(\bar{4}2m)$ into those of $M_P(\bar{4}m2)$.

(P6.24) Problem: Show that no change of basis matrix T exists that will transform the matrices of $M_P(3m1)$ into those of $M_P(31m)$.

To this point we have only considered the effect of reorienting the axes of a point group with respect to the primitive lattice. We will explore the effect of these reorientations on the centered lattices in the next two problems.

(P6.25) Problem: Show that 312 does not map R(*obv*) into self-coincidence.

(P6.26) Problem: Show that $\bar{4}m2$ and $\bar{4}2m$ map *I* into self-coincidence.

CHAPTER 7

THE CRYSTALLOGRAPHIC SPACE GROUPS

With the derivation of the 230 space groups it may be said that the problem of finding the possible symmetry groups of periodic media is completely solved." -- W.H. Zachariasen

INTRODUCTION

In Chapters 4 and 5 we derived the 32 crystallographic point groups. In Chapter 6 we derived the 14 Bravais lattice types. In this Chapter we will derive all of the crystallographic space groups and collect them into the 230 space group types. This is done by "gluing" the point groups to the appropriate lattice groups. We will accomplish this by constructing 4×4 matrices which can represent both the point isometries and the translations related to the lattices. At the end of Chapter 6 we found that the 32 point groups are represented by 37 distinct matrix groups. It is these matrix groups that we will use in our derivation.

TRANSLATIONS

In Chapter 1 we introduced the notion of geometric three-dimensional space S. All of our work with S to this point has been done with the help of a frame of reference imposed on S by a choice of an origin and a basis. In this section we shall view S in a more general context without the restriction of a choice of origin and basis. When an origin is undefined, there is no natural connection between the points in S and a vector space and so, in this case, we will not use boldface letters to denote the points in S. However, we assume that certain geometric information is available in S. For example, the distance between any two points in S is determined, angles between lines can be calculated and whether two lines are parallel or not can be ascertained. Since distances are determined in S, isometries, mappings that preserve distances, are defined in S regardless of a choice of origin. An important isometry we have yet to discuss is the translation.

(D7.1) Definition: Let p and q denote points in S. The *translation* from

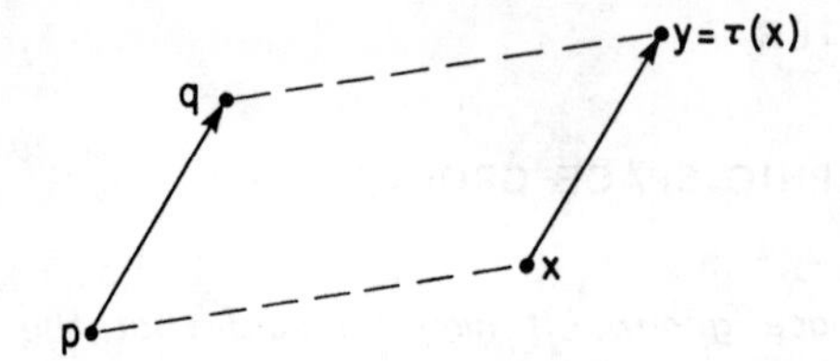

Figure 7.1: A directed line segment from point p to q defining a translational isometry τ that maps each point $x \varepsilon S$ to $\tau(x) = y$ such that pqyx is a parallelogram.

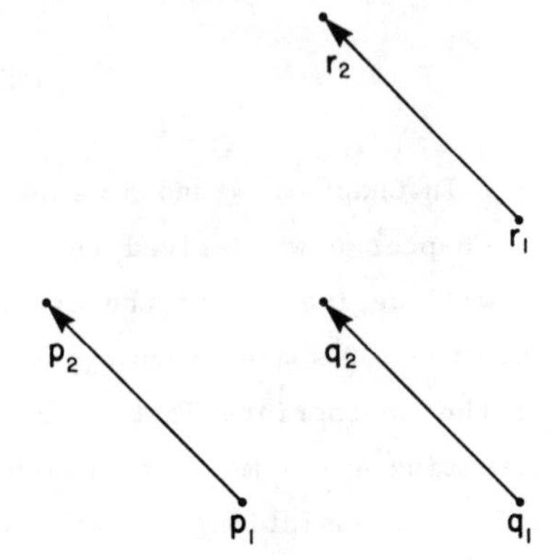

Figure 7.2: A translation τ displaces all points in space the same distance along parallel lines. Thus, if $\tau(p_1) = p_2$, then τ displaces q_1 to q_2 and r_1 to r_2

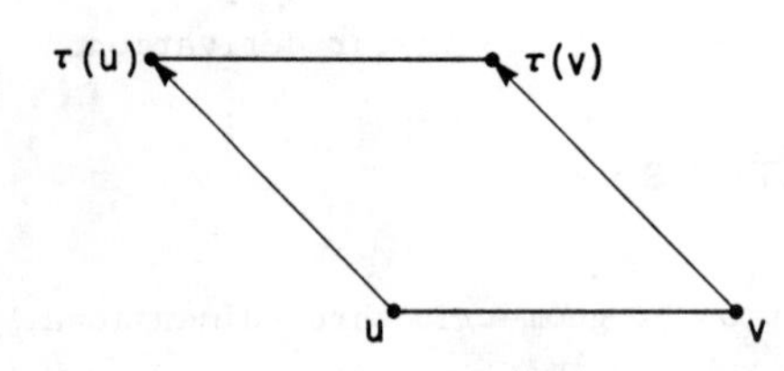

Figure 7.3: A translation τ is an isometry since if u and v are mapped to $\tau(u)$ and $\tau(v)$, then the distance between u and v equals that between $\tau(u)$ and $\tau(v)$ because $u\tau(u)\tau(v)v$ is a parallelogram.

p to q is the mapping τ that takes each $x \varepsilon S$ to $\tau(x) = y$, where y is the endpoint of the parallel copy of the directed line segment (arrow) from p to q positioned at x. That is, pqyx is a parallelogram (see Figure 7.1).

Note that any two points that define the same arrow as do p and q can be used to define τ. In Figure 7.2, if τ is the translation that maps p_1 to p_2, then it also maps q_1 to q_2 and r_1 to r_2 and τ is defined by any one of these pairs. The fact that τ is an isometry is verified by observing that if $u, v \varepsilon S$, then $u\tau(u)\tau(v)v$ is a parallelogram and so the distance between u and v equals that between $\tau(u)$ and $\tau(v)$ (see Figure 7.3). Hence τ is an isometry. If τ_1 and τ_2 are translations and $p \varepsilon S$, then $\tau_1 = \tau_2$ if and only if $\tau_1(p) = \tau_2(p)$. Consequently, two translations are equal if and only if they agree on any single point in S.

(T7.2) Theorem: The set of all translations Γ forms a group under composition.

The proof of this theorem is contained in the next three problems.

(P7.1) Problem: Show that the composition of two translations τ_1 followed by τ_2 is a translation. (Hint: Let p and q define τ_1 and q and r define τ_2. Then for an arbitrary $x \varepsilon S$, $\tau_3(x)$ where $\tau_3 = \tau_2\tau_1$ is found by taking $\tau_2(\tau_1(x))$. Hence the triangles pqr and $x\tau_1(x)\tau_3(x)$ are congruent. Use this observation to conclude that the arrows pr and $x\tau_3(x)$ are of the same length and direction. Hence τ_3 is the translation from p to r.)

(P7.2) Problem: Show that the identity mapping **1** is the translation from p to p and plays the role of the identity element with respect to composition.

(P7.3) Problem: Show that the composition of the translation from p to q and that from q to p is the identity mapping.

Since we showed that the composition of mappings is associative in Appendix 1, P7.1, P7.2 and P7.3 constitute a proof of T7.2. Note that the identity mapping **1** can be viewed as either a null translation (associated with an arrow of length 0) or a rotation of 0°. □

For the remainder of this chapter, we will denote the group of all translations by Γ. While the definition of a translation τ does not require the choice of an origin, τ can be more easily described within the framework provided by choosing an origin. Furthermore, within this framework, Γ is closely related to the vector space S of all vectors emanating from the origin. Let o denote the origin. Then any point $t \varepsilon S$ simultaneously defines a vector $\mathbf{t}$ in S emanating from o and the translation τ that maps o to t. In fact, the mapping τ can be described by

$$\tau(\mathbf{x}) = \mathbf{x} + \mathbf{t} \tag{7.1}$$

for all $\mathbf{x} \varepsilon S$. Here $\mathbf{t} = \tau(0)$ is called the *translational vector* associated with τ. Consequently there is a one-to-one correspondence between Γ and

S where each translation τ is associated with its translational vector $\tau(o)$. This close relationship between Γ and S extends into the algebraic nature of these sets as we shall see in the discussion below.

(E7.3) Example - The composition of two translational isometries: Show that if τ_1 and τ_2 are translations with translational vectors $\mathbf{t}_1$ and $\mathbf{t}_2$, respectively, then

$$\tau_1\tau_2(\mathbf{x}) = \mathbf{x} + (\mathbf{t}_1 + \mathbf{t}_2)$$

for all $\mathbf{x} \in S$.

Solution: Applying (7.1) twice we have

$$\begin{aligned} \tau_1\tau_2(\mathbf{x}) &= \tau_1(\tau_2(\mathbf{x})) \\ &= \tau_1(\mathbf{x} + \mathbf{t}_2) \\ &= (\mathbf{x} + \mathbf{t}_2) + \mathbf{t}_1 \\ &= \mathbf{x} + (\mathbf{t}_1 + \mathbf{t}_2) \text{ for all } \mathbf{x} \in S \quad . \end{aligned}$$ □

(P7.4) Problem: Show that if τ_1 and τ_2 are translations, then $\tau_1\tau_2 = \tau_2\tau_1$. That is, Γ is commutative under composition.

(D7.4) Definition: A group G is said to be an *abelian group* if G is commutative, that is, if

$$\mathbf{ab} = \mathbf{ba}$$

for all $\mathbf{a},\mathbf{b} \in G$.

Hence Γ is an abelian group.

(P7.5) Problem: Show that the cyclic groups derived in Chapter 4 are abelian.

(P7.6) Problem: Show that *322* is a *non-abelian group*, i.e., find two elements $\alpha,\beta \in$ *322* such that $\alpha\beta \neq \beta\alpha$.

(P7.7) Problem: Show that the translational vector for $\mathbf{1}$ is $\mathbf{0}$.

(E7.5) Example: Show that if $\mathbf{t}$ is the translational vector for τ, then $-\mathbf{t}$ is the translational vector for τ^{-1}. That is

$$\tau^{-1}(\mathbf{x}) = \mathbf{x} - \mathbf{t} \text{ for all } \mathbf{x} \in S \quad . \tag{7.2}$$

Solution: Let $\mathbf{s}$ denote the translational vector for τ^{-1}. Then

$$\tau^{-1}(\mathbf{x}) = \mathbf{x} + \mathbf{s} \text{ for all } \mathbf{x} \in S \quad .$$

Hence

$$\tau(\tau^{-1}(\mathbf{x})) = (\mathbf{x} + \mathbf{s}) + \mathbf{t}$$

and so

$$1(\mathbf{x}) = \mathbf{x} + (\mathbf{s} + \mathbf{t}) \quad .$$

By P7.7, $\mathbf{s} + \mathbf{t} = \mathbf{0}$ and consequently $\mathbf{s} = -\mathbf{t}$ as desired. □

(E7.6) Example: Show that if $\mathbf{t}_1$ and $\mathbf{t}_2$ are the translational vectors for τ_1, and τ_2, respectively, then

$$\tau_1^{-3}\tau_2^{2}(\mathbf{x}) = \mathbf{x} + (-3\mathbf{t}_1 + 2\mathbf{t}_2) \quad .$$

Solution: By repeated applications of (7.1) and (7.2), we have

$$\begin{aligned}
\tau_1^{-3}\tau_2^{2}(\mathbf{x}) &= \tau_1^{-3}\tau_2(\tau_2(\mathbf{x})) \\
&= \tau_1^{-3}\tau_2(\mathbf{x} + \mathbf{t}_2) \\
&= \tau_1^{-3}(\mathbf{x} + 2\mathbf{t}_2) \\
&= \tau_1^{-2}(\mathbf{x} + 2\mathbf{t}_2 - \mathbf{t}_1) \\
&= \mathbf{x} + (2\mathbf{t}_2 - 3\mathbf{t}_1) \\
&= \mathbf{x} + (-3\mathbf{t}_1 + 2\mathbf{t}_2) \quad .
\end{aligned}$$

□

Extending the logic used in E7.6, we see that if $\mathbf{t}$ is the translational vector for τ, then for any $n \in Z$,

$$\tau^{n}(\mathbf{x}) = \mathbf{x} + n\mathbf{t} \text{ for all } \mathbf{x} \in S \quad . \tag{7.3}$$

We generalize the definition of τ^{r} for the case where r is any real number by

$$\tau^{r}(\mathbf{x}) = \mathbf{x} + r\mathbf{t} \text{ for all } \mathbf{x} \in S \quad . \tag{7.4}$$

(P7.8) Problem: Using (7.4) show that

$$\tau^{r}\tau^{s} = \tau^{r+s} \text{ for } \tau \in \Gamma \text{ and } r,s \in R \quad .$$

(E7.7) Example: Show that

$$(\tau^r)^s = \tau^{rs} \text{ for } \tau \in \Gamma \text{ and } r,s \in R \quad .$$

Solution: By (7.4), we see that if $\mathbf{t}$ is the translation vector for τ, then $r\mathbf{t}$ is the translational vector for τ^r. Hence

$$\begin{aligned}(\tau^r)^s(\mathbf{x}) &= \mathbf{x} + s(r\mathbf{t}) \\ &= \mathbf{x} + (rs)\mathbf{t} \\ &= (\tau)^{rs}(\mathbf{x}) \quad .\end{aligned}$$

Hence $(\tau^r)^s = \tau^{rs}$. □

(P7.9) Problem: Let $\tau_1, \tau_2 \in \Gamma$ and $r \in R$. Show that

$$(\tau_1\tau_2)^r = \tau_1^r\tau_2^r \quad .$$

(Hint: Use E7.3.)

The results we have established above show that Γ is a vector space where vector "addition" is composition and scalar "multiplication" is exponentiation.

(P7.10) Problem: Rewrite each of the rules in D1.6 replacing $\mathbf{r} + \mathbf{s}$ by the composition $\mathbf{rs}$ and the scalar product $x\mathbf{r}$ by $\mathbf{r}^x$. Observe that we have verified all of these rules.

Let $\mathbf{o}$ denote an origin and $D = \{\mathbf{a},\mathbf{b},\mathbf{c}\}$ denote a basis for S. Let τ_x, τ_y, $\tau_z \in \Gamma$ be such that $\mathbf{a}$, $\mathbf{b}$. and $\mathbf{c}$ are their translational vectors, respectively. That is $\tau_x(\mathbf{o})$, $\tau_y(\mathbf{o})$ and $\tau_z(\mathbf{o})$ are the endpoints of $\mathbf{a}$, $\mathbf{b}$ and $\mathbf{c}$, respectively. If r, s, $t \in R$, then

$$\tau_x^r\tau_y^s\tau_z^t(\mathbf{x}) = \mathbf{x} + r\mathbf{a} + s\mathbf{b} + t\mathbf{c} \quad . \tag{7.5}$$

Since D is a basis, every vector in S can be written in the form $r\mathbf{a} + s\mathbf{b} + t\mathbf{c}$ and so every translation can be written in the form $\tau_x^r\tau_y^s\tau_z^t$. Consequently, $\Gamma = \{\tau_x^r\tau_y^s\tau_z^t \mid r,s,t \in R\}$. Therefore, $\{\tau_x,\tau_y,\tau_z\}$ is a basis for Γ and so Γ is a three dimensional vector space. Furthermore (7.5) gives the natural correspondence between Γ and S:

$$\tau_x^r\tau_y^s\tau_z^t \leftrightarrow r\mathbf{a} + s\mathbf{b} + t\mathbf{c} \quad , \tag{7.6}$$

which can be used to show that Γ and S are isomorphic. Note that viewing Γ as a group of mappings on S,

$$\mathrm{orb}_{\Gamma}(\mathbf{o}) = S \quad .$$

Now suppose that $D = \{\tau_x, \tau_y, \tau_z\}$ denotes some basis for Γ and that o_1 denotes a point in S. Then the basis of S associated with D and the origin o_1 is the set of vectors emanating from o_1 to $\tau_x(o_1)$, $\tau_y(o_1)$ and $\tau_z(o_1)$, respectively. We shall denote this basis by

$$D(o_1) = \{\tau_x(o_1) = \mathbf{a}_1,\ \ \tau_y(o_1) = \mathbf{b}_1,\ \ \tau_z(o_1) = \mathbf{c}_1\} \quad .$$

By identifying each $\tau \varepsilon \Gamma$ with the vector whose endpoint is $\tau(o_1)$ in S, we obtain the correspondence in (7.6) between Γ and the vectors emanating from o_1. Similarly, using

$$D(o_2) = \{\tau_x(o_2) = \mathbf{a}_2,\ \ \tau_y(o_2) = \mathbf{b}_2,\ \ \tau_z(o_2) = \mathbf{c}_2\} \quad ,$$

we obtain the correspondence in (7.6) between Γ and the vectors emanating from o_2. The bases $D(o_1) = \{\mathbf{a}_1, \mathbf{b}_1, \mathbf{c}_1\}$ and $D(o_2) = \{\mathbf{a}_2, \mathbf{b}_2, \mathbf{c}_2\}$ are similar in the sense that $\mathbf{a}_2$ is a translated parallel copy of $\mathbf{a}_1$, $\mathbf{b}_2$ is a parallel copy of $\mathbf{b}_1$ and $\mathbf{c}_2$ a parallel copy of $\mathbf{c}_1$. In view of (7.6), if $\tau \varepsilon \Gamma$, then

$$[\tau(o_1)]_{D(o_1)} = [\tau(o_2)]_{D(o_2)} \quad . \tag{7.7}$$

Hence corresponding to each translation τ and basis D there is a triple $\mathbf{t} \varepsilon R^3$ such that $\mathbf{t}$ represents the translational vector for τ regardless of origin.

Let $D_1 = \{\tau_x, \tau_y, \tau_z\}$ and D_2 denote bases for Γ. Let o_1 denote a point in S. Then the change of basis matrix T from $D_1(o_1)$ to $D_2(o_1)$ is

$$\mathrm{T} = \left[[\tau_x(o_1)]_{D_2(o_1)} \;\middle|\; [\tau_y(o_1)]_{D_2(o_1)} \;\middle|\; [\tau_z(o_1)]_{D_2(o_1)} \right] \quad .$$

By (7.7),

$$\begin{aligned} [\tau_x(o_1)]_{D_2(o_1)} &= [\tau_x(o_2)]_{D_2(o_2)} \\ [\tau_y(o_1)]_{D_2(o_1)} &= [\tau_y(o_2)]_{D_2(o_2)} \\ [\tau_z(o_1)]_{D_2(o_1)} &= [\tau_z(o_2)]_{D_2(o_2)} \quad . \end{aligned}$$

Hence T is also the change of basis matrix from $D_1(o_2)$ to $D_2(o_2)$. Hence we call T the *change of basis matrix from D_1 to D_2.*

(T7.8) Theorem: Let $D = \{\tau_x, \tau_y, \tau_z\}$ denote a basis for Γ. Let o_1, $o_2 \in S$ and $\tau \in \Gamma$ such that $\tau(o_1) = o_2$. Let $\mathbf{x}$, $\mathbf{y} \in S$. Then

$$[\mathbf{x}]_{D(o_1)} = [\mathbf{y}]_{D(o_2)}$$

if and only if $\tau(\mathbf{x}) = \mathbf{y}$.

Proof: Let $D(o_1) = \{\mathbf{a}_1, \mathbf{b}_1, \mathbf{c}_1\}$ and $D(o_2) = \{\mathbf{a}_2, \mathbf{b}_2, \mathbf{c}_2\}$ be defined as above. Suppose $[\mathbf{x}]_{D(o_1)} = [\mathbf{y}]_{D(o_2)}$. Then there exists p, q, $r \in R$ such that, simultaneously,

$$\mathbf{x} = p\mathbf{a}_1 + q\mathbf{b}_1 + r\mathbf{c}_1 \text{ and } \mathbf{y} = p\mathbf{a}_2 + q\mathbf{b}_2 + r\mathbf{c}_2 \quad .$$

Hence the translation $\tau_1 = \tau_x^p \tau_y^q \tau_z^r$ is associated with both $\mathbf{x}$ and $\mathbf{y}$ in the sense that $\tau_1(o_1) = \mathbf{x}$ and $\tau_1(o_2) = \mathbf{y}$. Then

$$\begin{aligned} \tau(\mathbf{x}) &= \tau(\tau_1(o_1)) \\ &= \tau_1(\tau(o_1)) \quad (\Gamma \text{ is abelian}) \\ &= \tau_1(o_2) \\ &= \mathbf{y} \quad . \end{aligned}$$

Conversely, suppose $\tau(\mathbf{x}) = \mathbf{y}$. Let p, q, $r \in R$ be such that

$$\mathbf{x} = p\mathbf{a}_1 + q\mathbf{b}_1 + r\mathbf{c}_1 \quad .$$

Then $\tau_x^p \tau_y^q \tau_z^r(o_1) = \mathbf{x}$. Hence

$$\begin{aligned} \mathbf{y} &= \tau(\mathbf{x}) \\ &= \tau(\tau_x^p \tau_y^q \tau_z^r(o_1)) \\ &= \tau_x^p \tau_y^q \tau_z^r(\tau(o_1)) \quad (\Gamma \text{ is abelian}) \\ &= \tau_x^p \tau_y^q \tau_z^r(o_2) \quad . \end{aligned}$$

Hence

$$\mathbf{y} = p\mathbf{a}_2 + q\mathbf{b}_2 + r\mathbf{c}_2 \quad ,$$

and so $[\mathbf{x}]_{D(o_1)} = [\mathbf{y}]_{D(o_2)} = [pqr]^t$. □

Thus far we have discussed various types of isometries. We have completely characterized the point isometries and found that only rotations and rotoinversions are possible. In the previous section we studied translations. We shall now characterize all isometries.

(T7.9) Theorem: Let α denote an isometry, then α can be expressed as

$$\alpha = \tau\beta$$

where τ is a translation and β is a point isometry.

Proof: Let α denote an isometry and let o denote a point in S. If $\alpha(o) = o$, then α is a point isometry and so, $\alpha = 1\alpha$ is in the desired form when 1 is the identity mapping (viewed as a translation). Now suppose $\alpha(o) = p \neq o$. Let τ be the translation such that $\tau(p) = o$. Then $\tau\alpha$ is an isometry such that

$$\begin{aligned} \tau\alpha(o) &= \tau(\alpha(o)) \\ &= \tau(p) \\ &= o \ . \end{aligned}$$

Hence $\tau\alpha$ is a point isometry. Let $\beta = \tau\alpha$. Then $\alpha = \tau^{-1}\beta$. Since τ is a translation, by T7.2, τ^{-1} is a translation and so α is the composition of a point isometry followed by a translation. □

The proof of T7.9 shows that corresponding to each choice of o, the isometry α can be written in the form $\tau\beta$ where β is not only a point isometry but one that maps o to itself. Hence if we choose a point o to be the origin, then α can be written as the composition of a point isometry that fixes the origin followed by a translation. If we change the origin to another point, then the point isometry used may be different. However, for a fixed origin, the translation τ and the point isometry β are unique as we shall show in the following theorem.

(T7.10) Theorem: Let α denote an isometry and let **o** denote an origin. If $\alpha = \tau_1\beta_1 = \tau_2\beta_2$ where $\beta_1(\mathbf{o}) = \mathbf{o} = \beta_2(\mathbf{o})$, then $\tau_1 = \tau_2$ and $\beta_1 = \beta_2$.

Proof: Since $\tau_1\beta_1 = \tau_2\beta_2$, we have

$$\tau_2^{-1}\tau_1\beta_1 = \beta_2 \quad .$$

Then

$$\tau_2^{-1}\tau_1\beta_1(\mathbf{o}) = \beta_2(\mathbf{o})$$

$$\tau_2^{-1}\tau_1(\mathbf{o}) = \mathbf{o}.$$

Hence $\tau_2^{-1}\tau_1 = \mathbf{1}$ and so $\tau_1 = \tau_2$. Furthermore, by the left cancellation law, $\beta_1 = \beta_2$. □

(T7.11) Theorem: Let α denote an isometry and let o_1 and o_2 denote points in S. Suppose $\alpha = \tau_1\beta_1$ where $\beta_1(o_1) = o_1$. If $\alpha = \tau_2\beta_2$ where $\beta_2(o_2) = o_2$, then $\tau_2 = \tau_1\tau^{-1}$ and $\beta_2 = \tau\beta_1$ where τ is the translation such that $\tau(\beta_1(o_2)) = o_2$.

Proof: Note that

$$\tau\beta_1(o_2) = \tau(\beta_1(o_2)) = o_2 \quad ,$$

and so $\tau\beta_1$ is a point isometry. Since $\tau_1\tau^{-1}$ is a translation and $\tau\beta_1$ is a point isometry such that $\alpha = (\tau_1\tau^{-1})(\tau\beta_1)$, by T7.10, $\tau_2 = \tau_1\tau^{-1}$ and $\beta_2 = \tau\beta_1$. □

Suppose the origin $\mathbf{o}$ is fixed and α is an isometry. Then by T7.10 there exists a unique translation τ and point isometry β such that $\beta(\mathbf{o}) = \mathbf{o}$ and $\alpha = \tau\beta$. We call τ the ***translational component*** of α and β the ***linear component*** of α (since a point isometry is a linear mapping) with respect to $\mathbf{o}$. Since $\alpha = \tau\beta$,

$$\begin{aligned} \alpha(\mathbf{r}) &= \tau(\beta(\mathbf{r})) \\ &= \beta(\mathbf{r}) + \mathbf{t} \quad . \end{aligned} \tag{7.8}$$

where $\mathbf{t} = \tau(\mathbf{o})$ is the translational vector for τ. Now suppose a basis $D = \{\mathbf{a},\mathbf{b},\mathbf{c}\}$ is chosen. Then β can be represented by the 3×3 matrix $M_D(\beta)$ and the translational vector $\mathbf{t}$ of τ can be represented as the triple $[\mathbf{t}]_D$. Creating a 4×4 matrix, denoted $R_D(\alpha) = \{M_D(\beta) \mid [\mathbf{t}]_D\}$, from these in the following manner,

$$R_D(\alpha) = \{M_D(\beta) \mid [\mathbf{t}]_D\} = \left[\begin{array}{c|c} M_D(\beta) & [\mathbf{t}]_D \\ & \\ 0\ 0\ 0 & 1 \end{array}\right] \tag{7.9}$$

we obtain a matrix representation for α. In order to accomplish this representation, we use 4-tuples, elements of R^4, to represent the vectors, $\mathbf{r}$ in S. The 4-tuples are constructed by putting a dummy 1 in the fourth position below $[\mathbf{r}]_D$.

(E7.12) Example: Verify that $R_D(\alpha) = \{M_D(\beta) \mid [\mathbf{t}]_D\}$ represents $\alpha = \tau\beta$ where $\mathbf{t} = \tau(\mathbf{o})$ in the sense that

$$R_D(\alpha)\begin{bmatrix} [\mathbf{r}]_D \\ \\ 1 \end{bmatrix} = \begin{bmatrix} [\alpha(\mathbf{r})]_D \\ \\ 1 \end{bmatrix}.$$

Solution: Let

$$M_D(\beta) = \begin{bmatrix} \ell_{11} & \ell_{12} & \ell_{13} \\ \ell_{21} & \ell_{22} & \ell_{23} \\ \ell_{31} & \ell_{32} & \ell_{33} \end{bmatrix},$$

$$[\mathbf{t}]_D = \begin{bmatrix} t_1 \\ t_2 \\ t_3 \end{bmatrix}, \quad [\mathbf{r}]_D = \begin{bmatrix} x \\ y \\ z \end{bmatrix}.$$

Then

$$\{M_D(\beta) \mid [\mathbf{t}]_D\}\begin{bmatrix} [\mathbf{r}]_D \\ \\ 1 \end{bmatrix} = \begin{bmatrix} \ell_{11} & \ell_{12} & \ell_{13} & t_1 \\ \ell_{21} & \ell_{22} & \ell_{23} & t_2 \\ \ell_{31} & \ell_{32} & \ell_{33} & t_3 \\ 0 & 0 & 0 & 1 \end{bmatrix}\begin{bmatrix} x \\ y \\ z \\ 1 \end{bmatrix}$$

$$= \begin{bmatrix} \ell_{11}x + \ell_{12}y + \ell_{13}z + t_1 \\ \ell_{21}x + \ell_{22}y + \ell_{23}z + t_2 \\ \ell_{31}x + \ell_{32}y + \ell_{33}z + t_3 \\ 1 \end{bmatrix}$$

$$= \begin{bmatrix} \ell_{11}x + \ell_{12}y + \ell_{13}z \\ \ell_{21}x + \ell_{22}y + \ell_{23}z \\ \ell_{31}x + \ell_{32}y + \ell_{33}z \\ 1 \end{bmatrix} + \begin{bmatrix} t_1 \\ t_2 \\ t_3 \\ 0 \end{bmatrix}$$

$$= \begin{bmatrix} M_D(\beta)[\mathbf{r}]_D \\ \\ 1 \end{bmatrix} + \begin{bmatrix} [\mathbf{t}]_D \\ \\ 0 \end{bmatrix}$$

$$= \begin{bmatrix} [\beta(\mathbf{r})]_D \\ \\ 1 \end{bmatrix} + \begin{bmatrix} [\mathbf{t}]_D \\ \\ 0 \end{bmatrix}$$

$$= \begin{bmatrix} [\beta(\mathbf{r})]_D + [\mathbf{t}]_D \\ \\ 1 \end{bmatrix}$$

$$= \begin{bmatrix} [\beta(\mathbf{r}) + \mathbf{t}]_D \\ \\ 1 \end{bmatrix}$$

$$= \begin{bmatrix} [\alpha(\mathbf{r})]_D \\ \\ 1 \end{bmatrix} \qquad \text{(by (7.8))}$$

□

In E7.12 we showed that the 4 × 4 matrix given in (7.9) represents α where $\alpha = \tau\beta$. The notation $\{M_D(\beta) \mid [\mathbf{t}]_D\}$ is called the *Seitz notation* (Seitz, 1935) for α with respect to $\mathbf{o}$ and D.

(D7.13) Definition: Let $\alpha = \tau\beta$ denote an isometry where τ is the translational component and β is the linear component of α with respect to the origin $\mathbf{o}$. Let D denote a basis. Then the matrix $R_D(\alpha) = \{M_D(\beta) \mid [\tau(\mathbf{o})]_D\}$ is called the 4 × 4 *matrix representation of* α *with respect to* D.

Let $\mathbf{o}$ denote a choice of origin and D a basis. If α and γ are isometries, then so is $\alpha\gamma$ and since

$$\alpha\gamma(\mathbf{r}) = \alpha(\gamma(\mathbf{r})) \quad ,$$

$$R_D(\alpha\gamma)\begin{bmatrix} [\mathbf{r}]_D \\ \\ 1 \end{bmatrix} = R_D(\alpha)\left(R_D(\gamma)\begin{bmatrix} [\mathbf{r}]_D \\ \\ 1 \end{bmatrix}\right) \quad .$$

Since matrix multiplication is associative, this equals

$$\left(R_D(\alpha)R_D(\gamma)\right)\begin{bmatrix} [\mathbf{r}]_D \\ \\ 1 \end{bmatrix} \quad .$$

Hence

$$R_D(\alpha\gamma) = R_D(\alpha)R_D(\gamma) \quad . \tag{7.10}$$

The composition $\alpha\gamma$ can also be viewed in terms of the decomposition described in T7.9. Suppose $\gamma = \tau_1\beta_1$ and $\alpha = \tau_2\beta_2$, then

$$\begin{aligned}(\alpha\gamma)(\mathbf{r}) &= (\tau_2\beta_2\tau_1\beta_1)(\mathbf{r}) \\ &= (\tau_2\beta_2\tau_1)(\beta_1(\mathbf{r})) \\ &= (\tau_2\beta_2)(\beta_1(\mathbf{r}) + \tau_1(\mathbf{o})) \\ &= \tau_2(\beta_2(\beta_1(\mathbf{r})) + \beta_2(\tau_1(\mathbf{o}))) \\ &= \beta_2\beta_1(\mathbf{r}) + \beta_2(\tau_1(\mathbf{o})) + \tau_2(\mathbf{o}) \quad .\end{aligned}$$

Hence, $\alpha\gamma = \tau\beta$ where $\beta = \beta_2\beta_1$ and τ is the translation whose translational vector is $\beta_2(\tau_1(\mathbf{o})) + \tau_2(\mathbf{o})$. Therefore, in the Seitz notation, given an origin $\mathbf{o}$ and a basis D,

$$R_D(\alpha\gamma) = \{M_2M_1 \mid M_2\mathbf{t}_1 + \mathbf{t}_2\} \quad , \tag{7.11}$$

where $M_1 = M_D(\beta_1)$, $M_2 = M_D(\beta_2)$, $\mathbf{t}_1 = [\tau_1(\mathbf{o})]_D$, $\mathbf{t}_2 = [\tau_2(\mathbf{o})]_D$. Note that $\mathbf{t}_1$ and $\mathbf{t}_2$ are defined here to be vectors in R^3. Since $R_D(\alpha\gamma) = R_D(\alpha)R_D(\gamma)$,

$$\{M_2 \mid \mathbf{t}_2\}\{M_1 \mid \mathbf{t}_1\} = \{M_2M_1 \mid M_2\mathbf{t}_1 + \mathbf{t}_2\} \quad . \tag{7.12}$$

(P7.11) Problem: Using matrix multiplication, confirm that (7.12) holds for any 3×3 matrices M_1 and M_2 and any two vectors $\mathbf{t}_1$ and $\mathbf{t}_2$ in R^3.

Suppose that $R_D(\alpha) = \{M \mid \mathbf{t}\}$. Then we can find $R_D(\alpha^{-1})$ by noting that $R_D(\alpha^{-1}) = (R_D(\alpha))^{-1}$ and observing that if $R_D(\alpha^{-1}) = \{N \mid \mathbf{r}\}$ then

$$\{M \mid \mathbf{t}\}\{N \mid \mathbf{r}\} = \{I_3 \mid \mathbf{0}\} .$$

Hence, by (7.11)

$$\{MN \mid M\mathbf{r} + \mathbf{t}\} = \{I_3 \mid \mathbf{0}\} .$$

Hence $N = M^{-1}$ and $\mathbf{r}$ is such that $M\mathbf{r} + \mathbf{t} = \mathbf{0}$. That is,

$$\mathbf{r} = -M^{-1}\mathbf{t} .$$

Consequently,

$$\{M \mid \mathbf{t}\}^{-1} = \{M^{-1} \mid -M^{-1}\mathbf{t}\} . \tag{7.13}$$

(P7.12) Problem: Use (7.11) to confirm (7.13) by showing that

$$\{M \mid \mathbf{t}\}\{M^{-1} \mid -M^{-1}\mathbf{t}\} = \{I_3 \mid \mathbf{0}\} .$$

(P7.13) Problem: Show that

$$\{M \mid \mathbf{t}\}^{-1} = \{M^{-1} \mid -M^{-1}\mathbf{t}\}$$

by extending the procedure for calculating the inverse of a 3 × 3 matrix given in Appendix 2 to the 4 × 4 case.

(T7.14) Theorem: Let I denote the group of all isometries. Then Γ is a normal subgroup of I.

Proof: Let $\alpha \in I$ and $\tau \in \Gamma$. According to the remark following DA7.11, we need only show that $\alpha\tau\alpha^{-1} \in \Gamma$. Using the Seitz notation with respect to an origin $\mathbf{o}$ and a basis D, we can write $R_D(\alpha) = \{M \mid \mathbf{t}_1\}$ and $R_D(\tau) = \{I_3 \mid \mathbf{t}_2\}$. By (7.13), $R_D(\alpha^{-1}) = \{M^{-1} \mid -M^{-1}\mathbf{t}_1\}$. Hence

$$\begin{aligned} R_D(\alpha\tau\alpha^{-1}) &= \{M \mid \mathbf{t}_1\}\{I_3 \mid \mathbf{t}_2\}\{M^{-1} \mid -M^{-1}\mathbf{t}_1\} \\ &= \{M \mid M\mathbf{t}_2 + \mathbf{t}_1\}\{M^{-1} \mid -M^{-1}\mathbf{t}_1\} \\ &= \{I_3 \mid M\mathbf{t}_2 + \mathbf{t}_1 - \mathbf{t}_1\} \\ &= \{I_3 \mid M\mathbf{t}_2\} . \end{aligned} \tag{7.14}$$

Hence $\alpha\tau\alpha^{-1}$ is a translation. Therefore $\alpha\tau\alpha^{-1} \in \Gamma$ for all $\alpha \in I$ and $\tau \in \Gamma$ and so Γ is a normal subgroup of I. □

This result says that we can almost commute translations and isometries. Since $\alpha\tau\alpha^{-1} \in \Gamma$ where $\alpha \in I$ and $\tau \in \Gamma$, we have $\alpha\tau\alpha^{-1} = \tau'$ for some $\tau' \in \Gamma$. Hence $\alpha\tau = \tau'\alpha$. That is, if we start with $\alpha\tau$ we can move the translation τ to the other side of α if we trade τ for a different translation τ'. This will be used to simplify some of our expressions.

(T7.15) Theorem: Let α denote an isometry, let o_1 and o_2 denote points in S and let $D = \{\tau_x, \tau_y, \tau_z\}$ denote a basis of Γ. Then

$$M_{D(o_1)}(\beta_1) = M_{D(o_2)}(\beta_2) \quad ,$$

where β_1 and β_2 are the linear components of α with respect to o_1 and o_2, respectively.

Proof: Let $\{\mathbf{a}_1, \mathbf{b}_1, \mathbf{c}_1\}$ denote the vectors in $D(o_1)$ and $\{\mathbf{a}_2, \mathbf{b}_2, \mathbf{c}_2\}$ those in $D(o_2)$. The first column of $M_{D(o_1)}(\beta_1)$ is $[\beta_1(\mathbf{a}_1)]_{D(o_1)}$ and the first column of $M_{D(o_2)}(\beta_2)$ is $[\beta_2(\mathbf{a}_2)]_{D(o_2)}$. By T7.11 $\beta_2 = \tau\beta_1$ where $\tau(\beta_1(o_2)) = o_2$. Hence

$$\begin{aligned} \beta_2(\mathbf{a}_2) &= \tau\beta_1(\mathbf{a}_2) \\ &= \tau\beta_1(\tau_x(o_2)) \\ &= \tau(\beta_1\tau_x)(o_2) \quad . \end{aligned}$$

Since Γ is a normal subgroup of I there exists a τ_x' such that

$$\beta_1\tau_x = \tau_x'\beta_1$$

and $\tau_x' = \beta_1\tau_x\beta_1^{-1} \in \Gamma$. Hence

$$\begin{aligned} \beta_2(\mathbf{a}_2) &= \tau(\tau_x'\beta_1)(o_2) \\ &= \tau_x'(\tau(\beta_1(o_2)) \qquad (\Gamma \text{ is abelian}) \\ &= \tau_x'(o_2) \qquad (\text{since } \tau(\beta_1(o_2)) = o_2) \\ &= \beta_1\tau_x\beta_1^{-1}(o_2) \quad . \end{aligned}$$

Also

$$\begin{aligned}\beta_1\tau_x\beta_1^{-1}(o_1) &= \beta_1(\tau_x(\beta_1^{-1}(o_1))) \\ &= \beta_1(\tau_x(o_1)) \qquad \text{(since } \beta_1 \text{ fixes } o_1\text{)} \\ &= \beta_1(\mathbf{a}_1) \quad .\end{aligned}$$

Since $\beta_1\tau_x\beta_1^{-1} \varepsilon \Gamma$, there exists $p,q,r \varepsilon R$ such that

$$\beta_1\tau_x\beta_1^{-1} = \tau_x^p\tau_y^q\tau_z^r \quad ,$$

and so $[\beta_1(\mathbf{a}_1)]_{D(o_1)} = [pqr]^t = [\beta_2(\mathbf{a}_2)]_{D(o_2)}$. Hence the first column of $M_{D(o_1)}(\beta_1)$ equals that of $M_{D(o_2)}(\beta_2)$. Similarly, the second and third columns of $M_{D(o_1)}(\beta_1)$ equal those of $M_{D(o_2)}(\beta_2)$. Therefore, $M_{D(o_1)}(\beta_1) = M_{D(o_2)}(\beta_2)$. □

Theorem T7.15 tells us a remarkable fact: *the nature of the linear component of α is the same regardless of the choice of origin*. For example, if β_1 and β_2 denote the linear components of α with respect to o_1 and o_2, respectively, and if β_1 is, say, a half-turn about an axis ℓ through o_1, then β_2 is a half-turn about an axis ℓ' through o_2 where ℓ' parallels ℓ.

Besides being important to our derivation of the crystallographic space groups, the next several results will enable us to transfer information about a crystal structure that is described in terms of a basis $D_1(o_1)$ to a new basis and origin $D_2(o_2)$.

(T7.16) Theorem: Let α denote an isometry, let $D = \{\tau_x,\tau_y,\tau_z\}$ denote a basis for Γ and let o_1 and o_2 denote points in S. Then if

$$R_{D(o_1)}(\alpha) = \{M \mid \mathbf{t}\} \quad ,$$

where M is a 3 × 3 matrix and $\mathbf{t} \varepsilon R^3$, then

$$R_{D(o_2)}(\alpha) = \{M \mid (M - I_3)\mathbf{p} + \mathbf{t}\} \quad ,$$

where $\mathbf{p} = [o_2]_{D(o_1)}$.

Proof: Let β_1 and β_2 denote the linear components of α with respect tc o_1 and o_2. Since $M = M_{D(o_1)}(\beta_1)$, T7.15 implies that $M = M_{D(o_2)}(\beta_2)$.

Let τ_1 and τ_2 be the translational components of α with respect to o_1 and o_2, respectively. Then $R_{D(o_2)}(\alpha) = \{M \mid [\tau_2(o_2)]\}_{D(o_2)}$. By T7.11, $\tau_2 = \tau_1\tau^{-1}$ where $\tau^{-1}(o_2) = \beta_1(o_2)$. We shall next determine the translational vector for τ_2 with respect to o_1. Let τ_p denote the translation such that $\tau_p(o_1) = o_2$. Then

$$\begin{aligned}\tau_2 &= \tau_1\tau^{-1} \\ &= \tau_1\tau^{-1}\tau_p^{-1}\tau_p \\ &= (\tau_1\tau_p^{-1})(\tau^{-1}\tau_p) \qquad (\Gamma \text{ is abelian}) \quad .\end{aligned}$$

Since $\tau^{-1}\tau_p(o_1) = \beta_1(o_2)$, $[\beta_1(o_2)]_{D(o_1)}$ represents the translational vector of $\tau^{-1}\tau_p$ with respect to o_1. The translational vector of τ_p^{-1}, with respect to o_1, is represented by $-[o_2]_{D(o_1)}$. Since $\mathbf{t}$ represents the translational vector of τ_1,

$$\begin{aligned}\tau_2(o_1) &= [\beta_1(o_2)]_{D(o_1)} - [o_2]_{D(o_1)} + \mathbf{t} \\ &= M[o_2]_{D(o_1)} - I_3[o_2]_{D(o_1)} + \mathbf{t} \\ &= (M - I_3)[o_2]_{D(o_1)} + \mathbf{t} \quad ,\end{aligned}$$

$(M - I_3)[o_2]_{D(o_1)} + \mathbf{t}$ is the translational vector of τ_2 with respect to $D(o_1)$. But the translational vector of τ_2 with respect to $D(o_2)$ is then also $(M - I_3)[o_2]_{D(o_1)} + \mathbf{t}$ by (7.7). Consequently,

$$R_{D(o_2)}(\alpha) = \{M \mid (M - I_3)\mathbf{p} + \mathbf{t}\} \quad ,$$

where $\mathbf{p} = [o_2]_{D(o_1)}$. □

(T7.17) Theorem: Let D_1 and D_2 denote bases for Γ, $\mathbf{o}$ denote the origin, T denote the change of basis matrix from D_1 to D_2 and let $\mathbf{r} \in S$. Then, using the R^4 representation of $\mathbf{r}$, we have

$$\{T \mid [000]^t\} \begin{bmatrix} [\mathbf{r}]_{D_1(\mathbf{o})} \\ \\ 1 \end{bmatrix} = \begin{bmatrix} [\mathbf{r}]_{D_2(\mathbf{o})} \\ \\ 1 \end{bmatrix} .$$

The proof of T7.17 is straightforward and is left to the reader.

(T7.18) Theorem: Let D denote a basis for Γ and let o_1 and o_2 denote points in S. Let $\mathbf{p} = [o_2]_{D(o_1)}$. Then if $r \in S$,

$$\begin{bmatrix} [r]_{D(o_2)} \\ \\ 1 \end{bmatrix} = \{I_3 \mid -\mathbf{p}\} \begin{bmatrix} [r]_{D(o_1)} \\ \\ 1 \end{bmatrix}$$

Again, the proof of this theorem is left to the reader.

(P7.14) Problem: Show that if we move the origin from o_1 to o_2 and if we then change the basis from D_1 to D_2, we have

$$\begin{bmatrix} [\mathbf{r}]_{D_2(o_2)} \\ \\ 1 \end{bmatrix} = \{T \mid -T\mathbf{p}\} \begin{bmatrix} [\mathbf{r}]_{D_1(o_1)} \\ \\ 1 \end{bmatrix}$$

where T is the change of basis matrix from D_1 to D_2, $\mathbf{p} = [o_2]_{D(o_1)}$ and $\mathbf{r} \in S$.

(P7.15) Problem: A structural analysis of coesite by Zoltai and Buerger (1959) shows it to be monoclinic with space group symmetry *B2/b*. The coordinates of the atoms determined in the analysis (Table 7.1) define a framework structure of silicate tetrahedra with double-crankshaft chains like those in sanidine, $KAlSi_3O_8$, a monoclinic feldspar with space group symmetry *C2/m*. The structures differ in that the chains in sanidine are related by a mirror plane whereas those in coesite are related by a glide plane. Figure 7.4 displays the sanidine structure between 0 and $\frac{1}{2}$. The origin $\mathbf{o}_1$ and basis vectors $D_1 = \{\mathbf{a}_1, \mathbf{b}_1, \mathbf{c}_1\}$ define the setting of the coesite unit cell and $\mathbf{o}_2$ and $D_2 = \{\mathbf{a}_2, \mathbf{b}_2, \mathbf{c}_2\}$ define the setting of the sanidine cell. The connection between these cells was established by Megaw (1970) who wrote a matrix for transforming the coesite coordinates to match those of sanidine. Following a comparison of the resulting coordinates with those of sanidine, she concluded that the coesite structure is impossible for $KAlSi_3O_8$ because there are no cavities in it large enough to accommodate K. She also concluded that the feldspar structure is impossible for coesite because it would require a bridging SiOSi angle of 110°, which theoretical evidence indicates is too narrow for stability (Gibbs, 1982).

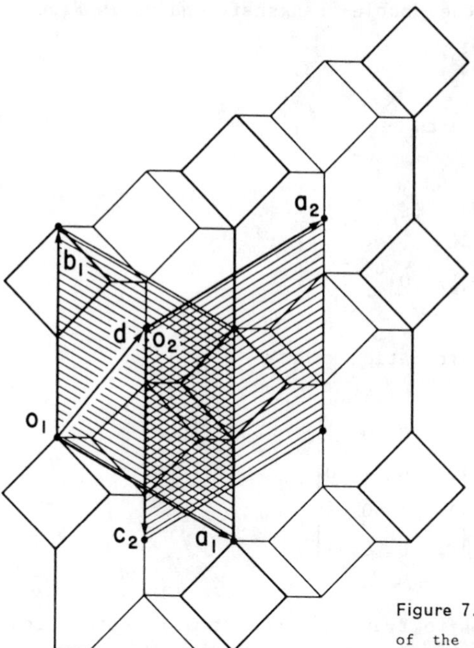

Figure 7.4: A drawing of the tetrahedral nodes of the double–crankshaft chains in sanidine spanning the structure along [100]. The origin o_1 and the D_1-basis vectors (c_1 is perpendicular to the plane forming a right-handed system) define the setting defined by Zoltai and Buerger (1959) for the coesite unit cell, and o_2 and the D_2-basis vectors (b_2 is perpendicular to the plane forming a right-handed system) define the setting for the sanidine unit cell. The vector **d** is the change of origin vector. This drawing is modified after Megaw (1970).

Table 7.1: Comparison of the atomic coordinates of coesite and sanidine.

	Zoltai & Buerger (1959)			*Megaw* (1970)				*Sanidine*		
Atom	x_1	y_1	z_1	x_2	y_2	z_2	*Atom*	x	y	z
Si_1	.1403	.0735	.1084	.6403	.1084	.3168	T_2	.7089	.1178	.3444
Si_2	.5063	.5388	.1576	.0063	.1576	.2175	T_1	.0097	.1850	.2233
O_1	0	0	0	−1/2	0	1/4	$O_{A(2)}$	−.3653	0	.2858
O_2	1/2	3/4	.1166	0	.1166	0	$O_{A(1)}$	0	.1472	0
O_3	.7306	.5595	.1256	.2306	.1256	.4211	O_D	.1793	.1269	.4025
O_4	.3080	.3293	.1030	−.1920	.1030	.2287	O_B	−.1722	.1469	.2244
O_5	.4877	.5274	.2878	−.0123	.2878	.2103	O_C	.0341	.3100	.2574

(1) Examine the drawing of the double-crankshaft chains in Figure 7.4 and show that

$$\mathbf{a}_2 = \mathbf{a}_1 + \mathbf{b}_1$$
$$\mathbf{b}_2 = \mathbf{c}_1$$
$$\mathbf{c}_2 = -\mathbf{b}_1$$

and $[\mathbf{o}_2]_{D_1} = [\mathbf{p}]_{D_1} = [\tfrac{1}{2}, 3/4, 0]^t$.

(2) Formulate the 4 × 4 transformation matrix

$$\{T \mid -T\mathbf{p}\} = \begin{bmatrix} 1 & 0 & 0 & \tfrac{1}{2} \\ 0 & 0 & 1 & 0 \\ 1 & -1 & 0 & \tfrac{1}{4} \\ 0 & 0 & 0 & 1 \end{bmatrix} .$$

(3) Use the matrix $\{T \mid -T\mathbf{p}\}$ to transform the atomic coordinates of coesite (Zoltai and Buerger, 1959) to match those in sanidine and compare your results with those obtained by Megaw (Table 7.1, Column 3).

(T7.19) Theorem: Let D_1 and D_2 denote bases of Γ, let o_1 and o_2 denote points in S and let α denote an isometry. Then

$$R_{D_2(o_2)}(\alpha) = \{TMT^{-1} \mid T[(M - I_3)\mathbf{p} + \mathbf{t}]\} \quad ,$$

where $R_{D_1(o_1)}(\alpha) = \{M \mid \mathbf{t}\}$, T is the change of basis matrix from D_1 to D_2 and $\mathbf{p} = [o_2]_{D_1(o_1)}$.

Proof: Consider the following circuit diagram where $\mathbf{r} \in S$:

$$\begin{array}{ccc} \begin{bmatrix} [\mathbf{r}]_{D_1(o_1)} \\ 1 \end{bmatrix} & \xrightarrow{\{T \mid -T\mathbf{p}\}} & \begin{bmatrix} [\mathbf{r}]_{D_2(o_2)} \\ 1 \end{bmatrix} \\ \{M \mid \mathbf{t}\} \downarrow & & \downarrow R_{D_2(o_2)}(\alpha) \\ \begin{bmatrix} [\alpha(\mathbf{r})]_{D_1(o_1)} \\ 1 \end{bmatrix} & \xrightarrow{\{T \mid -T\mathbf{p}\}} & \begin{bmatrix} [\alpha(\mathbf{r})]_{D_2(o_2)} \\ 1 \end{bmatrix} \end{array} .$$

It follows from the circuit diagram that

$$\begin{aligned} R_{D_2(o_2)}(\alpha) &= \{T \mid -T\mathbf{p}\}\{M \mid \mathbf{t}\}\{T \mid -T\mathbf{p}\}^{-1} \\ &= \{T \mid -T\mathbf{p}\}\{M \mid \mathbf{t}\}\{T^{-1} \mid \mathbf{p}\} \\ &= \{TMT^{-1} \mid -T\mathbf{p} + T[\mathbf{t} + M\mathbf{p}]\} \\ &= \{TMT^{-1} \mid T[(M - I_3)\mathbf{p} + \mathbf{t}]\} \quad . \end{aligned} \tag{7.15}$$

□

Note that (7.15) expresses the fact that $R_{D_2(o_2)}(\alpha)$ and $R_{D_1(o_1)}(\alpha)$ are similar matrices in the realm of all 4 × 4 real matrices (see Appendix 7).

CRYSTALLOGRAPHIC SPACE GROUPS

Not all groups of isometries are suitable for describing the symmetry of crystals. In Chapter 4, for example, we found the crystallographic restrictions that must be imposed on point isometries. In this section we study the restrictions that must be imposed on groups of isometries that leave the structure of some crystal invariant. The very nature of a crystal implies that there is three-dimensional periodicity in the structure. This periodicity is described in terms of a crystallographic translation group.

(D7.20) Definition: Let o denote a point in S. A translation group T is said to be a *crystallographic translation group* if and only if $\mathrm{orb}_T(o)$ is a three-dimensional lattice whose vectors emanate from o.

Note that, by (7.6), if T is a crystallographic translation group and if $\{\mathbf{a},\mathbf{b},\mathbf{c}\}$ is a basis for the lattice $\mathrm{orb}_T(o)$, then $D = \{\tau_x,\tau_y,\tau_z\}$ is a generating set for T where $\tau_x(o) = \mathbf{a}$, $\tau_y(o) = \mathbf{b}$ and $\tau_z(o) = \mathbf{c}$. That is, if

$$\mathrm{orb}_T(o) = \{u\mathbf{a} + v\mathbf{b} + w\mathbf{c} \mid u,v,w \in Z\} \quad ,$$

then

$$\mathrm{orb}_T(o) = \{\tau_x^u\tau_y^v\tau_z^w(o) \mid u,v,w \in Z\} \quad ,$$

and so

$$T = \{\tau_x^u\tau_y^v\tau_z^w \mid u,v,w \in Z\} \quad .$$

(D7.21) Definition: A set of translations $D = \{\tau_1,\tau_2,\tau_3\}$ of a crystallographic translation group T is said to be a generating set of T if and only if D is a basis for Γ and

$$T = \{\tau_1^u \tau_2^v \tau_3^w \mid u,v,w \in Z\} \quad .$$

Let $D = \{\tau_1, \tau_2, \tau_3\}$ denote a generating set for the crystallographic translation group T and let $o \in S$. Then $D(o)$ is a basis for the lattice $orb_T(o)$ as shown in the following problem.

(P7.16) Problem: Let T denote a crystallographic translation group and let $D = \{\tau_1, \tau_2, \tau_3\}$ be a generating set for T. Show that $L_{D(o)}$ is the $orb_T(o)$ where $o \in S$.

We shall now turn our attention to groups of isometries.

(P7.17) Problem: Show that if G denotes a group of isometries, then the set T of all of the translational isometries in G is a subgroup of G.

The subgroup T in P7.17 is called the *translation group* of G.

(D7.22) Definition: A group of isometries G is called a *crystallographic space group* if the translation group of G is a crystallographic translation group.

(T7.23) Theorem: Let G denote a crystallographic space group. Then the translation group of G is a normal subgroup of G.

(P7.18) Problem: Prove T7.23. (Hint: See the proof of T7.14 where G will play the role of I and the translation group of G will play the role of Γ.)

Let G denote a crystallographic space group and let T denote its translation group. If $D = \{\tau_x, \tau_y, \tau_z\}$ is a generating set for T, then, as shown in P7.16, $L_{D(\mathbf{o})}$ is the lattice equal to $orb_T(\mathbf{o})$. If $\alpha \in G$, then, by T7.15, the matrix representation for the linear component with respect to $D(\mathbf{o})$ is the same for all choices of origin $\mathbf{o}$. Let $M_D(\alpha)$ denote the 3×3 matrix representation of the linear component of α with respect to D. Also, let $M_D(G)$ denote the set

$$M_D(G) = \{M_D(\alpha) \mid \alpha \in G\} \quad ,$$

where G is any groups of isometries.

(P7.19) Problem: Let G denote a group of isometries and let $D = \{\tau_x, \tau_y, \tau_z\}$ denote a basis for Γ. Prove that $M_D(G)$ is a group. (Hint: See (7.12) and (7.13)).

Hence associated with each crystallographic space group G there is a group of 3×3 matrices $M_{D(G)}$ and a lattice $L_{D(\mathbf{o})}$. We will see how all possible crystallographic space groups G that share the same associated group of 3×3 matrices and lattice can be constructed. Before we can do this we must learn more about $M_D(G)$ and $L_{D(\mathbf{o})}$. Since $M_D(G)$ represents the linear components of the isometries in G, we will study these linear components first.

(D7.24) Definition: Let α denote an isometry and let $\mathbf{o} \in S$ denote the origin. Then we denote the linear component of α by $\Lambda_{\mathbf{o}}(\alpha)$. If G is a group of isometries, then we define $\Lambda_{\mathbf{o}}(G)$ to be

$$\Lambda_{\mathbf{o}}(G) = \{\Lambda_{\mathbf{o}}(\alpha) \mid \alpha \in G\} \quad .$$

(P7.20) Problem: Prove that $\Lambda_{\mathbf{o}}(G)$ is a group.

By T7.15, note that for any points $o_1, o_2 \in S$, $\Lambda_{o_1}(G)$ and $\Lambda_{o_2}(G)$ are isomorphic since they are both represented by $M_D(G)$. Using $\Lambda_{\mathbf{o}}(G)$, we will discover the relationship between $M_D(G)$ and $L_{D(\mathbf{o})}$.

(T7.25) Theorem: Let G denote a crystallographic space group, $D = \{\tau_x, \tau_y, \tau_z\}$ a generating set for its translation group, and $\mathbf{o}$ the origin. Then $\Lambda_{\mathbf{o}}(G)$ leaves $L_{D(\mathbf{o})}$ invariant. Hence $\Lambda_{\mathbf{o}}(G)$ is one of the 32 crystallographic point groups.

Proof: Recall that if $\mathbf{v}$ is a vector in S emanating from $\mathbf{o}$, then $\mathbf{v} \in L_{D(\mathbf{o})}$ if and only if $[\mathbf{v}]_{D(\mathbf{o})} \in Z^3$. Let $\beta \in \Lambda_{\mathbf{o}}(G)$ and $\mathbf{v} \in L_{D(\mathbf{o})}$. Let τ denote the translation such that $\tau(\mathbf{o}) = \mathbf{v}$. Then

$$R_{D(\mathbf{o})}(\beta) = \{M_D(\beta) \mid \mathbf{0}\}$$

and

$$R_{D(\mathbf{o})}(\tau) = \{I_3 \mid [\mathbf{v}]_{D(\mathbf{o})}\} \quad .$$

Hence

$$R_{D(\mathbf{o})}(\beta\tau\beta^{-1}) = \{I_3 \mid M_D(\beta)[\mathbf{v}]_{D(\mathbf{o})}\}$$
$$= \{I_3 \mid [\beta(\mathbf{v})]_{D(\mathbf{o})}\} .$$

Since the translation group T of G is a normal subgroup, $\beta\tau\beta^{-1} \in T$. Hence $[\beta(\mathbf{v})]_{D(\mathbf{o})} \in Z^3$ and so $\beta(\mathbf{v}) \in L_{D(\mathbf{o})}$. Hence $\beta(L_{D(\mathbf{o})})$ is a subset of $L_{D(\mathbf{o})}$. Since β is an isometry, $\beta(L_{D(\mathbf{o})}) = L_{D(\mathbf{o})}$. Hence $L_{D(\mathbf{o})}$ is invariant under $\Lambda_{\mathbf{o}}(G)$. □

Note that, in the proof of T7.25, we found that the isometry $\beta\tau\beta^{-1}$ is associated with the lattice vector $[\beta(\mathbf{v})]_{D(\mathbf{o})} = [\beta(\tau(\mathbf{o}))]_{D(\mathbf{o})}$. Hence, the fact that $\beta\tau\beta^{-1} \in T(G)$ is equivalent to the fact that $\beta(\mathbf{v}) \in L_{D(\mathbf{o})}$. Consequently an abstract approach to the study of space groups can be conducted from the point of view of the action of elements of G on elements of $T(G)$ where the action is defined by $\tau \to \alpha\tau\alpha^{-1}$ for a given $\alpha \in G$. This approach (see Farkas, 1981) has the advantage of being origin free and such that the invariance of $\text{orb}_{T(G)}(\mathbf{o})$ under G becomes the normality of $T(G)$ in G. In our treatment, however, we will continue to distinguish between a crystallographic translation group and its associated lattice with respect to an origin. This will enable us to relate our results directly to the geometry and symmetry of crystals.

Theorem T7.25 implies that if G is a crystallographic point group and if D is a basis for Γ, then $M_D(G) = M_D(H)$ for some point group H. Consequently we will build the crystallographic space groups from the matrix representations of the point groups and the lattices that fall within the appropriate Bravais lattice types. For each point group, H, we use the matrix representation of H with respect to the basis P of the primitive lattice belonging to H In Chapter 6 we found that there are 37 such matrix groups. The point groups for which there are two such matrix groups are listed in Table 6.4 where we have symbolized the corresponding point groups according to their orientations. We shall refer to these groups as the 37 *oriented point groups*.

In order to correctly build the space group G from its associated oriented point group and lattice, we must know how each fits inside G. The next several results will accomplish this for us.

Suppose G is a crystallographic space group associated with the oriented point group H and lattice L. Let $\{M \mid \mathbf{t}\} \in R_{P(\mathbf{o})}(G)$. Then we know that $M \in M_P(H)$. But we know much less about $\mathbf{t}$. If $M = I_3$, we know that $\{M \mid \mathbf{t}\}$ represents an element in $T(G)$ and so $\mathbf{t}$ represents a vector in L. Furthermore, all of the elements of $R_{P(\mathbf{o})}(G)$ of the form $\{M \mid \mathbf{t}\}$, where M is fixed, are related in a special way as described by the following theorem.

(T7.26) Theorem: Let G denote a crystallographic space group and let M be a matrix in $M_D(G)$. Let $\mathbf{t}_1 \in R^3$ be such that $\{M \mid \mathbf{t}_1\} \in R_{D(\mathbf{o})}(G)$. Then $\{M \mid \mathbf{t}_2\} \in R_{D(\mathbf{o})}(G)$ if and only if $\{I_3 \mid \mathbf{t}_2 - \mathbf{t}_1\} \in R_{D(\mathbf{o})}(T(G))$.

Proof: Suppose $\{M \mid \mathbf{t}_2\} \in R_{D(\mathbf{o})}(G)$, then $\{M \mid \mathbf{t}_2\}\{M \mid \mathbf{t}_1\}^{-1} \in R_{D(\mathbf{o})}(G)$. But

$$\begin{aligned}\{M \mid \mathbf{t}_2\}\{M \mid \mathbf{t}_1\}^{-1} &= \{M \mid \mathbf{t}_2\}\{M^{-1} \mid -M^{-1}\mathbf{t}_1\} \\ &= \{I_3 \mid \mathbf{t}_2 - \mathbf{t}_1\}.\end{aligned}$$

Conversely, if $\{I_3 \mid \mathbf{t}_2 - \mathbf{t}_1\} \in R_{D(\mathbf{o})}(T(G))$, then

$$\{I_3 \mid \mathbf{t}_2 - \mathbf{t}_1\}\{M \mid \mathbf{t}_1\} = \{M \mid \mathbf{t}_2\} \in R_{D(\mathbf{o})}(G). \qquad \square$$

Theorem T7.26 tells us that if we are constructing elements in $R_{D(\mathbf{o})}(G)$ of the form $\{M \mid \mathbf{t}_1\}$ and $\{M \mid \mathbf{t}_2\}$, then the points represented by $\mathbf{t}_1$ and $\mathbf{t}_2$ must be L-equivalent where $L = \mathrm{orb}_{T(G)}(\mathbf{o})$.

(E7.27) Example: Suppose G is a crystallographic space group and that $\Lambda_{\mathbf{o}}(G) = 2$. Suppose

$$\{M_D({}^{[010]}2) \mid \mathbf{t}_1\} = \begin{bmatrix} -1 & 0 & 0 & 2.5 \\ 0 & 1 & 0 & -1.5 \\ 0 & 0 & -1 & 3 \\ 0 & 0 & 0 & 1 \end{bmatrix}$$

is in $R_{D(\mathbf{o})}(G)$ and that $T(G) = P = L_{D(\mathbf{o})}$ is a monoclinic primitive lattice type in the second setting (Y unique axis - see Table 6.3). Find a $\mathbf{t} \in R^3$ that is as simple as possible such that $\{M_D({}^{[010]}2) \mid \mathbf{t}_1\} \in R_{D(\mathbf{o})}(G)$.

Solution: Since P is primitive, $\{[\mathbf{v}]_{D(\mathbf{o})} \mid \mathbf{v} \in P\} = Z^3$. Hence, by T7.26, $\{M_D({}^{[010]}2) \mid \mathbf{t}_2\} \in R_{D(\mathbf{o})}(G)$ if and only if $\mathbf{t}_2 - \mathbf{t}_1 = \mathbf{t}$ where $\mathbf{t} \in Z^3$. That is, any $\mathbf{t}_2 = \mathbf{t}_1 + \mathbf{t}$ where $\mathbf{t} \in Z^3$ forms an element $\{M_D({}^{[010]}2) \mid \mathbf{t}_2\}$ in

$R_{D(\mathbf{o})}(G)$. Since $[-2,2,-3]^t \in Z^3$, $\{M_D({}^{[010]}2) \mid [.5,.5,0]^t\} \in R_{D(\mathbf{0})}(G)$. This is the simplest form for elements of this type. □

(E7.28) Example: Solve E7.27 where $L_{D(\mathbf{o})}$ is taken to be a monoclinic C-centered lattice type C.

Solution: Since C is C-centered,

$$\{[\mathbf{v}]_{D(\mathbf{o})} \mid \mathbf{v} \in C\} = Z^3 \cup \{Z^3 + [\tfrac{1}{2},\tfrac{1}{2},0]^t\} \quad .$$

Hence

$$\{M_D({}^{[010]}2) \mid \mathbf{t}_2\} \in G$$

if and only if $\mathbf{t}_2 = \mathbf{t}_1 + \mathbf{t}$ where $\mathbf{t} \in Z^3$ or $\mathbf{t} \in Z^3 + [\tfrac{1}{2},\tfrac{1}{2},0]^t$. Since $[-2.5,1.5,-3]^t = [-3,1,-3]^t \in Z^3 + [\tfrac{1}{2},\tfrac{1}{2},0]^t$, $\{M_D({}^{[010]}2 \mid [000]^t\} \in R_{D(\mathbf{o})}(G)$. □

The next theorem yields more information on what types of elements $\mathbf{t}$ of R^3 can occur in matrices $\{M \mid \mathbf{t}\}$ in $R_{D(\mathbf{o})}(G)$.

(T7.29) Theorem: Let G denote a crystallographic space group, let $\mathbf{h}_1, \mathbf{h}_2,\ldots,\mathbf{h}_n \in \Lambda_{\mathbf{o}}(G)$ be such that

$$\mathbf{h}_1\mathbf{h}_2\ldots\mathbf{h}_k = \mathbf{h}_{k+1}\mathbf{h}_{k+2}\ldots\mathbf{h}_n \quad . \tag{7.16}$$

Let D denote a basis for Γ and let $\mathbf{o} \in S$. Let $M_i = M_D(\mathbf{h}_i)$. If $\{M_1 \mid \mathbf{t}_1\},\ldots,\{M_n \mid \mathbf{t}_n\} \in R_{D(\mathbf{o})}(G)$, then there exists $\{I_3 \mid \mathbf{s}\} \in R_{D(\mathbf{o})}(T(G))$ such that

$$\{M_1 \mid \mathbf{t}_1\}\{M_2 \mid \mathbf{t}_2\}\ldots\{M_k \mid \mathbf{t}_k\} = \{I_3 \mid \mathbf{s}\}\{M_{k+1} \mid \mathbf{t}_{k+1}\}\{M_{k+2} \mid \mathbf{t}_{k+2}\}\ldots\{M_n \mid \mathbf{t}_n\} \quad . \tag{7.17}$$

(P7.21) Problem: Use T7.26 to prove T7.29.

(D7.30) Definition: Let H denote a point group, let T denote a translation group and let D denote a basis for Γ. If $\mathbf{h}_1,\ldots,\mathbf{h}_n \in H$ are such that the relation

$$\mathbf{h}_1\mathbf{h}_2\ldots\mathbf{h}_k = \mathbf{h}_{k+1}\mathbf{h}_{k+2}\ldots\mathbf{h}_n \tag{7.18}$$

holds and if $M_i = M_D(\mathbf{h}_i)$, then we say that $\mathbf{t}_1,\ldots,\mathbf{t}_n \in R^3$ are *consistent* with the relation (7.18) with respect to T and D if $\{M_1 \mid \mathbf{t}_1\},\ldots,\{M_n \mid \mathbf{t}_n\}$ satisfy condition (7.17) for some $\{I_3 \mid \mathbf{s}\} \in R_D(T)$. In the case where the relation is in the form $\mathbf{h}_1\mathbf{h}_2\ldots\mathbf{h}_k = \mathbf{1}$, we say that $\mathbf{t}_1,\mathbf{t}_2,\ldots,\mathbf{t}_k$ is consistent with the relation if $\mathbf{t}_1,\mathbf{t}_2,\ldots,\mathbf{t}_k,\mathbf{0}$ is consistent. That is, if
$\{M_1 \mid \mathbf{t}_1\}\{M_2 \mid \mathbf{t}_2\}\ldots\{M_k \mid \mathbf{t}_k\} = \{I_3 \mid \mathbf{s}\}\{I_3 \mid \mathbf{0}\} = \{I_3 \mid \mathbf{s}\} \in R_D(T)$.

In the terminology of D7.30, T7.29 says that if the relation (7.18) occurs where $\mathbf{h}_1,\ldots,\mathbf{h}_n \in \Lambda_{\mathbf{o}}(G)$ and if $\mathbf{t}_1,\ldots,\mathbf{t}_n \in R^3$ such that $\{M_i \mid \mathbf{t}_i\} \in R_{D(\mathbf{o})}(G)$ where $M_i = M_D(\mathbf{h}_i)$, then $\mathbf{t}_1,\ldots,\mathbf{t}_n$ must be consistent with the relation (7.18) with respect to $T(G)$ and D. This result will enable us to transfer relations from a point group to its corresponding space groups. See Grossman and Magnus (1964) for a discussion of the generators and relations approach to group theory.

By T7.26, we see that each matrix M such that $\{M \mid \mathbf{t}_1\} \in R_{D(\mathbf{o})}(G)$, where G is a crystallographic space group, is associated with a collection of elements of G characterized by

$$\left\{\{M \mid \mathbf{t}_2\} \;\middle|\; \{I_3 \mid \mathbf{t}_2 - \mathbf{t}_1\} \in T(G)\right\} =$$
$$\left\{\{M \mid \mathbf{t}_2\} \;\middle|\; \{I_3 \mid \mathbf{t}\}\{M \mid \mathbf{t}_1\} = \{M \mid \mathbf{t}_2\} \text{ where } \{I_3 \mid \mathbf{t}\} \in T(G)\right\}$$
$$= R_D(T(G))\{M \mid \mathbf{t}_1\} \quad , \tag{7.19}$$

the right coset of $R_D(T(G))$ with respect to $\{M \mid \mathbf{t}_1\}$. Hence each element of $\Lambda_{\mathbf{o}}(G)$ corresponds to a right coset of $T(G)$. This correspondence yields an isomorphism from $\Lambda_{\mathbf{o}}(G)$ to $G/T(G)$.

(P7.22) Problem: Prove that $\theta:\Lambda_{\mathbf{o}}(G)$ to $R_{D(\mathbf{o})}(G)/R_D(T(G))$ defined by $\theta(\beta) = R_D(T(G))\{M \mid \mathbf{t}\}$ where $M = M_D(\beta)$ and $\{M \mid \mathbf{t}\} \in R_{D(\mathbf{o})}(G)$ is an isomorphism (you may choose to skip this problem on first reading). Conclude that $\Lambda_{\mathbf{o}}(G)$ is isomorphic to $G/T(G)$.

We are now ready to give a clear statement of how we will construct all of the crystallographic space groups G. Let H denote an oriented point group and let L denote a lattice left invariant by H. Let P denote that basis for the primitive sublattice of L associated with H and let $\mathbf{o}$ denote the origin. Then we denote the translation group that corresponds to L by T_L. That is

$$T_L = \{\tau \in \Gamma \mid \tau(\mathbf{o}) \in L\} \ . \tag{7.20}$$

Let $M_P(H) = \{M_1, M_2, \ldots, M_n\}$. Then we seek all possible $\mathbf{t}_1, \mathbf{t}_2, \ldots, \mathbf{t}_n \in R^3$ such that

$$R_P(G) = R_P(T_L)[\{M_1 \mid \mathbf{t}_1\}, \{M_2 \mid \mathbf{t}_2\}, \ldots, \{M_n \mid \mathbf{t}_n\}] \tag{7.21}$$

for some crystallographic space group G such that

$$R_P(T(G)) = R_P(T_L)$$

and (7.22)

$$M_P(G) = M_P(H) \quad .$$

By finding all such sets $\{\mathbf{t}_1, \ldots, \mathbf{t}_n\}$ we will find all of the crystallographic space groups (through their matrix representations). The strategy that we will use to obtain these sets is to determine a sufficient number or relations on H so that a set $\{\mathbf{t}_1, \mathbf{t}_2, \ldots, \mathbf{t}_n\}$ yields an $R_{P(\mathbf{o})}(G)$ if and only if the set is consistent with all of these relations.

Once all of the crystallographic space groups satisfying (7.22) for some choice of H and L are found, we will need to classify them into space group types. To do this, many different equivalence relations have been employed. Some of these are discussed by Hans Wondratschek in Chapter 8 of ITFC (Hahn, 1983). We shall use an equivalence relation that gives rise to the traditional 230 crystallographic space group types. The basic idea behind this relation is relatively simple. Suppose G_1 and G_2 are two crystallographic space groups. If there exist points o_1 and o_2 and generating sets D_1 and D_2 of $T(G_1)$ and $T(G_2)$ such that $D_1(o_1)$ and $D_2(o_2)$ are both right handed and $R_{D_1(\mathbf{o}_1)}(G_1)$ and $R_{D_2(\mathbf{o}_2)}(G_2)$ are the same set of matrices, then G_1 and G_2 are equivalent. Hence to show that G_1 is equivalent to G_2 we show that, by a change of origin and generating sets, the matrix representation of one can be converted into that of the other. In view of T7.16 we have the following definition.

(D7.31) Definition: Let G_1 and G_2 denote crystallographic space groups. Then G_1 is equivalent to G_2 if and only if there exists a vector $\mathbf{r} \in R^3$ and an integral matrix T with det(T) = 1 such that the matrix representations $\{M \mid \mathbf{t}\}$ of the elements of G_1 with respect to D, a generating set for $T(G)$, become those of G_2 by the transformation

$$\{M \mid \mathbf{t}\} \to \{TMT^{-1} \mid T[(M - I_3)\mathbf{p} + \mathbf{t}]\} \quad .$$

When discussing a space group G where $\mathrm{orb}_{T(G)}(\mathbf{o})$ is a centered-lattice, we will usually write the matrices representing $\Lambda_o(G)$ with respect to the basis P of the primitive lattice. However, in order to test for equivalence, D7.31 requires that $\{M \mid \mathbf{t}\}$ be written with respect to a basis D that generates $T(G)$. While this is a serious problem relative to the change of basis matrix T, we need not change to the basis D in order to detect equivalence that occurs because of a change of origin as the following theorem shows.

(T7.32) Theorem: Let α and β denote isometries, let D and P denote bases of Γ, o_1, $o_2 \in S$ and T be the change of basis matrix from D to P. Then

$$R_{D(o_1)}(\alpha) = R_{D(o_2)}(\beta)$$

if and only if

$$R_{P(o_1)}(\alpha) = R_{P(o_2)}(\beta) \quad .$$

Proof: Suppose $R_{D(o_1)}(\alpha) = R_{D(o_2)}(\beta)$. If $R_{D(o_1))}(\alpha) = \{M \mid [\mathbf{v}]_{D(o_1)}\}$ where $\mathbf{v} \in S$, then $R_{D(o_1)}(\beta) = \{M \mid [\mathbf{w}]_{D(o_1)}\}$ where $\mathbf{w}$ is such that

$$R_{D(o_2)}(\beta) = \{M \mid (M - I_3)[o_2]_{D(o_1)} + [\mathbf{w}]_{D(o_1)}\} = \{M \mid [\mathbf{v}]_{D(o_1)}\} \quad .$$

Hence

$$R_{P(o_1)}(\alpha) = \{TMT^{-1} \mid T[\mathbf{v}]_{D(o_1)}\} \quad \text{and} \quad R_{P(o_1)}(\beta) = \{TMT^{-1} \mid T[\mathbf{w}]_{D(o_1)}\} \quad .$$

Then

$$\begin{aligned}
R_{P(o_2)}(\beta) &= \{TMT^{-1} \mid (TMT^{-1} - I_3)[o_2]_{P(o_1)} + T[\mathbf{w}]_{D(o_1)}\} \\
&= \{TMT^{-1} \mid T(M - I_3)T^{-1}[o_2]_{P(o_1)} + T[\mathbf{w}]_{D(o_1)}\} \\
&= \{TMT^{-1} \mid T[(M - I_3)[o_2]_{D(o_1)} + [\mathbf{w}]_{D(o_1)}]\} \\
&\qquad (\text{Since } T^{-1}[o_2]_{D(o_1)} = [o_2]_{D(o_1)}) \\
&= \{TMT^{-1} \mid T[\mathbf{v}]_{D(o_1)}\} \\
&= R_{P(o_1)}(\alpha) \quad .
\end{aligned}$$

The converse is proved in a completely similar fashion. □

By T7.32, we may use the basis P of the primitive lattice to detect equivalence due to change of origin. However, one must change to the basis of the lattice under consideration to detect equivalence due to a change of basis.

CRYSTALLOGRAPHIC SPACE GROUP OPERATIONS

When an oriented point group H and a lattice L are given, to find the space groups satisfying (7.22) we must search for sets of elements in R^3 that are consistent with the relations in H. Let $\mathbf{h} \in H$. Then $\mathbf{h}^{o(\mathbf{h})} = \mathbf{1}$ is one relation that is always present. The next theorem gives the constraint that is imposed by this relation.

(T7.33) Theorem: Let $\mathbf{h}$ be a point isometry, let T denote a translation group and D be a basis for Γ. Then $\mathbf{t}$ is consistent with the relation $\mathbf{h}^{o(\mathbf{h})}$ with respect to T and D if and only if $\{I_3 \mid N\mathbf{t}\} \in R_D(T)$ where

$$N = M + M^2 + \ldots + M^{o(\mathbf{h})} \text{ and } M = M_D(\mathbf{h}).$$

Proof: Suppose $\mathbf{t}$ is consistent with the relation $\mathbf{h}^{o(\mathbf{h})} = \mathbf{1}$ with respect to T. Then $\{M \mid \mathbf{t}\}^{o(\mathbf{h})} = \{I_3 \mid \mathbf{s}\} \in R_D(T)$. But note that

$$\{M \mid \mathbf{t}\}^2 = \{M^2 \mid M\mathbf{t} + \mathbf{t}\}$$

$$\{M \mid \mathbf{t}\}^3 = \{M^3 \mid M^2\mathbf{t} + M\mathbf{t} + \mathbf{t}\}$$

$$\vdots$$

$$\{M \mid \mathbf{t}\}^{o(\mathbf{h})} = \{M^{o(\mathbf{h})} \mid M^{(o(\mathbf{h})-1)}\mathbf{t} + \ldots + M\mathbf{t} + \mathbf{t}\} \quad .$$

But $\{M \mid \mathbf{t}\}^{o(\mathbf{h})} = \{I_3 \mid \mathbf{s}\}$. Since $M^{o(\mathbf{h})} = I_3$, $\mathbf{t} = M^{o(\mathbf{h})}\mathbf{t}$ and so, by rearranging terms, we have

$$\mathbf{s} = (M + M^2 + \ldots + M^{o(\mathbf{h})})\mathbf{t} = N\mathbf{t} \quad .$$

By reversing this argument the converse is also proved. □

The constraint described by T7.33 gives rise to two new types of isometries. The screw and glide operations.

(E7.34) Example: Consider a space group G such that $M_P(G) = M_P(4)$ and $R_P(T(G)) = R_P(T_P)$. Let $P = \{\mathbf{a},\mathbf{b},\mathbf{c}\}$ denote the basis chosen in Chapter 6 with **c** along the axis of 4, **a** and **b** perpendicular to **c** with $\gamma = 90°$ such that D is a right-handed system. Then

$$M = M_P(\mathbf{4}) = \begin{bmatrix} 0 & -1 & 0 \\ 1 & 0 & 0 \\ 0 & 0 & 1 \end{bmatrix}.$$

Furthermore, since the lattice type is primitive, $\{[\mathbf{v}]_P \mid \mathbf{v} \in \text{orb}_{T(G)}(\mathbf{o})\}$ is Z^3. Then by T7.29 and T7.33, if $\{M \mid \mathbf{t}\} \in G$, then $N\mathbf{t} \in Z^3$ where $N = M + M^2 + M^3 + M^4$. Calculating N we obtain

$$N = \begin{bmatrix} 0 & 0 & 0 \\ 0 & 0 & 0 \\ 0 & 0 & 4 \end{bmatrix}.$$

Hence $N\mathbf{t} \in Z^3$ implies that $4t_3 \in Z$ Hence to be consistent with $\mathbf{4}^4 = \mathbf{1}$, $\mathbf{t}$ must be chosen such that $t_3 = 0, \frac{1}{4}, \frac{1}{2}$ or 3/4. □

The operations found in E7.34 are $\{M \mid [000]^t\}$, $\{M \mid [0,0,\frac{1}{4}]^t\}$, $\{M \mid [0,0,\frac{1}{2}]^t\}$ and $\{M \mid [0,0,3/4]^t\}$ where $M = M_P(\mathbf{4})$. These matrices represent *quarter-turn screw operations*. The translational component of each of these is directed along the axis of the rotation which in this case is along **c**. Hence, under $\{M \mid [0,0,\frac{1}{4}]^t\}$ a point $[xyz]^t$ is rotated a quarter turn about **c** and then displaced a distance of **c**/4 to the point $[-y,x,\frac{1}{4} + z]^t$. Consequently, this operation is called a quarter-turn screw with a *screw translation* of **c**/4 and will be denoted $\mathbf{4_1}$. Since the screw translation of $\{M \mid [000]^t\}$ is **0**, it is merely a quarter-turn and will be denoted by the usual symbol **4**. The operation $\{M \mid [00\frac{1}{2}]^t\}$ maps $[xyz]^t$ to $[-y,x,\frac{1}{2} + z]^t$. Since the screw translation is **c**/2, this quarter-turn screw will be denoted $\mathbf{4_2}$. Similarly, $\{M \mid [0,0,3/4]^t\}$ represents a quarter-turn screw denoted $\mathbf{4_3}$. In general, we make the notational convention that ${}^{[uvw]}\mathbf{n_m}$ represents an nth-turn screw about the vector $u\mathbf{a} + v\mathbf{b} + w\mathbf{c}$ with a screw translation of $m\mathbf{r}/n$ where **r** is the shortest (nonzero) vector in the lattice in the $[uvw]$ direction. The *symmetry element of a screw operation* will be taken to be that of its associated rotation. Hence in the case of a primitive cubic lattice,

${}^{[111]}3_2$ denotes a third-turn screw about [111] with a screw translation of 2/3(**a** + **b** + **c**) since **a** + **b** + **c** is the shortest nonzero vector in the [111] direction. However, in the body-centered cubic lattice, ${}^{[111]}3_2$ represents a third-turn screw about [111] with a screw translation of $(2/3)(\frac{1}{2}\mathbf{a} + \frac{1}{2}\mathbf{b} + \frac{1}{2}\mathbf{c}) = (1/3)\mathbf{a} + (1/3)\mathbf{b} + (1/3)\mathbf{c}$ since $\frac{1}{2}\mathbf{a} + \frac{1}{2}\mathbf{b} + \frac{1}{2}\mathbf{c}$ is the shortest nonzero lattice vector in the direction [111].

(P7.23) Problem: Show that with respect to the primitive cubic lattice *P*, the Seitz notation for ${}^{[111]}3_2$ is

$$\{M \mid [2/3,2/3,2/3]^t\} \quad ,$$

where $M = M_P({}^{[111]}3)$ and *P* is the basis for the primitive lattice given in Table 6.3.

(P7.24) Problem: Find the Seitz notation for ${}^{[111]}3_1$ with respect to the body-centered cubic lattice *I* described in Table 6.3.

(P7.25) Problem: Write the full 4 × 4 matrix for the screw operations discussed in P7.23 and P7.24.

(P7.26) Problem: Find the Seitz notation for ${}^{[100]}4_1$ with respect to a primitive cubic lattice *P*.

(P7.27) Problem: Find the Seitz notation for ${}^{[100]}4_1$ with respect to a body-centered cubic lattice *I*. Note that your answer here should be the same as in P7.26.

(P7.28) Problem: Find the Seitz notation for ${}^{[210]}2_1$ with respect to a primitive hexagonal lattice.

(E7.35) Example: Consider the space group *G* of E7.34 based on *4* but with a body-centered lattice type *I*. The basis *D* is the same here as in E7.34 and so M and N are the same. However $\{[\mathbf{v}]_D \mid \mathbf{v} \in \text{orb}_{T(G)}(\mathbf{o})\}$ is now $Z^3 + Z[\frac{1}{2},\frac{1}{2},\frac{1}{2}]^t$. Since $N\mathbf{t} = [0,0,4t_3]^t$, there must exist a $\mathbf{v} = [v_1 v_2 v_3]^t \in Z^3$ and an $n \in Z$ such that

$$[0,0,4t_3]^t = [v_1 v_2 v_3]^t + n[\tfrac{1}{2},\tfrac{1}{2},\tfrac{1}{2}]^t \quad .$$

Since $v_1 + n\frac{1}{2} = 0$ and $v_1, n \in Z$, $n = 0$. Hence $4t_3 \in Z$. Consequently, as in E7.34, we obtain only the operation **4**, $\mathbf{4}_1$, $\mathbf{4}_2$ and $\mathbf{4}_3$ for a body-centered lattice left invariant under *4*. □

(E7.36) Example: Consider the space group *G* based on the point group *m* with lattice type *P* (in the second setting with *Y* as the unique axis). Then

$$M = M_P({}^{[010]}\mathrm{m}) = \begin{bmatrix} 1 & 0 & 0 \\ 0 & -1 & 0 \\ 0 & 0 & 1 \end{bmatrix}.$$

$$N\mathbf{t} = [2t_1, 0, 2t_3] \in Z^3$$

and so $t_1 = 0$ or $\frac{1}{2}$ and $t_3 = 0$ or $\frac{1}{2}$. If we take $t_1 = \frac{1}{2}$ and $t_3 = 0$, then a point $[x,y,z]^t$ is reflected across the plane (010) and then translated parallel to it a distance of $\mathbf{a}/2$ along [100]. Such an operation is called a *glide operation*, (010) is called the *glide plane* and $\mathbf{a}/2$ is called the *glide translation*. The reflection plane of the mirror part of the glide operation will be taken to be its symmetry element. Because the glide translation is along **a** this glide is called an *a-glide* denoted by ${}^{[010]}\mathbf{a}$. In the case where $t_1 = t_3 = \frac{1}{2}$, the glide translation is $\frac{1}{2}\mathbf{a} + \frac{1}{2}\mathbf{c}$ which again parallels the (010) plane. This type of glide (where the glide translation is not in the direction of a basis vector) will be called an *n-glide*. The glide planes that arise in our example are listed below (note that a mirror operation is a glide with a **0** glide translation modulo the translation group).

t_1	t_3	*Symbol of Glide Operation*	*Type*
0	0	${}^{[010]}\mathrm{m}$	*reflection*
$\frac{1}{2}$	0	${}^{[010]}\mathrm{a}$	*a-glide*
0	$\frac{1}{2}$	${}^{[010]}\mathrm{c}$	*c-glide*
$\frac{1}{2}$	$\frac{1}{2}$	${}^{[010]}\mathrm{n}$	*n-glide*

□

(P7.29) Problem: Consider the space group *G* described in E7.36 where the lattice type *P* is taken to be in the first setting (*Z* the unique axis). In this case the mirror plane is perpendicular to **c**. Show that the following glide operations are possible: **m**, **a**, **b** and **n** (recall that when no orientation symbol is given, [001] is assumed) where the glide translation of the n-glide is $\frac{1}{2}\mathbf{a} + \frac{1}{2}\mathbf{b}$.

(P7.30) Problem: Show that the glide translations of the glide operations in P7.29 are unchanged if an *A*-centered or a *B*-centered lattice is used instead of a primitive lattice. Show that, in E7.36, the result is unchanged when *P* is replaced by an *A*-centered or *C*-centered lattice.

For a beautifully illustrated discussion of space group operations, see Bloss (1971) Chapter 7.

THE CRYSTALLOGRAPHIC SPACE GROUP TYPES DERIVED FROM THE ONE-GENERATOR POINT GROUPS

Given an oriented point group *H*, a lattice *L* left invariant under *H* and the basis *P* for the primitive lattice, the results stated thus far give us conditions that must hold for an $\{M \mid \mathbf{t}\}$ to be in $R_{P(\mathbf{o})}(G)$, where G is a space group satisfying (7.22). Theorem T7.37 below states necessary and sufficient conditions for a $\mathbf{t}$ to be such that $\{M \mid \mathbf{t}\}$ and $R_P(T_L)$ generate $R_P(G)$ for some G satisfying (7.22) when *H* is a one-generator point group. By a one-generator group we mean a cyclic group. Such a group *H* has an element α such that $H = \{\alpha, \alpha^2, \ldots, \alpha^{o(\alpha)}\}$. These groups are *1*, *2*, *3*, *4*, *6*, $\bar{1}$, *m*, $\bar{3}$, $\bar{4}$ and $\bar{6}$.

(T7.37) Theorem: Let *L* denote a lattice left invariant by the one-generator point group *H* generated by α and let *P* denote the basis for the primitive sublattice of *L* that is associated with *H*. Let $M = M_P(\alpha)$. There exists a crystallographic space group *G* satisfying (7.22) such that each element of $R_{P(\mathbf{o})}(G)$ can be written in the form

$$\{I_3 \mid \mathbf{v}\}\{M \mid \mathbf{r}\}^i \quad ,$$

where $\{I_3 \mid \mathbf{v}\} \in R_P(T_L)$ and $1 \le i \le O(\alpha)$ if and only if $\mathbf{r}$ is consistent with the relation $\alpha^{o(\alpha)} = 1$ with respect to T_L and *P*.

Proof: If there is such a crystallographic space group *G*, then, since

$\{M \mid \mathbf{r}\} \in R_{P(\mathbf{o})}(G)$, by T7.29, $\mathbf{r}$ must be consistent with the relation $\alpha^{o(\alpha)} = 1$ with respect to T_L and P. Now suppose that $\mathbf{r}$ is consistent with $\alpha^{o(\alpha)} = 1$ with respect to T_L and P. By (D7.30),

$$\{M \mid \mathbf{r}\}^{o(\alpha)} = \{I_3 \mid \mathbf{s}\} \quad , \tag{7.23}$$

where $\{I_3 \mid \mathbf{s}\} \in R_P(T_L)$. We will use (7.23) to show that the set

$$K = R_P(T_L)\{\{M \mid \mathbf{r}\}^i \mid i = 1,\ldots,o(\alpha)\}$$

is closed. Let $A = \{I_3 \mid \mathbf{s}_1\}\{M \mid \mathbf{r}\}^i$ and $B = \{I_3 \mid \mathbf{s}_2\}\{M \mid \mathbf{r}\}^j$ be elements of K where $\mathbf{s}_1$ and $\mathbf{s}_2$ are triples representing lattice vectors with respect to P and where $1 \le i,j \le o(\alpha)$. Then $\{M \mid \mathbf{r}\}^i = \{M^i \mid \mathbf{p}\}$ for some $\mathbf{p} \in R^3$ and so

$$\begin{aligned}
\{M \mid \mathbf{r}\}^i\{I_3 \mid \mathbf{s}_2\} &= \{M^i \mid \mathbf{p}\}\{I_3 \mid \mathbf{s}_2\} \\
&= \{M^i \mid \mathbf{p} + M^i\mathbf{s}_2\} \\
&= \{I_3 \mid M^i\mathbf{s}_2\}\{M^i \mid \mathbf{p}\} \\
&= \{I_3 \mid M^i\mathbf{s}_2\}\{M \mid \mathbf{r}\}^i \quad .
\end{aligned} \tag{7.24}$$

Since L is invariant under H, $M^i\mathbf{s}_2$ is a triple representing a lattice vector in L with respect to P. Hence

$$\begin{aligned}
AB &= \{I_3 \mid \mathbf{s}_1\}\{M \mid \mathbf{r}\}^i\{I_3 \mid \mathbf{s}_2\}\{M \mid \mathbf{r}\}^j \\
&= \{I_3 \mid \mathbf{s}_1\}\{I_3 \mid M^i\mathbf{s}_2\}\{M \mid \mathbf{r}\}^i\{M \mid \mathbf{r}\}^j = \\
&= \{I_3 \mid \mathbf{s}_1 + M^i\mathbf{s}_2\}\{M \mid \mathbf{r}\}^{i+j} \quad .
\end{aligned}$$

where $\{I_3 \mid \mathbf{s}_1 + M^i\mathbf{s}_2\} \in R_P(T_L)$. If $i + j \le o(\alpha)$, then $\{M \mid \mathbf{r}\}^{i+j}$ is in $\{\{M \mid \mathbf{r}\}^i \mid i = 1,\ldots,o(\alpha)\}$ and so then AB is in K. If $o(\alpha) < i + j$, then setting $k = i + j - o(\alpha)$

$$\{M \mid \mathbf{t}\}^{i+j} = \{M \mid \mathbf{t}\}^{o(\alpha)}\{M \mid \mathbf{t}\}^k \quad .$$

By (7.23), we have,

$$AB = \{I_3 \mid \mathbf{s}_1 + M^i\mathbf{s}_2\}\{I_3 \mid \mathbf{s}\}\{M \mid \mathbf{r}\}^k$$

$$= \{I_3 \mid \mathbf{s}_3\}\{M \mid \mathbf{r}\}^k \quad ,$$

where $\{I_3 \mid \mathbf{s}_3\} \in R_P(T_L)$ (since $R_P(T_L)$ is closed) and $1 \le k \le o(\alpha)$. Hence $AB \in K$.

(P7.31) Problem: Suppose that $A = \{I_3 \mid \mathbf{s}_1\}\{M \mid \mathbf{r}\}^i$ where $\{I_3 \mid \mathbf{s}_1\} \in R_P(T_L)$. Show that

$$A^{-1} = \{I_3 \mid -M^{o(\alpha) - i}(\mathbf{s} + \mathbf{s}_1)\}\{M \mid \mathbf{r}\}^{o(\alpha) - i}$$

by confirming that $AA^{-1} = \{I_3 \mid \mathbf{0}\}$. (Hint: use (7.23) to help with the simplification). Now show that A^{-1} is in K.

Using P7.31, we have completed the proof that K is a group. Hence $K = R_P(G)$ for some space group G. Now we need to show that (7.22) is satisfied. By the way K was constructed, $M_P(G) = M_P(4)$. To show that $R_P(T(G)) = R_P(T_L)$ we shall show that the only elements of the form $\{I_3 \mid \mathbf{t}\}$ in $R_P(G)$ are those where $\mathbf{t}$ represents a lattice vector in L.

(P7.32) Problem: Suppose $\{I_3 \mid \mathbf{t}\} = \{I_3 \mid \mathbf{s}_1\}\{M \mid \mathbf{r}\}^i$ where $1 \le i \le o(\alpha)$. Show that $M^i = I_3$ and so $i = o(\alpha)$. Then show that $\{I_3 \mid \mathbf{t}\} \in R_P(T_L)$ and that $R_P(T(G)) = R_P(T_L)$.

Since we now have $T(G)$ isomorphic to L, we have also shown that G is a crystallographic space group. □

(E7.38) Example: Find all of the crystallographic space group types G such that $M_P(G) = M_P(6)$.

Solution: From Table 6.3, the only lattice type left invariant under 6 is the primitive lattice P with basis P such that

$$M = M_P(\mathbf{6}) = \begin{bmatrix} 1 & -1 & 0 \\ 1 & 0 & 0 \\ 0 & 0 & 1 \end{bmatrix} .$$

Consider $\{M \mid \mathbf{r}\}$. By T7.33, we need to find all vectors $\mathbf{r}$ such that $N\mathbf{r}$ represents a lattice vector. Then by T7.37 we will have found all of the crystallographic space groups G that satisfy (7.22) with $H = 6$ and $L = P$. Since the lattice is primitive

$$R_P(T_P) = \{\{I_3 \mid \mathbf{s}\} \mid \mathbf{s} \in Z^3\}.$$

Since $N = M + M^2 + M^3 + M^4 + M^5 + M^6$, we have

$$N = \begin{bmatrix} 0 & 0 & 0 \\ 0 & 0 & 0 \\ 0 & 0 & 6 \end{bmatrix},$$

and so $N\mathbf{r} = [0\ 0\ 6r_3]^t$, which implies that $6r_3$ must be an integer. Hence, we have the generators

$$\{M \mid [r_1,r_2,0]^t\},\ \{M \mid [r_1,r_2,(1/6)]^t\},\ \{M \mid [r_1,r_2,(1/3)]^t\},$$

$$\{M \mid [r_1,r_2,\tfrac{1}{2}]^t\},\ \{M \mid \{r_1,r_2,(2/3)]^t\},\ \{M \mid [r_1,r_2,(5/6)]^t\}.$$

Since r_1 and r_2 can take on any values, we have an infinite number of crystallographic space groups associated with 6. However many of these are equivalent under D7.31. If the origin is shifted from its current position to some vector $\mathbf{p}$, then

$$\{M \mid \mathbf{r}\}$$

becomes

$$\{M \mid (M - I_3)\mathbf{p} + \mathbf{r}\}.$$

But

$$(M - I_3)\mathbf{p} + \mathbf{r} = \begin{bmatrix} 0 & -1 & 0 \\ 1 & -1 & 0 \\ 0 & 0 & 0 \end{bmatrix}\begin{bmatrix} p_1 \\ p_2 \\ p_3 \end{bmatrix} + \begin{bmatrix} r_1 \\ r_2 \\ r_3 \end{bmatrix}$$

$$= \begin{bmatrix} -p_2 \\ p_1 - p_2 \\ 0 \end{bmatrix} + \begin{bmatrix} r_1 \\ r_2 \\ r_3 \end{bmatrix}.$$

If we set $p_2 = r_1$ and $p_1 = r_1 - r_2$, then

$$(M - I_3)\mathbf{p} + \mathbf{r} = \begin{bmatrix} 0 \\ 0 \\ r_3 \end{bmatrix} .$$

Hence by an appropriate choice of origin, each of the generators we have found can be written in the form $\{M \mid [0,0,r_3]^t\}$. For example $\{M \mid [0,0,1/6]^t\}$ is the representative of the infinite number of space groups generated by the elements of the form $\{M \mid [r_1,r_2,1/6]^t\}$. Hence every space group G with $M_p(G) = M_p(6)$ is equivalent to one generated by one of

$$\{M \mid [0,0,0]^t\},\ \{M \mid [0,0,1/6]^t\},\ \{M \mid [0,0,1/3]^t\},$$

$$\{M \mid [0,0,\tfrac{1}{2})]^t\},\ \{M \mid [0,0,2/3]^t\},\ \{M \mid [0,0,5/6]^t\} \quad .$$

Since $(M - I_3)\mathbf{p}$ has a zero in the third component, a change of origin does not affect the third component of $\mathbf{r}$. Hence the only way two of these generators, say

$$\{M \mid [0,0,1/6]^t\} \text{ and } \{M \mid [0,0,1/3]^t\}$$

can lead to equivalent space groups is if there exists a proper unimodular matrix T such that

$$\{TMT^{-1} \mid T[0,0,1/6]^t\} = \{M \mid [0,0,1/3]^t\} .$$

It can be shown that no such matrix T exists. In fact all six of the generators lead to nonequivalent space group types. Hence we have found six nonequivalent space groups generated by $\mathbf{6}$, $\mathbf{6}_1$, $\mathbf{6}_2$, $\mathbf{6}_3$, $\mathbf{6}_4$ and $\mathbf{6}_5$, respectively. Since these all have the primitive lattice P, they are denoted $P6$, $P6_1$, $P6_2$, $P6_3$, $P6_4$ and $P6_5$. □

A brute force technique for showing that the six generators found in (E7.38) are nonequivalent can be developed along the lines of the argument used at the end of Chapter 6. More elegant approaches require the development of further algebraic tools. We do not have the space in this book to develop those tools. Hence we will leave this detail unexplored.

(E7.39) Example: Consider the space group $P6_1$. According to E7.38 and T7.37,

$$R_P(P6_1) = \bigcup_{i=1}^{6} R_P(T_P)\{M \mid \mathbf{r}\}^i$$

$$= R_P(T_P)\{M \mid \mathbf{r}\} \cup R_P(T_P)\{M \mid \mathbf{r}\}^2 \cup R_P(T_P)\{M \mid \mathbf{r}\}^3 \cup R_P(T_P)\{M \mid \mathbf{r}\}^4 \cup R_P(T_P)\{M \mid \mathbf{r}\}^5 \cup R_P(T_P)\{M \mid \mathbf{r}\}^6 ,$$

where $M = M_P(\mathbf{6})$ and $\mathbf{r} = [0,0,1/6]^t$. Consequently each element of $P6_1$ falls in exactly one of the six right cosets:

$R_P(T_P)\{M \mid [0,0,1/6]^t\}$, $R_P(T_P)\{M^2 \mid [0,0,1/3]^t\}$, $R_P(T_P)\{M^3 \mid [0,0,\tfrac{1}{2}]^t\}$,

$R_P(T_P)\{M^4 \mid [0,0,2/3]^t\}$, $R_P(T_P)\{M^5 \mid [0,0,5/6]^t\}$, $R_P(T_P)\{I_3 \mid [000]^t\}$,

where

$$M = \begin{bmatrix} 1 & -1 & 0 \\ 1 & 0 & 0 \\ 0 & 0 & 1 \end{bmatrix} .$$

We shall now describe the elements in each of these cosets. For example an element in $R_P(T_P)\{M^2 \mid [0,0,(1/3)]^t\}$ is of the form

$$\begin{bmatrix} 1 & 0 & 0 & u \\ 0 & 1 & 0 & v \\ 0 & 0 & 1 & w \\ 0 & 0 & 0 & 1 \end{bmatrix} \begin{bmatrix} 0 & -1 & 0 & 0 \\ 1 & -1 & 0 & 0 \\ 0 & 0 & 1 & 1/3 \\ 0 & 0 & 0 & 1 \end{bmatrix} = \begin{bmatrix} 0 & -1 & 0 & u \\ 1 & -1 & 0 & v \\ 0 & 0 & 1 & 1/3 + w \\ 0 & 0 & 0 & 1 \end{bmatrix} , \quad (7.25)$$

where $u,v,w \in Z$ since the lattice is primitive. Using the techniques for analyzing the matrix representation of a space group operation discussed by Wondratschek and Neubuser (1967) and Boisen and Gibbs (1978), we find that the matrix in (7.25) defines a third-turn screw operation with a screw translation of $((1/3) + w)\mathbf{c}$ occurring about an axis in the direction parallel to $\mathbf{c}$ passing through the point

$$(1/3)(2u - v)\mathbf{a} + (1/3)(u + v)\mathbf{b} \quad .$$

The points of this form that appear in the unit cell are those for which

$$0 \le (1/3)(2u - v) < 1 \quad \text{and} \quad 0 \le (1/3)(u + v) < 1 \quad .$$

Adding the corresponding terms of these inequalities we have

$$0 \le u < 2 \ .$$

Since u is an integer, u can only equal 0 or 1. When $u = 0$, the inequalities become $0 \le -(1/3)v < 1$ and $0 \le (1/3)v < 1$. The first implies that $v \le 0$ and the second that $v \ge 0$. Hence, when $u = 0$, v must equal 0. When $u = 1$, a similar analysis shows that v can only be 0 or 1. Hence, when $u = 1$, v can equal either 0 or 1. Thus, besides the $\mathbf{3_1}$ operation located at the origin, there are two others in the unit cell passing through the points $[1/3,2/3,0]^t$ and $[2/3,1/3,0]^t$, respectively. An element in the coset $R_P(T_P)\{M^4 \mid [0,0,2/3]^t\}$ is of the form

$$\begin{bmatrix} 1 & 0 & 0 & u \\ 0 & 1 & 0 & v \\ 0 & 0 & 1 & w \\ 0 & 0 & 0 & 1 \end{bmatrix} \begin{bmatrix} -1 & 1 & 0 & 0 \\ -1 & 0 & 0 & 0 \\ 0 & 0 & 1 & 2/3 \\ 0 & 0 & 0 & 1 \end{bmatrix} = \begin{bmatrix} -1 & 1 & 0 & u \\ -1 & 0 & 0 & v \\ 0 & 0 & 1 & 2/3 + w \\ 0 & 0 & 0 & 1 \end{bmatrix} .$$

An analysis of this matrix shows that it defines a negative third-turn screw operation $(\mathbf{3}^{-1})_2$ occurring about an axis passing through the point

$$(1/3)(u + v)\mathbf{a} + (1/3)(2v - u)\mathbf{b} \ .$$

Consequently, besides the $(\mathbf{3}^{-1})_2$ operation at the origin, there are two others in the unit cell passing through the points $[1/3,2/3,0]^t$ and $[2/3,1/3,0]^t$, respectively. This is as one would expect, since $(\mathbf{3}_1)^2 = (\mathbf{3}^{-1})_2$ and this composition does not move the axis of the $\mathbf{3}_1$ operation.

(P7.33) Problem: Analyze the matrices in each of the remaining cosets of $P6_1$ and show that the axes are positioned as shown in Figure 7.5.

(P7.34) Problem: Let τ denote the translational isometry defined by $\tau(\mathbf{o}) = \mathbf{t}$ where $\mathbf{t} = 2\mathbf{a} + 3\mathbf{b} + 5\mathbf{c}$ and $P = \{\mathbf{a},\mathbf{b},\mathbf{c}\}$. Show that

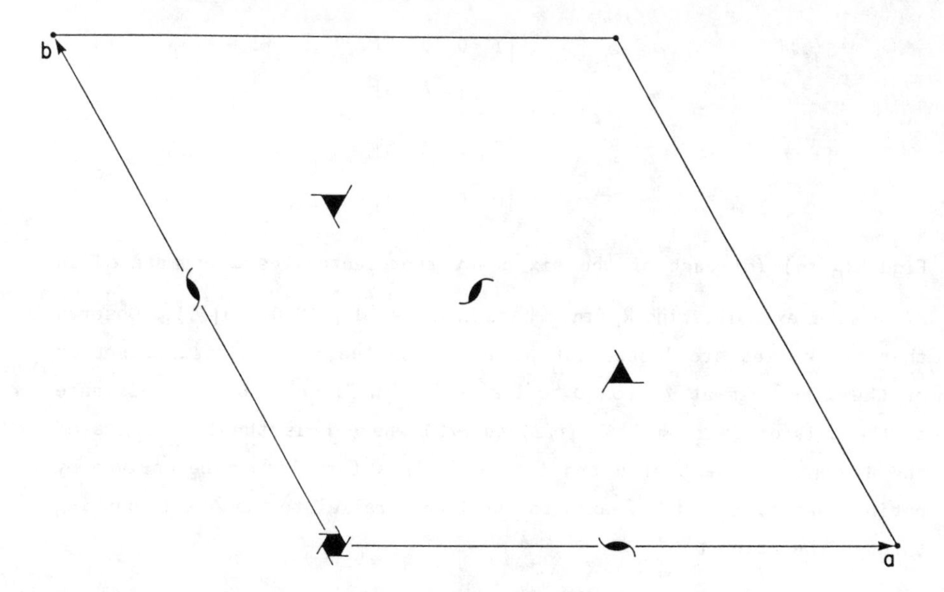

Figure 7.5: A diagram of a unit cell containing the screw axes of space group $P6_1$ each taking place about rotation axes paralleling **c**. The 6_1-screw axis is symbolized by a pronged hexad, the 3_1-screw axes by pronged triads and the 2_1-screw axes by pronged diad symbols.

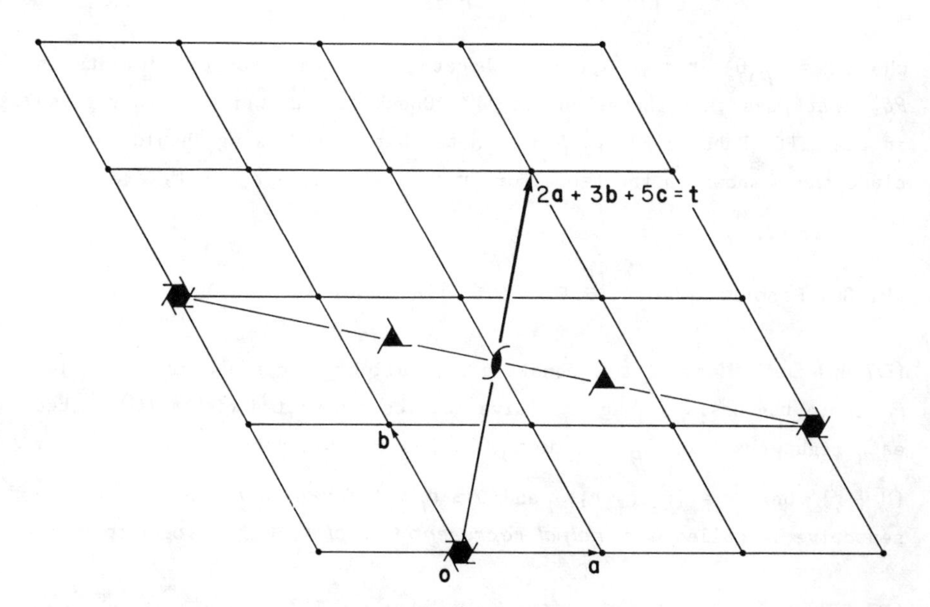

Figure 7.6: A diagram showing the composition of the translational isometry τ where $\tau(\mathbf{o}) = 2\mathbf{a} + 3\mathbf{b} + 5\mathbf{c} = \mathbf{t}$ with the rotational screw operations comprising a 6_1-screw axis that parallels **c** and passes through the origin. The resulting operations formed by this composition take place about rotation axes paralleling **c** and are located on the perpendicular bisector of **t** at a distance of $t/2\text{ctn}(\rho/2)$ from the line from **o** to **t** where ρ is the turn angle of the rotational component of the screw operation.

$$R_P(\tau) = \begin{bmatrix} 1 & 0 & 0 & 2 \\ 0 & 1 & 0 & 3 \\ 0 & 0 & 1 & 5 \\ 0 & 0 & 0 & 1 \end{bmatrix} .$$

Find $R_P(\tau\alpha)$ for each of the six coset representatives α enumerated in E7.39 (For example, find $R_P(\tau\alpha)$ where $R_P(\alpha) = \{M \mid [0,0,1/6]^t\}$). Observe that these axes are located at points along the perpendicular bisector of the line segment ℓ from **o** to **t** as shown in Figure 7.6. The distance of the axis of $\tau\alpha$ from ℓ is $(t/2)\text{ctn}(\rho/2)$ where ρ is the turn angle of the linear component of α and $t = \|\mathbf{t}\|$. This formula can be derived by noting that $\tau\alpha$ maps the Z axis to the line parallel to the Z axis passing through the point **t**.

(P7.35) Problem: Consider the space group $P6_2$. Then

$$R_P(P6_2) = \bigcup_{i=1}^{6} R_P(T_P)\{M \mid \mathbf{r}\}^i$$

where $M = M_P(\mathbf{6})$, $\mathbf{r} = [0,0,1/3]^t$. Locate all of the axes of elements in $P6_2$ that pass through the unit cell. Check your answer with that given in the ITFC (Hahn, 1983, p. 546). Note that your drawing should not include those shown in the table for $[\mathbf{r}_1,\mathbf{r}_2,\mathbf{r}_3]^t$ where $\mathbf{r}_1$ or $\mathbf{r}_2$ are equal to 1.

(P7.36) Problem: Do P7.35 for $P6_3$. (In this case $\mathbf{r} = [0,0,\frac{1}{2}]^t$.)

(D7.40) Definition: Let G denote a crystallographic space group and let P denote the basis of the primitive lattice associated with $\Lambda_o(G)$. Then each right coset of $R_P(T_P)$ in $R_P(G)$ has a representative of the form $\{M \mid \mathbf{r}\}$ where $\mathbf{r} = [\mathbf{r}_1,\mathbf{r}_2,\mathbf{r}_3]^t$ and $0 \le \mathbf{r}_i < 1$ for each i. Such a representative is called a ***principal representative of G with respect to P***.

(P7.37) Problem: Use (7.24) to prove that $R_P(T_P)$ is a normal subgroup of $R_P(G)$.

(P7.38) Problem: Show that the principal representative defined in D7.40 exists for each coset of $R_P(T_P)$ in $R_P(G)$ and that there is only one such representative for each coset.

(P7.39) Problem: Using E7.39, find the principal representatives of $P6_1$, with respect to P, the basis for the primitive lattice associated with G.

(D7.41) Definition: Let G denote a crystallographic space group and let $[xyz]^t$ denote a point that is moved by each isometry of G. Then the set of images of $[xyz]^t$ under the principal representatives of G with respect to P is called the *general position of G* (cf. Hahn, 1983). The triples in the set are called the *general equivalent positions* of G (Henry and Lonsdale, 1952).

Suppose that we know the triple that is the image of $[x,y,z]^t$ with respect to a basis P under an isometry α and wish to find the 4×4 matrix representation R of α with respect to P. Suppose $\mathbf{v} = [f_1(x,y,z), f_2(x,y,z), f_3(x,y,z)]^t$ is the image. Then $f_i(x,y,z)$ is a linear polynomial for each $1 \le i \le 3$ and

$$\mathrm{R}\begin{bmatrix} x \\ y \\ z \\ 1 \end{bmatrix} = \begin{bmatrix} r_{11} & r_{12} & r_{13} & t_1 \\ r_{21} & r_{22} & r_{23} & t_2 \\ r_{31} & r_{32} & r_{33} & t_3 \\ 0 & 0 & 0 & 1 \end{bmatrix}\begin{bmatrix} x \\ y \\ z \\ 1 \end{bmatrix} = \begin{bmatrix} f_1(x,y,z) \\ f_2(x,y,z) \\ f_3(x,y,z) \\ 1 \end{bmatrix} .$$

Hence we obtain the following set of polynomial identities:

$$r_{11}x + r_{12}y + r_{13}z + t_1 = f_1(x,y,z)$$
$$r_{21}x + r_{22}y + r_{23}z + t_2 = f_2(x,y,z)$$
$$r_{31}x + r_{32}y + r_{33}z + t_3 = f_3(x,y,z) .$$

Therefore, r_{i1}, r_{i2}, r_{i3} and t_i are the coefficients of $f_i(x,y,z)$. For example, if the image of $[x,y,z]^t$ under α is

$$[(3/4) + z, (1/4) - y, (3/4) - x]^t ,$$

then

$$r_{11}x + r_{12}y + r_{13}z + t_1 = 0x + 0y + 1z + (3/4)$$

implying that $r_{11}=r_{12}=0$, $r_{13}=1$ and $t_1=3/4$. Continuing this reasoning, we see that

$$R_{P(\mathbf{o})}(\alpha) = \begin{bmatrix} 0 & 0 & 1 & 3/4 \\ 0 & -1 & 0 & 1/4 \\ -1 & 0 & 0 & 3/4 \\ 0 & 0 & 0 & 1 \end{bmatrix} .$$

Consequently, the 4 × 4 representative of α can be found by an inspection of the corresponding image of $[x,y,z]^t$.

(E7.42) Example: Find the general equivalent positions of $P6_1$.

Solution: In P7.39, the principal representatives of $P6_1$ with respect to P are

$$\{M \mid [0,0,1/6]^t\},\ \{M^2 \mid [0,0,1/3]^t\},\ \{M^3 \mid 0,0,\tfrac{1}{2}]^t\},$$
$$\{M^4 \mid [0,0,2/3]^t\},\ \{M^5 \mid [0,0,5/6]^t\},\ \{I_3 \mid [000]^t\},$$

where

$$M = \begin{bmatrix} 1 & -1 & 0 \\ 1 & 0 & 0 \\ 0 & 0 & 1 \end{bmatrix} .$$

The images of $[x,y,z]^t$ under these representations are

$$[x-y,x,(1/6)+z]^t,\ [-y,x-y,(1/3)+z]^t,\ [-x,-y,\tfrac{1}{2}+z]^t,$$
$$[y-x,-x,(2/3)+z]^t,\ [y,y-x,(5/6)+z]^t,\ [x,y,z]^t \quad .$$

These triples are the general equivalent positions of $P6_1$. □

(P7.40) Problem: Find the general equivalent positions for $P6_2$ and $P6_3$. Verify your answers with the results given in (Hahn, 1983).

(E7.43) Example: Find all of the crystallographic space groups types G such that $\Lambda_{\mathbf{o}}(G)$ is isomorphic to 2.

Solution: From Table 6.3, the lattice types left invariant under 2 are P and C (where we have chosen the setting where Y is the unique axis). Hence we seek crystallographic space groups G such that $M_P(G) = M_P({}^{[010]}2)$. With respect to P, we have

$$M = M_P({}^{[010]}2) = \begin{bmatrix} -1 & 0 & 0 \\ 0 & 1 & 0 \\ 0 & 0 & -1 \end{bmatrix} .$$

We will find the space groups G such that $R_P(T(G)) = R_P(T_P)$ first. By T7.37, since

$$R_P(T_P) = \left\{ \{I_3 \mid \mathbf{s}\} \mid \mathbf{s} \in Z^3 \right\} , \tag{7.26}$$

we seek those elements of $\mathbf{r} \in R^3$ such that $N\mathbf{r} \in Z^3$ where

$$N = M + M^2 = \begin{bmatrix} 0 & 0 & 0 \\ 0 & 2 & 0 \\ 0 & 0 & 0 \end{bmatrix} .$$

Hence $N\mathbf{r} = [0, 2r_2, 0]$ must be in Z^3. Therefore, modulo P, we have the generators

$$\{M \mid [r_1, 0, r_3]^t\} \quad \text{and} \quad \{M \mid [r_1, \tfrac{1}{2}, r_3]^t\} , \tag{7.27}$$

where r_1, $r_3 \in R$. By T7.37, the generators in (7.27) yield all of the space groups G that satisfy (7.22) for $H = {}^{[010]}2$ and $L = P$.

To find a list of the nonequivalent space groups from all of the space groups based on (7.27), we determine the impact of moving the origin to $\mathbf{p}$. If this shift is made, $\{M \mid \mathbf{t}\}$ becomes

$$\{M \mid (M - I_3)\mathbf{p} + \mathbf{r}\} = \{M \mid [-2p_1, 0, -2p_3]^t + \mathbf{r}] .$$

Choosing p_1 and p_2 so that $2p_1 = r_1$ and $2p_3 = r_3$, we see that each of the generators in (7.27) is equivalent to either

$$\{M \mid [000]^t\} \quad \text{or} \quad \{M \mid [0, \tfrac{1}{2}, 0]^t\} .$$

It can be shown that these are nonequivalent. Thus we have found two crystallographic space group types where $M_P(G) = M_P({}^{[010]}2)$ and $R_P(T(G)) = R_P(T_P)$. We call the space group generated by $\{M \mid [000]^t\}$, $P2$, and that generated by $\{M \mid [0, \tfrac{1}{2}, 0]^t\}$, $P2_1$.

(P7.41) Problem: Consider the space group $P2_1$. Locate all of the symmetry axes in this space group that pass through the unit cell, find the principal representatives of $P2_1$ with respect to P, and calculate its set of general equivalent positions. Compare your results to those in ITFC (Hahn, 1983).

Continuing with E7.43, we consider the case where the lattice left invariant under 2 is taken to be C. Even though the basis of this lattice is $D = \{\frac{1}{2}\mathbf{a} + \frac{1}{2}\mathbf{b}, \mathbf{b}, \mathbf{c}\}$, we continue to write all of our matrices in terms of the basis P. Hence

$$\mathrm{R}_P(T_C) =$$

$$\left\{\{I_3 \mid \mathbf{s}\} \;\middle|\; \mathbf{s} = \mathbf{m} + u[\tfrac{1}{2},\tfrac{1}{2},0]^t \text{ where } \mathbf{m} \in Z^3 \text{ and } u = 0 \text{ or } 1\right\} . \quad (7.28)$$

Consequently, we seek those elements $\mathbf{r} = [r_1,r_2,r_3]^t \in R^3$ that satisfy the condition that $N\mathbf{r}$ represents a lattice vector. That is, such that

$$[0,2r_2,0]^t = \mathbf{m} + u[\tfrac{1}{2},\tfrac{1}{2},0]^t \quad ,$$

where $u = 0$ or 1. Then $0 = m_1 + u\frac{1}{2}$ where $m_1 \in Z$. Hence $u = 0$ and so, as in the case of P, we have the generators

$$\{M \mid [r_1,0,r_3]^t\} \quad \text{and} \quad \{M \mid [r_1,\tfrac{1}{2},r_3]^t\} \quad .$$

As before, these are equivalent to

$$\{M \mid [000]^t\} \quad \text{and} \quad \{M \mid [0,\tfrac{1}{2},0]^t\} \; ,$$

respectively. In particular $\{M \mid [0,\frac{1}{2},0]^t\}$ is equivalent to $\{M \mid [\frac{1}{2},\frac{1}{2},0]^t\}$ where we have chosen $r_1 = \frac{1}{2}$ and $r_3 = 0$. Note that since

$$\{M \mid [\tfrac{1}{2},\tfrac{1}{2},0]^t\} = \{I_3 \mid [\tfrac{1}{2},\tfrac{1}{2},0]^t\}\{M \mid [000]^t\} \quad ,$$

and since $\{I_3 \mid [\frac{1}{2},\frac{1}{2},0]^t\} \in \mathrm{R}_P(T_C)$,

$$\mathrm{R}_P(T_C)\{I_3 \mid [000]^t\} \cup \mathrm{R}_P(T_C)\{M \mid [000]^t\} =$$
$$\mathrm{R}_P(T_C)\{I_3 \mid [000]^t\} \cup \mathrm{R}_P(T_C)\{M \mid [\tfrac{1}{2},\tfrac{1}{2},0]^t\}$$

and so $\{M \mid [000]^t\}$ together with $\mathrm{R}_P(T_C)$ generates the same group of matrices as that generated by $\{M \mid [\frac{1}{2},\frac{1}{2},0]^t\}$ and $R_P(T_C)$.

Hence $\{M \mid [000]^t\}$ and $\{M \mid [0,\frac{1}{2},0]^t\}$ generate equivalent space groups. Consequently we obtain only one space group denoted $C2$. Hence there are three distinct crystallographic space groups based on the monoclinic point group ${}^{[010]}2$. They are $P2$, $P2_1$ and $C2$. □

(E7.44) Example: Find the principal coset representatives of $C2$ with respect to the P basis and the set of general equivalent positions of $C2$.

Solution: By (7.26) and (7.28) we see that

$$\begin{aligned} R_P(T_C) &= R_P(T_P)\{I_3 \mid [000]^t\} \cup R_P(T_P)\{I_3 \mid [\tfrac{1}{2},\tfrac{1}{2},0]^t\} \\ &= R_P(T_P) \cup R_P(T_P)\ )\{I_3 \mid [\tfrac{1}{2},\tfrac{1}{2},0]^t\} \quad . \end{aligned}$$

But

$$R_P(C2) = R_P(T_C)\{I_3 \mid [000]^t \cup R_P(T_C)\{M \mid [000]^t\} \quad .$$

where $M = M_P({}^{[010]}2)$. Hence

$$\begin{aligned} R_P(C2) &= \left[R_P(T_P) \cup R_P(T_P)\{I_3 \mid [\tfrac{1}{2},\tfrac{1}{2},0]^t\}\right]\{M \mid [000]^t\} \\ &\quad \cup \left[R_P(T_P) \cup R_P(T_P)\{I_3 \mid [\tfrac{1}{2},\tfrac{1}{2},0]^t\}\right]\{I_3 \mid [000]^t\} \\ &= R_P(T_P)\{M \mid [000]^t\} \cup R_P(T_P)\{M \mid [\tfrac{1}{2},\tfrac{1}{2},0]^t\} \\ &\quad \cup R_P(T_P)\{I_3 \mid [000]^t\} \cup R_P(T_P)\{I_3 \mid [\tfrac{1}{2},\tfrac{1}{2},0]\}^t \quad . \end{aligned}$$

The principal coset representatives of $C2$ with respect to P are $\{M \mid [000]^t\}$ $\{M \mid [\frac{1}{2},\frac{1}{2},0]^t\}$, $\{I_3 \mid [000]^t\}$, $\{I_3 \mid [\frac{1}{2},\frac{1}{2},0]^t\}$. Since

$$M_1 = \begin{bmatrix} -1 & 0 & 0 \\ 0 & 1 & 0 \\ 0 & 0 & -1 \end{bmatrix} \quad ,$$

the general equivalent positions for $C2$ are

$$[-x,y,-z]^t,\ [\tfrac{1}{2} - x,\tfrac{1}{2} + y,-z]^t,[x,y,z]^t \text{ and } [\tfrac{1}{2} + x,\tfrac{1}{2} + y,z]^t.$$

(P7.42) Problem: Show that the set of all crystallographic space group types G such that $\Lambda_o(G)$ is isomorphic to 3 consist of $P3$, $P3_1$, $P3_2$ and $R3$.

Applying the techniques illustrated in this section one can verify that the one-generator point groups give rise to the following list of 30 crystallographic space groups.

1:	$P1$
2:	$P2, P2_1, C2$
3:	$P3, P3_1, P3_2, R3$
4:	$P4, P4_1, P4_2, P4_3, I4, I4_1$
6:	$P6, P6_1, P6_2, P6_3, P6_4, P6_5$
$\bar{1}$:	$P\bar{1}$
m:	Pm, Pc, Cm, Cc
$\bar{3}$:	$P\bar{3}, R\bar{3}$
$\bar{4}$:	$P\bar{4}, I\bar{4}$
$\bar{6}$:	$P\bar{6}$

THE CRYSTALLOGRAPHIC SPACE GROUP TYPES DERIVED FROM THE TWO-GENERATOR POINT GROUPS

The two generator point groups H contain two elements α and β such that each element of H is expressed uniquely in the form $\alpha^i\beta^j$ where $1 \leq i \leq o(\alpha)$ and $1 \leq j \leq o(\beta)$. In order to do this, the elements α and β must be related in a special way. For each point group, we will choose α and β such that

$$\beta = \alpha\beta\alpha \tag{7.29}$$

and such that $\langle\alpha\rangle \cap \langle\beta\rangle = \{1\}$. A listing of the two-generator point groups together with their generators is given in Table 7.2.

(P7.43) Problem: Show that the α and β listed for each of the two-generator point groups satisfies (7.29).

(P7.44) Problem: Show that

$$\langle\alpha\rangle \cap \langle\beta\rangle = \{1\}$$

for each α and β pair given in Table 7.2 where $\langle\gamma\rangle$ denotes the cyclic group generated by γ.

Table 7.2: The two-generator oriented point groups and their generators.

Group	α	β	Group	α	β
$2/m$	$[010]_2$	i	$\bar{3}1m$	$\bar{3}$	$[210]_2$
$4/m$	4	i	422	4	$[100]_2$
$6/m$	6	i	$4mm$	4	$[100]_m$
222	2	$[100]_2$	$\bar{4}2m$	$\bar{4}$	$[100]_2$
$mm2$	2	$[100]_m$	$\bar{4}m2$	$\bar{4}$	$[100]_m$
321	3	$[100]_2$	622	6	$[100]_2$
312	3	$[210]_2$	$6mm$	6	$[100]_m$
$3m1$	3	$[100]_m$	$\bar{6}2m$	$\bar{6}$	$[100]_2$
$31m$	3	$[210]_m$	$\bar{6}m2$	$\bar{6}$	$[100]_m$
$\bar{3}m1$	$\bar{3}$	$[100]_2$			

(E7.45) Example: Let α and β be the generators listed in Table 7.2 for a two-generator point group. Show that if

$$\alpha^n \beta^m = 1,$$

then $\alpha^n = 1$ and $\beta^m = 1$.

Solution: Suppose $\alpha^n\beta^m = 1$. Then

$$\beta^m = \alpha^{-n}.$$

But $\beta^m \in \langle\beta\rangle$ and $\alpha^{-n} \in \langle\alpha\rangle$. Since $\beta^m = -\alpha^n$, $\beta^m \in \langle\beta\rangle \cap \langle\alpha\rangle$. By P7.44, $\beta^m = 1$. Therefore, $\alpha^n\beta^m = 1$ becomes $\alpha^n 1 = 1$ and so $\alpha^n = 1$. □

(P7.45) Problem: Show that, if $\alpha^i\beta^j = \alpha^k\beta^m$ where $1 \le i,\ k \le o(\alpha)$ and $1 \le j,\ m \le o(\beta)$ then $i = k$ and $j = m$. Use E7.45 together with the fact that $\{\gamma, \gamma^2, \ldots, \gamma^{o(\gamma)}\}$ are always distinct elements for any element of finite order.

(E7.46) Example: Use (7.29) to show that

$$H = \{\alpha^i\beta^j \mid i,j \in Z \text{ where } 1 \le i \le o(\alpha) \text{ and } 1 \le j \le o(\beta)\}$$

where H is a two-generator point group and α and β is as defined in Table 7.2.

Solution: We begin by showing that

$$K = \{\alpha^i\beta^j \mid i,j \in Z \quad \text{where } 1 \le i \le o(\alpha) \text{ and } 1 \le j \le o(\alpha)\}$$

is a group. Since K is contained in the finite group H, we need only show that K is nonempty and closed under composition. Since $\alpha \in K$, K is nonempty. To show that it is closed, we must show that elements like $(\alpha^2\beta)(\alpha^3\beta^2)$ can be rewritten in the form $\alpha^i\beta^j$. This is accomplished by noting that since $\alpha\beta\alpha = \beta$, we have $\beta\alpha = \alpha^{-1}\beta$. Hence if we move a β from the left side of an α to the right side we must replace α with α^{-1}. This is much like what we did in (7.24) to move the translations to the left. Hence,

$$\begin{aligned}(\alpha^2\beta)(\alpha^3\beta^2) &= \alpha^2\beta\alpha\alpha\alpha\beta^2 \\ &= \alpha^2\alpha^{-1}\beta\alpha\alpha\beta^2 \\ &= \alpha^2\alpha^{-1}\alpha^{-1}\beta\alpha\beta^2 \\ &= \alpha^2\alpha^{-1}\alpha^{-1}\alpha^{-1}\beta\beta^2 \\ &= \alpha^{-1}\beta^3 .\end{aligned}$$

Since $\alpha^{-1} \in \langle\alpha\rangle$, $\alpha^{-1} = \alpha^i$ for some $1 \le i \le o(\alpha)$. Similarly $\beta^3 = \beta^j$ for some $1 \le j \le o(\beta)$. Using this process, any product of the form $(\alpha^k\beta^\ell)(\alpha^m\beta^n)$ can be written in the form $\alpha^i\beta^j$ where $1 \le i \le o(\alpha)$ and $1 \le j \le o(\beta)$. Hence K is a subgroup of H (recall that $\alpha, \beta \in H$). But, by P7.45, the elements in K are all distinct and so $\#(K) = o(\alpha)o(\beta)$. Note that in each case in Table 7.2 $\#(H) = o(\alpha)o(\beta)$ and so $H = K$. □

(T7.47) Theorem: Let L denote a lattice left invariant by the two-generator point group H and let α and β be as given in Table 7.2. Let P denote the basis for the primitive sublattice of L associated with H. Let $M_1 = M_P(\alpha)$ and $M_2 = M_P(\beta)$. There exists a crystallographic space group G satisfying (7.22) such that each element of $R_{P(\mathbf{o})}(G)$ can be written in the form

$$\{I^3 \mid \mathbf{v}\}\{M_1 \mid \mathbf{r}\}^i\{M_2 \mid \mathbf{s}\}^j$$

where $\{I_3 \mid \mathbf{v}\} \in R_P(T_L)$, $1 \le i \le o(\alpha)$ and $1 \le j \le o(\beta)$ if and only if $\mathbf{r}$ and $\mathbf{s}$ are consistent with the relations $\alpha^{o(\alpha)} = 1$, $\beta^{o(\beta)} = 1$ and $\beta = \alpha\beta\alpha$ with respect to T_L and P.

Proof: By T7.29, if such a space group G exists, $\mathbf{r}$ and $\mathbf{s}$ must be consistent with the relations stated in the theorem. Suppose $\mathbf{r}$ and $\mathbf{s}$ are consistent with the stated relations. Consider

$$K = \left\{\{I_3 \mid \mathbf{v}\}\{M_1 \mid \mathbf{r}\}^i\{M_2 \mid \mathbf{s}\}^j \;\middle|\; \{I_3 \mid \mathbf{v}\} \in R_P(T_L),\right.$$

$$\left. 1 \leq i \leq o(\alpha),\ 1 \leq j \leq o(\beta)\right\}$$

To show that K is a group we employ the same process as discussed in the solution to E7.46. That is, we can move the elements of $R_P(T_L)$ to the left and the $\{M_2 \mid \mathbf{s}\}$'s to the right. However, by D7.30, each time we apply a relation, an element of $R_P(T_L)$ is created which can then be moved (with some modification) to the left. Hence K can be shown to be closed. Using the fact that

$$(g_1 g_2 g_3)^{-1} = g_3^{-1} g_2^{-1} g_1^{-1}$$

and then sorting, as described above, it is straight forward to show that each element in K has an inverse in K. Then K is a group. Let G denote the group of isometries that are represented by K with respect to P. Then $T(G)$ corresponds to the elements of K of the form $\{I_3 \mid \mathbf{w}\}$. Consider such an element. Then

$$\{I_3 \mid \mathbf{w}\} = \{I_3 \mid \mathbf{v}\}\{M_1 \mid \mathbf{r}\}^i\{M_2 \mid \mathbf{s}\}^j$$

where $\{I_3 \mid \mathbf{v}\} \in R_P(T_L)$, $1 \leq i \leq o(\alpha)$ and $1 \leq j \leq o(\beta)$. Then $M_1^i M_2^j = I_3$. By E7.45, $M_1^i = I_3$ and $M_2^j = I_3$ and so $i = o(\alpha)$ and $j = o(\beta)$. Since $\mathbf{r}$ and $\mathbf{s}$ were chosen to be consistent with $\alpha^{o(\alpha)} = 1$ and $\beta^{o(\beta)} = 1$ with respect to T_L and P, we have (D7.30)

$$\{I_3 \mid \mathbf{w}\} = \{I_3 \mid \mathbf{v}\}\{I_3 \mid \mathbf{u}\}\{I_3 \mid \mathbf{x}\}$$

where each of the factors on the right is an element of $R_P(T_L)$. Therefore $\{I_3 \mid \mathbf{w}\} \in R_P(T_L)$ and so $R_P(T(G)) = R_P(T_L)$. Also by E7.46, $M_P(G) = M_P(H)$. □

(E7.48) Example: Let $H = 422$ and $L = P$. By Table 7.2, $\alpha = 4$ and $\beta = [100]_2$. Let $M_1 = M_P(4)$ and $M_2 = M_P([100]_2)$. Show that $\mathbf{r} = [0,0,\frac{1}{4}]^t$ and

$\mathbf{s} = [\frac{1}{2},\frac{1}{2},0]^t$ are consistent with $\alpha^4 = \mathbf{1}$, $\beta^2 = \mathbf{1}$ and $\beta = \alpha\beta\alpha$.

Solution: Since $o(\alpha) = 4$, $o(\beta) = 2$, $N_1 = M_1 + M_1^2 + M_1^3 + M_1^4$ and $N_2 = M_2 + M_2^2$. Since

$$M_1 = \begin{bmatrix} 0 & -1 & 0 \\ 1 & 0 & 0 \\ 0 & 0 & 1 \end{bmatrix} \text{ and } M_2 = \begin{bmatrix} 1 & 0 & 0 \\ 0 & -1 & 0 \\ 0 & 0 & -1 \end{bmatrix},$$

we have

$$N_1 = \begin{bmatrix} 0 & 0 & 0 \\ 0 & 0 & 0 \\ 0 & 0 & 4 \end{bmatrix} \text{ and } N_2 = \begin{bmatrix} 2 & 0 & 0 \\ 0 & 0 & 0 \\ 0 & 0 & 0 \end{bmatrix}.$$

Since $N_1\mathbf{r} = [001]^t$ and $N_2\mathbf{s} = [100]^t$, by T7.33, $\mathbf{r}$ is consistent with $\alpha^4 = \mathbf{1}$ and $\mathbf{s}$ is consistent with $\beta^2 = \mathbf{1}$ with respect to T_P and P. To show that $\mathbf{r}$ and $\mathbf{s}$ are consistent with $\beta = \alpha\beta\alpha$, we calculate the product

$$\{M_1 \mid [0,0,\tfrac{1}{4}]^t\}\{M_2 \mid [\tfrac{1}{2},\tfrac{1}{2},0]^t\}\{M_1 \mid [0,0,\tfrac{1}{4}]^t\} = \{M_2 \mid [-\tfrac{1}{2},\tfrac{1}{2},0]^t\}.$$

But

$$\{M_2 \mid [-\tfrac{1}{2},\tfrac{1}{2},0]^t\} = \{I_3 \mid [-1,0,0]^t\}\{M_2 \mid [\tfrac{1}{2},\tfrac{1}{2},0]^t\}.$$

Since $\{I_3 \mid [-1,0,0]^t\} \in R_P(T_P)$, $\mathbf{r}$ and $\mathbf{s}$ are consistent with $\beta = \alpha\beta\alpha$ with respect to T_P and P. By T7.47, the set of matrices of the form

$$\{I_3 \mid \mathbf{v}\}\{M_1 \mid \mathbf{r}\}^i\{M_2 \mid \mathbf{s}\}^j,$$

where $\mathbf{v} \in Z^3$, $1 \le \mathbf{i} \le 4$ and $1 \le j \le 2$ represents the elements of a crystallographic space group G satisfying (7.22) where $L = P$ and $H = 422$. This space group is denoted $P4_12_12$. □

(E7.49) Example: Find all of the space groups G derived from $H = 321$ and $L = P$. Then determine all of the corresponding space group types.

Solution: By Table 7.2, $\alpha = \mathbf{3}$ and $\beta = [100]_2$. Then

$$M_1 = M_P(\mathbf{3}) = \begin{bmatrix} 0 & -1 & 0 \\ 1 & -1 & 0 \\ 0 & 0 & 1 \end{bmatrix} \text{ and } M_2 = M_P([100]_2) = \begin{bmatrix} 1 & -1 & 0 \\ 0 & -1 & 0 \\ 0 & 0 & -1 \end{bmatrix}.$$

We must now find all $\mathbf{r}$ and $\mathbf{s}$ that are consistent with the relations $\alpha^3 = 1$, $\beta^2 = 1$ and $\beta = \alpha\beta\alpha$. Then $N_1 = M_1 + M_1^2 + M_1^3$ and $N_2 = M_2 + M_2^2$ are

$$N_1 = \begin{bmatrix} 0 & 0 & 0 \\ 0 & 0 & 0 \\ 0 & 0 & 3 \end{bmatrix} \quad \text{and} \quad N_2 = \begin{bmatrix} 2 & -1 & 0 \\ 0 & 0 & 0 \\ 0 & 0 & 0 \end{bmatrix} .$$

By T7.33, since the lattice is primitive, $N_1\mathbf{r} = [0,0,3r_3]^t \in Z^3$. Hence $\mathbf{r} = [r_1,r_2,n/3]^t$ for some integer n. Since $N_2\mathbf{s} = [2s_1 - s_2,0,0]^t \in Z^3$, $\mathbf{s} = [s_1,2s_1 - m,s_3]^t$ for some integer m. Modulo P, $\mathbf{s}$ can be taken to be $\mathbf{s} = [s_1, 2s_1, s_3]^t$. To be consistent with $\beta = \alpha\beta\alpha$, $\mathbf{r}$ and $\mathbf{s}$ must be such that

$$\{M_1 \mid \mathbf{r}\}\{M_2 \mid \mathbf{s})\}\{M_1 \mid \mathbf{r}\} = \{I_3 \mid \mathbf{v}\}\{M_2 \mid \mathbf{s}\}$$

for some $\{I_3 \mid \mathbf{v}\} \in R_P(T_P)$. Hence

$$\{M_2 \mid [r_1 + r_2 - 2s_1, r_1 + r_2 - s_1, s_3]^t\} = \{I_3 \mid \mathbf{v}\}\{M_2 \mid \mathbf{s}\} .$$

Solving for $\mathbf{v}$, we have

$$\mathbf{v} = \begin{bmatrix} r_1 + r_2 - 3s_1 \\ r_1 + r_2 - 3s_1 \\ 0 \end{bmatrix} .$$

Since the lattice is primitive, $\mathbf{v} \in Z^3$, $r_1 + r_2 - 3s_1$ must be an integer. Hence any choice of $r_1,r_2,s_1,s_3 \in R$ and $n \in Z$ such that

$$r_1 + r_2 - 3s_1 \in Z \tag{7.30}$$

yields $\mathbf{r} = [r_1,r_2,n/3]^t$ and $\mathbf{s} = [s_1,2s_1,s_3]^t$ that are consistent with $\alpha^3 = 1$, $\beta^2 = 1$ and $\beta = \alpha\beta\alpha$. Conversely, any $\mathbf{r}$ and $\mathbf{s}$ that are consistent with these relations can be put (modulo P) into this form. By T7.47, each space group G that satisfies (7.22) with $H = 321$ and $L = P$ is represented by a matrix group whose elements are of the form

$$\{I_3 \mid \mathbf{v}\}\{M_1 \mid \mathbf{r}\}^i\{M_2 \mid \mathbf{s}\}^j ,$$

where $\{I_3 \mid \mathbf{v}\} \in R_P(T_P)$, $1 \le i \le 3$ and $1 \le j \le 2$ for some choice of $\mathbf{r}$ and $\mathbf{s}$ satisfying the conditions associated with (7.30).

We must now determine representatives of the different space group types in this collection. If the origin is moved to the vector $\mathbf{p}$, $\{M_1 \mid \mathbf{r}\}$ becomes

$$\{M_1 \mid (M_1 - I_3)\mathbf{p} + \mathbf{r}\} \quad . \tag{7.31}$$

Since $(M_1 - I_3)\mathbf{p} = [-p_1 - p_2, p_1 - 2p_2, 0]^t$, to simplify the first two components of $\mathbf{r}$, we wish to find $\mathbf{p}$ such that $-p_1 - p_2 = -r_1$ and $p_1 - 2p_2 = -r_2$. Solving this system of equations we find that if $p_1 = (2/3)r_1 - (1/3)r_2$ and $p_2 = (1/3)(r_1 + r_2)$, then (7.31) becomes

$$\{M_1 \mid [0,0,n/3]^t\} \quad .$$

Moving the origin in this manner changes the value of $\mathbf{s}$ as well. However the conditions associated with (7.30) must still hold. That is, for our new $\mathbf{r}$ and $\mathbf{s}$, $\mathbf{r} = [0,0,n/3]^t$ and $\mathbf{s} = [s_1, 2s_1, s_3]^t$ where $-3s_1 \in Z$. Hence $s_1 = m/3$ for some integer m. Therefore, $\mathbf{s} = [m/3, 2m/3, 0]^t$ for some $m \in Z$. Note that if a subsequent change of origin is such that $\{M_1 \mid \mathbf{r}\}$ becomes

$$\{M_1 \mid [u,v,n/3]^t\} \quad ,$$

then $\mathbf{r}$ can still be taken, modulo P, to be $[0,0,n/3]^t$. Hence, if we move the origin to a new $\mathbf{p}$, where p_1 and p_2 are chosen so that $-p_1 - p_2$ and $p_1 - 2p_2$ are integers then $\mathbf{r}$ is undisturbed. Note that p_3 can be any value we wish. Under such a change in origin, $\{M_2 \mid \mathbf{s}\}$ becomes

$$\{M_2 \mid [-p_2, -2p_2, -2p_3] + [m/3, 2m/3, s_3]^t\} \quad .$$

Setting $p_2 = m/3$, $p_3 = s_3/2$ and $p_1 = -m/3$, then $-p_1 - p_2 = 0$, $p_1 - 2p_2 = -3m/3 \in Z$ and $\{M_2 \mid \mathbf{s}\}$ becomes

$$\{M_2 \mid [0,0,0]^t\} \quad .$$

Consequently, we have the following choices of $\mathbf{r}$ and $\mathbf{s}$:

$\mathbf{r}$	$\mathbf{s}$
$[000]^t$	$[000]^t$
$[0,0,1/3]^t$	$[000]^t$
$[0,0,2/3]^t$	$[000]^t$.

The three pairs of $\{M_1 \mid \mathbf{r}\}$, $\{M_2 \mid \mathbf{s}\}$ formed using the vectors above lead to nonequivalent space groups. The three space groups obtained in this manner are denoted $P321$, $P3_121$ and $P3_221$, respectively. □

(E7.50) Example: Find the principle representatives of space group $P3_221$ and determine its general equivalent positions.

Solution: According to T7.47 and E7.49 the elements of $P3_221$ are of the form

$$\{I_3 \mid \mathbf{v}\}\{M_1 \mid \mathbf{r}\}^i\{M_2 \mid \mathbf{s}\}^j$$

where $\mathbf{v} \in Z^3$, $1 \le i \le 3$, $1 \le j \le 2$ and

$$M_1 = M_P(\mathbf{3}) = \begin{bmatrix} 0 & -1 & 0 \\ 1 & -1 & 0 \\ 0 & 0 & 1 \end{bmatrix} \text{ and } M_2 = M_P({}^{[100]}2) = \begin{bmatrix} 1 & -1 & 0 \\ 0 & -1 & 0 \\ 0 & 0 & -1 \end{bmatrix} .$$

$\mathbf{r} = [0,0,2/3]^t$ and $\mathbf{s} = [0,0,0]^t$. Thus,

$$\begin{aligned} P3_221 = {} & R_P(T_P)\{M_1 \mid [0,0,2/3]^t\}\{M_2 \mid [000]^t\} \\ & \cup R_P(T_P)\{M_1^2 \mid [0,0,1/3]^t\}\{M_2 \mid [000]^t\} \\ & \cup R_P(T_P)\{I_3 \mid [000]^t\}\{M_2 \mid [000]^t\} \\ & \cup R_P(T_P)\{M_1 \mid [0,0,2/3]^t\}\{I_3 \mid [000]^t\} \\ & \cup R_P(T_P)\{M_1^2 \mid [0,0,1/3]^t\}\{I_3 \mid [000]^t\} \\ & \cup R_P(T_P)\{I_3 \mid [000]^t\}\{I_3 \mid [000]^t\} . \end{aligned}$$

That is, each element in $P3_221$ resides in exactly one of the following six right cosets:

$$\begin{aligned} & R_P(T_P)\{M_1M_2 \mid [0,0,2/3]^t\}, \ R_P(T_P)\{M_1^2M_2 \mid [0,0,1/3]^t\} \\ & R_P(T_P)\{M_2 \mid [000]^t\}, \ R_P(T_P)\{M_1 \mid [0,0,2/3]^t\}, \\ & R_P(T_P)\{M_1^2 \mid [0,0,1/3]^t\}, \ R_P(T_P)\{I_3 \mid [000]^t\} . \end{aligned}$$

Hence, the principal representatives of $P3_221$ with respect to P are

$$\begin{aligned} & \{I_3 \mid [000]^t\}, \ \{M_1 \mid [0,0,2/3]^t\}, \ \{M_1^2 \mid [0,0,1/3]^t\} \\ & \{M_2 \mid [000]^t\}, \ \{M_1M_2 \mid [0,0,2/3]^t\}, \ \{M_1^2M_2 \mid [0,0,1/3]^t\} . \end{aligned}$$

Since

$$M_1M_2 = \begin{bmatrix} 0 & 1 & 0 \\ 1 & 0 & 0 \\ 0 & 0 & -1 \end{bmatrix} \quad \text{and} \quad M_1^2M_2 = \begin{bmatrix} -1 & 0 & 0 \\ -1 & 1 & 0 \\ 0 & 0 & -1 \end{bmatrix} ,$$

the general equivalent positions for $P3_221$ are

$$[x,y,z]^t,\ [-y,x-y,\ (2/3)+z]^t,\ [y-x,\ -x,\ (1/3)+z]^t,$$
$$[x-y,\ -y,\ -z]^t,\ [y,\ x,\ (2/3)-z]^t,\ [-x,\ y-x,\ (1/3)-z]^t \ . \quad \square$$

(P7.46) Problem: The ITFC (Hahn, 1983) gives the following general equivalent positions for $P3_221$:

$$[x,\ y,\ z]^t,\ [-y,\ x-y,\ (2/3)+z]^t,\ [y-x,\ -x,\ (1/3)+z]^t,$$
$$[y,x,-z]^t,\ [x-y,\ -y,\ (1/3)-z]^t,\ [-x,\ y-x,\ (2/3)-z]^t \ .$$

Show that if we move our origin to $\mathbf{p} = [0,0,-1/6]^t$, then $\{M_1 \mid \mathbf{r}\}$ and $\{M_2 \mid \mathbf{s}\}$ are changed so that the general equivalent positions as given in the ITFC are obtained.

(P7.47) Problem: Determine the principal representatives of space group $P4_12_12$ (E7.48) and confirm that its general equivalent positions are:

$$[x,\ y,\ z]^t,\ [-y,\ x,\ \tfrac{1}{4}+z]^t,\ [-x,\ -y,\ \tfrac{1}{2}+z]^t,$$
$$[y,\ -x,\ (3/4)+z]^t,\ [\tfrac{1}{2}+x,\ \tfrac{1}{2}-y,\ -z]^t,\ [y-\tfrac{1}{2},\ \tfrac{1}{2}+x,\ \tfrac{1}{4}-z]^t,$$
$$[-\tfrac{1}{2}-x,\ y-\tfrac{1}{2},\ \tfrac{1}{2}-z]^t,\ [\tfrac{1}{2}-y,\ -\tfrac{1}{2}-x,\ (3/4)-z]^t \ .$$

(P7.48) Problem: The ITFC (Hahn, 1983) gives the following general equivalent positions for $P4_12_12$:

$$[x,\ y,\ z]^t,\ [-x,\ -y,\ \tfrac{1}{2}+z]^t,\ [\tfrac{1}{2}-y,\ \tfrac{1}{2}+x,\ \tfrac{1}{4}+z]^t,$$
$$[\tfrac{1}{2}+y,\ \tfrac{1}{2}-x,\ (3/4)+z]^t,\ [\tfrac{1}{2}-x,\ \tfrac{1}{2}+y,\ \tfrac{1}{4}-z]^t,$$
$$[\tfrac{1}{2}+x,\ \tfrac{1}{2}-y,\ (3/4)-z]^t,\ [y,\ x,\ -z]^t,\ [-y,\ -x,\ \tfrac{1}{2}-z]^t \ .$$

Find $\mathbf{p}$ such that if our origin is moved to $\mathbf{p}$, these general equivalent positions occur.

(E7.51) Example: An element of the right coset $R_p(T_p)\{M_2 \mid [000]^t\}$ of $P3_221$ is of the form

$$\begin{bmatrix}1 & 0 & 0 & \mathbf{u}\\0 & 1 & 0 & \mathbf{v}\\0 & 0 & 1 & \mathbf{w}\\0 & 0 & 0 & 1\end{bmatrix}\begin{bmatrix}1 & -1 & 0 & 0\\0 & -1 & 0 & 0\\0 & 0 & -1 & 0\\0 & 0 & 0 & 1\end{bmatrix} = \begin{bmatrix}1 & -1 & 0 & \mathbf{u}\\0 & -1 & 0 & \mathbf{v}\\0 & 0 & -1 & \mathbf{w}\\0 & 0 & 0 & 1\end{bmatrix},$$

where, because the lattice is primitive, $u, v, w \in Z$. Analyze the matrix on the right, using the techniques discussed by Boisen and Gibbs (1978) and determine the name of the space group operation α that it represents and a point on its symmetry element.

Solution: As $tr(M_2) = -1$ and $det(M_2) = +1$, we know that the linear part of α defines a half-turn. We also know that its rotation axis parallels [100]. The translational component, $\mathbf{d}$, of α is found by solving

$$\mathbf{d} = \tfrac{1}{2} \sum_{i=1}^{2} M_1^i\mathbf{t} = \tfrac{1}{2}(M_1\mathbf{t} + M_1^2\mathbf{t}) \quad ,$$

where $\mathbf{t} = [u,v,w]^t$ and so

$$\mathbf{d} = \tfrac{1}{2}\left(\begin{bmatrix}1 & -1 & 0\\0 & -1 & 0\\0 & 0 & -1\end{bmatrix}\begin{bmatrix}u\\v\\w\end{bmatrix} + \begin{bmatrix}1 & 0 & 0\\0 & 1 & 0\\0 & 0 & 1\end{bmatrix}\begin{bmatrix}u\\v\\w\end{bmatrix}\right) = \begin{bmatrix}u - v/2\\0\\0\end{bmatrix}.$$

Hence, $\alpha = {}^{[100]}2_1$ modulo the translation group when v is odd and $\alpha = {}^{[100]}2$ modulo the translation group when v is even.

The set of points defining the symmetry element of α is found using the equation

$$\mathbf{t} - \mathbf{d} = (I_3 - M_1)\mathbf{e}$$

where $\mathbf{e} = [e_1,e_2,e_3]^t$ denotes a point on the rotation axis of α. Thus,

$$\begin{bmatrix}u\\v\\w\end{bmatrix} - \begin{bmatrix}u - v/2\\0\\0\end{bmatrix} = \left(\begin{bmatrix}1 & 0 & 0\\0 & 1 & 0\\0 & 0 & 1\end{bmatrix} - \begin{bmatrix}1 & -1 & 0\\0 & -1 & 0\\0 & 0 & -1\end{bmatrix}\right)\begin{bmatrix}e_1\\e_2\\e_3\end{bmatrix}$$

$$\begin{bmatrix}v/2\\v\\w\end{bmatrix} = \begin{bmatrix}0 & 1 & 0\\0 & 2 & 0\\0 & 0 & 2\end{bmatrix}\begin{bmatrix}e_1\\e_2\\e_3\end{bmatrix}.$$

With the matrix methods described in Appendix 2, we find that e_1 is indeterminant, $e_2 = v/2$ and $e_3 = w/2$. Hence, the rotation axes of ${}^{[100]}2$ and ${}^{[100]}2_1$ both pass through the point $(v/2)\mathbf{b} + (w/2)\mathbf{c}$. For the case where v is even and $\alpha = {}^{[100]}2$, we see that there are exactly two such axes in the unit cell passing through the points $[0,0,0]^t$ and $[0,0,\frac{1}{2}]^t$. When v is odd and $\alpha = {}^{[100]}2_1$, again there are two axes in the unit cell passing through the points $[0,\frac{1}{2},0]^t$ and $[0,\frac{1}{2},\frac{1}{2}]^t$ (Figure 7.7). □

(P7.49) Problem: Analyze an element from each of the remaining right cosets of $P3_221$ and confirm the symmetry elements displayed in Figure 7.7.

(E7.52) Example: The space group symmetry of the α-quartz structure displayed in Figure 1.4 is $P3_221$. As a verification of this observation, show that the atomic coordinates of the oxygen atoms in the unit cell (Table 1.3) are $P3_221$-equivalent.

Solution: This will be done by showing that the triple representing O_1 is $P3_221$-equivalent to those of O_2, O_3, O_4, O_5 and O_6. Note that the basis used in Table 1.3 is $D = P$. The triple representing $\mathbf{r}_1$ radiating from the origin to O_1 in Table 1.3 is

$$[\mathbf{r}_1]_P = [0.4141,\ 0.2681,\ 0.1188]^t .$$

Hence, we have (modulo P)

$$R_P(1)\begin{bmatrix}\mathbf{r}_1\\ 1\end{bmatrix} = \begin{bmatrix}1&0&0&0\\0&1&0&0\\0&0&1&0\\0&0&0&1\end{bmatrix}\begin{bmatrix}0.4141\\0.2681\\0.1188\\1\end{bmatrix} = \begin{bmatrix}0.4141\\0.2681\\0.1188\\1\end{bmatrix},$$

$$R_P(3_2)\begin{bmatrix}\mathbf{r}_1\\ 1\end{bmatrix} = \begin{bmatrix}0&-1&0&0\\1&-1&0&0\\0&0&1&2/3\\0&0&0&1\end{bmatrix}\begin{bmatrix}0.4141\\0.2681\\0.1188\\1\end{bmatrix} = \begin{bmatrix}-0.2681\\0.1460\\0.7855\\1\end{bmatrix} = \begin{bmatrix}0.7319\\0.1460\\0.7855\\1\end{bmatrix},$$

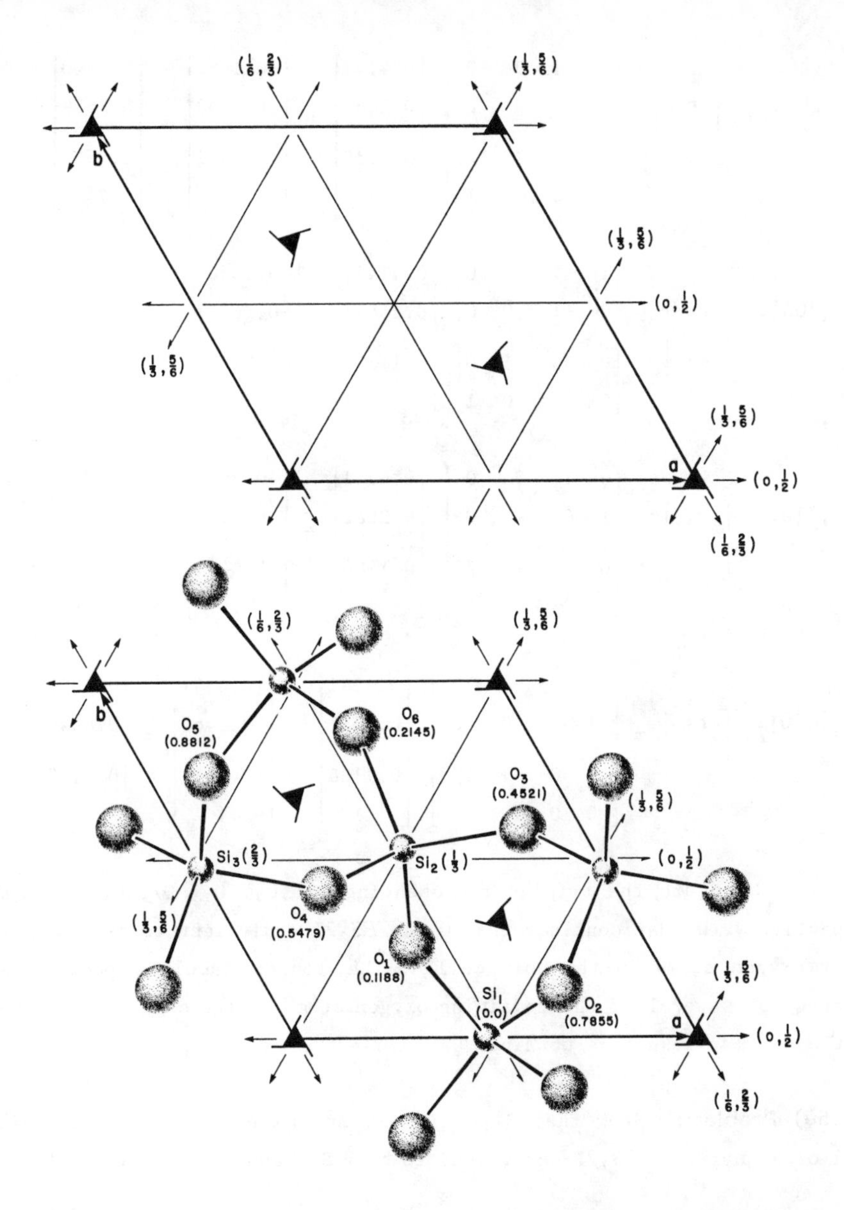

Figure 7.7 (top): A unit cell displaying the symmetry elements of space group $P3_221$ viewed down **c**. The rotation axes of the vertical 3_2-screw axes are symbolized by pronged triads and those of the horizontal 2-fold screw and 2-fold rotation axes are symbolized by single-barbed and double-barbed arrows, respectively, paralleling [100], [110] and [010]. The numbers enclosed between the parentheses define the Z-coordinates of these axes. For example, $(0,\frac{1}{2})$ placed at the ends of the two-fold rotation axes paralleling **a** indicates that two such axes occur in the cell, one at a height $Z = 0$ and the other at height $Z = \frac{1}{2}$.

Figure 7.8 (bottom): A diagram of the symmetry elements of space group $P3_221$ inserted into a unit cell of the α-quartz structure viewed down **c**. The atoms are labelled as in Table 1.3. The numbers next to each atom in the cell specify its Z-coordinate.

$$R_p(3_2^2)\begin{bmatrix}\mathbf{r}_1\\1\end{bmatrix}=\begin{bmatrix}-1&1&0&0\\-1&0&0&0\\0&0&1&1/3\\0&0&0&1\end{bmatrix}\begin{bmatrix}0.4141\\0.2681\\0.1188\\1\end{bmatrix}=\begin{bmatrix}-0.1460\\-0.4141\\0.4521\\1\end{bmatrix}=\begin{bmatrix}0.8540\\0.5859\\0.4521\\1\end{bmatrix}$$

$$R_p({}^{[100]}2)\begin{bmatrix}\mathbf{r}_1\\1\end{bmatrix}=\begin{bmatrix}1&-1&0&0\\0&-1&0&0\\0&0&-1&0\\0&0&0&1\end{bmatrix}\begin{bmatrix}0.4141\\0.2681\\0.1188\\1\end{bmatrix}=\begin{bmatrix}0.1460\\-0.2681\\-0.1188\\1\end{bmatrix}=\begin{bmatrix}0.1460\\0.7319\\0.8812\\1\end{bmatrix},$$

$$R_p({}^{[110]}2)\begin{bmatrix}\mathbf{r}_1\\1\end{bmatrix}=\begin{bmatrix}0&1&0&0\\1&0&0&0\\0&0&-1&2/3\\0&0&0&1\end{bmatrix}\begin{bmatrix}0.4141\\0.2681\\0.1188\\1\end{bmatrix}=\begin{bmatrix}-0.2681\\0.4141\\0.5479\\1\end{bmatrix},$$

$$R_p({}^{[010]}2)\begin{bmatrix}\mathbf{r}_1\\1\end{bmatrix}=\begin{bmatrix}-1&0&0&0\\-1&1&0&0\\0&0&-1&1/3\\0&0&0&1\end{bmatrix}\begin{bmatrix}0.4141\\0.2681\\0.1188\\1\end{bmatrix}=\begin{bmatrix}-0.4141\\-0.1460\\0.2145\\1\end{bmatrix}=\begin{bmatrix}0.5859\\0.8540\\0.2145\\1\end{bmatrix}.$$

Because these are the triples representing O_1, O_2, O_3, O_5, O_4 and O_6, respectively, we may conclude that O_1 is $P3_221$-equivalent to each of the other oxygen atoms in the unit cell. The reader should repeat this process for several of the remaining oxygen atoms in the cell and observe that a similar result is obtained. □

(P7.50) Problem: Show that Si_1 which is at the end of $\mathbf{r}_2$ in the unit cell of α-quartz is $P3_221$-equivalent to both Si_2 and Si_3 (see Table 1.3).

Note that Si_1 resides on the rotation axis of ${}^{[100]}2$. Hence ${}^{[100]}2$ maps Si_1 to itself. As we learned earlier, whenever an atom is mapped to itself by at least one operation of the group in addition to the identity, we say that the atom is on a special position. Thus, the Si atoms in α-quartz are on special positions (each resides on a 2-fold axis). As a consequence, there are only three Si atoms in the unit cell. On the other hand, because O_1 is on a general position, there are six oxygen atoms in the unit cell. Therefore, the unit cell of α-quartz contains three SiO_2 formula units. A drawing of the structure of α-quartz and its symmetry elements is given in Figure 7.8 (above).

(P7.51) Problem: Analyze an element from each of the eight right cosets of $P4_12_12$ and construct a diagram of its unit cell showing all its symmetry elements viewed down **c**.

(P7.52) Problem: In a structural analysis of α-cristobalite, Pluth *et al* (1985) found an oxygen atom at $[0.7392, 0.0637, 0.8036]^t$ and a Si atom at $[0.8005, 0.3005, 5/8]^t$.

(1) Given that the space group of α-cristobalite is $P4_12_12$ described in E7.48, find the coordinates of the O atoms in the unit cell that are $P4_12_12$-equivalent. Then find the coordinates of the Si atoms that are $P4_12_12$-equivalent.

(2) Make a drawing of the α-cristobalite structure viewed down **c** and compare with that given in Figure 1.7.

(3) Calculate the SiO bond lengths and SiOSi angles and compare your results with those calculated in E1.13 for α-quartz.

(E7.53) Example: Find all of the space groups G such that $M_p(G) = M_p(2/m)$ and $R_p(T(G)) = R_p(T_C)$. Then determine all of the corresponding space group types.

Solution: By Table 7.2, $\alpha = {}^{[010]}2$ and $\beta = i$. Hence

$$M_1 = M_p({}^{[010]}2) = \begin{bmatrix} -1 & 0 & 0 \\ 0 & 1 & 0 \\ 0 & 0 & -1 \end{bmatrix} \quad \text{and} \quad M_2 = M_p(i) = \begin{bmatrix} -1 & 0 & 0 \\ 0 & -1 & 0 \\ 0 & 0 & -1 \end{bmatrix}.$$

We now find all **r** and **s** that are consistent with the relations $\alpha^2 = \mathbf{1}$, $\beta^2 = \mathbf{1}$ and $\beta = \alpha\beta\alpha$. Since $N_1\mathbf{r} = [0,2r_2,0]^t$ and the lattice is C-centered, $[0,2r_2,0]^t = [uvw]^t + k[\frac{1}{2},\frac{1}{2},0]^t$ where $u,v,w \in Z$ and $k = 0$ or 1. Therefore, $0 = u + k/2$ which implies that $k = 0$. Hence $\mathbf{r}_2 = n/2$ for some $n \in Z$. Since $N_2\mathbf{s} = [000]^t$, no conditions are imposed by $\beta^2 = \mathbf{1}$. To be consistent with $\beta = \alpha\beta\alpha$,

$$\{M_1 \mid \mathbf{r}\}\{M_2 \mid \mathbf{s}\}\{M_1 \mid \mathbf{r}\} = \{I_3 \mid \mathbf{v}\}\{M_2 \mid \mathbf{s}\} .$$

where $\{I_3 \mid \mathbf{v}\} \varepsilon R_P(T_C)$. By taking the product and solving, we find that

$$[2r_1 - 2s_1,\ 0,\ 2r_3 - 2s_3]^t = [uvw]^t + k[\tfrac{1}{2},\tfrac{1}{2},0]^t ,$$

where u, v, $w \varepsilon Z$ and $k = 0$ or 1. Since $0 = v + k/2$, $k = 0$. Hence, modulo C, $\mathbf{r} = [s_1 + m/2,\ n/2,\ s_3 + \ell/2]^t$ where m, $\ell \varepsilon Z$. Hence any choice of s_1, s_2, $s_3 \varepsilon R$, m, n, $\ell \varepsilon Z$ yields

$$\mathbf{r} = [s_1 + m/2,\ n/2,\ s_3 + \ell/_2]^t \text{ and } \mathbf{s} = [s_1,s_2,s_3]^t \qquad (7.32)$$

that are consistent with the relations.

By T7.47, each space group G such that $M_P(G) = M_P(2/m)$ and $R_P(T(G)) = R_P(T_C)$ is represented by a matrix group whose elements are of the form

$$\{I_3 \mid \mathbf{v}\}\{M_1 \mid \mathbf{r}\}^i\{M_2 \mid \mathbf{s}\}^j ,$$

where $\{I_3 \mid \mathbf{v}\} \varepsilon R_P(T_C)$, $1 \le i \le 2$, $1 \le j \le 2$ for some choice of $\mathbf{r}$ and $\mathbf{s}$ satisfying the conditions assiciated with (7.32).

We now determine the different space group types. Suppose the origin is moved to $\mathbf{p}$. Since $(M_2 - I_3)\mathbf{p} = [-2p_1,\ -2p_2,\ -2p_3]^t$, if we set $p_1 = s_1/2$ and $p_2 = s_3/2$, then $\{M_2 \mid \mathbf{s}\}$ becomes

$$\{M_2 \mid [000]^t\}$$

and the conditions in (7.32) yield $\{M_1 \mid [m/2,\ n/2,\ \ell/2]^t\}$. Since any subsequent change of origin so that

$$[-2p_1,\ -2p_2,\ -2p_3]^t \ \varepsilon\ Z^3 + k[-\tfrac{1}{2},\tfrac{1}{2},0]^t$$

where $k = 0$ or 1 will leave $\{M_2 \mid [000]^t\}$ unchanged. Since $(M_1 - I_3)\mathbf{p} = [-2p_1,\ 0,\ -2p_3]^t$, setting $p_1 = m/4$, $p_2 = m/4$, $p_3 = 0$, yields $[-2p_1,\ -2p_2,\ -2p_3]^t = [-m/2,-m/2,0]^t \ \varepsilon\ Z^3 + k[\tfrac{1}{2},\tfrac{1}{2},0]^t$ where $k =$ 0 or 1 and changes $\{M_1 \mid [m/2,n/2,\ell/2]^t\}$ into $\{M_1 \mid [0,n/2,\ell/2]^t\}$. Hence it appears that we have four choices for $\mathbf{r}$. Since

$$\{M_1 \mid [0,\tfrac{1}{2},0]^t\} = \left[\{I_3 \mid [-1,0,0]^t\}\{I_3 \mid [\tfrac{1}{2},\tfrac{1}{2},0]^t\}\right]\{M_1 \mid [\tfrac{1}{2},0,0]^t\} ,$$

$\{M_1 \mid [0,\frac{1}{2},0]^t\}$ and $\{M_1 \mid [\frac{1}{2},0,0]^t\}$ generate the same matrix group with $R_P(T_C)$. Since $\{M_1 \mid [\frac{1}{2},0,0]^t\}$ is equivalent to $\{M_1 \mid [000]^t\}$ under the change of origin to $[\frac{1}{4},\frac{1}{4},0]^t$, $\{M_1 \mid [0,\frac{1}{2},0]^t\}$ and $\{M_1 \mid [000]^t\}$ generate equivalent space groups with $R_P(T_C)$. Similarly $[0,\frac{1}{2},\frac{1}{2}]^t$ and $[0,0,\frac{1}{2}]^t$ yield equivalent space groups. Therefore the space group types associated with *2/m* and *C* are *C2/m* (when $\mathbf{r} = [000]^t$ and $\mathbf{s} = [000]^t$) and *C2/c* (when $\mathbf{r} = [0,0,\frac{1}{2}]^t$ and $\mathbf{s} = [000]^t$). □

(P7.53) Problem: Find all of the space groups G that satisfy (7.22) with $H = 222$ and $L = P$. Find one representative from each space group type. The following steps may be helpful in your derivation.

(1) Consult Table 7.2 and let

$$M_1 = M_P(2) = \begin{bmatrix} -1 & 0 & 0 \\ 0 & -1 & 0 \\ 0 & 0 & 1 \end{bmatrix} \quad \text{and} \quad M_2 = M_P({}^{[100]}2) = \begin{bmatrix} 1 & 0 & 0 \\ 0 & -1 & 0 \\ 0 & 0 & -1 \end{bmatrix}.$$

(2) Show that to be consistent with $\alpha^2 = 1$ and $\beta^2 = 1$,

$$\{M_1 \mid \mathbf{r}\} = \{M_1 \mid [r_1, r_2, m/2]^t\}$$

and

$$\{M_2 \mid \mathbf{s}\} = \{M_2 \mid [n/2, s_2, s_3]^t\} \quad ,$$

where $m,n \in Z$.

(3) Show that to be consistent with $\beta = \alpha\beta\alpha$, $2(r_2 - s_2) \in Z$. Hence, modulo P, $\mathbf{s} = [n/2,\ r_2 + k/2,\ s_3]^t$ where $k \in Z$. Then by T7.47 the set of all crystallographic space groups G satisfying (7.22) with $H = 222$ and $L = P$ consists of those represented by $R_{P(\mathbf{o})}(G)$ which are generated by $R_P(T_P)$, $\{M_1 \mid \mathbf{r}\}$ and $\{M_2 \mid \mathbf{s}\}$ where $\mathbf{r} = [r_1, r_2, m/2]^t$ and $\mathbf{s} = [n/2,\ r_2 + k/2,\ s_3]^t$ with $r_1,\ r_2,\ s_3 \in R$, $m,\ n,\ k \in Z$.

(4) Show that by an appropriate change of origin, $\{M_1 \mid \mathbf{r}\}$ becomes

$$\{M_1 \mid [0,0,m/2]^t\}$$

and $\{M_2 \mid \mathbf{s}\}$ becomes

$$\{M_2 \mid [n/2,k/2,0]^t\} ,$$

where $k,m,n \in Z$.

(5) Note that $\{M_2 \mid \mathbf{r}\}\{M_1 \mid \mathbf{s}\} = \{M_3 \mid \mathbf{r} + M_2\mathbf{s}\}$ where

$$M_3 = \begin{bmatrix} -1 & 0 & 0 \\ 0 & 1 & 0 \\ 0 & 0 & -1 \end{bmatrix} .$$

Show that $\{M_2 \mid [\frac{1}{2},0,0]^t\}\{M_1 \mid [000]^t\} = \{M_3 \mid [\frac{1}{2},0,0]^t\}$.

(6) Show that if we change the basis using

$$T = \begin{bmatrix} 0 & 0 & 1 \\ 1 & 0 & 0 \\ 0 & 1 & 0 \end{bmatrix}$$

and subsequently change the origin to $\mathbf{p} = [0,-\frac{1}{4},0]^t$, then $\{M_1 \mid [000]^t\}$, $\{M_2 \mid [\frac{1}{2},0,0]^t\}$,$\{M_3 \mid [\frac{1}{2},0,0]^t\}$ becomes

$$\{M_1 \mid [0,\tfrac{1}{2},0]^t\},\{M_2 \mid [000]^t\},\{M_3 \mid [0,\tfrac{1}{2},0]^t\} .$$

Hence the matrix group generated by $R_P(T_P)$, $\{M_1 \mid [000]^t\}$ and $\{M_2 \mid [\frac{1}{2},0,0]^t\}$ is equivalent to the matrix group generated by $R_P(T_P)$, $\{M_1 \mid [0,\frac{1}{2},0]^t\}$ and $\{M_2 \mid [000]^t\}$.

(7) Using the matrix T of part 6 and making appropriate origin shifts, show that the matrix groups generated by each of the following set of generators are equivalent:

$$\{M_1 \mid [000]^t\}\{M_2 \mid [\tfrac{1}{2},0,0]^t\},\ \{M_1 \mid [000]^t\}\{M_2 \mid [0,\tfrac{1}{2},0]^t\},$$

and $\{M_1 \mid [0,0,\frac{1}{2}]^t\}\{M_2 \mid [000]^t\}$.

(8) As in (7), show that the matrix groups generated by the following set of generators are equivalent

$$\{M_1 \mid [000]^t\}\{M_2[\tfrac{1}{2},\tfrac{1}{2},0]^t\},\ \{M_1 \mid [0,0,\tfrac{1}{2}]^t\}\{M_2 \mid [\tfrac{1}{2},0,0]^t\},$$

$$\text{and}\ \{M_1 \mid [0,0,\tfrac{1}{2}]^t\}\{M_2[0,\tfrac{1}{2},0]^t\}\ .$$

(9) The space group generated from

$$\{M_1 \mid [000]^t\}\{M_2 \mid [000]^t\}$$

is denoted *P222* the while one generated from

$$\{M_1 \mid [0,0,\tfrac{1}{2}]^t\}\{M_2 \mid [\tfrac{1}{2},\tfrac{1}{2},0]^t\}$$

is denoted $P2_12_12_1$. That generated according to (7) is denoted $P222_1$ and that from (8) $P2_12_12$. □

Applying the methods in this section one can verify that the two-generator point groups give rise to the following 116 space groups:

2/m: $P2/m$, $P2_1/m$, $C2/m$, $P2/c$, $P2_1/c$, $C2/c$
4/m: $P4/m$, $P4_2/m$, $P4/n$, $P4_2/n$, $I4/m$, $I4_1/a$
6/m: $P6/m$, $P6_3/m$
222: $P222$, $P222_1$, $P2_12_12$, $P2_12_12_1$,
$C222_1$, $C222$, $F222$,
$I222$, $I2_12_12_1$
321: $P321$, $P3_121$, $P3_221$, $R32$
312: $P312$, $P3_112$, $P3_212$ *(see P6.25 for why R312 does not exist)*
$\bar{3}m1$: $P\bar{3}m1$, $P\bar{3}c1$, $R\bar{3}m$, $R\bar{3}c$
$\bar{3}1m$: $P\bar{3}1m$, $P\bar{3}1c$
422: $P422$, $P42_12$, $P4_122$, $P4_12_12$, $P4_222$, $P4_22_12$, $P4_322$,
$P4_32_12$, $I422$, $I4_122$
622: $P622$, $P6_122$, $P6_522$, $P6_222$, $P6_422$, $P6_322$
mm2: $Pmm2$, $Pmc2_1$, $Pcc2$, $Pma2$, $Pca2_1$, $Pnc2$, $Pmn2_1$,
$Pba2$, $Pna2_1$, $Pnn2$, $Cmm2$, $Cmc2_1$, $Ccc2$, $Amm2$,
$Abm2$, $Ama2$, $Aba2$, $Fmm2$, $Fdd2$, $Imm2$, $Iba2$,
$Ima2$ *(Note that the A-lattice type is equivalent to B but not C.)*
3m1: $P3m1$, $P3c1$, $R3m$, $R3c$
31m: $P31m$, $P31c$,

4mm: $P4mm$, $P4bm$, $P4_2cm$, $P4_2nm$, $P4cc$, $P4nc$, $P4_2mc$, $P4_2bc$, $I4mm$, $I4cm$, $I4_1md$, $I4_1cd$

6mm: $P6mm$, $P6cc$, $P6_3cm$, $P6_3mc$

$\bar{4}2m$: $P\bar{4}2m$, $P\bar{4}2c$, $P\bar{4}2_1m$, $P\bar{4}2_1c$, $I\bar{4}2m$, $I\bar{4}2d$,

$\bar{4}m2$: $P\bar{4}m2$, $P\bar{4}c2$, $P\bar{4}b2$, $P\bar{4}n2$, $I\bar{4}m2$, $I\bar{4}c2$

$\bar{6}m2$: $P\bar{6}m2$, $P\bar{6}c2$

$\bar{6}2m$: $P\bar{6}2m$, $P\bar{6}2c$

(E7.54) Example - Space group symmetry of coesite in the sanidine setting: The space group determined by Zoltai and Buerger (1959) for coesite is $B2/b$ (see P7.15) with general equivalent positions

$$[x,y,z]^t,\ [\tfrac{1}{2}+x,y,\tfrac{1}{2}+z]^t,\ [-x,-y,-z]^t,$$
$$[\tfrac{1}{2}-x,-y,\tfrac{1}{2}-z]^t,\ [-x,\tfrac{1}{2}-y,z]^t,\ [\tfrac{1}{2}-x,\tfrac{1}{2}-y,\tfrac{1}{2}+z]^t,$$
$$[x,\tfrac{1}{2}+y,-z]^t,\ [\tfrac{1}{2}+x,\tfrac{1}{2}+y,\tfrac{1}{2}-z]^t.$$

(1) Find the matrix representation $R_{D_1(o_1)}(\alpha)$ and Seitz's notation of the space group operation α in $B2/b$ that maps the triple $[x,y,z]^t$ to $[\tfrac{1}{2}-x,-y,\tfrac{1}{2}-z]^t$ in the $D_1(o_1)$ setting.

Solution: By inspection we see that

$$R_{D_1(o_1)}(\alpha) = \begin{bmatrix} -1 & 0 & 0 & \tfrac{1}{2} \\ 0 & -1 & 0 & 0 \\ 0 & 0 & -1 & \tfrac{1}{2} \\ 0 & 0 & 0 & 1 \end{bmatrix}.$$

Casting $R_{D_1(o_1)}(\alpha)$ into Seitz's notation, we get $\{M \mid \mathbf{t}\}$ where

$$M = \begin{bmatrix} -1 & 0 & 0 \\ 0 & -1 & 0 \\ 0 & 0 & -1 \end{bmatrix} \quad \text{and} \quad \mathbf{t} = \begin{bmatrix} \tfrac{1}{2} \\ 0 \\ \tfrac{1}{2} \end{bmatrix}.$$

(2) Find $R_{D_2(o_2)}(\alpha)$, the matrix representation for α for coesite in the $D_2(o_2)$ sanidine setting as described in P7.15 where it is shown that

$$T = \begin{bmatrix} 1 & 0 & 0 \\ 0 & 0 & 1 \\ 1 & -1 & 0 \end{bmatrix} \quad \text{and} \quad [\mathbf{p}]_{D_1} = \begin{bmatrix} \frac{1}{2} \\ 3/4 \\ 0 \end{bmatrix} .$$

Solution: According to (7.15),

$$R_{D_2(o_2)}(\alpha) = \{TMT^{-1} \mid T[(M - I_3)\mathbf{p} + \mathbf{t}]\} .$$

Using the information provided in part 1, we have

$$TMT^{-1} = \begin{bmatrix} 1 & 0 & 0 \\ 0 & 0 & 1 \\ 1 & -1 & 0 \end{bmatrix} \begin{bmatrix} -1 & 0 & 0 \\ 0 & -1 & 0 \\ 0 & 0 & -1 \end{bmatrix} \begin{bmatrix} 1 & 0 & 0 \\ 0 & 0 & 1 \\ 1 & -1 & 0 \end{bmatrix}^{-1}$$

$$= \begin{bmatrix} -1 & 0 & 0 \\ 0 & -1 & 0 \\ 0 & 0 & -1 \end{bmatrix}$$

and

$$T[(M - I_3)\mathbf{p} + \mathbf{t}] = \begin{bmatrix} 1 & 0 & 0 \\ 0 & 0 & 1 \\ 1 & -1 & 0 \end{bmatrix} \left(\left(\begin{bmatrix} -1 & 0 & 0 \\ 0 & -1 & 0 \\ 0 & 0 & -1 \end{bmatrix} - \begin{bmatrix} 1 & 0 & 0 \\ 0 & 1 & 0 \\ 0 & 0 & 1 \end{bmatrix} \right) \begin{bmatrix} \frac{1}{2} \\ 3/4 \\ 0 \end{bmatrix} + \begin{bmatrix} \frac{1}{2} \\ 0 \\ \frac{1}{2} \end{bmatrix} \right)$$

$$= \begin{bmatrix} -\frac{1}{2} \\ \frac{1}{2} \\ 1 \end{bmatrix} = \begin{bmatrix} \frac{1}{2} \\ \frac{1}{2} \\ 0 \end{bmatrix} \quad \text{(modulo } P\text{, which is contained in the lattice)}$$

Hence,

$$R_{D_2(o_2)}(\alpha) = \begin{bmatrix} -1 & 0 & 0 & \frac{1}{2} \\ 0 & -1 & 0 & \frac{1}{2} \\ 0 & 0 & -1 & 0 \\ 0 & 0 & 0 & 1 \end{bmatrix}$$

□

(P7.54) Problem: Determine the principle representatives, $R_{D_2(o_2)}(\alpha)$, given in Table 7.3 for the space group of coesite in the $D_2(o_3)$ sanidine setting as done in (2) of E7.54.

(P7.55) Problem: Analyze each of the principle representatives $R_{D_2(o_2)}(\alpha)$ in Table 7.3 and show that the space group of coesite is

Table 7.3: The principal representatives $R_{D_1(o_1)}(\alpha)$ and $R_{D_2(o_2)}(\alpha)$ of the space group of coesite in the Zoltai-Buerger setting, $D_1(o_1)$, and the sanidine setting, $D_2(o_2)$.

$R_{D_1(o_1)}(\alpha)$	$R_{D_2(o_2)}(\alpha)$	$R_{D_1(o_1)}(\alpha)$	$R_{D_2(o_2)}(\alpha)$
$\begin{bmatrix}1&0&0&0\\0&1&0&0\\0&0&1&0\\0&0&0&1\end{bmatrix}$	$\begin{bmatrix}1&0&0&0\\0&1&0&0\\0&0&1&0\\0&0&0&1\end{bmatrix}$	$\begin{bmatrix}-1&0&0&0\\0&-1&0&\frac{1}{2}\\0&0&1&0\\0&0&0&1\end{bmatrix}$	$\begin{bmatrix}-1&0&0&0\\0&1&0&0\\0&0&-1&0\\0&0&0&1\end{bmatrix}$
$\begin{bmatrix}1&0&0&\frac{1}{2}\\0&1&0&0\\0&0&1&\frac{1}{2}\\0&0&0&1\end{bmatrix}$	$\begin{bmatrix}1&0&0&\frac{1}{2}\\0&1&0&\frac{1}{2}\\0&0&1&\frac{1}{2}\\0&0&0&1\end{bmatrix}$	$\begin{bmatrix}-1&0&0&\frac{1}{2}\\0&-1&0&\frac{1}{2}\\0&0&1&\frac{1}{2}\\0&0&0&1\end{bmatrix}$	$\begin{bmatrix}-1&0&0&\frac{1}{2}\\0&1&0&\frac{1}{2}\\0&0&-1&\frac{1}{2}\\0&0&0&1\end{bmatrix}$
$\begin{bmatrix}-1&0&0&0\\0&-1&0&0\\0&0&-1&0\\0&0&0&1\end{bmatrix}$	$\begin{bmatrix}-1&0&0&0\\0&-1&0&0\\0&0&-1&\frac{1}{2}\\0&0&0&1\end{bmatrix}$	$\begin{bmatrix}1&0&0&0\\0&1&0&\frac{1}{2}\\0&0&-1&0\\0&0&0&1\end{bmatrix}$	$\begin{bmatrix}1&0&0&0\\0&-1&0&0\\0&0&1&\frac{1}{2}\\0&0&0&1\end{bmatrix}$
$\begin{bmatrix}-1&0&0&\frac{1}{2}\\0&-1&0&0\\0&0&-1&\frac{1}{2}\\0&0&0&1\end{bmatrix}$	$\begin{bmatrix}-1&0&0&\frac{1}{2}\\0&-1&0&\frac{1}{2}\\0&0&-1&0\\0&0&0&1\end{bmatrix}$	$\begin{bmatrix}1&0&0&\frac{1}{2}\\0&1&0&\frac{1}{2}\\0&0&-1&\frac{1}{2}\\0&0&0&1\end{bmatrix}$	$\begin{bmatrix}1&0&0&\frac{1}{2}\\0&-1&0&\frac{1}{2}\\0&0&1&0\\0&0&0&1\end{bmatrix}$

$I2/a$ when its structure is described in terms of the $D_2(o_2)$ setting of sanidine.

THE CRYSTALLOGRAPHIC SPACE GROUPS BASED ON THE THREE-GENERATOR POINT GROUPS

The strategy for deriving these space groups will be similar to that used in the previous section. The first step in the two-generator case was to find two generators α and β for the oriented point group H such that every element in H can be expressed uniquely in the form $\alpha^i\beta^j$ where $1 \le i \le o(\alpha)$ and $1 \le j \le o(\beta)$. The relations needed to rearrange an expression, such as $\alpha^2\beta\alpha\beta^3\alpha^2$ into the $\alpha^i\beta^j$ form imposed conditions on the **r** and **s** vectors so that they are consistent with the relations with respect to the chosen lattice L and basis P. In particular, the necessary relations were $\alpha^{o(\alpha)} = 1$, $\beta^{o(\beta)} = 1$ and $\beta = \alpha\beta\alpha$. In this section, we will determine three generators α, β, γ of H together with relations such that every element of H can be expressed uniquely in the form $\alpha^i\beta^j\gamma^k$ when

Table 7.4: The generators and relations for the three-generator oriented point groups.

Point Group	α	β	γ	Additional Relations
mmm	2	$[100]_2$	i	$\alpha\gamma = \gamma\alpha, \beta\gamma = \gamma\beta$
$4/mmm$	4	$[100]_2$	i	$\alpha\gamma = \gamma\alpha,\ \beta\gamma = \gamma\beta$
$6/mmm$	6	$[100]_2$	i	$\alpha\gamma = \gamma\alpha,\ \beta\gamma = \gamma\beta$
23	2	$[100]_2$	$[111]_3$	$\gamma\alpha = \beta\gamma,\ \gamma\beta = \alpha\beta\gamma$
432	4	$[110]_2$	$[111]_3$	$\gamma\alpha = \alpha^{-1}\gamma^2,\ \gamma\beta = \alpha\gamma^2$
$\bar{4}3m$	$\bar{4}$	$[110]_m$	$[111]_3$	$\gamma\alpha = \alpha^{-1}\gamma^2,\ \gamma\beta = \alpha\gamma^2$
$2/m\bar{3}$	2	$[100]_2$	$[111]_{\bar{3}}$	$\gamma\alpha = \beta\gamma,\ \gamma\beta = \alpha\beta\gamma$
$4/m\bar{3}2/m$	4	$[110]_2$	$[111]_{\bar{3}}$	$\gamma\alpha = \alpha^{-1}\gamma^2,\ \gamma\beta = \alpha\gamma^2$

$1 \le i \le o(\alpha)$, $1 \le j \le o(\beta)$, $1 \le k \le o(\gamma)$. Then **r**, **s** and **t** will be found so that they are consistent with these relations on H with respect to L and P. A theorem that parallels T7.47 will then guarantee that all space groups based on H and L can be found by constructing the matrix groups generated by $R_P(T_L)$, $\{M_P(\alpha) \mid \mathbf{r}\}$, $\{M_P(\beta) \mid \mathbf{s}\}$ and $\{M_P(\gamma) \mid \mathbf{t}\}$. The relations we will use for H are $\alpha^{o(\alpha)} = 1$, $\beta^{o(\beta)} = 1$, $\gamma^{o(\gamma)} = 1$, $\beta = \alpha\beta\alpha$ and two additional relations involving γ. These additional relations together with the choices of α, β and γ are given for each three-generator groups H given in Table 7.4.

(P7.56) Problem: Verify the relations given in Table 7.4 for each of the point groups listed. Also verify that $\beta = \alpha\beta\alpha$ holds for each.

(E7.55) Example: Consider $H = 23$ and let α, β, γ be as in Table 7.4. Write $\alpha\gamma^2\beta\alpha$ in the form $\alpha^i\beta^j\gamma^k$ where $1 \le i \le o(\alpha)$, $1 \le j \le o(\beta)$, $1 \le k \le o(\gamma)$. Explain how any product of α's, β's and γ's can be written in this form.

Solution: The relations we have to work with are $\alpha^2 = 1$, $\beta^2 = 1$, $\gamma^3 = 1$, $\beta\alpha = \alpha^{-1}\beta$ (from $\beta = \alpha\beta\alpha$), $\gamma\alpha = \beta\gamma$ and $\gamma\beta = \alpha\beta\gamma$. Hence

$$
\begin{aligned}
\alpha\gamma^2\beta\alpha &= \alpha\gamma\gamma\beta\alpha \\
&= \alpha\gamma\alpha\beta\gamma\alpha && (\gamma\beta = \alpha\beta\gamma) \\
&= \alpha\beta\gamma\beta\gamma\alpha && (\gamma\alpha = \beta\gamma) \\
&= \alpha\beta\alpha\beta\gamma\gamma\alpha && (\gamma\beta = \alpha\beta\gamma) \\
&= \alpha\beta\alpha\beta\gamma\beta\gamma \\
&= \alpha\beta\alpha\beta\alpha\beta\gamma\gamma \\
&= \alpha\alpha^{-1}\beta\beta\alpha\beta\gamma\gamma
\end{aligned}
$$

$$= \alpha\alpha^{-1}\beta\alpha^{-1}\beta\beta\gamma\gamma$$
$$= \alpha\alpha^{-1}\beta\alpha\beta\beta\gamma\gamma \qquad (\alpha^{-1} = \alpha)$$
$$= \alpha\alpha^{-1}\alpha^{-1}\beta\beta\beta\gamma\gamma$$
$$= \alpha^{-1}\beta^3\gamma^2$$
$$= \alpha\beta\gamma^2 \quad .$$

As illustrated in this simplification, the relations $\gamma\alpha = \beta\gamma$ and $\gamma\beta = \alpha\beta\gamma$ enable us to move all of the γ's to the right. The α's and β's can be sorted using $\beta\alpha = \alpha^{-1}\beta$. We then obtain an expression where all of the α's are on the left, the β's are in the middle and the γ's are on the right. The relations $\alpha^2 = \beta^2 = 1$ and $\gamma^3 = 1$ are then used for the final simplification. Recall that $\mathbf{g}^{-1} = \mathbf{g}^{n-1}$ where $n = o(\mathbf{g})$. This is useful in putting a term in the correct form for a relation as was shown in the above illustration when α^{-1} was replaced by α. □

(P7.57) Problem: Consider $H = 4/mmm$ and let α, β, γ be as in Table 7.4. Write $\gamma\alpha\beta\gamma\alpha^3$ in the form $\alpha^i\beta^j\gamma^k$ where $1 \le i \le o(\alpha)$, $1 \le j \le o(\beta)$ and $1 \le k \le o(\gamma)$. Explain how any product of α's, β's and γ's can be so written.

(P7.58) Problem: Consider $H = 4/m\bar{3}2/m$ and let α, β, γ be as in Table 7.4. Write $\beta\alpha^2\gamma^5\beta\alpha$ in the form $\alpha^i\beta^j\gamma^k$ where $1 \le i \le o(\alpha)$, $1 \le j \le o(\beta)$, $1 \le k \le o(\gamma)$. Explain how any such product can be written in this form.

In general, the generators α, β, γ given for the point groups listed in Table 7.4 are such that each element of H can be written in the form $\alpha^i\beta^j\gamma^k$ where $1 \le i \le o(\alpha)$, $1 \le j \le o(\beta)$, $1 \le k \le o(\gamma)$.

(E7.56) Example: Let $H = mmm$ and let α, β, γ be as in Table 7.4. Show that each element of H appears exactly once in the list of elements of the form $\alpha^i\beta^j\gamma^k$ where $1 \le i \le o(\alpha)$, $1 \le j \le o(\beta)$ and $1 \le k \le o(\gamma)$.

Solution: Since $o(\alpha) = o(\beta) = o(\gamma) = 2$, we have

$$\alpha^1\beta^1\gamma^1 = [010]_m \qquad \alpha^1\beta^1\gamma^2 = [010]_2$$
$$\alpha^2\beta^1\gamma^1 = [100]_m \qquad \alpha^2\beta^1\gamma^2 = [100]_2$$
$$\alpha^1\beta^2\gamma^1 = m \qquad \alpha^1\beta^2\gamma^2 = 2$$
$$\alpha^2\beta^2\gamma^1 = i \qquad \alpha^2\beta^2\gamma^2 = 1 \quad .$$

□

(P7.59) Problem: Let $H = 4/mmm$ and let α, β, γ be as in Table 7.4. Show that each element of H appears exactly once in the list of elements of the form $\alpha^i\beta^j\gamma^k$ where $1 \le i \le o(\alpha)$, $1 \le j \le o(\beta)$ $1 \le k \le o(\gamma)$.

In general, the generators α, β, γ given for the point groups listed in Table 7.4 are such that each element of H appears exactly once in the list of elements of the form $\alpha^i\beta^j\gamma^k$ where $1 \le i \le o(\alpha)$, $1 \le j \le o(\beta)$, $1 \le k \le o(\gamma)$. This fact will be shown in the following problems.

(P7.60) Problem: For each H in Table 7.4, show that $\langle\alpha\rangle \cap \langle\beta\rangle = \{1\}$.

(P7.61) Problem: For each H in Table 7.4, show that $K \cap \langle\gamma\rangle = \{1\}$ where $K = \{\alpha^i\beta^j \mid 1 \le i \le o(\alpha) \text{ and } 1 \le j \le o(\beta)\}$. Hint: If $\gamma = i$ and α and β are proper point isometries, use the fact that the set of proper isometries is closed. If γ involves a third-turn, check to see if α and β are members of a group that does not contain an isometry involving a third-turn.

(P7.62) Problem: Use the result of P7.60 and P7.61 to show that if $\alpha^i\beta^j\gamma^k = 1$ for any H in Table 7.4, then $\alpha^i = 1$, $\beta^j = 1$ and $\gamma^k = 1$.

(P7.63) Problem: Use P7.62 to show that if

$$\alpha^i\beta^j\gamma^k = \alpha^\ell\beta^m\gamma^n \quad ,$$

where $1 \le i, \ell \le o(\alpha)$, $1 \le j, m \le o(\beta)$, $1 \le k, n \le o(\gamma)$, then $i = \ell$, $j = m$, $k = n$.

(P7.64) Problem: Use P7.63 to show that each element of H can be written uniquely in the form $\alpha^i\beta^j\gamma^k$ where $1 \le i \le o(\alpha)$, $1 \le j \le o(\beta)$ and $1 \le k \le o(\gamma)$. (Note that in each case $\#(H) = o(\alpha)o(\beta)o(\gamma)$.)

(T7.57) Theorem: Let L denote a lattice left invariant by the three-generator point group H and let α, β, γ be as given in Table 7.4. Let P denote the basis for the primitive sublattice of L associated with H. Let $M_1 = M_P(\alpha)$, $M_2 = M_P(\beta)$ and $M_3 = M_P(\gamma)$. There exists a crystallographic space group G satisfying (7.22) such that each element of $R_P(G)$ can be written in the form

$$\{I_3 \mid \mathbf{v}\}\{M_1 \mid \mathbf{r}\}^i\{M_2 \mid \mathbf{s}\}^j\{M_3 \mid \mathbf{t}\}^k \quad ,$$

where $\{I_3 \mid \mathbf{v}\} \varepsilon R_P(T_L)$, $1 \le i \le o(\alpha)$, $1 \le j \le o(\beta)$ and $1 \le k \le o(\gamma)$ if and only if $\mathbf{r}$, $\mathbf{s}$, and $\mathbf{t}$ are consistent with the relations $\alpha^{o(\alpha)} = 1$, $\beta^{o(\beta)} = 1$, $\gamma^{o(\gamma)} = 1$, $\beta = \alpha\beta\alpha$ and the two relations listed for H in Table 7.4 with respect to T_L and P.

In view of the examples and problems preceding T7.57, the proof of T7.57 is a straightforward extension of the proof given for T7.47.

(E7.58) Example: Find all crystallographic groups G that satisfy (7.22) with $H = 432$ and $L = P$. Then determine the different space group types.

Solution: By Table 7.4, $\alpha = 4$, $\beta = [110]_2$ and $\gamma = [111]_3$ then

$$M_1 = M_P(4) = \begin{bmatrix} 0 & -1 & 0 \\ 1 & 0 & 0 \\ 0 & 0 & 1 \end{bmatrix} \quad , \quad M_2 = M_P([110]_2) = \begin{bmatrix} 0 & 1 & 0 \\ 1 & 0 & 0 \\ 0 & 0 & -1 \end{bmatrix}$$

and

$$M_3 = M_P([111]_3) = \begin{bmatrix} 0 & 0 & 1 \\ 1 & 0 & 0 \\ 0 & 1 & 0 \end{bmatrix} \quad .$$

Show that the relation $\alpha^4 = 4^4 = 1$ implies that $\mathbf{r}$ can be cast in the form $[r_1, r_2, k/4]^t$ where $k \varepsilon Z$; that $\beta^2 = [110]_2{}^2 = 1$ implies $\mathbf{s} = [s_1, -s_1, s_3]^t$ (modulo P); and that $\gamma^3 = [111]_3{}^3 = 1$ implies that $\mathbf{t} = [t_1, t_2, -t_1 - t_2]^t$. The relation $\beta\alpha = \alpha^{-1}\beta$ implies that $r_2 = m/2 - s_1$. The relation $\gamma\alpha = \alpha^{-1}\gamma^2$ implies $s_1 = k/4 + m/2 - t_2$. The relation $\gamma\beta = \alpha\gamma^2$ implies that $r_1 = 2t_1 + t_2 + s_3$ and that $k/2 = m/2$ (plus or minus an integer). Combining these results, we see that

$$\mathbf{r} = [r_1, t_2 - k/4, k/4]^t \quad ,$$

$$\mathbf{s} = [(3k/4) - t_2, t_2 - (3k/4), s_3]^t \quad ,$$

$$\mathbf{t} = [t_1, t_2, -t_1 - t_2]^t \quad ,$$

where $r_1, t_1, t_2, s_3 \in R$, $2t_1 + t_2 + s_3 - r_1 \in Z$ and $k \in Z$. From these choices of $\mathbf{r}$, $\mathbf{s}$ and $\mathbf{t}$, using T7.57, all space groups based on *432* and *P* can be constructed.

Now we wish to determine the space group types represented in this set of space groups. We first note that since $(M_3 - I_3)\mathbf{p} = [-p_1 + p_3, p_1 - p_2, p_2 - p_3]^t$, if we set $p_1 = 0$, $p_2 = t_2$ and $p_3 = -t_1$, then $(M_3 - I_3)\mathbf{p} + t = [0,0,0]^t$. Hence by this change of origin $\{M_3 \mid \mathbf{t}\}$ becomes $\{M_3 \mid [0,0,0]^t\}$. While this change of origin alters some of the entries in $\mathbf{r}$ and $\mathbf{s}$, the basic relationship that we discovered earlier still must hold. That is,

$$\mathbf{r} = [r_1, -k/4, k/4,]^t \text{ and } \mathbf{s} = [(3k/4), -3k/4, s_3]^t ,$$

where $s_3 - r_1 \in Z$. Since $s_3 - r_1 \in Z$, r_1 can be taken to be equal to s_3. Any subsequent change of origin that we will consider will be such that $\mathbf{t}$ remains $[000]^t$. That is, such that $-p_1 + p_3, p_1 - p_2$ and $p_2 - p_3$ are all zero. Hence any change of origin to a $\mathbf{p}$ such that $p_1 = p_2 = p_3$ will be acceptable. Since

$$(M_2 - I_3)\mathbf{p} = [-p_1 + p_2,\ p_1 - p_2,\ -2p_3]^t ,$$

if we set $p_1 = p_2 = p_3 = s_3/2$, $\{M_2 \mid \mathbf{s}\}$ becomes $\{M_2 \mid [3k/4, -3k/4, 0]^t\}$. Since $(M_1 - I_3)\mathbf{p} = [-p_1 - p_2,\ p_1 - p_2,\ 0]^t$, and since $r_1 = s_3$, when $p_1 = p_2 = p_3 = s_3/2 = r_1/2$, $\{M \mid r\}$ becomes $\{M_1 \mid [0, -k/4, k/4]^t\}$. We have listed all of the possible $\mathbf{r}$, $\mathbf{s}$ and $\mathbf{t}$ vectors for the choices of k in Table 7.5, where the space group symbol is listed for each. □

Table 7.5: The space group types based on *432* and *P*.

k	$\mathbf{r}^t$	$\mathbf{s}^t$	$\mathbf{t}^t$	Space Group Symbol
0	[000]	[000]	[000]	$P432$
1	$[0, -\frac{1}{4}, \frac{1}{4}]$	[3/4,−3/4,0]	[000]	$P4_132$
2	$[0, -\frac{1}{2}, \frac{1}{2}]$	$[\frac{1}{2}, -\frac{1}{2}, 0]$	[000]	$P4_232$
3	[0,−3/4,3/4]	$[\frac{1}{4}, -\frac{1}{4}, 0]$	[000]	$P4_332$

A list of all of the space group types that can be derived from the three-generator point groups appears below. Counting the space group types found for the one-, two-, and three-generator point groups, we have found a total of 230 crystallographic space group types.

mmm: *Pmmm*, *Pnnn*, *Pccn*, *Pban*, *Pmma*, *Pnna*, *Pmna*, *Pcca*, *Pbam*, *Pccn*, *Pbcm*, *Pnnm*, *Pmmn*, *Pbcn*, *Pbca*, *Pnma*, *Cmcm*, *Cmca*, *Cmmm*, *Cccm*, *Cmma*, *Ccca*, *Fmmm*, *Fddd*, *Immm*, *Ibam*, *Ibca*, *Imma*.

$4/mmm$: $P4/mmm$, $P4/mcc$, $P4/nbm$, $P4/nnc$, $P4/mbm$, $P4/mnc$, $P4/nmm$, $P4/ncc$, $P4_2/mmc$, $P4_2/mcm$, $P4_2/nbc$, $P4_2/nnm$, $P4_2/mbc$, $P4_2/mnm$, $P4_2/nmc$, $P4_2/ncm$, $I4/mmm$, $I4/mcm$, $I4_1/amd$, $I4_1/acd$.

$6/mmm$: $P6/mmm$, $P6/mcc$, $P6_3/mcm$, $P6_3/mmc$.

23: $P23$, $P2_13$, $I23$, $I2_13$, $F23$

$2/m\bar{3}$: $Pm3$, $Pa3$, $Pn3$, $Im3$, $Ia3$, $Fm3$, $Fd3$

432: $P432$, $P4_132$, $P4_232$, $P4_332$, $I432$, $I4_132$ $F432$, $F4_132$

$\bar{4}3m$: $P\bar{4}3m$, $P\bar{4}3n$, $I\bar{4}3m$, $I\bar{4}3d$, $F\bar{4}3m$, $F\bar{4}3c$

$4/m\bar{3}2/m$: $Pm3m$, $Pn3n$, $Pm3n$, $Pn3m$, $Im3m$, $Ia3d$ $Fm3m$, $Fm3c$, $Fd3m$, $Fd3c$.

APPENDIX 1

MAPPINGS

When a cartographer draws a map of an area of the earth's surface, he constructs a miniature representation of the area so that each point in that area has exactly one point on the map assigned to it by the mapping. In its uses in crystallography, the notion of a mapping never wanders far from this basic idea of assigning the elements of some object with those of a representation or a map of the object.

(DA1.1) Definition: A *mapping* α of a set A into a set B is a rule which assigns to each element $\mathbf{a} \varepsilon A$ a unique element $\mathbf{b}$ in B.

To illustrate this definition, possible mappings α between two sets $A = \{\mathbf{a}_1,\mathbf{a}_2,\mathbf{a}_3,\mathbf{a}_4\}$ and $B = \{\mathbf{b}_1,\mathbf{b}_2,\mathbf{b}_3,\mathbf{b}_4,\mathbf{b}_5\}$ are examined. In each case, a set of arrows defines the rule for the mapping α. In Figure A1.1(a), α qualifies as a mapping because each element in A has exactly *one* arrow emanating from it. For example, α assigns the element $\mathbf{b}_3$ to $\mathbf{a}_4$. That is, $\mathbf{a}_4$ is mapped to $\mathbf{b}_3$ and $\mathbf{b}_3$ is said to be the *image* of $\mathbf{a}_4$ under the mapping. As no other arrow emanates from $\mathbf{a}_4$, no other element of B is assigned to it. The fact that both $\mathbf{a}_1$ and $\mathbf{a}_4$ are mapped to the common element $\mathbf{b}_3$ does not violate the definition of a mapping. On the other hand, the two remaining rules described in the figure are not mappings since $\mathbf{a}_3$ in Figure A1.1(b) does not have an element of B assigned to it and $\mathbf{a}_4$ in Figure A1.1(c) has two elements of B ($\mathbf{b}_3$ and $\mathbf{b}_5$) assigned to it. If α is a mapping from A and B, we write $\alpha:A\rightarrow B$ and if α maps $\mathbf{a}$ to $\mathbf{b}$, we write $\alpha(\mathbf{a}) = \mathbf{b}$.

So that we may deal effectively with the concept of mappings, we require a precise definition of what it means for two mappings to be equal.

(DA1.2) Definition: Let α and β denote mappings from A to B. Then α and β are *equal mappings* if and only if $\alpha(\mathbf{a}) = \beta(\mathbf{a})$ for all $\mathbf{a} \varepsilon A$.

While the rules used to define the mappings α and β may be quite different, if the image of $\mathbf{a}$ is the same for both rules for all $\mathbf{a} \varepsilon A$, then the mappings are considered to be equal. For example, if α is a rotation of space of 180° about a line and if β is a rotation of 540° about

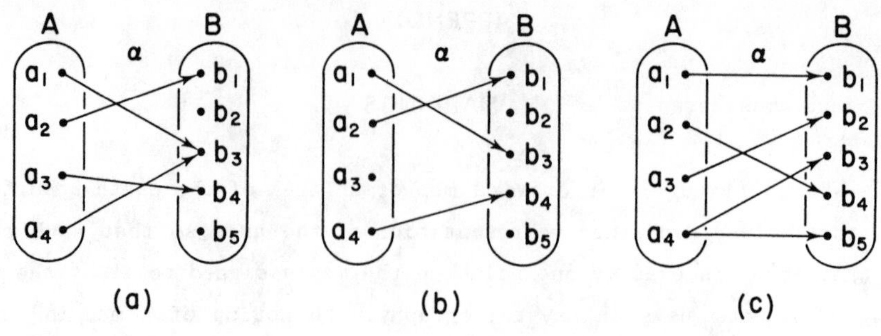

Figure A1.1

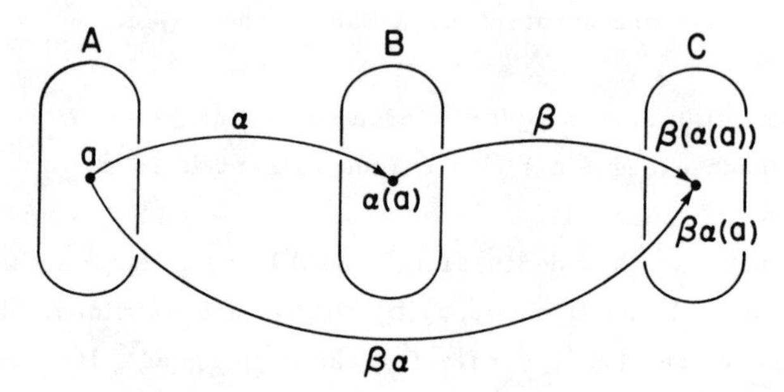

Figure A1.2

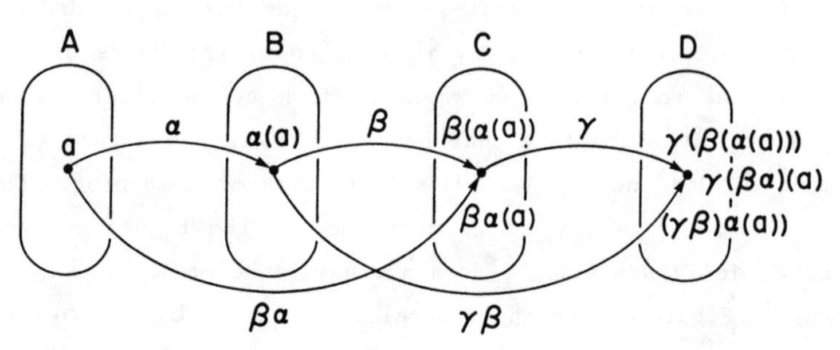

Figure A1.3

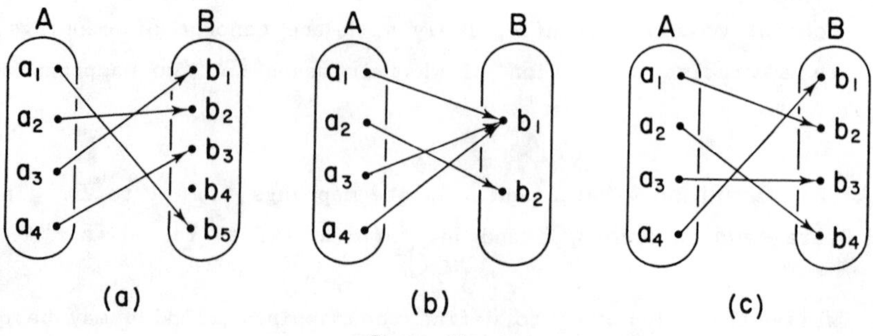

Figure A1.4

the same line, then since the final destination of each point in space is the same under both rotations, by our definition, the two mappings α and β are considered to be equal, i.e., $\alpha = \beta$.

The method for combining operations is called composition.

(DA1.3) Definition: Let $\alpha: A \to B$ and $\beta: B \to C$. Then the *composition* of β and α denoted by $\beta\alpha$ is defined by $\beta\alpha(\mathbf{a}) = \beta(\alpha(\mathbf{a}))$ as is shown in Figure A1.2.

An example of DA1.3 is the composition of two successive clockwise rotations of space α and β about the same line. If α denotes a rotation of 90° and β denotes a rotation of 180°, then the composition $\beta\alpha$ of these two rotations is a third-turn clockwise rotation of 270° about the same axis.

The composition of two mappings α and β is not a commutative operation. That is, in general $\alpha\beta \neq \beta\alpha$. However, the familiar algebraic rules of associativity does hold for composition as evinced by the following theorem.

(TA1.4) Theorem: Let $\alpha: A \to B$, $\beta: B \to C$ and $\gamma: C \to D$, then $\gamma(\beta\alpha) = (\gamma\beta)\alpha$.

Proof: Let $\mathbf{a} \in A$, then by DA1.3

$$\gamma(\beta\alpha)(\mathbf{a}) = \gamma(\beta(\alpha(\mathbf{a}))) \text{ and}$$

$$(\gamma\beta)\alpha(\mathbf{a}) = \gamma\beta(\alpha)) = \gamma(\beta(\alpha(\mathbf{a}))).$$

Hence by DA1.2, $\gamma(\beta\alpha) = (\gamma\beta)\alpha$. This proof is illustrated by Figure A1.3. □

There are three special kinds of mappings that will be particularly of important to us. These are called (1) one-to-one mappings, (2) onto mappings and (3) one-to-one and onto mappings.

(DA1.5) Definition: Let $\alpha: A \to B$ denote a mapping.

(1) If each element of B is an image of *no more than one* element of A, then α is said to be *one-to-one*.
(2) If each element in B is the image of *at least one* element of A, then α is said to be *onto*

(3) If α is both one-to-one and onto, then α is said to be *one-to-one and onto*.

Each of these types of mappings is illustrated in Figure A1.4. The mapping illustrated by Figure A1.4(a) is one-to-one because each element in B has at most one element from A mapped to it, i.e., $\mathbf{b}_1$, $\mathbf{b}_2$, $\mathbf{b}_3$ and $\mathbf{b}_5$ each have one element from A mapped to them whereas $\mathbf{b}_4$ has no elements from A mapped to it. The mapping in Figure A1.4(b) is onto because each element in B has one or more elements mapped to it: $\mathbf{b}_1$ has three elements of A $\{\mathbf{a}_1, \mathbf{a}_3, \mathbf{a}_4\}$ mapped to it and $\mathbf{b}_2$ has one element, $\mathbf{a}_2$, from A mapped to it. Figure A1.4(c) illustrates a one-to-one and onto mapping. In this mapping each element in B has exactly one element in A mapped to it. When the sets are finite, this requires that the number of elements in A equal the number of elements in B. Stated another way, we say that the order of A, denoted $\#(A)$, is equal to the order of B, denoted $\#(B)$, i.e., $\#(A) = \#(B)$.

When α: $A \rightarrow A$ is a one-to-one and onto mapping of a set A onto itself, then α is called a *permutation* on A. For example, a rotation of geometric space S is a permutation on S. Specifically, S, the identity mapping $1(\mathbf{r}) = (\mathbf{r})$ for all $\mathbf{r} \in S$, is a permutation on S. An important property of permutations on a set S is that each permutation α has an inverse.

(DA1.5) Definition: Let $\alpha: A \rightarrow B$ denote a mapping. A mapping $\beta: B \rightarrow A$ is said to be the *inverse of* α in case

$$\beta(x) = y \text{ if and only if } \alpha(y) = x \text{ .}$$

(TA1.6) Theorem: The mapping $\alpha: A \rightarrow B$ has an inverse if and only if α is one-to-one and onto.

We will not give a formal proof of TA1.6. Note, however, that if α is not one-to-one, then some element of B is the image of two or more elements of A. Hence, defining β as in DA1.5 would require mapping $\mathbf{b}$ to more than one element which violates the definition of a mapping. Similarly, if α is not onto, there exists an element $\mathbf{b} \in B$ that is not an image under α and so a β, defined as in DA1.5, would not map $\mathbf{b}$ anywhere. Conversely, it is straightforward to show that if α is one-to-one and onto then its inverse exists. We denote the inverse of α by α^{-1}.

The technique for showing that a mapping $\alpha: A \rightarrow B$ is one-to-one is to assume that $\mathbf{a}_1, \mathbf{a}_2 \in A$ such that $\alpha(\mathbf{a}_1) = \alpha(\mathbf{a}_2)$ and prove that $\mathbf{a}_1 = \mathbf{a}_2$. This will show that no two *different* elements of A map to the same element of B.

(EA1.7) Example: Let $\alpha: Z \to R$ be defined by $\alpha(z) = 2z$. Show that α is one-to-one.

Solution: Let $z_1, z_2 \in Z$ such that $\alpha(z_1) = \alpha(z_2)$. By the definition of α, $2z_1 = 2z_2$. Hence, $z_1 = z_2$, and so α is one-to-one. □

Note that the mapping α of EA1.7 is not onto since there does not exist an integer $z \in Z$ such that $\alpha(z) = 5$. The technique for showing that a mapping $\alpha: A \to B$ is onto is to choose an arbitrary $\mathbf{b} \in B$ and find an element $\mathbf{a} \in A$ such that $\alpha(\mathbf{a}) = \mathbf{b}$.

(EA2.8) Example: Let $\alpha: R \to R^+$ be defined by $\alpha(r) = e^r$ where R^+ is the set of positive real numbers. Prove that α is onto.

Solution: Let $p \in R^+$. Then we must find an element $r \in R$ such that $\alpha(r) = p$. By the definition of α, this would mean that $e^r = p$. Solving this equation we have $r = \ell n(p)$. Since $\ell n(p)$ is defined for all positive real numbers, $r = \ell n(p)$ exists, and

$$\begin{aligned}\alpha(r) &= \alpha(\ell n(p)) \\ &= e^{\ell n(p))} \\ &= p .\end{aligned}$$

Hence α is onto. □

(PA1.1) Problem: Show that the α defined in EA1.7 is one-to-one.

(PA1.2) Problem: Show that $\alpha: R \to R$ defined by $\alpha(r) = 3r + 5$ is one-to-one and onto. Find α^{-1}.

APPENDIX 2

MATRIX METHODS

(DA2.1) Definition: A *matrix* is a rectangular array of numbers enclosed between brackets and arranged in rows and columns as displayed below:

$$\begin{array}{rc} & \text{First column} \downarrow \quad\quad m\text{th column} \downarrow \\ \text{First row} \rightarrow & \left[\begin{array}{cccc} a_{11} & a_{12} & \cdots & a_{1m} \\ a_{21} & a_{22} & \cdots & a_{2m} \\ \cdot & \cdot & & \cdot \\ \cdot & \cdot & & \cdot \\ \cdot & \cdot & & \cdot \\ a_{n1} & a_{n2} & \cdots & a_{nm} \end{array}\right] \\ n\text{th row} \rightarrow & \end{array}.$$

In this matrix, we have n rows and m columns. The number at the intersection of the ith row and jth column is referred to as the (i,j)-entry. Note that we have denoted it by a_{ij}. A matrix with n rows and m columns is called an $n \times m$ matrix (read n by m matrix).

(DA2.2) Definition: Two $n \times m$ matrices A and B are said to be *equal* if $a_{ij} = b_{ij}$ for all $1 \leq i \leq n$ and $1 \leq j \leq m$.

This definition enables us to solve matrix equations such as

$$\begin{bmatrix} c\rho & -s\rho & 0 \\ s\rho & c\rho & 0 \\ 0 & 0 & 1 \end{bmatrix} = \begin{bmatrix} 0 & -1 & 0 \\ 1 & 0 & 0 \\ 0 & 0 & 1 \end{bmatrix}$$

for ρ where $c\rho = \cos\rho$ and $s\rho = \sin\rho$. In this example, by the definition of equality, $c\rho = 0$, $s\rho = 1$. Hence, $\rho = 90°$. In fact, the matrix on the right represents a 90° counterclockwise rotation with respect to a cartesian basis $C = \{i,j,k\}$ about k (see Appendix 3).

OPERATIONS

Two $n \times m$ matrices A and B can be added according to the following rule:

$$A + B = \begin{bmatrix} a_{11} & a_{12} \cdots a_{1m} \\ a_{21} & a_{22} \cdots a_{2m} \\ \vdots & \vdots \quad\ \ \vdots \\ a_{n1} & a_{n2} \cdots a_{nm} \end{bmatrix} + \begin{bmatrix} b_{11} & b_{12} \cdots b_{1m} \\ b_{21} & b_{22} \cdots b_{2m} \\ \vdots & \vdots \quad\ \ \vdots \\ b_{n1} & b_{n2} \cdots b_{nm} \end{bmatrix}$$

$$= \begin{bmatrix} a_{11}+b_{11} & a_{12}+b_{12} \cdots a_{1m}+b_{1m} \\ a_{21}+b_{21} & a_{22}+b_{22} \cdots a_{2m}+b_{2m} \\ \vdots & \vdots \qquad\qquad \vdots \\ a_{n1}+b_{n1} & a_{n2}+b_{n2} \cdots a_{nm}+b_{nm} \end{bmatrix} .$$

If two matrices are not the same size, then no rule of addition is defined for them. But when they are the same size, we say that they are *conformable for addition*. A more compact definition for A + B is

$$A + B = C \quad \text{where } c_{ij} = a_{ij} + b_{ij} .$$

(EA2.3) Example: The sum of the following two 2×3 matrices is shown:

$$\begin{bmatrix} 2 & -1 & 4 \\ 3 & 1 & 8 \end{bmatrix} + \begin{bmatrix} -5 & 2 & 14 \\ -3 & 6 & 4 \end{bmatrix} = \begin{bmatrix} -3 & 1 & 18 \\ 0 & 7 & 12 \end{bmatrix} .$$

□

If A is an $n \times m$ matrix and B is a $m \times t$ matrix (that is, A has the same number of columns as B has rows), then we can multiply A times B to form the product

$$AB = C \quad \text{where } c_{ij} = \sum_{k=1}^{m} a_{ik}b_{kj} .$$

If A is 3×2 and B is 2×3, we have

$$\begin{bmatrix} a_{11} & a_{12} \\ a_{21} & a_{22} \\ a_{31} & a_{32} \end{bmatrix} \begin{bmatrix} b_{11} & b_{12} & b_{13} \\ b_{21} & b_{22} & b_{23} \end{bmatrix} = \begin{bmatrix} a_{11}b_{11}+a_{12}b_{21} & a_{11}b_{12}+a_{12}b_{22} & a_{11}b_{13}+a_{12}b_{23} \\ a_{21}b_{11}+a_{22}b_{21} & a_{21}b_{12}+a_{22}b_{22} & a_{21}b_{13}+a_{22}b_{23} \\ a_{31}b_{11}+a_{32}b_{21} & a_{31}b_{12}+a_{32}b_{22} & a_{31}b_{13}+a_{32}b_{23} \end{bmatrix}$$

If A is $n \times m$ and B is $s \times t$ where $m \neq s$, then we cannot form the product AB. When $m = s$, we say that A and B are *conformable for multiplication*.

(EA2.4) Example: If $C = \{\mathbf{i}, \mathbf{j}, \mathbf{k}\}$ is a cartesian basis, then the matrices

$$M_C(\rho) = \begin{bmatrix} c\rho & s\rho & 0 \\ -s\rho & c\rho & 0 \\ 0 & 0 & 1 \end{bmatrix} \quad \text{and} \quad M_C(\phi) = \begin{bmatrix} c\phi & s\phi & 0 \\ -s\phi & c\phi & 0 \\ 0 & 0 & 1 \end{bmatrix}$$

describe clockwise rotations of space through an angle of ρ° and ϕ° about the same vector $\mathbf{k}$ (see Appendix 3). Moreover, the product of these two matrices $M_C(\phi)M_C(\rho) = M_C(\phi\rho)$ describes the combined effect of the two rotations both about $\mathbf{k}$ through a turn angle of $(\phi+\rho)$, that is

$$M_C(\phi)M_C(\rho) = \begin{bmatrix} c\phi & s\phi & 0 \\ -s\phi & c\phi & 0 \\ 0 & 0 & 1 \end{bmatrix} \begin{bmatrix} c\rho & s\rho & 0 \\ -s\rho & c\rho & 0 \\ 0 & 0 & 1 \end{bmatrix} = \begin{bmatrix} c\phi c\rho - s\phi s\rho & c\phi s\rho + s\phi c\rho & 0 \\ -s\phi c\rho - c\phi s\rho & -s\phi s\rho + c\phi c\rho & 0 \\ 0 & 0 & 1 \end{bmatrix}$$

$$= \begin{bmatrix} \cos(\phi+\rho) & \sin(\phi+\rho) & 0 \\ -\sin(\phi+\rho) & \cos(\phi+\rho) & 0 \\ 0 & 0 & 1 \end{bmatrix} .$$

□

(DA2.5) Definition: The *transpose* of a matrix A, denoted A^t, is formed by interchanging the entries of the rows and columns of A. Thus, if a_{ij} is the (i,j)-entry of A, then a_{ij} becomes the (j,i)-entry of A^t.

Thus, if $A = \begin{bmatrix} a_{11} & a_{12} \\ a_{21} & a_{22} \\ a_{31} & a_{32} \end{bmatrix}$, then $A^t = \begin{bmatrix} a_{11} & a_{21} & a_{31} \\ a_{12} & a_{22} & a_{32} \end{bmatrix}$.

(PA2.1) Problem: Calculate the matrices $(AB)^t$ and B^tA^t given that

$$A = \begin{bmatrix} -1 & 0 & 0 \\ 1 & 1 & 0 \\ 0 & 0 & -1 \end{bmatrix} \quad \text{and} \quad B = \begin{bmatrix} 0 & 1 & -1 \\ 0 & 0 & -1 \\ 1 & 0 & -1 \end{bmatrix} .$$

Note that $(AB)^t = B^tA^t$. This is an example of the property that if A and B are conformable for multiplication, then $(AB)^t = B^tA^t$.

An $n \times n$ matrix B is said to be a *symmetric matrix* if $B = B^t$. If A and B are symmetric matrices, then $(AB)^t = B^tA^t = BA$. Hence, the product of symmetric matrices is symmetric.

SOLVING SYSTEMS OF LINEAR EQUATIONS

The system of equations

$$\begin{aligned} a_{11}x_1 + a_{12}x_2 + \dots + a_{1m}x_m &= b_1 \\ a_{21}x_1 + a_{22}x_2 + \dots + a_{2m}x_m &= b_2 \\ &\vdots \\ a_{n1}x_1 + a_{n2}x_2 + \dots + a_{nm}x_m &= b_n \end{aligned} \tag{A2.1}$$

can be solved using matrices where the a_{ij}'s and b_j's are given numbers and the x_i's are the unknowns. In solving for the unknowns, we first form the *augmented matrix*

$$\left[\begin{array}{cccc|c} a_{11} & a_{12} & \dots & a_{1m} & b_1 \\ a_{21} & a_{22} & \dots & a_{2m} & b_2 \\ \vdots & & & \vdots & \vdots \\ a_{n1} & a_{n2} & \dots & a_{nm} & b_n \end{array}\right] . \tag{A2.2}$$

We then change this matrix into an equivalent matrix of a simpler form whose system of equations can be easily solved. For this to work, it is important that the solution set of the changed system be identical with that of the original system. As shown in any standard text on linear algebra, the solution set of a matrix is unchanged by the following three elementary row operations:

(1) Any two rows of the matrix can be switched;
(2) Each entry of a given row can be multiplied by a nonzero constant;

(3) A given row can be replaced by the sum of that row and a multiple of another.

The augmented matrix of the system of linear equations in (A2.1) is given in (A2.2). If we apply any one of the three row operations given above to this matrix and obtain

$$\left[\begin{array}{cccc|c} c_{11} & c_{12} & \cdots & c_{1m} & d_1 \\ c_{21} & c_{22} & \cdots & c_{2m} & d_2 \\ \cdot & & & \cdot & \cdot \\ \cdot & & & \cdot & \cdot \\ \cdot & & & \cdot & \cdot \\ c_{n1} & c_{n2} & \cdots & c_{nm} & d_n \end{array}\right],$$

then the system of linear equations

$$\begin{array}{l} c_{11}x_1 + c_{12}x_2 + \ldots + c_{1m}x_m = d_1 \\ c_{21}x_1 + c_{22}x_2 + \ldots + c_{2m}x_m = d_2 \\ \cdot \qquad\qquad \cdot \quad \cdot \\ \cdot \qquad\qquad \cdot \quad \cdot \\ \cdot \qquad\qquad \cdot \quad \cdot \\ c_{n1}x_1 + c_{n2}x_2 + \ldots + c_{nm}x_m = d_n \end{array}$$

has precisely the same solution set as the original system.

Our objective in this procedure is to obtain, through consecutive applications of row operations, a matrix whose system of equations is easy to solve. The simplest system of equations to solve consists of the following set of n equations and n unknowns

$$\begin{array}{l} 1x_1 + 0x_2 + \ldots + 0x_n = b_1 \\ 0x_1 + 1x_2 + \ldots + 0x_n = b_2 \\ \cdot \qquad\qquad \cdot \quad \cdot \\ \cdot \qquad\qquad \cdot \quad \cdot \\ \cdot \qquad\qquad \cdot \quad \cdot \\ 0x_1 + 0x_2 + \ldots + 1x_n = b_n \end{array}$$

because it has the solution $x_1 = b_1$, $x_2 = b_2$, ..., $x_n = b_n$. The augmented matrix for this system is

$$\begin{bmatrix} 1 & 0 & \ldots & 0 & | & b_1 \\ 0 & 1 & \ldots & 0 & | & b_2 \\ . & & & . & | & . \\ . & & & . & | & . \\ . & & & . & | & . \\ 0 & 0 & \ldots & 1 & | & b_n \end{bmatrix} . \tag{A2.3}$$

(EA2.6) Example - Determining whether a set of vectors is a basis: Three vectors **r**, **v**, **s** are chosen from the α-quartz lattice whose triples with respect to a natural basis D = {**a**,**b**,**c**} are $[\mathbf{r}]_D = [2\ 1\ 2]^t$, $[\mathbf{v}]_D = [-1\ 2\ 2]^t$ and $[\mathbf{s}]_D = [2\ 1\ -1]^t$. Can **d** be written as a linear combination of the set H = {**r**,**v**,**s**} where $[\mathbf{d}]_D = [4\ 1\ 3]^t$? Show that any vector $[\mathbf{e}]_H = [e_1\ e_2\ e_3]^t$ can be written as a linear combination of H. Show that H is a basis of S. (Bases are discussed in Chapter 1 after equation (1.2).)

Solution: In order for **d** to be written as a linear combination of H = {**r**,**v**,**s**}, there must exist real numbers x_1, x_2 and x_3 such that

$$x_1\mathbf{r} + x_2\mathbf{v} + x_3\mathbf{s} = \mathbf{d}$$

or equivalently

$$x_1[\mathbf{r}]_D + x_2[\mathbf{v}]_D + x_3[\mathbf{s}]_D = [\mathbf{d}]_D$$

Hence we want to find a solution to the system of equations

$$x_1 \begin{bmatrix} 2 \\ 1 \\ 2 \end{bmatrix} + x_2 \begin{bmatrix} -1 \\ 2 \\ 2 \end{bmatrix} + x_3 \begin{bmatrix} 2 \\ 1 \\ -1 \end{bmatrix} = \begin{bmatrix} 4 \\ 1 \\ 3 \end{bmatrix} .$$

Written in matrix form, we have

$$\begin{bmatrix} 2 & -1 & 2 \\ 1 & 2 & 1 \\ 2 & 2 & -1 \end{bmatrix} \begin{bmatrix} x_1 \\ x_2 \\ x_3 \end{bmatrix} = \begin{bmatrix} 4 \\ 1 \\ 3 \end{bmatrix} .$$

The augmented matrix is

$$\begin{bmatrix} 2 & -1 & 2 & | & 4 \\ 1 & 2 & 1 & | & 1 \\ 2 & 2 & -1 & | & 3 \end{bmatrix} .$$

We shall use the elementary row operations to convert our augmented matrix into the form shown in (A2.3). Our first objective is to put a 1 in the (1,1) entry of the matrix. In this case we will do this by switching the first two rows obtaining the following matrix.

$$\begin{bmatrix} 1 & 2 & 1 & | & 1 \\ 2 & -1 & 2 & | & 4 \\ 2 & 2 & -1 & | & 3 \end{bmatrix} .$$

Our next objective is to put zeros in the remaining entries of the first column. This is called "sweeping out the first column." We accomplish this by performing two elementary row operations consecutively. We first replace the second row by the sum of it plus (-2) times the first and then we replace the third row by the sum of it plus (-2) times the first, obtaining

$$\begin{bmatrix} 1 & 2 & 1 & | & 1 \\ 0 & -5 & 0 & | & 2 \\ 0 & -2 & -3 & | & 1 \end{bmatrix} .$$

The next objective is to put a 1 in the (2,2) entry. This is done by multiplying the second row by (-1/5), obtaining

$$\begin{bmatrix} 1 & 2 & 1 & | & 1 \\ 0 & 1 & 0 & | & -2/5 \\ 0 & -2 & -3 & | & 1 \end{bmatrix} .$$

To sweep out the second row (putting zero everywhere but in the (2,2) entry), we replace the first row by the sum of it plus (-2) times the second and replace the third row by the sum of it plus 2 times the second, obtaining

$$\begin{bmatrix} 1 & 0 & 1 & | & 9/5 \\ 0 & 1 & 0 & | & -2/5 \\ 0 & 0 & -3 & | & 1/5 \end{bmatrix} .$$

Moving to the third column, we wish to have a 1 in the (3,3) entry which is accomplished by multiplying the third row by (-1/3), obtaining

$$\left[\begin{array}{ccc|c} 1 & 0 & 1 & 9/5 \\ 0 & 1 & 0 & -2/5 \\ 0 & 0 & 1 & -1/15 \end{array}\right] .$$

We now sweep out the third column by replacing the first row by the sum of it plus to (-1) times the third, obtaining

$$\left[\begin{array}{ccc|c} 1 & 0 & 0 & 28/15 \\ 0 & 1 & 0 & -2/5 \\ 0 & 0 & 1 & -1/15 \end{array}\right] .$$

Hence, the solution is $x_1 = 28/15$, $x_2 = -2/5$ and $x_3 = -1/15$. Therefore,

$$\mathbf{d} = (28/15)\mathbf{r} + (-2/5)\mathbf{v} - (1/15)\mathbf{s} .$$

To show that $\mathbf{e}$ is a linear combination of H where $[\mathbf{e}]_H = [e_1\ e_2\ e_3]^t$, we seek a solution to the system of equations whose augmented matrix is

$$\left[\begin{array}{ccc|c} 2 & -1 & 2 & e_1 \\ 1 & 2 & 1 & e_2 \\ 2 & 2 & -1 & e_3 \end{array}\right] . \tag{A2.4}$$

Performing the elementary row operations exactly as above, we obtain

$$\left[\begin{array}{ccc|c} 1 & 0 & 0 & (4e_1-3e_2+5e_3)/15 \\ 0 & 1 & 0 & (-e_1+2e_2)/5 \\ 0 & 0 & 1 & (2e_1+6e_2-5e_3)/15 \end{array}\right] . \tag{A2.5}$$

Given any vector, we can use matrix (A2.5) to find the linear combination of H that yields the given vector by simply substituting the coordinates of the given vector with respect to D for e_1, e_2 and e_3, respectively. Consequently, we have shown that every vector in S can be written as a linear combination of H, and so H spans S. Furthermore, since the solution sets for (A2.4) and (A2.5) are identical, and since there is only one solution to (A2.5), each vector can only be expressed one way as a linear combination of H. Therefore, H is a basis. □

(PA2.2) Problem: Starting with the augmented matrix (A2.4) use row operations to obtain (A2.5). Then use (A2.5) to express the vectors **f** and **t** as linear combinations of H where $[\mathbf{f}]_D = [5,1,6]^t$ and $[\mathbf{t}]_D = [-4,3,7]^t$. Confirm that you have found the right answer by calculating the appropriate linear combinations.

(PA2.3) Problem: Consider the set of vectors $T = \{\mathbf{x},\mathbf{y},\mathbf{z}\}$ where $[\mathbf{x}]_D = [2\ -2\ 3]^t$, $[\mathbf{y}]_D = [4\ -3\ 9]^t$ and $[\mathbf{z}]_D = [20,-17,40]^t$. Show that $[\mathbf{w}]_D = [2\ 1\ 8]^t$ can be written as a linear combination of T and that T forms a basis for S.

Answer: $11\mathbf{x} + 15\mathbf{y} - 4\mathbf{z} = \mathbf{w}$ and the augmented matrix for a linear combination equal to an arbitrary vector $\mathbf{e} = [e_1\ e_2\ e_3]^t$ is

$$\left[\begin{array}{ccc|c} 1 & 0 & 0 & (33e_1+20e_2-8e_3)/2 \\ 0 & 1 & 0 & (29e_1+20e_2-6e_3)/2 \\ 0 & 0 & 1 & (-9e_1-6e_2+2e_3)/2 \end{array}\right].$$

Reduced Row Echelon Matrices: It is not always possible to transform a matrix into the ideal form given in (A2.3). For example, suppose after a few row operations the following matrix is obtained,

$$\left[\begin{array}{ccc|c} 1 & 2 & -1 & 3 \\ 0 & 0 & 4 & 2 \\ 0 & 0 & -1 & 8 \end{array}\right]. \qquad \text{(A2.6)}$$

As there is no row operation that allows us to put a 1 in the (2,2) entry without disrupting the first column, we cannot transform this coefficient matrix into the identity matrix. Instead, we shall transform it into a more general matrix form called the reduced row echelon form.

(DA2.7) Definition: A matrix is in *reduced row echelon* form if

(1) All of the nonzero rows occur above the first occurrence of a row with all zeros;
(2) The first nonzero entry in a row (if there is one) is 1 and, except for the first row, appears to the right of the first nonzero entry of the previous row;

(3) If the first nonzero entry of a given row is in the *jth* column, then every other entry in the *jth* column must be zero.

Unlike the form given in (A2.3), any matrix can be placed into reduced row echelon form by a sequence of row operations.

(EA2.8) Example: Place the matrix given in (A2.6) into reduced row echelon form and find the solution set.

Solution: By switching rows 2 and 3 and multiplying the new row 2 by -1, the matrix in (A2.6) becomes

$$\left[\begin{array}{ccc|c} 1 & 2 & -1 & 3 \\ 0 & 0 & 1 & -8 \\ 0 & 0 & 4 & 2 \end{array}\right].$$

Next, replacing the first row by the sum of itself and 1 times the second and the third row by the sum of itself and (-4) times the second, we have

$$\left[\begin{array}{ccc|c} 1 & 2 & 0 & -5 \\ 0 & 0 & 1 & -8 \\ 0 & 0 & 0 & 34 \end{array}\right].$$

The system

$$\begin{aligned} x_1 + 2x_2 + 0x_3 &= -5 \\ 0x_1 + 0x_2 + x_3 &= -8 \\ 0x_1 + 0x_2 + 0x_3 &= 34 \end{aligned}$$

has no solution since the left side of the third equation will always be zero. Hence we may conclude that the original system had no solution. □

We will illustrate how solution sets are determined once the augmented matrix has been placed into reduced row echelon form in the following three examples.

(EA2.9) Example: Suppose the reduced row echelon matrix is

$$\left[\begin{array}{ccc|c} 1 & 0 & 0 & 1/4 \\ 0 & 1 & 0 & 1/4 \\ 0 & 0 & 1 & -1/4 \end{array}\right],$$

then there is *exactly one solution*: $x_1 = 1/4$, $x_2 = 1/4$, $x_3 = -1/4$. □

(EA2.10) Example: Suppose the reduced row echelon matrix is

$$\left[\begin{array}{ccc|c} 1 & 0 & -1 & 1/4 \\ 0 & 1 & -1 & 1/2 \\ 0 & 0 & 0 & 3/4 \end{array}\right].$$

Since $0x_1 + 0x_2 + 0x_3 = 3/4$ is one of the equations and its left side is always zero, there is no solution. □

(EA2.11) Example: Suppose the reduced row echelon matrix is

$$\left[\begin{array}{cccccc|c} 1 & 2 & 0 & 3 & 0 & 5 & 1 \\ 0 & 0 & 1 & 4 & 0 & 2 & -3 \\ 0 & 0 & 0 & 0 & 1 & 4 & 6 \\ 0 & 0 & 0 & 0 & 0 & 0 & 0 \end{array}\right].$$

To find the solution set, first circle the columns of the matrix that have the first nonzero entry of some row in it, i.e.,

$$\left[\begin{array}{cccccc|c} \textcircled{1} & 2 & \textcircled{0} & 3 & \textcircled{0} & 5 & 1 \\ \textcircled{0} & 0 & \textcircled{1} & 4 & \textcircled{0} & 2 & -3 \\ \textcircled{0} & 0 & \textcircled{0} & 0 & \textcircled{1} & 4 & 6 \\ \textcircled{0} & 0 & \textcircled{0} & 0 & \textcircled{0} & 0 & 0 \end{array}\right].$$

The unknowns corresponding to the *uncircled* columns are x_2, x_4, and x_6. Note that given any values for x_2, x_4 and x_6, values can be found for x_1, x_3 and x_5 so that the system is satisfied. Hence, if we set $x_2 = \ell$, $x_4 = m$ and $x_6 = n$, we obtain the solution set

$$\begin{aligned} x_1 &= 1 - 2\ell - 3m - 5m \\ x_2 &= \ell \\ x_3 &= -3 - 4m - 2n \end{aligned}$$

$$x_4 = m$$
$$x_5 = 6 - 4n$$
$$x_6 = n.$$

So we obtain an infinite number of solutions. □

The three examples above exhaust the types of solutions possible. The three types of solution sets that can occur are (1) a unique solution, (2) an infinite number of solutions, or (3) no solution. A more detailed discussion of solutions of equations can be found in most linear algebra texts (e.g., Johnson and Riess, 1980).

The next example explores how to deal with the situation where a relatively large number of experimental equations determines a relatively small number of variables and the best approximate solution is sought.

(EA2.12) Example - A determination of a crystal's cell dimensions by least-squares methods: In this example, we shall consider a set of X-ray diffraction data used to construct a set of linear equations that requires solution by least-squares methods. The least-squares method is a powerful tool devised by Johann Gauss at the age of 18 to find a close fit between a set of experimental data and some theoretical model that provides a connection between the data and a set of parameters characterizing the model. Because of the relatively large number of linear equations that are typically supplied in such an experiment, matrix theory provides a neat and compact way of deriving the appropriate least-squares equations. To illustrate the use of matrices in the method, we shall analyze a set of measured Q-values ($Q = 1/d^2$ where d is the interplanar spacing between diffracting planes $(hk\ell)$) for a crystal called protoamphibole and show by example how the method can be used to find a set of reciprocal cell parameters for the crystal that provides a close fit between the observed and calculated Q-values.

The d-spacings and Q-values for planes with indices $(hk\ell)$ as measured for the crystal are given in Table A2.1. We observe that the indices $(hk\ell)$ of these planes must be known before a least-squares analysis can be completed, i.e., the diffraction pattern must be indexed.

A study of the optical properties and the diffraction symmetry of protoamphibole indicates that it is orthorhombic which means that $a^* \neq b^* \neq c^*$ and $\alpha^* = \beta^* = \gamma^* = 90°$. Setting $Q = 1/d^2$, we can write (see (2.7) and C2.7).

Table A2.1. Selected diffraction data for protoamphibole.

h	k	ℓ	d	Q	h	k	ℓ	d	Q
1	1	0	8.276	0.01460	3	3	0	2.759	0.13137
0	4	0	4.469	0.05007	1	0	2	2.543	0.15463
1	3	1	3.645	0.07527	2	6	1	2.268	0.19441
2	2	1	3.260	0.09409	4	6	1	1.735	0.33220
3	1	0	3.063	0.10659	5	6	1	1.515	0.43569
1	5	1	2.822	0.12557	6	6	1	1.334	0.56194

$$Q = [h\ k\ \ell] \begin{bmatrix} a^{*2} & 0 & 0 \\ 0 & b^{*2} & 0 \\ 0 & 0 & c^{*2} \end{bmatrix} \begin{bmatrix} h \\ k \\ \ell \end{bmatrix}$$

$$= h^2 a^{*2} + k^2 b^{*2} + \ell^2 c^{*2} \ .$$

Hence,

$$Q = HA + KB + LC \ ,$$

where $H = h^2$, $K = k^2$, $L = \ell^2$, $A = a^{*2}$, $B = b^{*2}$ and $C = c^{*2}$. As each Q-value, Q_i, is a random variable subject to a measurement error, ε_i, a set of n measurements provides the following set of n linear equations:

$$\begin{aligned} Q_1 &= H_1 A + K_1 B + L_1 C + \varepsilon_1 \\ Q_2 &= H_2 A + K_2 B + L_2 C + \varepsilon_2 \\ &\ \ \vdots \\ Q_n &= H_n A + K_n B + L_n C + \varepsilon_n \ . \end{aligned}$$

With matrix theory, these n equations can be written succinctly as $\mathbf{Q} = \mathrm{D}\mathbf{v} + \varepsilon$ where

$$\mathbf{Q} = \begin{bmatrix} Q_1 \\ Q_2 \\ \vdots \\ Q_n \end{bmatrix}, \quad D = \begin{bmatrix} H_1 & K_1 & L_1 \\ H_2 & K_2 & L_2 \\ \vdots & & \vdots \\ H_n & K_n & L_n \end{bmatrix}, \quad \mathbf{v} = \begin{bmatrix} A \\ B \\ C \end{bmatrix} \text{ and } \varepsilon = \begin{bmatrix} \varepsilon_1 \\ \varepsilon_2 \\ \vdots \\ \varepsilon_n \end{bmatrix}.$$

Consistent with Gauss' postulate of minimum variance, we minimize V where

$$V = \varepsilon^t \varepsilon .$$

As $\mathbf{Q} = D\mathbf{v} + \varepsilon$, we can write the equality

$$\begin{aligned} V &= \varepsilon^t \varepsilon = (\mathbf{Q}-D\mathbf{v})^t(\mathbf{Q}-D\mathbf{v}) \\ &= (\mathbf{Q}^t-\mathbf{v}^t D^t)(\mathbf{Q}-D\mathbf{v}) \\ &= \mathbf{Q}^t\mathbf{Q} - \mathbf{Q}^t D\mathbf{v} - \mathbf{v}^t D^t \mathbf{Q} + \mathbf{v}^t D^t D\mathbf{v} . \end{aligned}$$

Since $\mathbf{v}^t D^t \mathbf{Q} = \mathbf{Q}^t D\mathbf{v}$,

$$V = \mathbf{Q}^t\mathbf{Q} - 2\mathbf{Q}^t D\mathbf{v} + \mathbf{v}^t D^t D\mathbf{v} .$$

Without going into the details, it can be shown that when the derivative $\partial V/\partial \mathbf{v}$ is equated with zero that

$$D^t D\mathbf{v} = D^t \mathbf{Q} . \tag{A2.7}$$

Using the data in Table A2.1, we can construct the matrices

$$D = \begin{bmatrix} 1 & 1 & 0 \\ 0 & 16 & 0 \\ 1 & 9 & 1 \\ 4 & 4 & 1 \\ 9 & 1 & 0 \\ 1 & 25 & 1 \\ 9 & 9 & 0 \\ 1 & 0 & 4 \\ 4 & 36 & 1 \\ 16 & 36 & 1 \\ 25 & 36 & 1 \\ 36 & 36 & 1 \end{bmatrix}, \quad \mathbf{Q} = \begin{bmatrix} 0.01460 \\ 0.05007 \\ 0.07527 \\ 0.09409 \\ 0.10659 \\ 0.12557 \\ 0.13137 \\ 0.15463 \\ 0.19441 \\ 0.33220 \\ 0.43569 \\ 0.56194 \end{bmatrix},$$

$$D^t = \begin{bmatrix} 1 & 0 & 1 & 4 & 9 & 1 & 9 & 1 & 4 & 16 & 25 & 36 \\ 1 & 16 & 9 & 4 & 1 & 25 & 9 & 0 & 36 & 36 & 36 & 36 \\ 0 & 0 & 1 & 1 & 0 & 1 & 0 & 4 & 1 & 1 & 1 & 1 \end{bmatrix},$$

$$D^tD = \begin{bmatrix} 2375 & 3057 & 91 \\ 3057 & 6245 & 182 \\ 91 & 182 & 23 \end{bmatrix} \quad \text{and} \quad D^t\mathbf{Q} = \begin{bmatrix} 40.1030 \\ 61.1703 \\ 2.43769 \end{bmatrix}.$$

Next, we form the normal equations $(D^tD)\mathbf{v} = D^t\mathbf{Q}$

$$\begin{bmatrix} 2375 & 3057 & 91 \\ 3057 & 6245 & 182 \\ 91 & 182 & 23 \end{bmatrix} \begin{bmatrix} A \\ B \\ C \end{bmatrix} = \begin{bmatrix} 40.1030 \\ 61.1703 \\ 2.43769 \end{bmatrix},$$

The solution of these equations will provide an unbiased set of least-squares estimates of A, B and C. As a first step in the solution of these equations for A, B and C, we form the augmented matrix

$$\left[\begin{array}{ccc|c} 2375 & 3057 & 91 & 40.1030 \\ 3057 & 6245 & 182 & 61.1703 \\ 91 & 182 & 23 & 2.43769 \end{array}\right].$$

Our first objective is to get a 1 in the (1,1) entry. We accomplish this by first switching the first and third rows and then multiplying the resulting first row by (1/91),

$$\left[\begin{array}{ccc|c} 1 & 2 & 0.252747 & 0.0267878 \\ 3057 & 6245 & 182 & 61.1703 \\ 2375 & 3057 & 91 & 40.1030 \end{array}\right].$$

Now we would like zeros in the remaining entries of the first column. This can be accomplished by replacing the second row by the sum of itself and (-3057) times the first row and then replacing the third row by the sum of itself and (-2375) times the first,

$$\left[\begin{array}{ccc|c} 1 & 2 & 0.252747 & 0.0267878 \\ 0 & 131 & -590.64758 & -20.72000 \\ 0 & -1693 & -509.27412 & -23.518025 \end{array}\right].$$

We next wish to get a 1 at the (2,2) entry. We can accomplish this by multiplying the second row by the nonzero constant (1/131),

$$\begin{bmatrix} 1 & 2 & 0.252747 & | & 0.0267878 \\ 0 & 1 & -4.508760 & | & -0.158168 \\ 0 & -1693 & -509.27413 & | & -23.518025 \end{bmatrix} .$$

Now we want to obtain zeros in the remaining entries of the second columns. We do this by replacing the first row by the sum of itself and (-2) times the second row and replacing the third row by the sum of itself and 1693 times the second row,

$$\begin{bmatrix} 1 & 0 & 9.270267 & | & 0.343124 \\ 0 & 1 & -4.508760 & | & -0.158168 \\ 0 & 0 & -8142.6048 & | & -291.29645 \end{bmatrix} .$$

Next we want to obtain a 1 in the (3,3) entry. If the third row is multiplied by the constant (-1/8142.6048), we obtain

$$\begin{bmatrix} 1 & 0 & 9.27067 & | & 0.343124 \\ 0 & 1 & -4.508760 & | & -0.158168 \\ 0 & 0 & 1 & | & 0.035774 \end{bmatrix} .$$

To obtain zeros in the remaining entries of the third column of the matrix, we replace the second row with the sum of itself and (4.50876) times the third row and replace the first row by the sum of itself and (-9.27067) times the third row,

$$\begin{bmatrix} 1 & 0 & 0 & | & 0.011475 \\ 0 & 1 & 0 & | & 0.003128 \\ 0 & 0 & 1 & | & 0.035774 \end{bmatrix} .$$

Hence the solution set to the normal equations is

$$A = a^{*2} = 0.011475$$
$$B = b^{*2} = 0.003128$$
$$C = c^{*2} = 0.035774 .$$

Since, for a crystal with orthorhombic geometry $a = 1/a^*$, $b = 1/b^*$ and $c = 1/c^*$, we see that $a = 9.335$A, $b = 17.880$A and $c = 5.287$A. □

(PA2.4) Problem: A measurement of the X-ray diffraction powder pattern of α-quartz provides the interplanar spacings given in Table A2.2 for the 10 largest angle reflections observed. Since α-quartz is hexagonal ($a^* = b^* \neq c^*$, $\alpha^* = \beta^* = 90°$, $\gamma^* = 60°$),

Table A2.2. Selected interplanar spacings and Q-values for α-quartz.

h	k	ℓ	d	Q	h	k	ℓ	d	Q
4	0	4	0.8395	1.41892	2	2	5	0.8115	1.51853
2	0	6	0.8295	1.45334	3	3	1	0.8096	1.52566
4	1	3	0.8254	1.46781	4	2	0	0.8041	1.54661
3	3	0	0.8189	1.49121	3	1	5	0.7971	1.57389
5	0	2	0.8117	1.51778	4	2	1	0.7952	1.58142

its reciprocal lattice parameters are related to $1/d^2 = Q$ by

$$Q = [h\ k\ \ell] \begin{bmatrix} a^{*2} & a^{*2}/2 & 0 \\ a^{*2}/2 & a^{*2} & 0 \\ 0 & 0 & c^{*2} \end{bmatrix} \begin{bmatrix} h \\ k \\ \ell \end{bmatrix}$$

$$= h(ha^{*2} + ka^{*2}/2) + k(ha^{*2}/2 + ka^{*2}) + \ell^2 c^{*2} .$$

Hence

$$Q = (h^2 + hk + k^2)a^{*2} + \ell^2 c^{*2} .$$

Therefore,

$$HA + LC = Q ,$$

where $H = (h^2 + hk + k^2)$, $L = \ell^2$, $A = a^{*2}$ and $C = c^{*2}$. Construct the matrices $\mathbf{Q}$, D and D^t using the data in Table A2.2 and show that

$$D^tD = \begin{bmatrix} 4677 & 1369 \\ 1369 & 2901 \end{bmatrix}$$

and

$$D^t\mathbf{Q} = \begin{bmatrix} 305.00785 \\ \\ 174.72195 \end{bmatrix}.$$

Then solve the normal equations $D^tD\mathbf{v} = D^t\mathbf{Q}$ for the vector

$$\mathbf{v} = \begin{bmatrix} 0.055212 \\ \\ 0.034174 \end{bmatrix}$$

of the least-squares estimates of A and C and show that $a^* = 0.23497$ and $c^* = 0.18486$. Since for a hexagonal crystal, $a = 2/(\sqrt{3}a^*)$ and $c = 1/c^*$, the least-squares estimates of the unit cell dimensions of α-quartz are a = 4.914A and c = 5.409A.

DETERMINANTS

Given an $n \times n$ matrix A, a number called the determinant of A is assigned. The determinant of such a matrix is denoted either as det(A) or |A|. We shall first see how to calculate det(A) and then explore how it is applied.

The determinant of a 2 × 2 matrix is defined as

$$\det(A) = \begin{vmatrix} a_{11} & a_{12} \\ \\ a_{21} & a_{22} \end{vmatrix} = a_{11}a_{22} - a_{12}a_{21} .$$

(EA2.13) Example: Suppose that $A = \begin{bmatrix} 2 & -1 \\ 5 & 4 \end{bmatrix}$, find det(A).

Solution: $\det(A) = \begin{vmatrix} 2 & -1 \\ 5 & 4 \end{vmatrix} = (2)(4) - (-1)(5) = 8 + 5 = 13$.

The determinant of an $n \times n$ matrix A where $n > 2$ can be expressed in terms of determinants of smaller related matrices which can in turn be expressed in terms of determinants of smaller matrices until finally det(A) is expressed in terms of determinants of 2 × 2 matrices. The related matrices in this process are called the minors of A. The *minor* A_{ij} of A is the matrix formed by deleting the *ith* row and *jth* column of A.

(EA2.14) Example: If

$$A = \begin{bmatrix} -3 & 5 & 14 \\ 2 & 1 & 8 \\ -2 & 4 & 6 \end{bmatrix},$$

then $A_{11} = \begin{bmatrix} 1 & 8 \\ 4 & 6 \end{bmatrix}$, $A_{23} = \begin{bmatrix} -3 & 5 \\ -2 & 4 \end{bmatrix}$ and $A_{31} = \begin{bmatrix} 5 & 14 \\ 1 & 8 \end{bmatrix}$.

(DA2.15) Definition: The *determinant* of an $n \times n$ matrix A is defined to be

$$\det(A) = |A| = \sum_{j=1}^{n} (-1)^{1+j} a_{1j} |A_{1j}| .$$

(EA2.16) Example: If

$$A = \begin{bmatrix} a_{11} & a_{12} & a_{13} \\ a_{21} & a_{22} & a_{23} \\ a_{31} & a_{32} & a_{33} \end{bmatrix},$$

then

$$|A| = (-1)^{1+1} a_{11}|A_{11}| + (-1)^{1+2} a_{12}|A_{12}| + (-1)^{1+3} a_{13}|A_{13}|$$

$$= a_{11} \begin{vmatrix} a_{22} & a_{23} \\ a_{32} & a_{33} \end{vmatrix} - a_{12} \begin{vmatrix} a_{21} & a_{23} \\ a_{31} & a_{33} \end{vmatrix} + a_{13} \begin{vmatrix} a_{21} & a_{22} \\ a_{31} & a_{32} \end{vmatrix} .$$

Note that the calculation of the determinant of a 3 × 3 matrix can be accomplished by calculating the determinants of three 2 × 2 matrices. If A is a 4 × 4 matrix, then you obtain an expression involving four 3 × 3 matrices, each of which leads to an expression involving three 2 × 2 matrices. Hence, 12 determinants of 2 × 2 matrices would be required.

(EA2.17) Example: If

$$A = \begin{bmatrix} 1 & -2 & 5 \\ 3 & 1 & 8 \\ -4 & 5 & 2 \end{bmatrix},$$

then

$$|A| = (-1)^{1+1}(1)\begin{vmatrix} 1 & 8 \\ 5 & 2 \end{vmatrix} + (-1)^{1+2}(-2)\begin{vmatrix} 3 & 8 \\ -4 & 2 \end{vmatrix} + (-1)^{1+3}(5)\begin{vmatrix} 3 & 1 \\ -4 & 5 \end{vmatrix}$$

$$= (1)(2 - 40) - (-2)(6 + 32) + (5)(15 + 4)$$
$$= -38 + 76 + 95 = 133 \ .$$

Hence, $|A| = 133$. □

The definition of a determinant (DA2.15) uses the entries and minors of only the first row. However, the determinant can be found by expanding about any row:

$$|A| = \sum_{j=1}^{n} (-1)^{i+j} a_{ij} |A_{ij}| \quad \text{for any } 1 \le i \le n$$

or about any column:

$$|A| = \sum_{i=1}^{n} (-1)^{i+j} a_{ij} |A_{ij}| \quad \text{for any } 1 \le j \le n \ .$$

(EA2.18) Example: Using the matrix A of EA2.17, the determinant of A expanded about the third row becomes

$$|A| = (-1)^{3+1}(-4)\begin{vmatrix} -2 & 5 \\ 1 & 8 \end{vmatrix} + (-1)^{3+2}(5)\begin{vmatrix} 1 & 5 \\ 3 & 8 \end{vmatrix} + (-1)^{3+3}(2)\begin{vmatrix} 1 & -2 \\ 3 & 1 \end{vmatrix}$$

$$= (-4)(-16 - 5) - (5)(8 - 15) + (2)(1 + 6)$$
$$= 133 \ .$$ □

(EA2.19) Example: Expand the A matrix given in EA2.17 about its second column:

$$|A| = (-1)^{1+2}(-2)\begin{vmatrix} 3 & 8 \\ -4 & 2 \end{vmatrix} + (-1)^{2+2}(1)\begin{vmatrix} 1 & 5 \\ -4 & 2 \end{vmatrix} + (-1)^{3+2}(5)\begin{vmatrix} 1 & 5 \\ 3 & 8 \end{vmatrix}$$

$$= 133 \ .$$ □

One property of determinants that will be particularly useful to us is that if A and B are $n \times n$ matrices, then

$$\det(AB) = \det(A)\det(B) \ . \tag{A2.8}$$

The proof of this fact can be found in any standard linear algebra text (Johnson and Riess, 1981).

In Chapter 1, we showed that for any basis $D = \{\mathbf{a},\mathbf{b},\mathbf{c}\}$, the volume $V = \mathbf{a} \bullet (\mathbf{b} \times \mathbf{c})$. One way to calculate V is to form a cartesian basis C and find the change of basis matrix A from D to C. That is,

$$A = \left[\begin{array}{c|c|c} [\mathbf{a}]_C & [\mathbf{b}]_C & [\mathbf{c}]_C \end{array} \right] .$$

Then $V = \mathbf{a} \bullet \mathbf{b} \times \mathbf{c} = \det(A)$. The next example shows how to find the volume of the unit cell of one basis knowing the volume of the unit cell of another.

(EA2.20) Example: Suppose that the vectors of a new basis $D_2 = \{\mathbf{a}_2,\mathbf{b}_2,\mathbf{c}_2\}$ is given in terms of an old basis $D_1 = \{\mathbf{a}_1,\mathbf{b}_1,\mathbf{c}_1\}$ whose unit cell volume V_1 is known. Show that $V_2 = V_1\det(B)$ where B is the change of basis matrix from D_2 to D_1.

Solution: Let C denote a cartesian basis and let A denote the change of basis matrix from D_1 to C. Then, as noted above, $V_1 = \det(A)$. Let B denote the change of basis matrix from D_2 to D_1. That is,

$$B = \left[\begin{array}{c|c|c} [\mathbf{a}_2]_{D_1} & [\mathbf{b}_2]_{D_1} & [\mathbf{c}_2]_{D_1} \end{array} \right] .$$

Then

$$AB = \left[\begin{array}{c|c|c} A[\mathbf{a}_2]_{D_1} & A[\mathbf{b}_2]_{D_1} & A[\mathbf{c}_2]_{D_1} \end{array} \right]$$

$$= \left[\begin{array}{c|c|c} [\mathbf{a}_2]_C & [\mathbf{b}_2]_C & [\mathbf{c}_2]_C \end{array} \right] ,$$

and so AB is the change of basis matrix from D_2 to C. Hence

$$\begin{aligned} V_2 &= \det(AB) \\ &= \det(A)\det(B) \\ &= V_1\det(B) \quad . \end{aligned} \qquad (A2.9)$$

□

(EA2.21) Example - A unit cell of kyanite contains 20 oxygen atoms: The volume of the face-centered sub-cell of the kyanite structure is $v_1 = 58.719$ A^3 (E2.19). Using the definitions of the basis vectors for the kyanite cell (E2.17), we can compute the volume of the kyanite cell v_2 using (A2.9), that is,

$$v_2 = v_1 \begin{vmatrix} 3/2 & 0 & -1 \\ -1/2 & 2 & 0 \\ 1 & 0 & 1 \end{vmatrix}$$

$$= 58.719 \left((-1)^{2+2}(2) \begin{vmatrix} 3/2 & -1 \\ 1 & 1 \end{vmatrix} \right)$$

$$= 58.719 \times 5 = 293.60A^3 \; .$$

Hence, the kyanite cell volume is five times larger than that of its face-centered subcell. Moreover, as the subcell contains four oxygen atoms per volume, the kyanite cell must contain 5 × 4 = 20 oxygen atoms. □

INVERSES

The matrix

$$I_3 = \begin{bmatrix} 1 & 0 & 0 \\ 0 & 1 & 0 \\ 0 & 0 & 1 \end{bmatrix}$$

is called the *identity matrix* because if A is any 3 × 3 matrix, then

$$AI_3 = \begin{bmatrix} a_{11} & a_{12} & a_{13} \\ a_{21} & a_{22} & a_{23} \\ a_{31} & a_{32} & a_{33} \end{bmatrix} \begin{bmatrix} 1 & 0 & 0 \\ 0 & 1 & 0 \\ 0 & 0 & 1 \end{bmatrix} = \begin{bmatrix} a_{11} & a_{12} & a_{13} \\ a_{21} & a_{22} & a_{23} \\ a_{31} & a_{32} & a_{33} \end{bmatrix} = A \; .$$

Similarly, $I_3A = A$. Hence, I_3 is called the *multiplicative identity* on the set of all 3 × 3 matrices. In general, the $n \times n$ matrix I_n whose (i,j)-entry equals 0 if $i \neq j$ and 1 if $i = j$ is the identity matrix for the set of all $n \times n$ matrices.

Let A denote an $n \times n$ matrix. If there exists a matrix B such that

$$AB = I_n \; ,$$

then B is said to be the *inverse* of A, and we write

$$A^{-1} = B \; .$$

When A^{-1} exists, A is said to be *invertible*.

Let A denote an invertible matrix. Since

$$AA^{-1} = \begin{bmatrix} 1 & 0 & 0 \\ 0 & 1 & 0 \\ 0 & 0 & 1 \end{bmatrix},$$

we have the equations

$$A \text{ (first column of } A^{-1}) = \begin{bmatrix} 1 \\ 0 \\ 0 \end{bmatrix}$$

$$A \text{ (second column of } A^{-1}) = \begin{bmatrix} 0 \\ 1 \\ 0 \end{bmatrix}$$

$$A \text{ (third column of } A^{-1}) = \begin{bmatrix} 0 \\ 0 \\ 1 \end{bmatrix}$$

where the entries of A^{-1} are the unknowns. Instead of solving these systems of equations individually, we can deal with them all at once in the following augmented matrix:

$$\left[\begin{array}{ccc|ccc} a_{11} & a_{12} & a_{13} & 1 & 0 & 0 \\ a_{21} & a_{22} & a_{23} & 0 & 1 & 0 \\ a_{31} & a_{32} & a_{33} & 0 & 0 & 1 \end{array}\right].$$

If this matrix can be reduced to a matrix of the following form using elementary row operations,

$$\left[\begin{array}{ccc|ccc} 1 & 0 & 0 & b_{11} & b_{12} & b_{13} \\ 0 & 1 & 0 & b_{21} & b_{22} & b_{23} \\ 0 & 0 & 1 & b_{31} & b_{32} & b_{33} \end{array}\right]$$

then the first column of A^{-1} is $\begin{bmatrix} b_{11} \\ b_{21} \\ b_{31} \end{bmatrix}$, the second is $\begin{bmatrix} b_{12} \\ b_{22} \\ b_{32} \end{bmatrix}$, and the third is $\begin{bmatrix} b_{13} \\ b_{23} \\ b_{33} \end{bmatrix}$. Therefore,

$$A^{-1} = \begin{bmatrix} b_{11} & b_{12} & b_{13} \\ b_{21} & b_{22} & b_{23} \\ b_{31} & b_{32} & b_{33} \end{bmatrix}$$

If the A matrix cannot be reduced to I_3 using row operations, then A has no inverse. Under these circumstances, A is said to be *singular* and to lack an inverse.

Finding A^{-1} can be helpful in solving many types of problems. For example, the matrix equation

$$AB = C$$

can be solved for B if A has an inverse by premultiplying both sides of the equation by A^{-1}; that is,

$$A^{-1}(AB) = A^{-1}C$$
$$(A^{-1}A)B = A^{-1}C$$
$$I_nB = A^{-1}C$$
$$B = A^{-1}C \ .$$

In our next example, we shall consider the problem of constructing and inverting a transformation matrix.

(EA2.22) Example: The oxygen atoms in the rock-forming mineral aenigmatite, $Na_2Fe_5TiSi_6O_{20}$, can be described as closest-packed defining a face-centered subcell with cell dimensions a_1 = 4.10A, b_1 = 4.33A, c_1 = 4.44A, α_1 = 103.1°, β_1 = 84.8°, γ_1 = 103.8°. The linear equations that relate the observed D_2 basis vectors (a_2 = 9.752A, b_2 = 10.406A, c_2 = 8.926A, α_2 = 83.130°, β_2 = 65.587°, γ_2 = 64.785°) to those of the distorted subcell (D_1 basis) are

$$\begin{aligned} \mathbf{a}_2 &= 2\mathbf{a}_1 - 1/2\mathbf{b}_1 - 1/2\mathbf{c}_1 \\ \mathbf{b}_2 &= 2\mathbf{a}_1 + 2\mathbf{b}_1 \\ \mathbf{c}_2 &= \mathbf{a}_1 + \mathbf{b}_1 + 2\mathbf{c}_1 \ . \end{aligned} \qquad (A2.10)$$

With these results, the transformation matrix

$$T^{-1} = \left[[\mathbf{a}_2]_{D_1} \mid [\mathbf{b}_2]_{D_1} \mid [\mathbf{c}_2]_{D_1} \right] = \begin{bmatrix} 2 & 2 & 1 \\ -1/2 & 2 & 1 \\ -1/2 & 0 & 2 \end{bmatrix}$$

can be constructed by simply writing the coordinates of $\mathbf{a}_2$, $\mathbf{b}_2$ and $\mathbf{c}_2$ in columns. With this matrix, the coordinates of any vector $[\mathbf{v}]_{D_2}$ can be expressed as a linear combination of $[\mathbf{v}]_{D_1}$ by the expression (see T2.10),

$$T^{-1}[\mathbf{v}]_{D_2} = [\mathbf{v}]_{D_1} , \tag{A2.11}$$

Our problem is to find the matrix T that defines any vector $[\mathbf{v}]_{D_1}$ as a linear combination of $[\mathbf{v}]_{D_2}$. If both sides of (A2.11) are premultiplied by T, then we have

$$[\mathbf{v}]_{D_2} = T[\mathbf{v}]_{D_1} .$$

This equation defines each vector $[\mathbf{v}]_{D_2}$, including the D_2 basis vectors, as a linear combination of the D_1 basis vectors. In fact, the columns of T are the coordinates of $\mathbf{a}_1$, $\mathbf{b}_1$ and $\mathbf{c}_1$ in the D_2 basis. We now find T. We begin by writing the augmented matrix,

$$\left[\begin{array}{ccc|ccc} 2 & 2 & 1 & 1 & 0 & 0 \\ -1/2 & 2 & 1 & 0 & 1 & 0 \\ -1/2 & 0 & 2 & 0 & 0 & 1 \end{array}\right] .$$

To get a 1 in the (1,1)-entry, the first row is multiplied by (1/2),

$$\left[\begin{array}{ccc|ccc} 1 & 1 & 1/2 & 1/2 & 0 & 0 \\ -1/2 & 2 & 1 & 0 & 1 & 0 \\ -1/2 & 0 & 2 & 0 & 0 & 1 \end{array}\right] .$$

To get zeros in the remaining entries of the first column, we replace the second row by the sum of itself and (1/2) times the first and replace the third row by the sum of itself and (1/2) times the first.

$$\left[\begin{array}{ccc|ccc} 1 & 1 & 1/2 & 1/2 & 0 & 0 \\ 0 & 5/2 & 5/4 & 1/4 & 1 & 0 \\ 0 & 1/2 & 9/4 & 1/4 & 0 & 1 \end{array}\right] .$$

To get a 1 in the (2,2)-entry, multiply the second row by 2/5.

$$\left[\begin{array}{ccc|ccc} 1 & 1 & 1/2 & 1/2 & 0 & 0 \\ 0 & 1 & 1/2 & 1/10 & 2/5 & 0 \\ 0 & 1/2 & 9/4 & 1/4 & 0 & 1 \end{array}\right].$$

To get zeros in the remaining entries of the second column, replace the first row by the sum of itself and (-1) times the second and replace the third by the sum of itself and (-1/2) times the second.

$$\left[\begin{array}{ccc|ccc} 1 & 0 & 0 & 2/5 & -2/5 & 0 \\ 0 & 1 & 1/2 & 1/10 & 2/5 & 0 \\ 0 & 0 & 2 & 1/5 & -1/5 & 1 \end{array}\right].$$

To get a 1 in the (3,3) entry, multiply the third row by (-1/2).

$$\left[\begin{array}{ccc|ccc} 1 & 0 & 0 & 2/5 & -2/5 & 0 \\ 0 & 1 & 1/2 & 1/10 & 2/5 & 0 \\ 0 & 0 & 1 & 1/10 & -1/10 & 1/2 \end{array}\right].$$

To get a zero at the (2,3) entry, replace the second row by the sum of itself and (-1/2) times the third row,

$$\left[\begin{array}{ccc|ccc} 1 & 0 & 0 & 2/5 & -2/5 & 0 \\ 0 & 1 & 0 & 1/20 & 9/20 & -1/4 \\ 0 & 0 & 1 & 1/10 & -1/10 & 1/2 \end{array}\right].$$

Hence

$$T = \begin{bmatrix} 2/5 & -2/5 & 0 \\ 1/20 & 9/20 & -1/4 \\ 1/10 & -1/10 & 1/2 \end{bmatrix} = \left[\begin{array}{c|c|c} [\mathbf{a}_1]_{D_2} & [\mathbf{b}_1]_{D_2} & [\mathbf{c}_1]_{D_2} \end{array}\right].$$

Hence, we can write the linear equations

$$\begin{aligned} \mathbf{a}_1 &= 2/5\mathbf{a}_2 + 1/20\mathbf{b}_2 + 1/10\mathbf{c}_2 \\ \mathbf{b}_1 &= -2/5\mathbf{a}_2 + 9/20\mathbf{b}_2 - 1/10\mathbf{c}_2 \\ \mathbf{c}_1 &= -1/4\mathbf{b}_2 + 1/2\mathbf{c}_2 \ . \end{aligned} \qquad \text{(A2.1}$$

Each of these equations can be checked by replacing $\mathbf{a}_1$, $\mathbf{b}_1$, and $\mathbf{c}_1$ in (A2.10) by the appropriate expressions in (A2.12).

(EA2.23) Example - Finding a transformation matrix from experimental data: We now consider an experimental method for finding a transformation matrix T. Suppose that the cell dimensions ($a_1 = b_1 = c_1 = 7.268$ A, $\alpha_1 = 84.84°$, $\beta_1 = 138.82°$, $\gamma_1 = 109.71°$) of hemimorphite, $Zn_4Si_2O_7(OH)_2 \bullet H_2O$, were determined in a diffraction experiment in which the indices $h_1k_1\ell_1$ given in Table A2.3 were assigned to the diffraction record. Suppose also that a later investigator re-indexed the record with the indices $h_2k_2\ell_2$. Find the transformation matrix T from D_1 to D_2 and the cell dimensions of the basis vectors $D_2 = \{\mathbf{a}_2, \mathbf{b}_2, \mathbf{c}_2\}$.

Table A2.3. Indexed X-ray (CuKα) diffraction record for hemimorphite.

h_1	k_1	ℓ_1	h_2	k_2	ℓ_2	2θ(obs)
0	1	1	1	2	1	26.36°
1	1	0	2	1	1	28.80°
1	0	1	1	1	2	37.79°

Solution: In Chapter 2, we found that perpendicular to each face on a crystal there exists a vector $[\mathbf{s}]_{D^*}$. We also found that the matrix T^t transforms a vector $[\mathbf{s}]_{D^*_2}$ into $[\mathbf{s}]_{D^*_1}$. Thus, T^t can be used to transform the indices of the planes $h_2k_2\ell_2$ assigned by the second investigator into those, $h_1k_1\ell_1$, assigned by the first. Since $[\mathbf{s}]_{D^*_1} = T^t[\mathbf{s}]_{D^*_2}$, we can select any three planes of the first set of indices (provided they are not parallel to one another), say (011), (110) and (101) and match them up with those of the second set as follows:

$$T^t \begin{bmatrix} 1 \\ 2 \\ 1 \end{bmatrix} = \begin{bmatrix} 0 \\ 1 \\ 1 \end{bmatrix},$$

$$T^t \begin{bmatrix} 2 \\ 1 \\ 1 \end{bmatrix} = \begin{bmatrix} 1 \\ 1 \\ 0 \end{bmatrix},$$

$$T^t \begin{bmatrix} 1 \\ 1 \\ 2 \end{bmatrix} = \begin{bmatrix} 1 \\ 0 \\ 1 \end{bmatrix}.$$

That is,

$$T^t \begin{bmatrix} 1 & 2 & 1 \\ 2 & 1 & 1 \\ 1 & 1 & 2 \end{bmatrix} = \begin{bmatrix} 0 & 1 & 1 \\ 1 & 1 & 0 \\ 1 & 0 & 1 \end{bmatrix}.$$

Thus,

$$T^t = \begin{bmatrix} 0 & 1 & 1 \\ 1 & 1 & 0 \\ 1 & 0 & 1 \end{bmatrix} \begin{bmatrix} 1 & 2 & 1 \\ 2 & 1 & 1 \\ 1 & 1 & 2 \end{bmatrix}^{-1}.$$

Using elementary row operations, we find that

$$\begin{bmatrix} 1 & 2 & 1 \\ 2 & 1 & 1 \\ 1 & 1 & 2 \end{bmatrix}^{-1} = \begin{bmatrix} -1/4 & 3/4 & -1/4 \\ 3/4 & -1/4 & -1/4 \\ -1/4 & -1/4 & 3/4 \end{bmatrix}.$$

Then we have

$$T = \begin{bmatrix} 1/2 & 1/2 & -1/2 \\ -1/2 & 1/2 & 1/2 \\ 1/2 & -1/2 & 1/2 \end{bmatrix}.$$

Calculating $G_2 = T^{-t}G_1T^{-1}$, we get the cell dimensions $a_2 = 8.368$, $b_2 = 10.731$, $c_2 = 5.112$ A, $\alpha_2 = \beta_2 = \gamma_2 = 90.00°$ for hemimorphite with the second choice of basis vectors. □

Table A2.4. Selected indices for the diffraction record for coesite.

h_1	k_1	ℓ_1	h_2	k_2	ℓ_2
1	1	1	2	0	1
1	0	0	1	-1	0
0	1	1	1	1	1

(PA2.5) Problem: In a preliminary study of coesite, SiO_2, its diffraction record was indexed with indices $h_1k_1\ell_1$ and the cell dimensions $a_1 = b_1 =$ 7.141A, c_1 = 7.173A, $\alpha_1 = \beta_1$ = 104.626°, γ_1 = 120.055°. The pattern was then re-indexed with a new set of indices $h_2k_2\ell_2$. With the information in Table A2.4, show that the cell dimenions of coesite in the new setting are a_2 = 7.135A, b_2 = 12.372A, c_2 = 7.173A, $\alpha_2 = \gamma_2$ = 90.0°, β_2 = 120.36°.

We now discuss a few useful properties involving inverses and determinants.

(TA2.24) Theorem: Let A denote an $n \times n$ invertible matrix. Then,

$$\det(A^{-1}) = 1/\det(A) .$$

Proof: Since $AA^{-1} = I_n$, $\det(AA^{-1}) = \det(I_n) = 1$. By (A2.8), $\det(AA^{-1}) = \det(A)\det(A^{-1})$. Therefore, $\det(A^{-1}) = 1/\det(A)$. □

Note that TA2.24 implies that if det(A) = 0, it can have no inverse. In fact, the converse is also true, and so A has an inverse if and only if $\det(A) \neq 0$.

(PA2.6) Problem: Given that

$$T = \begin{bmatrix} 1/2 & -1/2 & 1/2 \\ 1/2 & 1/2 & -1/2 \\ -1/2 & 1/2 & 1/2 \end{bmatrix} \quad \text{and} \quad T^{-1} = \begin{bmatrix} 1 & 1 & 0 \\ 0 & 1 & 1 \\ 1 & 0 & 1 \end{bmatrix},$$

show that det(T) = 0.5 and $\det(T^{-1}) = 1/\det(T) = 2.0$ by calculating the determinants for these two matrices.

(TA2.25) Theorem: Let A denote a matrix consisting entirely of integer entries. If $\det(A) = \pm 1$, then A^{-1} consists entirely of integer entries.

Proof: Using the cofactor method of inverting A (see any standard linear algebra text), A^{-1} is written with entries of the form $(1/\det(A))(-1)^{i+j}\det(A_{ji})$ where A_{ji} is the (j,i) minor. Hence, if $\det(A) = \pm 1$ and if A consists entirely of integers, each entry of A^{-1} is also an integer. □

(PA2.7) Problem: Given that

$$T = \begin{bmatrix} 1 & 0 & 0 \\ 0 & 0 & 1 \\ 0 & -1 & 0 \end{bmatrix}$$

and $\det(T) = 1.0$, show that T^{-1} consists entirely of integers.

APPENDIX 3

CONSTRUCTION AND INTERPRETATION OF MATRICES REPRESENTING POINT ISOMETRIES

INTRODUCTION

In this appendix, we show how the matrix representation of a given rotation can be created when specific information about the images of the basis vectors is not given in a convenient form. We do this first with respect to a cartesian basis C and then extend these results to general bases. We then turn to the question of interpreting the matrix of an unknown rotation or rotoinversion. Again we begin with a cartesian basis and then develop a method that works for general bases. Central to the appendix is the general cartesian rotation matrix which is presented in the first section and is proved at the end of the appendix.

CONSTRUCTION OF MATRIX REPRESENTATIONS

Cartesian Bases: The *general cartesian rotation matrix*

$$M_C(\alpha) = \begin{bmatrix} \ell_1{}^2(1-c\rho)+c\rho & \ell_1\ell_2(1-c\rho)-\ell_3 s\rho & \ell_1\ell_3(1-c\rho)+\ell_2 s\rho \\ \ell_2\ell_1(1-c\rho)+\ell_3 s\rho & \ell_2{}^2(1-c\rho)+c\rho & \ell_2\ell_3(1-c\rho)-\ell_1 s\rho \\ \ell_3\ell_1(1-c\rho)-\ell_2 s\rho & \ell_3\ell_2(1-c\rho)+\ell_1 s\rho & \ell_3{}^2(1-c\rho)+c\rho \end{bmatrix} \quad \text{(A3.1)}$$

represents a rotation, α, of space through a turn angle of ρ about the unit vector $\mathbf{L} = \ell_1\mathbf{i} + \ell_2\mathbf{j} + \ell_3\mathbf{k}$ where $C = \{\mathbf{i},\mathbf{j},\mathbf{k}\}$ forms a cartesian basis, $c\rho = \cos\rho$ and $s\rho = \sin\rho$. The *general cartesian rotoinversion matrix*

$$M_C(\sigma) = \begin{bmatrix} \ell_1{}^2(c\rho-1)-c\rho & \ell_1\ell_2(c\rho-1)+\ell_3 s\rho & \ell_1\ell_3(c\rho-1)-\ell_2 s\rho \\ \ell_2\ell_1(c\rho-1)-\ell_3 s\rho & \ell_2{}^2(1-c\rho)-c\rho & \ell_2\ell_3(c\rho-1)+\ell_1 s\rho \\ \ell_3\ell_1(c\rho-1)+\ell_2 s\rho & \ell_3\ell_2(c\rho-1)-\ell_1 s\rho & \ell_3{}^2(c\rho-1)-c\rho \end{bmatrix} \quad \text{(A3.2)}$$

represents a rotoinversion, σ, of space with the same turn angle ρ and about the same vector $\mathbf{L}$. Note that $M_C(\sigma)$ can be obtained by simply multiplying (A3.1) by the inversion matrix

$$M_C(i) = \begin{bmatrix} -1 & 0 & 0 \\ 0 & -1 & 0 \\ 0 & 0 & -1 \end{bmatrix}$$

which is tantamount to multiplying each entry of $M_C(\alpha)$ by (-1).

(PA3.1) Problem: Show that the general cartesian rotation matrix becomes

$$M_C(\alpha) = \begin{bmatrix} 0 & 0 & 1 \\ 1 & 0 & 0 \\ 0 & 1 & 0 \end{bmatrix}$$

when α represents a third turn about the unit vector $\mathbf{L} = \sqrt{3}/3(\mathbf{i} + \mathbf{j} + \mathbf{k})$, i.e., $\ell_1 = \ell_2 = \ell_3 = \sqrt{3}/3$ and $\rho = 120°$. It is important to note that $\mathbf{L}$ must be a vector of unit length in order to apply the general cartesian rotation matrix. Since we are dealing with a rotation, the form given in (A3.1) should be used to compute the matrix.

(PA3.2) Problem: Show that the general cartesian rotoinversion matrix becomes

$$M_C(\sigma) = \begin{bmatrix} 0 & -1 & 0 \\ -1 & 0 & 0 \\ 0 & 0 & 1 \end{bmatrix}$$

when σ represents a half-turn rotoinversion about $\mathbf{L} = \sqrt{2}/2(\mathbf{i} + \mathbf{j})$. As a half-turn rotoinversion is equivalent to a reflection isometry, $M_C(\sigma)$ can also be viewed as representing a reflection of space over a plane (a mirror plane) perpendicular to $\mathbf{L}$. As we are calculating a matrix for a rotoinversion, the form given in (A3.2) should be used.

General Bases: In Chapter 3, we presented a method for constructing the matrix representation $M_D(\alpha)$ for a point isometry α. In the method, we first examined where the natural basis vectors $D = \{\mathbf{a},\mathbf{b},\mathbf{c}\}$ of the system are mapped by α, and then we formed $M_D(\alpha)$ by setting its three columns equal to the coordinate vectors of $\alpha(\mathbf{a})$, $\alpha(\mathbf{b})$ and $\alpha(\mathbf{c})$. We now study a method for constructing $M_D(\alpha)$ that does not require such specific information about $\alpha(\mathbf{a})$, $\alpha(\mathbf{b})$ and $\alpha(\mathbf{c})$. The idea is to translate all of the information about α into a cartesian basis C, find $M_C(\alpha)$ and then transform $M_C(\alpha)$ into $M_D(\alpha)$.

Suppose a, b, c, α, β, γ are given for the basis D. Then using the technique discussed in Chapter 2, we establish a cartesian basis C such that the change of basis matrix from D to C is (see equation (2.31))

$$A = \begin{bmatrix} a\sin\beta & -b\sin\alpha\cos\gamma^* & 0 \\ 0 & b\sin\alpha\sin\gamma^* & 0 \\ a\cos\beta & b\cos\alpha & c \end{bmatrix} \tag{A3.3}$$

Suppose we have a vector $\mathbf{r}$ along the rotation axis associated with α and we know $[\mathbf{r}]_D$. Then, since $A[\mathbf{r}]_D = [\mathbf{r}]_C$, we can find the direction cosines of the rotation axis with respect to C. Using (A3.1) or (A3.2), we can then obtain $M_C(\alpha)$. We then obtain $M_D(\alpha)$ from the following circuit diagram:

$$\begin{array}{ccccc} & [\mathbf{r}]_D & \xrightarrow{\quad A \quad} & [\mathbf{r}]_C & \\ M_D(\alpha) & \downarrow & & \downarrow & M_C(\alpha) \\ & [\alpha(\mathbf{r})]_D & \xrightarrow{\quad A \quad} & [\alpha(\mathbf{r})]_C & \end{array} \tag{A3.4}$$

which yields $M_D(\alpha) = A^{-1}M_C(\alpha)A$. This technique is illustrated in the next example.

(EA3.1) Example - Algebraic determination of the matrix representation of a point isometry: Spinel, $MgAl_2O_4$, which has a face-centered cubic structure can be described in terms of a structure with $a = b = c = 5.730$ A and $\alpha = \beta = \gamma = 60°$. As an example, we shall compute the matrix entries of $M({}^{[1\bar{1}1]}4)$ for a quarter-turn about the line $\mathbf{a} - \mathbf{b} + \mathbf{c}$.

Solution: Using (A3.3), we have

$$A = \begin{bmatrix} 4.962326 & 1.654109 & 0.0 \\ 0.0 & 4.678525 & 0.0 \\ 2.865 & 2.865 & 5.730 \end{bmatrix}. \tag{A3.5}$$

Since $A[\mathbf{r}]_D = [\mathbf{r}]_C$, we can calculate the components of a vector directed along the rotation axis using (A3.5) as follows:

$$\begin{bmatrix} 4.962326 & 1.654109 & 0.0 \\ 0.0 & 4.678525 & 0.0 \\ 2.865 & 2.865 & 5.730 \end{bmatrix} \begin{bmatrix} 1 \\ -1 \\ 1 \end{bmatrix} = \begin{bmatrix} 3.308217 \\ -4.678525 \\ 5.730 \end{bmatrix} .$$

Dividing this vector by its length, we obtain the direction cosines $\ell_1 = 0.40825$, $\ell_2 = -0.57735$ and $\ell_3 = 0.70711$ of the rotation axis defined in terms of C. Substituting these values along with $\rho = 90°$ into (A3.1), we get

$$M_C(\mathbf{g}) = \begin{bmatrix} 0.166667 & -0.942809 & -0.288675 \\ 0.471405 & 0.333333 & -0.816497 \\ 0.866025 & 0.0 & 0.5 \end{bmatrix}$$

where $\mathbf{g}$ is the quarter-turn $^{[1\bar{1}1]}4$. Since $M_D(\mathbf{g}) = A^{-1}M_C(\mathbf{g})A$, and since

$$A^{-1} = \begin{bmatrix} 0.201518 & -0.071248 & 0.0 \\ 0.0 & 0.213743 & 0.0 \\ -0.100759 & -0.071248 & 0.174520 \end{bmatrix} ,$$

$$M_D(\mathbf{g}) = M_D(^{[1\bar{1}1]}4) = \begin{bmatrix} 0 & -1 & 0 \\ 0 & 0 & -1 \\ 1 & 1 & 1 \end{bmatrix} .$$

□

INTERPRETATION OF MATRICES REPRESENTING POINT ISOMETRIES

Cartesian Bases: In the derivation of point groups, we are often confronted with the problem of determining the properties of point isometries. For example, suppose that we are given two matrices $M_C(\theta)$ and $M_C(\phi)$ representing two point isometries θ and ϕ with respect to a cartesian basis C and suppose that their product

$$M_C(\theta)\, M_C(\phi) = M_C(\omega)$$

represents some point isometry $\omega = \theta\phi$. From the matrix of ω, we wish to determine the properties of ω. That is, we need rules that can be applied to the matrix to discover (1) whether ω is a rotation or rotoinversion,

(2) the turn angle ρ and (3) the direction cosines ℓ_1, ℓ_2 and ℓ_3 of a unit vector **L** in the positive direction along the axis of ω.

(RA3.2) Rules for interpreting a general cartesian matrix M representing some point isometry, ω: Let ω denote some point isometry and ℓ_{ij} denote the (i,j) entry of its matrix representation M. Then

$$M = M_C(\omega) = \begin{bmatrix} \ell_{11} & \ell_{12} & \ell_{13} \\ \ell_{21} & \ell_{22} & \ell_{23} \\ \ell_{31} & \ell_{32} & \ell_{33} \end{bmatrix}.$$

The properties of ω are determined by the following rules:

(1) Calculate the determinant of M. If det(M) = +1, ω is a rotation. If det(M) = -1, ω is a rotoinversion.

(2) If det(M) = -1, multiply each entry of M by -1 and then go to (3), i.e., proceed by analyzing the properties of the rotation part of the rotoinversion.

(3) The turn angle ρ is given by

$$\rho = \cos^{-1}((\mathrm{tr}(M) - 1)/2)$$

where $\mathrm{tr}(M) = \ell_{11} + \ell_{22} + \ell_{33}$ (tr(M) is called the trace of M).

(4a) If $\rho = 180°$, go to step (4b). If not, the direction cosines ℓ_1, ℓ_2 and ℓ_3 are given by

$$\ell_1 = \frac{\ell_{32} - \ell_{23}}{2s_\rho}, \quad \ell_2 = \frac{\ell_{13} - \ell_{31}}{2s_\rho}, \quad \ell_3 = \frac{\ell_{21} - \ell_{12}}{2s_\rho}, \qquad (A3.6)$$

where the rotation axis of ω is along $\mathbf{L} = \ell_1\mathbf{i} + \ell_2\mathbf{j} + \ell_3\mathbf{k}$, ℓ_{ij} are the entries of M, and $0 < \rho < 180°$.

(4b) If $\rho = 180°$, then the direction cosines ℓ_1, ℓ_2, ℓ_3 are given by

$$\ell_1 = (\tfrac{1}{2}(\ell_{11}+1))^{\frac{1}{2}}, \quad \ell_2 = \pm(\tfrac{1}{2}(\ell_{22}+1))^{\frac{1}{2}}, \quad \ell_3 = \pm(\tfrac{1}{2}(\ell_{33}+1))^{\frac{1}{2}}. \qquad (A3.7)$$

When $\ell_1 \neq 0$, the sign of ℓ_2 is chosen to agree with that of ℓ_{21} and that of ℓ_3 to agree with ℓ_{31}. When $\ell_1 = 0$ and $\ell_2 \neq 0$, choose the positive sign for ℓ_2 and choose the sign of ℓ_3 to agree with ℓ_{23}.

If $\ell_1 = \ell_2 = 0$, then choose ℓ_3 to be positive.

In rule 2, each entry of M was multiplied by -1 when det(M) = -1. In this case, ω is the rotoinversion $i\beta$ where β is the rotation whose properties were determined in rules 3 and 4.

(PA3.3) Problem: Examine the matrix representing the composition $\omega = \theta\phi$ of the two point isometries θ and ϕ given that

$$M_C(\theta\phi) = M_C(\theta)M_C(\phi) = \begin{bmatrix} 0 & 0 & 1 \\ 1 & 0 & 0 \\ 0 & 1 & 0 \end{bmatrix} \begin{bmatrix} 0 & -1 & 0 \\ -1 & 0 & 0 \\ 0 & 0 & 1 \end{bmatrix} = \begin{bmatrix} 0 & 0 & 1 \\ 0 & -1 & 0 \\ -1 & 0 & 0 \end{bmatrix}$$

and show that ω is an inversion quarter-turn ($\rho = 90°$) about the unit vector $\mathbf{L} = -\mathbf{j}$.

(PA3.4) Problem: Suppose that the matrix representation of the point isometry α is

$$M_C(\alpha) = \begin{bmatrix} 0 & 0 & 1 \\ 0 & -1 & 0 \\ 1 & 0 & 0 \end{bmatrix}$$

Show that $\det(M(\alpha)_C) = +1$, that the turn angle of α is $\rho = 180°$ and that $\ell_1 = \sqrt{\tfrac{1}{2}}$, $\ell_2 = 0$ and $\ell_3 = \sqrt{\tfrac{1}{2}}$. Hence, α defines a half-turn rotation of space about the unit vector, $\mathbf{L} = \sqrt{2}/2(\mathbf{i} + \mathbf{k})$.

(PA3.5) Problem: Show that the trace of the general cartesian rotation matrix (A3.1) is $2c\rho + 1$. Note that since $\mathbf{L} = \ell_1\mathbf{i} + \ell_2\mathbf{j} + \ell_3\mathbf{k}$ is a unit vector written in terms of a cartesian basis, $\ell_1^2 + \ell_2^2 + \ell_3^2 = 1$.

(PA3.6) Problem: Consider the following equality:

$$\begin{bmatrix} \ell_{11} & \ell_{12} & \ell_{13} \\ \ell_{21} & \ell_{22} & \ell_{23} \\ \ell_{31} & \ell_{32} & \ell_{33} \end{bmatrix} = \begin{bmatrix} \ell_1^2(1-c\rho)+c\rho & \ell_1\ell_2(1-c\rho)-\ell_3 s\rho & \ell_1\ell_3(1-c\rho)+\ell_2 s\rho \\ \ell_2\ell_1(1-c\rho)+\ell_3 s\rho & \ell_2^2(1-c\rho)+c\rho & \ell_2\ell_3(1-c\rho)-\ell_1 s\rho \\ \ell_3\ell_1(1-c\rho)-\ell_2 s\rho & \ell_3\ell_2(1-c\rho)+\ell_1 s\rho & \ell_3^2(1-c\rho)+c\rho \end{bmatrix}.$$

Assuming that $\rho \neq 0°$ and $\rho \neq 180°$ (hence $s\rho \neq 0$), evaluate $\ell_{21} - \ell_{12}$ and solve for ℓ_3 obtaining the direction cosine

$$\ell_3 = \frac{\ell_{21} - \ell_{12}}{2s_\rho} .$$

Similarly evaluate the direction cosines ℓ_1 and ℓ_2 as in (A3.6).

(PA3.7) Problem: Show that when $\rho = 180°$, that (A3.1) becomes

$$\begin{bmatrix} \ell_{11} & \ell_{12} & \ell_{13} \\ \ell_{21} & \ell_{22} & \ell_{23} \\ \ell_{31} & \ell_{32} & \ell_{33} \end{bmatrix} = \begin{bmatrix} 2\ell_1^2-1 & 2\ell_1\ell_2 & 2\ell_1\ell_3 \\ 2\ell_2\ell_1 & 2\ell_2^2-1 & 2\ell_2\ell_3 \\ 2\ell_3\ell_1 & 2\ell_3\ell_2 & 2\ell_3^2-1 \end{bmatrix} .$$

Verify the equations given in (A3.7).

General bases: Before we can present rules for interpreting α from $M_D(\alpha)$, we need to establish what the matrix representations of α with respect to different bases have in common.

Let α denote an isometry that fixes the origin. Then for each basis, D, a matrix $M_D(\alpha)$ representing α exists. Since there are an infinite number of bases, there are an infinite number of distinct matrices representing α. However, the matrices in the class of all matrices representing α share certain important properties. For example, we shall show in this section that if D_1 and D_2 are two bases, then

$$\text{tr}(M_{D_1}(\alpha)) = \text{tr}(M_{D_2}(\alpha))$$

and

$$\det(M_{D_1}(\alpha)) = \det(M_{D_2}(\alpha)) .$$

Consider the following circuit diagram where D_1 and D_2 are bases and T is the change of basis matrix from D_1 and D_2.

$$\begin{array}{ccc} [v]_{D_1} & \xrightarrow{\ T\ } & [v]_{D_2} \\ M_{D_1}(\alpha)\ \downarrow & & \downarrow\ M_{D_2}(\alpha) \\ [\alpha(v)]_{D_1} & \xrightarrow{\ T\ } & [\alpha(v)]_{D_2} \end{array}$$

From the circuit diagram we see that

$$M_{D_2}(\alpha) = TM_{D_1}(\alpha)T^{-1} .$$

(DA3.3) Definition: If A and B are n × n matrices such that there exists an invertible matrix T such that

$$A = TBT^{-1} ,$$

then we say A is similar to B.

By the above discussion we have the following theorem.

(TA3.4) Theorem: If D_1 and D_2 are bases and α is an isometry, then $M_{D_2}(\alpha)$ is similar to $M_{D_1}(\alpha)$.

In fact, two matrices A and B represent the same isometry if and only if A is similar to B.

(TA3.5) Theorem: If matrix A is similar to matrix B, then tr(A) = tr(B).

Proof: Since A is similar to B, there exists an invertible matrix T such that

$$A = TBT^{-1} .$$

Hence tr(A) and $tr(TBT^{-1})$ are equal. By the result proved later in Lemma LA3.8, tr(MN) = tr(NM) for any two matrices M and N. Therefore,

$$\begin{aligned} tr(TBT^{-1}) &= tr((TB)T^{-1}) \\ &= tr((BT)T^{-1}) \\ &= tr(B(TT^{-1})) \\ &= tr(B) . \end{aligned}$$

Consequently, tr(A) = tr(B). □

(TA3.6) Theorem: If α is a point isometry with turn angle ρ and D is a basis, then

(i) if α is a rotation, $tr(M_D(\alpha)) = 1 + 2\cos\rho$

or

(ii) if α is a rotoinversion, $tr(M_D(\alpha)) = -(1 + 2c\rho)$.

Proof: Assume α is a rotation and let C denote a cartesian basis.

Then $M_C(\alpha)$ is given by (A3.1). Evaluating the trace of this matrix, we have

$$\begin{aligned} \mathrm{tr}(M_C(\alpha)) &= \ell_1^2(1-c\rho) + c\rho + \ell_2^2(1-c\rho) + c\rho + \ell_3^2(1-c\rho) + c\rho \\ &= (\ell_1^2+\ell_2^2+\ell_3^2)(1-c\rho) + 3c\rho \\ &= (1-c\rho) + 3c\rho \qquad (\text{since } \ell_1^2+\ell_2^2+\ell_3^2 = 1) \\ &= 1 + 2c\rho \ . \end{aligned}$$

Let D denote any basis. By theorem TA3.4, $M_D(\alpha)$ is similar to $M_C(\alpha)$. By theorem TA3.5,

$$\begin{aligned} \mathrm{tr}(M_D(\alpha)) &= \mathrm{tr}(M_C(\alpha)) \\ &= 1 + 2\cos\rho \ . \end{aligned}$$

□

(PA3.8) Problem: Show that part (ii) of theorem TA3.6 is true.

(TA3.7) Theorem: If A is similar to B, then

$$\det(A) = \det(B) \ .$$

Proof: Since A and B are similar, there exists an invertible matrix T such that

$$A = TBT^{-1} \ .$$

In Appendix 2, we showed that $\det(MN) = \det(M)\det(N)$ for matrices M and N and that $\det(M^{-1}) = 1/\det(M)$ for an invertible matrix M. Hence

$$\begin{aligned} \det(A) &= \det(TBT^{-1}) \\ &= \det(T)\det(B)\det(T^{-1}) \\ &= \det(B)\det(T)\det(T^{-1}) \\ &= \det(B) \ . \end{aligned}$$

□

(PA3.9) Problem: Use theorem TA3.7 to show that if D_1 and D_2 are bases and α is an isometry leaving the origin fixed, then $\det(M_{D_1}(\alpha)) = \det(M_{D_2}(\alpha))$.

(CA3.8) Corollary: If M is the matrix representation of a point isometry α, then α is a rotation if and only if $\det(M) = +1$ and α is a rotoinversion if and only if $\det(M) = -1$.

(PA3.10) Problem: Prove CA3.8.

Given the matrix representation of a point isometry α, we can determine its turn angle by TA3.6 and whether it is a rotation or rotoinversion by CA3.8. What remains is to find the axis associated with α and to determine which is its positive end. Note that if $\mathbf{r}$ is along the axis of α, then $\alpha(\mathbf{r}) = \mathbf{r}$. In matrix terms, we have

$$[M_D(\alpha)][\mathbf{r}]_D = [\mathbf{r}]_D \tag{A3.8}$$

Since $I_3[\mathbf{r}]_D = [\mathbf{r}]_D$, we can substitute this result into (A3.8) obtaining

$$[M_D(\alpha)][\mathbf{r}]_D = I_3[\mathbf{r}]_D .$$

Then

$$([M_D(\alpha)] - I_3)[\mathbf{r}]_D = [\mathbf{0}]_D \tag{A3.9}$$

where $[\mathbf{r}]_D{}^t = [\ell_1 \ell_2 \ell_3]$ are the components of a vector along the rotation axis of α. The equations given in (A3.9) form a system of linear equations that can be solved using techniques presented in Appendix 2. In linear algebra, this procedure is called finding the eigenvectors of $M(\alpha)_D$ associated with the eigenvalue 1.

Now that we can find the axis, we need to determine its positive end. Without loss of generality, we can do this for rotations since if α is a rotoinversion, $i\alpha$ is a rotation with the same axis. Hence, if $M_D(\alpha)$ has determinant -1, we multiply each of the elements of $M_D(\alpha)$ by -1 and then apply the following theorem.

(TA3.9) Theorem: Let α denote a rotation with turn angle $0° < \rho < 180°$ and let D denote a basis (as usual, right handed) for which we are given

$$M_D(\alpha) = \begin{bmatrix} \ell_{11} & \ell_{12} & \ell_{13} \\ \ell_{21} & \ell_{22} & \ell_{23} \\ \ell_{31} & \ell_{32} & \ell_{33} \end{bmatrix} .$$

Let $\mathbf{r}$ denote a nonzero vector along the axis ℓ of α for which we are given u,v,w such that $[\mathbf{r}]_D = [uvw]^t$. Then $\mathbf{r}$ is directed in the positive direction of ℓ (with the choice $0° < \rho < 180°$) if and only if in the case

(a) when $v = w = 0$,

$$\begin{vmatrix} u & \ell_{12} \\ w & \ell_{32} \end{vmatrix} > 0 ,$$

or (b) otherwise

$$\begin{vmatrix} \ell_{21} & v \\ \ell_{31} & w \end{vmatrix} > 0 .$$

The proof of TA3.9 is given at the end of this Appendix. We now present an example that will illustrate our approach.

(EA3.10) Example - Determination of the properties of a point isometry: Determine the properties of the point isometry α given that its matrix representation

$$M_D(\alpha) = \begin{bmatrix} 0 & 0 & -1 \\ -1 & 0 & 0 \\ 1 & 1 & 1 \end{bmatrix}$$

is defined in terms of some general bases $D = \{\mathbf{a},\mathbf{b},\mathbf{c}\}$.

Solution: In determining the properties of α, we begin by evaluating the determinant of the matrix. As

$$\det M_D(\alpha) = (-1)^{1+3}(-1) \begin{vmatrix} -1 & 0 \\ 1 & 1 \end{vmatrix} = +1 ,$$

we conclude that α is a rotation isometry. Next, the trace of $M_D(\alpha)$ = 1 is equated with $2c\rho + 1$ from which it follows that $c\rho = 0$. Thus, the turn angle is taken to be 90° and α is taken to be a quarter turn, 4. The rotation axis of the quarter turn is provided by replacing $M_D(\alpha)$ in (A3.8) by $M_D(4)$ and solving for the vectors along the axis $[\ell_1,\ell_2\ ,\ell_3]^t$:

$$\left(\begin{bmatrix} 0 & 0 & -1 \\ -1 & 0 & 0 \\ 1 & 1 & 1 \end{bmatrix} - \begin{bmatrix} 1 & 0 & 0 \\ 0 & 1 & 0 \\ 0 & 0 & 1 \end{bmatrix} \right) \begin{bmatrix} \ell_1 \\ \ell_2 \\ \ell_3 \end{bmatrix} = \begin{bmatrix} 0 \\ 0 \\ 0 \end{bmatrix}$$

$$\begin{bmatrix} -1 & 0 & -1 \\ -1 & -1 & 0 \\ 1 & 1 & 0 \end{bmatrix} \begin{bmatrix} \ell_1 \\ \ell_2 \\ \ell_3 \end{bmatrix} = \begin{bmatrix} 0 \\ 0 \\ 0 \end{bmatrix} .$$

The augmented matrix for this system of equations is

$$\left[\begin{array}{ccc|c} -1 & 0 & -1 & 0 \\ -1 & -1 & 0 & 0 \\ 1 & 1 & 0 & 0 \end{array}\right] .$$

When this matrix is transformed into reduced row echelon form by elementary row operations, it becomes

$$\left[\begin{array}{ccc|c} 1 & 0 & 1 & 0 \\ 0 & 1 & -1 & 0 \\ 0 & 0 & 0 & 0 \end{array}\right] .$$

In this example, since ℓ_3 can take on any value, we equate ℓ_3 with some scalar x and then solve for y and z: $\ell_1 = -x$, $\ell_2 = x$ and $\ell_3 = x$. Hence, the rotation axis of α is

$$\ell = \{-x\mathbf{a} + x\mathbf{b} + x\mathbf{c} | x \in R\} = \{x(-\mathbf{a} + \mathbf{b} + \mathbf{c}) | x \in R\} = \{x\mathbf{r} | x \in R\}$$

where $\mathbf{r} = -\mathbf{a} + \mathbf{b} + \mathbf{c}$. About one end of ℓ, α is a quarter turn and about the other it is a negative quarter turn. We now wish to determine whether $\mathbf{r}$ is directed toward the end of ℓ about which α is a quarter turn. This is accomplished by applying the rules described in theorem TA3.9. Because $\mathbf{r} = -\mathbf{a} + \mathbf{b} + \mathbf{c}$, $[\mathbf{r}]_D{}^t = [-111]^t = [uvw]^t$, $v \neq 0$, we apply rule (a) of the theorem and observe that

$$\begin{vmatrix} \ell_{21} & v \\ \ell_{31} & w \end{vmatrix} = \begin{vmatrix} -1 & 1 \\ 1 & 1 \end{vmatrix} = -2 < 0 .$$

Hence, $\mathbf{r}$ is not directed in the positive direction when α is taken to be a quarter turn. However, α is a quarter turn about ℓ when the positive direction of the axis is taken to be $\mathbf{t} = \mathbf{a} - \mathbf{b} - \mathbf{c}$. Hence, $\alpha = [1\bar{1}\bar{1}]_4$.

□

(PA3.11) Problem: Suppose for the spinel crystal described in EA3.1 that $M_D(\beta)$ represents a quarter-turn rotation about the vector $\mathbf{a} + \mathbf{b} - \mathbf{c}$ and $M_D(\alpha)$ represents a quarter-turn rotation about $\mathbf{a} - \mathbf{b} + \mathbf{c}$. Show that $\gamma = \beta\alpha$, represented by the product

$$M_D(\gamma) = M_D(\beta)\, M_D(\alpha) = \begin{bmatrix} 0 & 0 & -1 \\ 1 & 1 & 1 \\ 0 & -1 & 0 \end{bmatrix} \begin{bmatrix} 0 & -1 & 0 \\ 0 & 0 & -1 \\ 1 & 1 & 1 \end{bmatrix}$$

$$= \begin{bmatrix} -1 & -1 & -1 \\ 1 & 0 & 0 \\ 0 & 0 & 1 \end{bmatrix},$$

is $[\bar{1}\bar{1}1]_3$.

(LA3.11) Lemma: Let A and B denote $n \times n$ matrices. Then

$$\mathrm{tr}(AB) = \mathrm{tr}(BA)\ .$$

Proof: Since

$$(i,i) \text{ term of } AB = \sum_{j=1}^{n} a_{ij} b_{ji}$$

and since the trace of AB is the sum of the (i,i) terms,

$$\begin{aligned} \mathrm{tr}(AB) &= \sum_{i=1}^{n} \left(\sum_{j=1}^{n} a_{ij} b_{ji} \right) \\ &= \sum_{j=1}^{n} \left(\sum_{i=1}^{n} a_{ij} b_{ji} \right) \quad \text{(by switching the summations)} \\ &= \sum_{j=1}^{n} \left(\sum_{i=1}^{n} b_{ji} a_{ij} \right) \\ &= \sum_{j=1}^{n} ((j,j) \text{ term of } BA) \\ &= \mathrm{tr}(BA)\ . \end{aligned}$$

□

PROOFS OF (A3.1), (A3.2) AND TA3 .9

Proof of (A3.1) and (A3.2): To facilitate the derivation, we shall use two cartesian bases $C_1 = \{\mathbf{i}_1, \mathbf{j}_1, \mathbf{k}_1\}$ and $C_2 = \{\mathbf{i}_2, \mathbf{j}_2, \mathbf{k}_2\}$. Suppose that (A3.1) and (A3.2) are both defined in terms of the C_1 basis (see Figure A3.1(a) where the unit vector $\mathbf{L}$ is shown about which the rotation α takes place). A second cartesian basis C_2 is illustrated in Figure A3.1(b) with $\mathbf{k}_2 = \mathbf{L}$. The matrix representation of α with respect to C_2 can be easily calculated from the information in Figure A3.1(c) to be

$$M_{C_2}(\alpha) = \begin{bmatrix} c\rho & -s\rho & 0 \\ s\rho & c\rho & 0 \\ 0 & 0 & 1 \end{bmatrix} .$$

Let A denote a change of basis matrix such that $A[\mathbf{v}]_{C_1} = [\mathbf{v}]_{C_2}$. In deriving the general cartesian rotation matrix, we consider the following circuit diagram:

$$\begin{array}{ccc} [\mathbf{v}]_{C_1} & \xrightarrow{\quad A \quad} & [\mathbf{v}]_{C_2} \\ M_{C_1}(\alpha) \Big\downarrow & & \Big\downarrow M_{C_2}(\alpha) \\ [\alpha(\mathbf{v})]_{C_1} & \xrightarrow{\quad A \quad} & [\alpha(\mathbf{v})]_{C_2} \end{array}$$

By inspecting the diagram, we see that $M_{C_1}(\alpha) = A^{-1}M_{C_2}(\alpha)A$. Since C_1 and C_2 are cartesian bases, $C_1^* = C_1$ and $C_2^* = C_2$ and so by (2.22) $A = A^{-t}$. Hence $A^{-1} = A^t$. Upon expansion, these matrices become

$$M_{C_1}(\alpha) = \begin{bmatrix} \ell_{11} & \ell_{12} & \ell_{13} \\ \ell_{21} & \ell_{22} & \ell_{23} \\ \ell_{31} & \ell_{32} & \ell_{33} \end{bmatrix}$$

$$= \begin{bmatrix} a_{11} & a_{21} & a_{31} \\ a_{12} & a_{22} & a_{32} \\ a_{13} & a_{23} & a_{33} \end{bmatrix} \begin{bmatrix} c\rho & -s\rho & 0 \\ s\rho & c\rho & 0 \\ 0 & 0 & 1 \end{bmatrix} \begin{bmatrix} a_{11} & a_{12} & a_{13} \\ a_{21} & a_{22} & a_{23} \\ a_{31} & a_{32} & a_{33} \end{bmatrix} \qquad \text{(A3.10)}$$

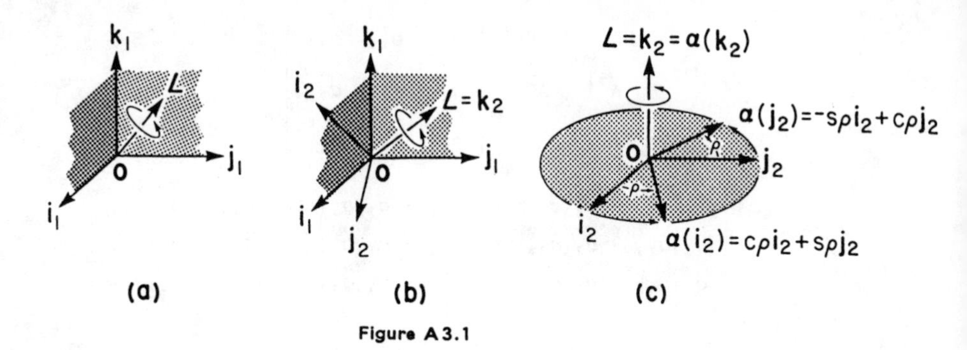

Figure A3.1

where ℓ_{ij} is the (i,j)-entry of $M_{C_1}(\alpha)$. When the three matrices on the right are multiplied together, we get nine linear equations which we equate with the nine entries of $M_{C_1}(\alpha)$.

Since A is the change of basis matrix from C_1 to C_2, the first column of A is $[\mathbf{i}_1]_{C_2}$. Hence, $a_{31} = \mathbf{i}_1 \bullet \mathbf{k}_2$, the direction cosine of the angle between $\mathbf{i}_1$ and $\mathbf{k}_2 = \mathbf{L}$. That is, $a_{31} = \ell_1$. Similarly, as the second and third columns of A are $[\mathbf{j}_1]_{C_2}$ and $[\mathbf{k}_1]_{C_2}$, respectively, $a_{32} = \ell_2$ and $a_{33} = \ell_3$. We shall use this information to determine the entries of $M_{C_1}(\alpha)$ in terms of $s\rho$, $c\rho$, ℓ_1, ℓ_2, and ℓ_3. From (A3.10), we have

$$\begin{aligned}\ell_{11} &= a_{11}^2 c\rho + a_{21}^2 c\rho + a_{31}^2 \\ &= (a_{11}^2 + a_{21}^2) c\rho + a_{31}^2 \ .\end{aligned}$$

Since the (1,1)-entry of A^tA is $a_{11}^2 + a_{21}^2 + a_{31}^2$ and since $A^tA = I_3$, we get the equality $a_{11}^2 + a_{21}^2 + a_{31}^2 = 1$. Hence, $a_{11}^2 + a_{21}^2 = 1 - a_{31}^2$. Because $a_{31} = \ell_1$, we have

$$\ell_{11} = \ell_1^2(1 - c\rho) + c\rho \ .$$

By similar arguments,

$$\ell_{22} = \ell_2^2(1 - c\rho) + c\rho \ ,$$

$$\ell_{33} = \ell_3^2(1 - c\rho) + c\rho \ .$$

The strategy used to derive the off-diagonal entries of (A3.1) requires a knowledge of several other properties of A. As the two cartesian basis sets C_1 and C_2 are both right-handed, both outline a unit volume $v = \det(A) = \det(A^t) = 1$. Using the cofactor method, A^{-1} is

$$A^{-1} = A^{t} = \begin{bmatrix} a_{11} & a_{21} & a_{31} \\ a_{12} & a_{22} & a_{32} \\ a_{13} & a_{23} & a_{33} \end{bmatrix}$$

$$= \begin{bmatrix} a_{22}a_{33} - a_{23}a_{32} & a_{13}a_{32} - a_{12}a_{33} & a_{12}a_{23} - a_{22}a_{13} \\ a_{31}a_{23} - a_{21}a_{33} & a_{11}a_{33} - a_{31}a_{13} & a_{21}a_{13} - a_{11}a_{23} \\ a_{21}a_{32} - a_{31}a_{22} & a_{31}a_{12} - a_{11}a_{32} & a_{11}a_{22} - a_{12}a_{21} \end{bmatrix} .$$

From this matrix equality, we obtain the following three identities:

$$\begin{aligned} a_{31} &= \ell_1 = a_{12}a_{23} - a_{22}a_{13} , \\ a_{32} &= \ell_2 = a_{21}a_{13} - a_{11}a_{23} , \\ a_{33} &= \ell_3 = a_{11}a_{22} - a_{12}a_{21} . \end{aligned} \tag{A3.11}$$

Returning to (A3.10) and recalling that, since $A^{t}A = I_3$, $a_{11}a_{12} + a_{21}a_{22} = -a_{31}a_{32}$,

$$\begin{aligned} \ell_{12} &= (a_{11}a_{12} + a_{21}a_{22})c\rho + (a_{21}a_{12} - a_{11}a_{22})s\rho + a_{31}a_{32} \\ &= (-a_{31}a_{32})c\rho + (-\ell_3)s\rho + a_{31}a_{32} \\ &= (-\ell_1\ell_2)c\rho - \ell_3 s\rho + \ell_1\ell_2 \\ &= \ell_1\ell_2(1 - c\rho) - \ell_3 s\rho . \end{aligned}$$

Similarly, the remaining ℓ_{ij} entries can be found using (A3.10), (A3.11) and the equality $A^{t}A = I_3$. The final matrix is shown in (A3.1).

Because a rotoinversion is the composition of an inversion and a rotation, it follows, as observed earlier, that (A3.2) can be obtained from (A3.1) by multiplying each of its entries by -1. □

Discussion and proof of theorem TA3.9: Let α denote a rotation with turn angle ρ about an axis ℓ. From one end of the axis, α is viewed as a rotation of ρ degrees (where a positive angle viewed down the positive end of ℓ is counterclockwise). From the other end, α is viewed as a rotation of 360-ρ°. Hence, if the orientation of the axis can be determined for angles $0 \le \rho \le 180°$, then the rotation is completely determined. Note that if ρ = 0°, then α is the identity map and so it makes no sense to "orient" the axis. If ρ = 180°, then α is a half turn about either end of ℓ and so there is no preferred positive end of the axis. Hence we shall need a procedure for orienting the axis only when $0° < \rho < 180°$. Suppose we have $\ell = \{x\mathbf{r} | x \in R\}$ and we wish to determine whether $\mathbf{r}$ is directed toward the positive end of ℓ or not. If $\mathbf{r}$ is so directed and if $\mathbf{t}$ is any vector not along ℓ, then, since $0° < \rho < 180°$, $\{\mathbf{t}, \alpha(\mathbf{t}), \mathbf{r}\}$ form a right-handed system. Suppose we know $M_D(\alpha)$ with respect to some right-handed coordinate system with basis $D = \{\mathbf{a},\mathbf{b},\mathbf{c}\}$. Then the determinant of the matrix that changes the basis from $\{\mathbf{t}, \alpha(\mathbf{t}), \mathbf{r}\}$ to D must be positive. Since the change of basis matrix is

$$T = \begin{bmatrix} [\mathbf{t}]_D & [\alpha(\mathbf{t})]_D & [\mathbf{r}]_D \end{bmatrix}$$

we have the result that $\mathbf{r}$ is in the positive direction along ℓ if and only $\det(T) > 0$. By judiciously choosing $\mathbf{t}$ we can simplify the calculation of $\det(T)$. If $\mathbf{a}$ is not along ℓ, let $\mathbf{t} = \mathbf{a}$. Then

$$T = \begin{bmatrix} 1 & \ell_{11} & u \\ 0 & \ell_{21} & v \\ 0 & \ell_{31} & w \end{bmatrix},$$

where $[\ell_{11},\ell_{21},\ell_{31}]^t$ is the first column of $M_D(\alpha)$ (hence $[\alpha(\mathbf{a})]_D$) and $[uvw]^t = [\mathbf{r}]_D$. Then

$$\det(T) = \begin{vmatrix} \ell_{21} & v \\ \ell_{31} & w \end{vmatrix}.$$

If $\mathbf{a}$ is directed along ℓ, then $v = w = 0$. In this case $\mathbf{b}$ cannot be along the axis because α is not the identity, hence at least one of ℓ_{12} or ℓ_{32} is nonzero. Letting $\mathbf{t} = \mathbf{b}$, we have

$$T = \begin{bmatrix} 0 & \ell_{12} & u \\ 1 & \ell_{22} & v \\ 0 & \ell_{32} & w \end{bmatrix}$$

where $[\ell_{12},\ell_{22},\ell_{32}]^t$ is the second column of $M_D(\alpha)$. Hence

$$\det(T) = -\begin{vmatrix} \ell_{12} & u \\ \ell_{32} & w \end{vmatrix} = \begin{vmatrix} u & \ell_{12} \\ w & \ell_{32} \end{vmatrix} .$$

□

APPENDIX 4

POTPOURRI

HANDEDNESS OF BASES

Let $D = \{\mathbf{u},\mathbf{v},\mathbf{w}\}$ denote a basis of S. A geometrical definition for the handedness of D is illustrated in Figure A4.1. If the basis is viewed so that the first two vectors, $\mathbf{u}$ and $\mathbf{v}$ in this case, are directed toward the viewer with $\mathbf{u}$ on the left and $\mathbf{v}$ on the right (Figure A4.1(a)), then

(1) if the third vector $\mathbf{w}$ is directed upward, D is right-handed (Figure A4.1(b)), or
(2) if $\mathbf{w}$ is directed downward, D is left-handed (Figure A4.1(c)).

Note that by referring to $\mathbf{u}$ as the first vector, $\mathbf{v}$ as the second and so forth, we are assuming that D is an ordered set. If, for example, we switched the order of $\mathbf{u}$ and $\mathbf{v}$, the handedness of the basis would change. However, by convention, the unordered set symbolism $\{\ldots\}$ is normally used to denote bases.

To use vector analysis to determine the handedness of D, we observe that $\{\mathbf{u},\mathbf{v},\mathbf{u} \times \mathbf{v}\}$ always forms a right-handed system. Hence the question becomes "Is $\mathbf{w}$ on the same side of the plane of $\mathbf{u}$ and $\mathbf{v}$ as is $\mathbf{u} \times \mathbf{v}$ or not?" If it is, then D is right-handed; if not, then D is left-handed. If $\mathbf{w}$ is on the same side of the plane as $\mathbf{u} \times \mathbf{v}$ then the projection of $\mathbf{w}$ onto $\mathbf{u} \times \mathbf{v}$ will be positive (Figure A4.2(a)); otherwise, it will be negative (Figure A4.2(b)). The projection of $\mathbf{w}$ onto $\mathbf{u} \times \mathbf{v}$ is

$$\frac{\mathbf{w} \cdot (\mathbf{u} \times \mathbf{v})}{\|\mathbf{u} \times \mathbf{v}\|}$$

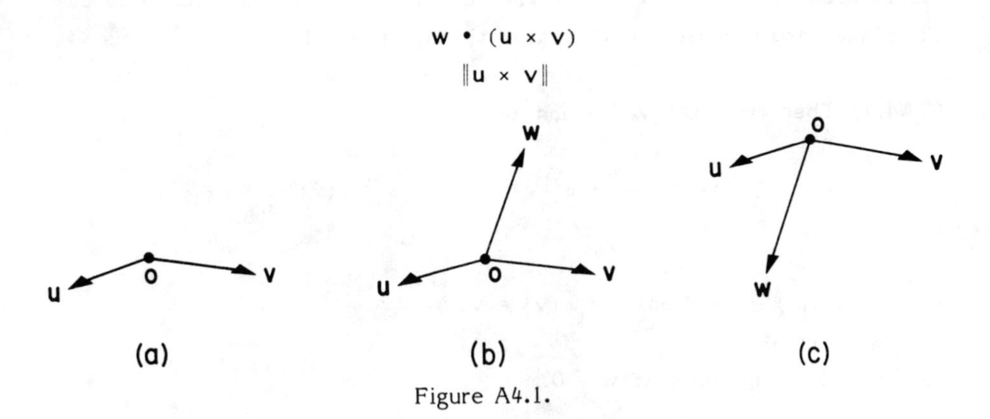

Figure A4.1.

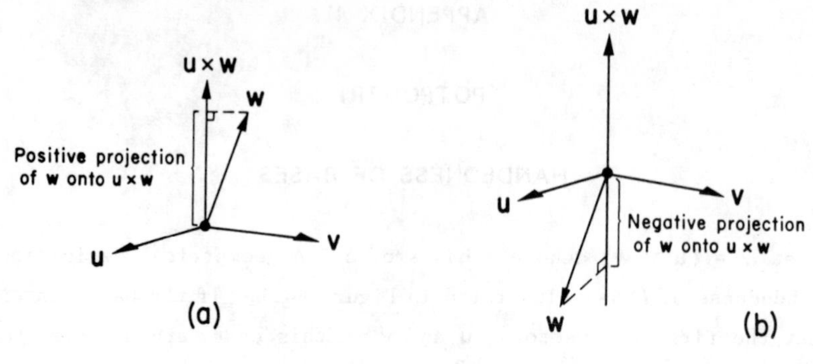

Figure A4.2.

as was shown in Chapter 1. Since $\|\mathbf{u} \times \mathbf{v}\| > 0$, we have the following rules:

(1) If $\mathbf{w} \cdot (\mathbf{u} \times \mathbf{v}) > 0$, then D is right-handed.

(2) If $\mathbf{w} \cdot (\mathbf{u} \times \mathbf{v}) < 0$, then D is left-handed.

Note that in Chapter 1, we showed that the volume of the parallelepiped outlined by the basis vectors $D = \{\mathbf{a},\mathbf{b},\mathbf{c}\}$ is $\mathbf{a} \cdot (\mathbf{b} \times \mathbf{c})$. This would have been incorrect if D had been a left-handed basis since, in that case, $\mathbf{a} \cdot (\mathbf{b} \times \mathbf{c}) < 0$. However, throughout the book our bases, unless otherwise indicated, are assumed to be right-handed.

DISCUSSION AND PROOF OF T6.15

Let L denote a lattice left invariant under a proper rotation α of finite order $o(\alpha)$. Let ℓ denote the rotation axis of α and let Q denote the plane perpendicular to ℓ passing through the origin.

(TA4.1) Theorem: Let $\mathbf{v} \varepsilon S$ and let

$$\mathbf{w} = \alpha(\mathbf{v}) + \alpha^2(\mathbf{v}) + \ldots + \alpha^{o(\alpha)}(\mathbf{v}).$$

Then

(1) $\mathbf{v}$ is on ℓ if and only if $\alpha(\mathbf{v}) = \mathbf{v}$,

(2) $\mathbf{w}$ is on ℓ,

(3) $\mathbf{v} \varepsilon Q$ if and only if $\mathbf{w} = \mathbf{0}$.

The proof of (1) follows directly from the definition of a rotation axis.

(PA4.1) Problem: Establish part (2) of TA4.1 by applying part (1) to **w**.

(PA4.2) Problem: Establish part (3) to TA4.1 by observing that if d is the distance from Q to **v**, then d is the distance from Q to $\alpha^i(\mathbf{v})$ for all i, and so the distance from Q to **w** is $do(\alpha)$. □

The proof of T6.15 will be conducted through the following discussion and problems.

Since L is a three-dimensional lattice, there must exist a vector $\mathbf{v} \in L$ that is not in Q. By TA4.1 part (2), $\mathbf{w} = \alpha(\mathbf{v}) + \ldots + \alpha^{o(\alpha)}(\mathbf{v})$ is on ℓ. By part (3), $\mathbf{w} \neq 0$. Furthermore, since L is left invariant under α, $\alpha^i(\mathbf{v})$ is in L for all i and so $\mathbf{w} \in L$. Hence, we have found a nonzero vector **w** in L on ℓ as required.

The fact that there is a two-dimensional lattice plane perpendicular to ℓ passing through the origin (that is, in Q) will now be established. Since L is three dimensional, nonzero lattice vectors $\mathbf{v}_1$ and $\mathbf{v}_2$ can be found such that $\mathbf{v}_1$ is not along ℓ and such that $\mathbf{v}_2$ is not in the plane of ℓ and $\mathbf{v}_1$. Let $\mathbf{u}_1 = \mathbf{v}_1 - \alpha(\mathbf{v}_1)$ and $\mathbf{u}_2 = \mathbf{v}_2 - \alpha(\mathbf{v}_2)$.

(PA4.3) Problem: Suppose α is a half-turn. Prove each of the following statements:

(1) $\mathbf{u}_1 \in Q$ (use part (3) of TA4.1).

(2) $\mathbf{u}_1$ is not on ℓ (assume it is, apply part (1) of TA4.1, show that this implies that $\mathbf{v}_1$ is along ℓ, which is a contradiction).

(3) Conclude that $\mathbf{u}_1$ and $\mathbf{u}_2$ are nonzero vectors in Q.

(4) Show that $\mathbf{u}_1$ and $\mathbf{u}_2$ are not collinear (assume $\mathbf{u}_1 = s\mathbf{u}_2$ for some $s \in R$, show that $\mathbf{v}_1 - s\mathbf{v}_2 = \alpha(\mathbf{v}_1 - s\mathbf{v}_2)$ and so $\mathbf{v}_1 - s\mathbf{v}_2$ is along ℓ. This implies that $\mathbf{v}_2$ is in the plane of $\mathbf{v}_1$ and ℓ, a contradiction).

(5) Conclude that there is a lattice plane of L lying in Q.

(PA4.4) Problem: Repeat PA4.3 where α is a third-turn. Note that in the second step of PA4.3, you may also need to appeal to part (2) of TA4.1.

APPENDIX 5

SOME PROPERTIES OF LATTICE PLANES

All of the equations and vectors in this appendix are written in terms of a basis $D = \{\mathbf{a},\mathbf{b},\mathbf{c}\}$ for S.

(TA5.1) Theorem: Let P denote a lattice plane. Then there exists integers h, k, ℓ, m such that

$$hx + ky + \ell z = m$$

is the equation for P.

Proof: According to the material in Chapter 2, an equation for P is of the form

$$t_1 x + t_2 y + t_3 z = w$$

where $t_1, t_2, t_3, w \in R$. If $w \neq 0$, then we have

$$(t_1/w)x + (t_2/w)y + (t_3/w)z = 1 \ .$$

Setting $v_i = t_i/w$ for the case where $w \neq 0$ and $v_i = t_i$ for the case where $w = 0$, this equation can be written as

$$v_1 x + v_2 y + v_3 z = w \tag{A5.1}$$

where $w = 1$ or 0. Since P is a lattice plane, there must exist three lattice vectors $\mathbf{p} = p_1\mathbf{a} + p_2\mathbf{b} + p_3\mathbf{c}$, $\mathbf{q} = q_1\mathbf{a} + q_2\mathbf{b} + q_3\mathbf{c}$, $\mathbf{r} = r_1\mathbf{a} + r_2\mathbf{b} + r_3\mathbf{c}$ whose end points are lattice points on P where p_i, q_i, r_i are all integers. As the end points of these vectors are on P, they satisfy (A5.1). Hence

$$\begin{aligned} v_1 p_1 + v_2 p_2 + v_3 p_3 &= w \\ v_1 q_1 + v_2 q_2 + v_3 q_3 &= w \\ v_1 r_1 + v_2 r_2 + v_3 r_3 &= w \ . \end{aligned}$$

We now solve this system of equations for v_1, v_2 and v_3 using the row operations described in Appendix 2. Since each row operation consists of only additions, subtractions, multiplications or divisions, the solution to this system must be rational numbers (recall that p_i, q_i, r_i and w are all integers). Hence, v_1, v_2 and v_3 are rational numbers. Multiplying (A5.1) on both sides by each of the denominators of v_1, v_2 and v_3 and simplifying, this equation can be written in the form

$$hx + ky + \ell z = m$$

where $h, k, \ell, m \in Z$. □

(EA5.2) Example: Find the equation of the plane P passing through the end points of $\mathbf{p} = 2\mathbf{a} - 3\mathbf{b} + \mathbf{c}$, $\mathbf{q} = -\mathbf{a} + \mathbf{b} - 2\mathbf{c}$ and $\mathbf{r} = -3\mathbf{a} + 2\mathbf{b} + 3\mathbf{c}$ with $w = 1$.

Solution: Since the end points of these vectors lie on P, their coefficients must satisfy the equation

$$v_1x + v_2y + v_3z = 1 .$$

Hence we have the three equations

$$\begin{aligned} v_1(2) + v_1(-3) + v_3(1) &= 1 \\ v_1(-1) + v_1(1) + v_3(-2) &= 1 \\ v_1(-3) + v_1(2) + v_3(3) &= 1 \end{aligned}$$

which can be rewritten in the form of the augmented matrix

$$\left[\begin{array}{ccc|c} 2 & -3 & 1 & 1 \\ -1 & 1 & -2 & 1 \\ -3 & 2 & 3 & 1 \end{array}\right] .$$

Using Gauss-Jordan elimination (see Appendix 2), we obtain the solution set $v_1 = -23/12$, $v_2 = -7/4$ and $v_3 = -5/12$. For this reason, the equation becomes

$$-23/12\ x - 7/4\ y - 5/12\ z = 1 .$$

Multiplying both sides of the equation by each of the denominators of v_1, v_2 and v_3 and then simplifying, we get

$$-23x - 21y - 5z = 12$$

as the equation of P where $h = -23$, $k = -21$ and $\ell = -5$. Thus the equation can be written in the form of (A5.1) where h,k,ℓ and m are all integers. Later we shall learn that P is parallel to the plane

$$-23x - 21y - 5z = 1$$

which intercepts the X-axis at $(-1/23)\mathbf{a}$, the Y-axis at $(-1/21)\mathbf{b}$ and the Z-axis at $(-1/5)\mathbf{c}$ and that P is the twelfth plane in the $(\overline{23},\overline{21},\overline{5})$ stack measured from the origin. □

If we had tried to find the equation of a plane passing through the points $\mathbf{p}$, $\mathbf{q}$ and $\mathbf{r}$ with $w = 0$, we would have found that $v_1 = v_2 = v_3 = 0$ which yields the equation $0x + 0y + 0z = 0$ of all space. The only time three points will determine a plane whose equation has $w = 0$ is when that plane passes through the origin.

(TA5.3) Theorem: The equation

$$hx + ky + \ell z = 0 \qquad \text{(A5.2)}$$

where h,k,ℓ are integers but not all equal to zero is the equation of a lattice plane passing through the origin.

Proof: We begin by assuming that at least one of the coefficients of (A5.2) is nonzero. Without loss of generality, we shall assume that $\ell \neq 0$. In our proof of the theorem, we shall show that the plane contains the end points of three lattice vectors, that these points are noncollinear and that the plane passes through the origin. Consider the lattice vectors $\mathbf{p} = 0\mathbf{a} + 0\mathbf{b} + 0\mathbf{c}$, $\mathbf{q} = 0\mathbf{a} - \ell\mathbf{b} + k\mathbf{c}$ and $\mathbf{r} = -\ell\mathbf{a} + 0\mathbf{b} + h\mathbf{c}$. Since the coefficients of these vectors satisfy (A5.2), we can conclude that the end points of these vectors are on the plane. Also, as the end point of the zero vector $\mathbf{p}$ is on (A5.2), the plane must pass through the origin. Finally, we show that $\mathbf{p}$, $\mathbf{q}$ and $\mathbf{r}$ are noncollinear.

We do this by showing that $(\mathbf{q} - \mathbf{p})$ and $(\mathbf{r} - \mathbf{p})$ are not multiples of each other. Since

$$[\mathbf{q} - \mathbf{p}]_D = [\mathbf{q}]_D - [\mathbf{p}]_D = [0,-\ell,k]^t ,$$
$$[\mathbf{r} - \mathbf{p}]_D = [\mathbf{r}]_D - [\mathbf{p}]_D = [-\ell,0,h]^t$$

and $\ell \neq 0$, we see that $(\mathbf{q} - \mathbf{p})$ and $(\mathbf{r} - \mathbf{p})$ are not multiples of each other.□

The definition of a lattice plane given in Chapter 2 states that we need to locate three noncollinear lattice points on a plane to conclude that the plane is a lattice plane. However, in the special case where the equation for the plane can be written as $hx + ky + \ell z = m$ where $h,k,\ell,m \in Z$, much less needs to be shown. The next theorem shows that we need only locate one lattice point on the plane.

(TA5.4) Theorem: If the plane with equation

$$hx + ky + \ell z = m \qquad \text{(A5.3)}$$

with $h,k,\ell,m \in Z$, not all h,k,ℓ equal to zero, has at least one lattice point on it, then it is a lattice plane.

Proof: Clearly if no lattice points satisfy the equation, then it is not the equation of a lattice plane. On the other hand, suppose that the coefficients of the vector $\mathbf{p} = p_1\mathbf{a} + p_2\mathbf{b} + p_3\mathbf{c}$ satisfies the equation. By the previous theorem, $hx + ky + \ell z = 0$ is a lattice plane which we shall label Q. Let $\mathbf{q} = q_1\mathbf{a} + q_2\mathbf{b} + q_3\mathbf{c}$ be a lattice vector whose end point is on Q and consider

$$\mathbf{p} + \mathbf{q} = (p_1 + q_1)\mathbf{a} + (p_2 + q_2)\mathbf{b} + (p_3 + q_3)\mathbf{c} .$$

Then

$$h(p_1 + q_1) + k(p_2 + q_2) + \ell(p_3 + q_3) =$$
$$hp_1 + kp_2 + \ell p_3 + hq_1 + kq_2 + \ell q_3 = m + 0 = m .$$

Hence, $\mathbf{p} + \mathbf{q}$ is on the plane defined by (A5.3) for all $\mathbf{q}$ on Q. Therefore, (A5.3) defines a lattice plane. □

(TA5.5) Theorem: The equation

$$hx + ky + \ell z = 1 \; , \qquad \text{(A5.4)}$$

where h, k, ℓ are nonzero integers is the equation of a lattice plane if and only if h, k and ℓ have no common integer factor larger than one (that is, there does not exist some integer $t > 1$ such that $h = tn_1$, $k = tn_2$, $\ell = tn_3$ where $n_1, n_2, n_3 \in Z$).

Proof: Suppose there exists an integer $t > 1$ that is an integer factor of h, k and ℓ. Then there exists integers n_1, n_2 and n_3 such that $h = n_1 t$, $k = n_2 t$ and $\ell = n_3 t$. But then (A5.4) becomes

$$n_1 tx + n_2 ty + n_3 tz = 1 \; ,$$

which yields

$$n_1 x + n_2 y + n_3 z = 1/t \; .$$

But if integers are substituted for x, y and z on the left, we would have an integer on the left and a non-integer $1/t$ on the right (since $t > 1$). Hence, in this case, because there are no integer solutions to (A5.4), it does not define a lattice plane.

Now suppose there does not exist an integer greater than one that is an integral divisor of h, k and ℓ. Note that this does *not* say that these integers are pairwise coprime. Let t denote the greatest common divisor of h and k. Then ℓ is relatively prime (i.e., 1 is the greatest common divisor) to t since otherwise t would be greater than one and be a common divisor of h, k and ℓ. Since ℓ and t are relatively prime, by the euclidean algorithm (see any standard algebra text), there exist integers m_1, m_2 such that

$$m_1 t + m_2 \ell = 1 \; .$$

Since t is the greatest common divisor of h and k, by the euclidean algorithm, there exist integers n_1 and n_2 such that

$$n_1 h + n_2 k = t \; .$$

Hence,

$$m_1(n_1h + n_2k) + m_2\ell = 1 \ .$$

Therefore,

$$h(m_1n_1) + k(m_1n_2) + m_2\ell = 1 \ .$$

For this reason, $[m_1n_1, m_1n_2, m_2]^t$ represents a lattice point at the end of the vector $m_1n_1\mathbf{a} + m_1n_2\mathbf{b} + m_2\mathbf{c} = \mathbf{p}$ that satisfies (A5.4). Thus, by TA5.4, (A5.4) is the equation of a lattice plane. □

(TA5.6) Theorem: The equation

$$hx + ky + 0z = 1$$

where $h,k \in Z$ with $h \neq 0$ and $k \neq 0$ is the equation of a lattice plane if and only if h and k are relatively prime.

Proof: Suppose that h and k are not relatively prime. Then there exists an integer $t > 1$ such that $h = tn$ and $k = tm$ for some $n,m \in Z$. Hence, we can write the equation

$$t(nx + my) = 1$$

and

$$nx + my = 1/t \ .$$

Since $1/t$ is not an integer, this equation can have no integral solutions. Hence, the equation does not define a lattice plane. Next, suppose that h and k are relatively prime. Then by the euclidean algorithm, there exist integers n and m such that

$$hn + km = 1 \ .$$

Hence, $[n,m,0]^t$ represents a lattice point satisfying the equation, and by TA5.4, it is the equation of a lattice plane. □

(TA5.7) Theorem: The equation

$$hx + 0y + 0z = 1$$

where $h \in Z$ and $h \neq 0$ is the equation of a lattice plane if and only if $h = \pm 1$.

Proof: Suppose that $h \neq \pm 1$, then $|h| > 1$ and so $1/h$ is not an integer. But any solution is of the form $[1/h, n, m]^t$ and so no lattice points satisfy the equation. We next suppose that $h = \pm 1$. Then $\mathbf{p} = h\mathbf{a} + 0\mathbf{b} + 0\mathbf{c}$ is a vector whose coefficients satisfy the equation. By TA5.4, this is the equation of a lattice plane. □

(TA5.8) Theorem: The equation

$$hx + ky + \ell z = n \tag{A5.5}$$

where $h, k, \ell, n \in Z$ with h, k, ℓ not all zero, $n > 0$ is the equation of a lattice plane if and only if n is a multiple of the largest common factor of h, k and ℓ.

Proof: First we shall establish that $hx + ky + \ell z = t$ defines a lattice plane where t denotes the largest common factor of h, k and ℓ. Then $n_1 = h/t$, $n_2 = k/t$ and $n_3 = \ell/t$ are integers such that the largest common factor of n_1, n_2 and n_3 is 1. Hence by TA5.5, TA5.6 and TA5.7

$$n_1x + n_2y + n_3z = 1 \tag{A5.6}$$

is the equation of a lattice plane. Let $p_1\mathbf{a} + p_2\mathbf{b} + p_3\mathbf{c}$ be a vector whose coefficients satisfy (A5.6). Hence

$$n_1p_1 + n_2p_2 + n_3p_3 = 1 .$$

Therefore,

$$tn_1p_1 + tn_2p_2 + tn_3p_3 = t$$

and so

$$hp_1 + kp_2 + \ell p_3 = t \ .$$

Consequently, the end point of **p** is a lattice point on the plane defined by

$$hx + ky + \ell z = t \tag{A5.7}$$

and so, by TA5.4, it is a lattice plane.

Next, suppose that n is a multiple of t, say $n = st$ where $s \in Z$. Since the end point of the lattice vector **p** is on the plane of (A5.7),

$$s(hp_1 + kp_2 + \ell p_3) = st = n \ .$$

Hence,

$$h(sp_1) + k(sp_2) + \ell(sp_3) = n \ .$$

Therefore, $[sp_1, sp_2, sp_3]^t$ represents a lattice point on the plane defined by (A5.5).

Now suppose that (A5.5) defines a lattice plane, and let t denote the largest common factor of h, k and ℓ. Again, let $n_1, n_2, n_3 \in Z$ be such that $h = n_1 t$, $k = n_2 t$ and $\ell = n_3 t$. Then (A5.7) becomes

$$n_1 tx + n_2 ty + n_3 tz = n \ .$$

Therefore, we have

$$n_1 x + n_2 y + n_3 z = n/t \ .$$

Since (A5.7) defines a lattice plane, there exist integer solutions to this equation (for x, y and z). Hence, n/t must be an integer, and so t is a factor of n. □

(TA5.9) Theorem: The distance between the origin and the plane defined by

$$hx + ky + \ell z = 1$$

is $1/([hk\ell]G^{-1}[hk\ell]^t)^{1/2}$ where G is the metrical matrix for $D = \{\mathbf{a},\mathbf{b},\mathbf{c}\}$. Furthermore, the distance between the origin and $hx + ky + \ell z = n$ is $n/([hk\ell]G^{-1}[hk\ell]^t)^{1/2}$.

Proof: In Chapter 2, we found that associated with the plane is a vector $\mathbf{s}$ such that $\mathbf{p}$ is on the plane if and only if $\mathbf{p} \bullet \mathbf{s} = 1$. It is shown that the distance from the origin to the plane is $1/s$ and that

$$[\mathbf{s}]_D^t = [hk\ell]G^{-1} .$$

Using the procedures described in Chapter 1, we see that

$$\begin{aligned} s^2 &= [\mathbf{s}]_D^t G[\mathbf{s}]_D \\ &= ([hk\ell]G^{-1})G([hk\ell]G^{-1})^t \\ &= ([hk\ell]G^{-1})G(G^{-1}[hk\ell]^t) \\ &= ([hk\ell]G^{-1}[hk\ell]^t) . \end{aligned}$$

Since the distance d between the origin and the plane equals $1/\mathbf{s}$, we have

$$d = 1/([hk\ell]G^{-1}[hk\ell]^t)^{1/2} .$$

The proof that the distance from $\mathbf{0}$ to the plane $hx + ky + \ell z = n$ is equal to $n/([hk\ell]G^{-1}[hk\ell]^t)^{1/2}$ is left to the reader. □

(TA5.10) Theorem: Let Q denote a lattice plane and let P denote a lattice plane that is closest to the origin of those that are parallel to Q and do not pass through the origin. Then there is an equation for P of the form

$$hx + ky + \ell z = 1$$

where $h,k,\ell \in Z$.

Proof: By our earlier discussions, we know that there exist integers n_1, n_2, n_3 and $m > 0$ such that

$$n_1x + n_2y + n_3z = m .$$

By TA5.8, the largest common factor t of n_1, n_2 and n_3 must be a factor of m. Hence, setting $h = n_1/t$, $k = n_2/t$, $\ell = n_3/t$ and $n = m/t$, we obtain the equation for P

$$hx + ky + \ell z = n$$

where the largest common factor of h, k and ℓ is 1. By TA5.5, TA5.6 and TA5.7,

$$hx + ky + \ell z = 1$$

is the equation of a lattice plane N. If $n \neq 1$, P would be further from the origin than N which contradicts the choice of P. Hence, $n = 1$. □

(TA5.11) Theorem: Let P denote any lattice plane. Then there exist integers h,k,ℓ,m such that the largest common integer factor of h, k and ℓ is 1 and

$$hx + ky + \ell z = m$$

is the equation for P.

Proof: By TA5.8, an equation exists of the form

$$n_1 x + n_2 y + n_3 z = n$$

where the largest integer factor of n_1, n_2 and n_3, denoted by t, is a factor of n. Hence, letting $h = n_1/t$, $k = n_2/t$ and $\ell = n_3/t$, we obtain an equation of the desired form. □

APPENDIX 6

INTERSECTION ANGLES BETWEEN ROTATION AXES

In Chapter 5 we showed that there are only six possible proper polyaxial crystallographic point groups. Four of these groups (*222*, *322*, *422*, *622*) are referred to as the dihedral groups, and the remaining two (*23* = *332*, *432*) are referred to as cubic axial groups. We shall now consider a method for finding the intersection angles among the rotation axes of both of these two types of groups. Not only can the methods developed in this appendix be used to find the intersection angles of the polyaxial crystallographic groups, but they can also be used to find the intersection angles for the noncrystallographic rotation groups.

Dihedral groups: We begin by considering the intersection angles that must exist for a dihedral group $\nu_1 22$. According to Table 5.1, there are precisely two pole points which we denote p_{11} and p_{12} in the equivalence class $C_1(\nu_1 22)$ belonging to the ν_1-fold axes of this group. Hence, there is exactly one ν_1-fold axis, and so p_{11} and p_{12} are antipodal points. We now need to determine the angle between the ν_1-fold axis and the rotation axis of α, one of the half-turns, whose pole points are not in $C_1(\nu_1 22)$. If $\alpha(p_{11}) = p_{11}$, then α must take place about the ν_1-fold axis, and its pole points are in $C_1(\nu_1 22)$. Hence, the other possibility is that α must permute the pole points of $C_1(\nu_1 22)$ such that $\alpha(p_{11}) = p_{12}$. In this case, let $\mathbf{p}$ denote a pole point for α. Since α is a point isometry, the distance between $\mathbf{p}$ and $\mathbf{p}_{11}$ must be the same as that between $\mathbf{p}$ (= $\alpha(\mathbf{p})$) and $\mathbf{p}_{12}$ (= $\alpha(\mathbf{p}_{11})$). Thus, $\mathbf{p}$ must lie on a plane perpendicular to the ν_1-axis which passes through the origin and pierces the unit ball at $\mathbf{p}_{11}$ and $\mathbf{p}_{12}$. Therefore, the rotation axis of the half-turn lies in a plane perpendicular to the ν_1-axis. A similar argument shows that the two-fold axes associated with the remaining pole points in $C_2(\nu_1 22)$ and $C_3(\nu_1 22)$ are also perpendicular to the ν_1-fold axis. Thus, the ν_1 two-fold axes associated with each pole point in $C_2(\nu_1 22)$ and $C_3(\nu_1 22)$ are all perpendicular to the ν_1-fold axis. The only angles left to determine are those between these two-fold axes.

We continue by defining a cartesian coordinate system C such that $\mathbf{k}$ is directed along the ν_1-fold axis and $\mathbf{i}$ is along the axis of one of the half-turns α. Let ρ denote the turn angle of the generating rotation ν_1 of the monoaxial group defining the ν_1-fold axis. Then the matrix representation of the composition $\nu_1\alpha$ with respect to C is given by

$$M_C(\nu_1\alpha) = M_C(\nu_1)\, M_C(\alpha)$$

$$= \begin{bmatrix} c\rho & -s\rho & 0 \\ s\rho & c\rho & 0 \\ 0 & 0 & 1 \end{bmatrix} \begin{bmatrix} 1 & 0 & 0 \\ 0 & -1 & 0 \\ 0 & 0 & -1 \end{bmatrix}$$

$$= \begin{bmatrix} c\rho & s\rho & 0 \\ s\rho & -c\rho & 0 \\ 0 & 0 & -1 \end{bmatrix} .$$

Applying the rules for analyzing a symmetric general cartesian rotation matrix (A3.4), we have

$$\ell_1 = \left(\frac{c\rho + 1}{2}\right)^{\frac{1}{2}} , \quad \ell_2 = \pm\left(\frac{1 - c\rho}{2}\right)^{\frac{1}{2}} \quad \text{and} \quad \ell_3 = 0 .$$

By the half-angle formulae and the sign convention of the direction cosines set forth in (A3.4), we obtain

$$\ell_1 = \cos(\rho/2), \quad \ell_2 = \sin(\rho/2) = \cos((90-\rho)/2) \quad \text{and} \quad \ell_3 = 0.$$

Hence, the rotation axis of $\nu_1\alpha$ makes an angle of $\rho/2$ with respect to that of α. Repeating this process by applying ν_1 to $\nu_1\alpha$ and so forth until we obtain α, we find that two-fold axes occur at intervals of $\rho/2$, thus generating all of the ν_1 two-fold axes. Note that since $\rho = 360/\nu_1$, the angle between adjacent two-fold axes is $180/\nu_1$. These results are summarized in the following theorem.

(TA6.1) Theorem: The ν_1-fold rotation axis of a dihedral $\nu_1 22$ group is perpendicular to each half-turn axis (except those contained in ν_1). Furthermore, if α_1 and α_2 are adjacent half-turns not contained in ν_1, then their rotation axes intersect at an angle of $180/\nu_1$.

In the case of *422*, for example, the angle between adjacent two-fold axes in the plane perpendicular to the four-fold axis is $180/4 = 45°$.

(PA6.1) Problem: Determine the angle between adjacent two-fold axes in the plane perpendicular to the ν_1-fold for the dihedral groups *222*, *322* and *422*.

Cubic axial groups: As noted above, there are two possible cubic axial crystallographic proper point groups, *432* and *332* (= *23*). Both of these are of the form $\nu_1\nu_2 2$. In Chapter 5 it is shown that **43** = **2** occurs in *432* and **33** = **2** occurs in *332*. We are left with the problem of determining the angles between the ν_1-fold axis and the ν_2-fold axis in both of these cases. A special case of a theorem which will be stated here but proved later in this section is of particular importance in resolving this problem.

(TA6.2) Theorem: Given that the composition of rotations ν_1 and ν_2 is a half-turn, the intersection angle $<\nu_1:\nu_2$ between the axes of ν_1 and ν_2 is

$$<(\nu_1:\nu_2) = \cos^{-1}(\text{ctn}(180/\nu_1)\ \text{ctn}(180/\nu_2)) \ .$$

(EA6.3) Example: In the case of *432*, we have $\nu_1 = 4$, $\nu_2 = 3$, and the composition **43** = **2**. Hence, by Theorem TA6.2, the angle <(**4**:**3**) is

$$<(\mathbf{4}:\mathbf{3}) = \cos^{-1}(\text{ctn}(45°)\ \text{ctn}(60°)) = \cos^{-1}(1/\sqrt{3}) \simeq 54.74°.$$

(PA6.2) Problem: In the case of *332*, find <(**3**:**3**). (Answer: $\cos^{-1}(1/3) = 70.53°$.)

(PA6.3) Problem: Consider the case of $\nu_1\nu_2 2$ where $\nu_2 = 2$ and show that $<(\nu_1:2) = 90°$. Hence, as observed in TA6.1, we see that the composition of ν_1 and **2** is a half-turn only when the rotation axis of the half-turn is perpendicular to the rotation axis of ν_1.

(PA6.4) Problem: A study of the herpes virus with the electron microscope indicates that the virus forms as a small particle with icosahedral *532* rotational symmetry. Determine the angle between the generating fifth-turn and third-turn, <(**5**:**3**), of the five- and three-fold axes of such a particle assuming that **53** = **2** (see R.W. Horne and P. Wildy, 1961). (Answer: 37.377°.)

Now that we have determined the angle between the ν_1 and ν_2 axes so that $\nu_1\nu_2$ is a half-turn, we can calculate the relative position of the two-fold axis using matrices. In the case of *432*, orient a cartesian basis C such that $\mathbf{k}$ is along the four-fold axis and $\mathbf{i}$ is in the plane of $\mathbf{k}$ and the three-fold axis where the angle between the four-fold and three-fold axes is $\cos^{-1}(1/\sqrt{3})$. Then the direction cosines

$$[\ell_1, \ell_2, \ell_3]^t$$

of these two axes with respect to C are $[001]^t$ for the four-fold and $[\sqrt{2}/\sqrt{3}, 0, 1/\sqrt{3}]^t$ for the three-fold. Using the general cartesian rotation matrix (see Appendix 3), we have

$$M_C(4) = \begin{bmatrix} 0 & -1 & 0 \\ 1 & 0 & 0 \\ 0 & 0 & 1 \end{bmatrix}$$

$$M_C(3) = \begin{bmatrix} 1/2 & -1/2 & -\sqrt{2}/2 \\ 1/2 & -1/2 & -\sqrt{2}/2 \\ \sqrt{2}/2 & \sqrt{2}/2 & 0 \end{bmatrix}.$$

Hence

$$M_C(2) = M_C(43) = M_C(4)\, M_C(3) = \begin{bmatrix} -1/2 & 1/2 & \sqrt{2}/2 \\ 1/2 & -1/2 & -\sqrt{2}/2 \\ \sqrt{2}/2 & \sqrt{2}/2 & 0 \end{bmatrix}.$$

Applying (A3.4) to this matrix, we see that the triple of a unit vector along the two-fold axis is

$$[\mathbf{v}]_C = [1/2, 1/2, \sqrt{2}/2]^t.$$

Hence the angle between the four-fold axis and the two-fold axis can be found by taking the inner product of $\mathbf{v}$ with $[001]^t$ and solving for the angle. Since the inner product is $\sqrt{2}/2$, we have $\angle(4{:}2) = \cos^{-1}(\sqrt{2}/2) = 45°$. The angle between the three-fold axis and the two-fold axis is found by forming the inner product of $[\mathbf{v}]_C$ and $[\sqrt{2}/\sqrt{3}, 0, 1/\sqrt{3}]^t$, the unit vector along the three-fold axis):

$$\angle(3{:}2) = \cos^{-1}((\sqrt{2}/\sqrt{3})(1/2) + (1/\sqrt{3})(\sqrt{2}/2)) \simeq 35.26\ .$$

(PA6.5) Problem: For *332*, let α and β denote two third-turns such that αβ = **2**. Define a Cartesian basis *C* with **k** along the rotation axis of α and **i** in the plane of **k** and the rotation axis of β. Find $M_C(\alpha)$ and $M_C(\beta)$ and then calculate $M_C(\mathbf{2})$. Analyze $M_C(\mathbf{2})$ to determine $<\alpha:\mathbf{2}$ and $<\beta:\mathbf{2}$.

Answer: The axis of α is along $[001]^t$ and that of β is along the unit vector $[\mathbf{v}]_C = [\sqrt{8}/3\ \ 0\ \ 1/3]^t$. Hence

$$M_C(\alpha) = \begin{bmatrix} -1/2 & -\sqrt{3}/2 & 0 \\ \sqrt{3}/2 & -1/2 & 0 \\ 0 & 0 & 1 \end{bmatrix} \quad \text{and} \quad M_C(\beta) = \begin{bmatrix} 5/6 & -\sqrt{3}/6 & \sqrt{2}/3 \\ \sqrt{3}/6 & -1/2 & -\sqrt{6}/3 \\ \sqrt{2}/3 & \sqrt{6}/3 & -1/3 \end{bmatrix}.$$

The product of these two matrices is

$$M_C(\mathbf{2}) = \begin{bmatrix} -2/3 & 1/\sqrt{3} & \sqrt{2}/3 \\ 1/\sqrt{3} & 0 & \sqrt{6}/3 \\ \sqrt{2}/3 & \sqrt{6}/3 & -1/3 \end{bmatrix}.$$

Using the techniques presented in Appendix 3, we found that the unit vector along the two-fold axis is $[\mathbf{v}]_C = [1/\sqrt{6},\ 1/\sqrt{2},\ 1/\sqrt{3}]^t$. Consequently, $<\alpha:\mathbf{2} = <\beta:\mathbf{2} = \cos^{-1}(1/\sqrt{3}) \simeq 54.74°$.

(PA6.6) Problem: In Problem PA6.4, we found that for *532*, the angle between a five-fold axis and a three-fold axis such that **53** = **2** is 37.377°. Using this information and the general cartesian rotation matrix (Appendix 3), find (1) $M_C(\mathbf{2})$, (2) the unit vector **v** along the two-fold axis, (3) $<(\mathbf{5}:\mathbf{2})$ and $<(\mathbf{3}:\mathbf{2})$.

Answers:

$$M_C(\mathbf{2}) = \begin{bmatrix} -0.638197 & 0.262866 & 0.723607 \\ 0.262866 & -0.809017 & 0.525731 \\ 0.723607 & 0.525731 & 0.447214 \end{bmatrix},$$

$$\mathbf{v} = 0.42532\mathbf{i} + 0.30902\mathbf{j} + 0.85065\mathbf{k},$$

$$<(\mathbf{5}:\mathbf{2}) = 31.717° \quad \text{and} \quad <(\mathbf{3}:\mathbf{2}) = 20.905°.$$

Theorem TA6.2 is a special case of Euler's theorem which we now are ready to state and prove. The proof of Euler's theorem is just a generalization of the technique used to find, for example, $\measuredangle(\mathbf{4}:\mathbf{3})$ in *432*.

(TA6.4) Theorem (Euler): Let α_1, α_2, α_3 denote nonidentity rotations with turn angles ρ_1,ρ_2,ρ_3 about axes passing through the origin such that $\alpha_1\alpha_2 = \alpha_3$. Then the angle $\measuredangle(\alpha_1:\alpha_2)$ between the rotation axes of α_1 and α_2 is given by

$$\cos\measuredangle(\alpha_1:\alpha_2) = [\cos(\rho_1/2)\cos(\rho_2/2) \pm \cos(\rho_3/2)]/(\sin(\rho_1/2)\sin(\rho_2/2)) .$$

Proof: A cartesian coordinate system is chosen such that $\mathbf{k}$ is directed in the positive direction of the α_1 axis and $\mathbf{i}$ is in the plane of $\mathbf{k}$ and the α_2 axis. The unit vectors along the axes of α_1 and α_2 are $[001]^t$ and $[\sin\phi, 0, \cos\phi]^t$ where $\phi = \measuredangle(\alpha_1:\alpha_2)$. Then the matrix representations are given by Equation (A3.1) of Appendix 3 as

$$M_C(\alpha_1) = \begin{bmatrix} c\rho_1 & -s\rho_1 & 0 \\ s\rho_1 & c\rho_1 & 0 \\ 0 & 0 & 1 \end{bmatrix} ,$$

$$M_C(\alpha_2) = \begin{bmatrix} s\phi^2(1-c\rho_2)+c\rho_2 & -c\phi s\rho_2 & s\phi c\phi(1-c\rho_2) \\ c\phi s\rho_2 & c\rho_2 & -s\phi s\rho_2 \\ s\phi c\phi(1-c\rho_2) & s\phi s\rho_2 & c\phi^2(1-c\rho_2)+c\rho_2 \end{bmatrix} .$$

Hence, the matrix representation for α_3 is

$$M_C(\alpha_3) = M_C(\alpha_1)\, M_C(\alpha_2) =$$

$$\begin{bmatrix} (c\rho_1 s\phi^2(1-c\rho_2)+c\rho_1 c\rho_2-s\rho_1 s\rho_2 c\phi) & (-c\phi c\rho_1 s\rho_2-s\rho_1 c\rho_2) & (c\rho_1 s\phi c\phi(1-c\rho_2)+s\phi s\rho_1 s\rho_2) \\ (s\rho_1 s\phi^2(1-c\rho_2)+s\rho_1 c\rho_2+c\rho_1 s\rho_2 c\phi) & (-s\rho_1 s\rho_2 c\phi+c\rho_1 c\rho_2) & (s\rho_1 s\phi c\phi(1-c\rho_2)-c\rho_1 s\rho_2 s\phi) \\ (s\phi c\phi(1-c\rho_2)) & (s\rho_2 s\phi) & (c\phi^2(1-c\rho_2)+c\rho_2) \end{bmatrix} \tag{A6.1}$$

But by rule 3 of RA3.1 in Appendix 3, the trace of $M_C(\alpha_3)$ must equal $1 + 2\cos\rho_3$. Hence, equating $1 + 2\cos\rho_3$ with the trace of the matrix in (A6.1), we obtain

$$\begin{aligned} 1 + 2c\rho_3 = {} & c\rho_1 s\phi^2 - c\rho_1 c\rho_2 s\phi^2 + c\rho_1 c\rho_2 - s\rho_1 s\rho_2 c\phi \\ & - s\rho_1 s\rho_2 c\phi + c\rho_1 c\rho_2 + c\phi^2 - c\rho_2 c\phi^2 + c\rho_2 . \end{aligned}$$

Collecting terms, this expression becomes

$$1 + 2\cos\rho_3 = (1 - \cos\rho_1)(1 - \cos\rho_2)\cos^2\phi$$
$$- 2\sin\rho_1\sin\rho_2\cos\phi + (1 + \cos\rho_1)(1 + \cos\rho_2) - 1 . \qquad (A6.2)$$

Since $2\cos^2(\alpha/2) = 1 + \cos\alpha$ and $2\sin^2(\alpha/2) = 1 - \cos\alpha$, (A6.2) can be rewritten as

$$4\sin^2(\rho_1/2)\sin^2(\rho_2/2)\cos^2\phi + (-2\sin\rho_1\sin\rho_2)\cos\phi$$
$$+ 4\cos^2(\rho_1/2)\cos^2(\rho_2/2) - 4\cos^2(\rho_3/2) = 0 .$$

Setting $a = 4\sin^2(\rho_1/2)\sin^2(\rho_2/2)$, $b = -2\sin\rho_1\sin\rho_2$, and $c = 4\cos^2(\rho_1/2)\cos^2(\rho_2/2) - 4\cos^2(\rho_3/2)$, we can use the quadratic formula,

$$\cos\phi = \frac{-b \pm (b^2 - 4ac)^{\frac{1}{2}}}{2a}$$

to find $\cos\phi$. Using the identity $\sin^2\rho_2 = 4\sin^2(\rho_2/2)\cos^2(\rho_2/2)$, the formula becomes

$$\cos\phi = \frac{2\sin\rho_1\sin\rho_2 \pm (64\sin^2(\rho_1/2)\sin^2(\rho_2/2)\cos^2(\rho_3/2))^{\frac{1}{2}}}{(8\sin^2(\rho_1/2)\sin^2(\rho_2/2))}$$

$$\cos\phi = \frac{2\sin\rho_1\sin\rho_2 \pm 8\sin(\rho_1/2)\sin(\rho_2/2)\cos(\rho_3/2)}{8\sin^2(\rho_1/2)\sin^2(\rho_2/2)}$$

Next, recalling that $\sin\rho_1 = 2\cos(\rho_1/2)\sin(\rho_1/2)$, etc., we get

$$\cos\phi = \frac{(8\sin(\rho_1/2)\sin(\rho_2/2))(\cos(\rho_1/2)\cos(\rho_2/2) \pm \cos(\rho_3/2))}{(8\sin(\rho_1/2)\sin(\rho_2/2))(\sin(\rho_1/2)\sin(\rho_2/2))}$$

and

$$\cos\angle(\alpha_1:\alpha_2) = \cos\phi = \frac{\cos(\rho_1/2)\cos(\rho_2/2) \pm \cos(\rho_3/2)}{\sin(\rho_1/2)\sin(\rho_2/2)}$$

Proof of Theorem TA6.2: Applying Euler's theorem to the case where $\rho_3 = 180°$, we have

$$\cos\angle(\alpha_1:\alpha_2) = \cos(\rho_1/2)\cos(\rho_2/2)/\sin(\rho_1/2)\sin(\rho_2/2)$$
$$= \mathrm{ctn}(\rho_1/2)\mathrm{ctn}(\rho_2/2)$$
$$= \mathrm{ctn}(180/\nu_1)\mathrm{ctn}(180/\nu_2)$$

since $\nu_i = 360/\rho_i$.

(PA6.6) Problem: Let α and β denote third-turns about the vectors $[\mathbf{v}]_C = (-1/\sqrt{3}, -1/\sqrt{3}, 1/\sqrt{3})^t$ and $[\mathbf{w}]_C = (1/\sqrt{3}, 1/\sqrt{3}, 1/\sqrt{3})^t$), respectively.

(1) Show that the intersection angle between the axes of these two rotations is $<(\alpha:\beta) = \cos^{-1}(-1/3) \approx 109.47°$.

(2) Evaluate the product $M_C(\gamma) = M_C(\alpha)M_C(\beta)$ and show that $\gamma = [111]_3$.

(3) Show that $<(\alpha:\gamma) = <(\beta:\gamma) = \cos^{-1}(1/3) \approx 70.53°$.

APPENDIX 7

EQUIVALENCE RELATIONS, COSETS AND FACTOR GROUPS

Equivalence Relations: In the section "Equivalent Points and Planes" in Chapter 4, the notion of G-equivalence was defined. This notion is an example of a binary relation defined on a set. Such a *relation* on a set T is a rule that determines for each pair of elements x and y whether x is related to y ($x \sim y$) or not ($x \not\sim y$). Let K denote the set of all crystals and define crystal x to be related to crystal y if and only if the density (mass per unit volume) of x, denoted $\rho(x)$, is greater than $\rho(y)$. Written another way,

$$x \sim y \iff \rho(x) > \rho(y) .$$

Then $\sim$ is a binary relation. Note that unless $\rho(x) = \rho(y)$, we cannot have both $x \sim y$ and $y \sim x$. In T4.33, three properties of G-equivalence were established. These three properties constitute the definition of an equivalence relation defined below.

(DA7.1) Definition: A binary relation $\sim$ defined on a set T is an *equivalence relation* if the following three properties are satisfied:

(1) $x \sim x$ for all $x \in T$ (*reflexive property*);
(2) For all $x,y \in T$, if $x \sim y$, then $y \sim x$ (*symmetric property*);
(3) For all $x,y,z \in T$, if $x \sim y$ and $y \sim z$, then $x \sim z$ (*transitive property*).

Besides G-equivalence, there are many important equivalence relations in the study of crystallography. For example, on the set of all crystals K, define $x \sim y$ if and only if x has the same crystal structure as y (that is, x and y are *isostructural*) is an equivalence relation. We shall now examine several other examples which will illustrate the notion.

(EA7.2) Example: Show that the relation defined on the set of all turn angles T defined by

$$\theta \sim \phi \iff \theta = \phi + 360°m \text{ for some } m \in Z$$

is an equivalence relation.

Solution: Since $\theta = \theta + 360°(0)$ and $0 \in Z$, $\theta \sim \theta$ and so $\sim$ is reflexive. Now suppose that $\theta \sim \phi$. Then there exists an integer m such that $\theta = \phi + 360m$. Hence $\phi = \theta + 360(-m)$. Since $m \in Z$, we have $-m \in Z$ and so $\phi \sim \theta$. Hence $\sim$ is symmetric. Finally, suppose $\theta \sim \phi$ and $\phi \sim \rho$. Then there exist integers m_1 and m_2 such that $\theta = \phi + 360m_1$ and $\phi = \rho + 360m_2$. Hence, $\theta = (\rho + 360m_2) + 360m_1 = \rho + 360(m_1 + m_2)$. Since $m_1, m_2 \in Z$, $m_1 + m_2 \in Z$ and so $\theta \sim \rho$. Hence $\sim$ is transitive. □

The manner in which turn angles are related in EA7.2 is a special case of an important concept in mathematics known as modular arithmetic which wil be introduced in the following definition.

(DA7.3) Definition: Let n denote an integer. Given two integers a and b, we say a is congruent to b modulo n if and only if there exists an integer m such that $a = b + nm$. In this case we write $a \equiv b$ modulo n.

In EA7.2, θ, ϕ, and 360 play the roles of a, b and n, respectively. Hence, the equivalence relation of this example can be restated as $\theta \sim \phi \iff \theta \equiv \phi$ modulo 360.

(PA7.1) Problem: Let n denote an integer. Prove that

$$a \sim b \iff a \equiv b \text{ modulo } n$$

is an equivalence relation.

Hint: Adapt the solution to EA7.2 where n replaces 360.

(EA7.4) Example: Let M denote the set of all $n \times n$ matrices with real entries. Define $\sim$ on M by

$$A \sim B \iff \text{there exists } P \in M \text{ such that } PAP^{-1} = B .$$

In Appendix 3 we called two matrices such that $A \sim B$ similar matrices. Show that $\sim$ is an equivalence relation on M.

Solution: Let $A \in M$. Then $I_n A I_n^{-1} = A$ where I_n is the $n \times n$ identity matrix. Hence, $A \sim A$ and so $\sim$ is reflexive. Let $A,B \in M$ such that $A \sim B$. Then there exists $P \in M$ such that $PAP^{-1} = B$. Hence, $A = P^{-1}BP = (P^{-1})B(P^{-1})^{-1}$. Since $P^{-1} \in M$, $B \sim A$. Hence, $\sim$ is symmetric. Let $A,B,C \in M$ such that $A \sim B$ and $B \sim C$. Then there exist matrices, P,Q in M such that $PAP^{-1} = B$ and $QBQ^{-1} = C$. Hence, substituting for B

$$Q(PAP^{-1})Q^{-1} = C$$
$$(QP)A(P^{-1}Q^{-1}) = C .$$

But $P^{-1}Q^{-1} = (QP)^{-1}$ and so

$$(QP)A(QP)^{-1} = C$$

where PQ ε M. Hence, A ~ C, and so ~ is transitive.

(PA7.2) Problem: Let M denote the set of all $n \times n$ matrices with real entries. Define ~ on M by

$$A \sim B \iff \text{there exists an invertible matrix } P \text{ such that } P^tAP = B .$$

Show that ~ is an equivalence relation on M.

Hint: Recall that $(P^{-1})^t = (P^t)^{-1}$ and $(PQ)^t = Q^tP^t$.

The following example enables us to study a lattice extending infinitely in all directions by considering only a finite portion of the lattice.

(PA7.3) Problem: Let $D = \{\mathbf{a},\mathbf{b},\mathbf{c}\}$ denote a basis for S and let L denote the lattice generated by D. That is,

$$L = \{u\mathbf{a} + v\mathbf{b} + w\mathbf{c} \mid u,v,w \in Z\} .$$

Define ~ on S by

$$\mathbf{x} \sim \mathbf{y} \iff \mathbf{y}-\mathbf{x} \in L .$$

show that ~ is an equivalence relation.

Hint: The necessary facts about L are listed below for each part of the proof:

Reflexive: $\mathbf{0} \in L$
Symmetric: If $\mathbf{v} \in L$, then $-\mathbf{v} \in L$
Transitive: If $\mathbf{u},\mathbf{v} \in L$, then $\mathbf{u} + \mathbf{v} \in L$.

Equivalence Classes: Let T denote a set on which the equivalence relation ~ has been defined. Then, in a natural way, ~ partitions T into disjoint sets. These sets are called *equivalence classes*.

(DA7.5) Definition: Let T denote a set on which the equivalence relation ~ is defined. Let $t \in T$. Then the equivalence class of t is denoted by $[t]$ and is defined to be

$$[t] = \{x \in T \mid t \sim x\} .$$

Given an equivalence relation ~ on T, then the set of all equivalence classes has two very important properties. First, T is the union of all of the equivalence classes. Second, two equivalence classes are either exactly the same set or completely disjoint. Proof of these facts can be found in many modern algebra texts (e.g., see Durbin, 1981). Moreover, if E denotes an equivalence class, then any element of E is said to be a *representative* of E.

(EA7.6) Example: Let G denote a group of isometries. In T4.33, it was shown that G-equivalence is an equivalence relation of S. Let $\mathbf{r} \in S$, then

$$\begin{aligned} [\mathbf{r}] &= \{\mathbf{x} \in S \mid \mathbf{r} \sim \mathbf{x}\} \\ &= \{\mathbf{g}(\mathbf{r}) \mid \mathbf{g} \in G\} \\ &= \mathrm{Orb}_G(\mathbf{r}) . \end{aligned}$$

Hence, the set of equivalence classes of the G-equivalence relation on S is the set of all orbits of points in S under G. In practice, to find the orbits of a point $\mathbf{r}$, we define a basis D and determine the matrix representations of the isometries in G, $\mathrm{M}_D(G)$. Then the triples in R^3 that represent the points in $\mathrm{Orb}_G(\mathbf{r})$ are

$$\mathrm{Orb}_{G,D}(\mathbf{r}) = \{\mathrm{M}_D(\alpha)[\mathbf{r}]_D \mid \alpha \in G\} .$$

(PA7.4) Problem: Find $\mathrm{Orb}_{322,P}(\mathbf{r})$ for each of the following choices of $\mathbf{r}$. (See Chapter 5 for $\mathrm{M}_P(322)$.)

(1) $[\mathbf{r}]_P = [111]^t$

(2) $[\mathbf{r}]_P = [001]^t$

(3) $[\mathbf{r}]_P = [100]^t$

(4) $[\mathbf{r}]_P = [110]^t$

(5) $[\mathbf{r}]_P = [000]^t$

(PA7.5) Problem: Consider the equivalence relation of EA7.2. List six elements of each of the following equivalence classes:

(1) [45] (4) [-72]

(2) [72] (5) [-3826]

(3) [4358]

In Chapter 3 we decided that the preferred turn angle θ for a given rotation should be selected such that $-180 < \theta \le 180$. The reason that we can always do this is that each equivalence class of the relation in EA7.2 has a representative in this range.

(PA7.6) Problem: For each of the following turn angles θ, find the representative ϕ of $[\theta]$ such that $-180 < \phi \le 180$.

(1) 810 (2) -240

(3) 240 (4) 2589

(PA7.7) Problem: List six elements of [5] using the equivalence relation "modulo n" defined in DA7.3 for each of the following values of n

(1) $n = 2$ (2) $n = 4$

(3) $n = 10$ (4) $n = 1$.

Answer: When $n = 2$, then 5, 7, -13, 19, 1 and 21 are all elements of [5]. This can be seen through the following equations:

$$5 = 5 + 2(0)$$
$$5 = 7 + 2(-1)$$
$$5 = -13 + 2(9)$$
$$5 = 19 + 2(-7), \text{ etc.}$$

(PA7.8) Problem: Using the equivalence relation defined in EA7.4, find two matrices that are similar to each of the following matrices:

$$(1)\begin{bmatrix} 0 & 0 & 1 \\ 1 & 0 & 0 \\ 0 & 1 & 0 \end{bmatrix} \qquad (2)\begin{bmatrix} 1 & 1 & 1 \\ 0 & -1 & 0 \\ 0 & 0 & -1 \end{bmatrix}$$

Solution: To solve this problem we need only find a 3×3 invertible matrix P and form PAP^{-1}. Recall that a matrix P is invertible if and only if $\det(P) \neq 0$. For example, if

$$P = \begin{bmatrix} 2/3 & -1/3 & -1/3 \\ 1/3 & 1/3 & 2/3 \\ 1/3 & 1/3 & 1/3 \end{bmatrix}$$

then

$$B = PAP^{-1} = \begin{bmatrix} 2/3 & -1/3 & -1/3 \\ 1/3 & 1/3 & -2/3 \\ 1/3 & 1/3 & 1/3 \end{bmatrix} \begin{bmatrix} 0 & 0 & 1 \\ 1 & 0 & 0 \\ 0 & 1 & 0 \end{bmatrix} \begin{bmatrix} 1 & 0 & 1 \\ -1 & 1 & 1 \\ 0 & -1 & 1 \end{bmatrix}$$

$$= \begin{bmatrix} 0 & -1 & 0 \\ 1 & -1 & 0 \\ 0 & 0 & 1 \end{bmatrix}$$

is related to A. The reader should finish the problem in a similar fashion for the other choice of A. The matrix A in this example represents a third-turn about [111] for a crystal with basis D_1, the matrix P is the change of basis matrix from a D_1 to a D_2 basis, and B represents the same third-turn but about [001] of the D_2-basis of the crystal.

(EA7.7) Example: Using the relation ~ defined in PA7.2, find a matrix that is related to

$$A = \begin{bmatrix} 291.829 & 0 & 0 \\ 0 & 94.829 & 0 \\ 0 & 0 & 87.142 \end{bmatrix}$$

Solution: Let

$$P = \begin{bmatrix} 1/2 & 0 & 0 \\ -1/2 & 1 & 0 \\ 0 & 0 & 1 \end{bmatrix},$$

then

$$B = \begin{bmatrix} 1/2 & -1/2 & 0 \\ 0 & 1 & 0 \\ 0 & 0 & 1 \end{bmatrix} \begin{bmatrix} 291.829 & 0 & 0 \\ 0 & 94.829 & 0 \\ 0 & 0 & 87.142 \end{bmatrix} \begin{bmatrix} 1/2 & 0 & 0 \\ -1/2 & 1 & 0 \\ 0 & 0 & 1 \end{bmatrix}$$

$$= \begin{bmatrix} 96.664 & -47.414 & 0 \\ -47.414 & 94.829 & 0 \\ 0 & 0 & 87.142 \end{bmatrix}.$$

(PA7.8) Problem: Find another matrix related to the matrix A of EA7.7, using the relation given in PA7.2.

The matrix A given in EA7.7 is the metrical matrix for the C-centered orthorhombic unit cell of low cordierite. The matrix P of the solution is the change of basis matrix from a pseudo-hexagonal primitive cell basis to the orthorhombic basis. Hence according to T2.12 the matrix $B = P^tAP$ is the metrical matrix for the pseudo-hexagonal basis. In analyzing B, we find that pseudo-hexagonal basis {**a**,**b**,**c**} is such that a = 9.832A, b = 9.738A, c = 9.335A, $\alpha = \beta = 90°$ and $\gamma = 119.68°$, which is close to being hexagonal. In fact, any basis D = {**a**,**b**,**c**} could be chosen, the change of basis matrix from D to the orthorhombic basis could be found and used for P to yield $B = P^tAP$, the metrical matrix for D. Hence the equivalence class of A with respect to this equivalence relation is the set of all metrical matrices.

(PA7.9) Problem: Define a basis D = {**a**,**b**,**c**} in terms of the orthorhombic basis of low cordierite. Find P, the change of basis matrix from D to the orthorhombic basis, and calculate the metrical matrix $B = P^tAP$ for D. Calculate a, b, c, α, β, γ.

Cosets: We shall now study an equivalence relation on a group G whose equivalence classes, called left or right cosets, are indispensable to our development of the crystallographic groups.

(EA7.8) Example: Let G denote a group and let H denote a subgroup of G. Define ~ on G by

$$a \sim b \iff ba^{-1} \varepsilon H .$$

Show that ~ is an equivalence relation.

Solution: Let $a \varepsilon G$. Then $aa^{-1} = e$ where e is the identity element of G. But the identity of H is the same as that of G (see Durbin, 1981), and so $aa^{-1} \varepsilon H$. Hence $a \sim a$. Let $a,b \varepsilon G$ such that $a \sim b$. Then ba^{-1} εH. Since H is a group $(ba^{-1})^{-1} \varepsilon H$. But $(ba^{-1})^{-1} = ab^{-1}$ and so $ab^{-1} \varepsilon H$. Hence $b \sim a$. Let $a,b,c \varepsilon G$ such that $a \sim b$ and $b \sim c$. Then $ba^{-1} \varepsilon H$ and $cb^{-1} \varepsilon H$. Since H is closed under the binary operation of G, $(cb^{-1})(ba^{-1}) \varepsilon H$. But $(cb^{-1})(ba^{-1}) = ca^{-1}$. Hence $a \sim c$. Therefore ~ is an equivalence relation. □

Let G denote a group, H a subgroup of G and $a \varepsilon G$. Then the equivalence class of a with respect to the equivalence relation of EA7.7 has a simple form:

$$\begin{aligned}
[a] &= \{b \varepsilon G \mid a \sim b\} \\
&= \{b \varepsilon G \mid ba^{-1} \varepsilon H\} \\
&= \{b \varepsilon G \mid ba^{-1} = h \text{ for some } h \varepsilon H\} \\
&= \{b \varepsilon G \mid b = ha \text{ for some } h \varepsilon H\} \\
&= \{ha \mid h \varepsilon H\} .
\end{aligned}$$

(DA7.8) Definition: Let G denote a group, H a subgroup of G and $a \varepsilon G$. Then the *right coset of H in G determined by a*, denoted Ha, is

$$Ha = \{ha \mid h \varepsilon H\} .$$

Hence the equivalence classes of a with respect to the relation of EA7.7 are the right cosets of H in G determined by a.

(EA7.9) Example: Let $G = 322$, $H = 3 = \{1,3,3^{-1}\}$. Find the right coset $3g$ for each $g \varepsilon 322$.

Solution:

$$31 = \{1,3,3^{-1}\}1 = \{1,3,3^{-1}\}$$

$$33 = \{1,3,3^{-1}\}3 = \{3,3^{-1},1\}$$

$$33^{-1} = \{1,3,3^{-1}\}3^{-1} = \{3^{-1},1,3\}$$

$$3[100]_2 = \{1,3,3^{-1}\}[100]_2 = \{[100]_2,[110]_2,[010]_2\}$$

$$3[110]_2 = \{1,3,3^{-1}\}[110]_2 = \{[110]_2,[010]_2,[100]_2\}$$

$$3[010]_2 = \{1,3,3^{-1}\}[010]_2 = \{[010]_2,[100]_2,[110]_2\} .$$

Note that we only have two distinct right cosets since $31 = 33 = 33^{-1}$ and $3[100]_2 = 3[010]_2 = 3[110]_2$. □

(PA7.10) Problem: Let $G = 422$, $H = 4 = \{1,4,2,4^{-1}\}$. Find the right coset $4g$ for each $g \in 422$.

(PA7.11) Problem: Let G denote a group and let H denote a subgroup of G. Define $\sim$ on G by

$$a \sim b \iff a^{-1}b \in H .$$

Show that $\sim$ is an equivalence relation.

Hint: The proof will be similar to that given in EA7.7.

(PA7.12) Problem: Using the relation of PA7.11, show that if $a \in G$, then $[a] = \{ah \mid h \in H\}$.

(DA7.10) Definition: Let G denote a group, H a subgroup of G, $a \in G$. Then the *left coset* of H in G determined by a, denoted aH, is

$$ah = \{ah \mid h \in H\}.$$

(PA7.13) Problem: Let $G = 322$ and $H = 3 = \{1,3,3^{-1}\}$. Find the left coset $g3$ for each $g \in 322$.

Note that the left cosets of H calculated in PA7.13 are such that

$\mathbf{g}H = H\mathbf{g}$ (see EA7.9) for all $\mathbf{g} \varepsilon G$. When this happens the subgroup is said to be a *normal subgroup*.

(DA7.11) Definition: A subgroup H of a group G is said to be *normal* if $\mathbf{g}H = H\mathbf{g}$ for all $\mathbf{g} \varepsilon G$.

It is straightforward to show that H is a normal subgroup of G if and only if $\mathbf{ghg}^{-1} \varepsilon H$ for all $\mathbf{g} \varepsilon G$ and $\mathbf{h} \varepsilon H$.

(EA7.12) Example: Let $G = 322$ and $H = {}^{[100]}2 = \{1, {}^{[100]}2\}$. Find the left and right cosets $\mathbf{g}^{[100]}2$ and ${}^{[100]}2\mathbf{g}$ for each $\mathbf{g} \varepsilon 322$.

Solution: Let $\mathbf{g} = \mathbf{3}$. Then

$$3^{[100]}2 = \{3, {}^{[110]}2\} \quad \text{and} \quad {}^{[100]}23 = \{3, {}^{[010]}2\} .$$

The reader can find the remaining left and right cosets.

Note that in EA7.12, we have shown that ${}^{[100]}2$ is *not* a normal subgroup of 322.

(PA7.14) Problem: Let $G = 422$. Show that 4 is a normal subgroup of 422 while ${}^{[100]}2$ is not.

(TA7.13) Theorem (Lagrange). Let G denote a finite group and H a subgroup of G. Then $\#(G)/\#(H)$ is an integer. This integer is the number of distinct cosets of H in G and is called the *index* of H in G.

Proof: Denote the distinct elements of H by $H = \{\mathbf{h}_1, \mathbf{h}_2, \ldots, \mathbf{h}_m\}$. Then $H\mathbf{a} = \{\mathbf{h}_1\mathbf{a}, \mathbf{h}_2\mathbf{a}, \ldots, \mathbf{h}_m\mathbf{a}\}$. Since there are m terms in the list for $H\mathbf{a}$, $H\mathbf{a}$ has at most m elements. To see that these m elements are distinct, suppose $\mathbf{h}_i\mathbf{a} = \mathbf{h}_j\mathbf{a}$. By the cancellation law, $\mathbf{h}_i = \mathbf{h}_j$, implying that $i = j$. Hence $\#(H\mathbf{a}) = \#(H)$ for all $\mathbf{a} \varepsilon G$. Recall that the right cosets are equivalence classes of the equivalence relation given in EA7.7. Hence by the discussion following DA7.5, G is the union of the distinct right cosets of H, and these cosets are disjoint. Choosing one representative $\mathbf{a}$ from each coset, we have

$$G = \bigcup_{a \varepsilon G} H\mathbf{a} .$$

Since the right cosets are disjoint,

$$\#(\bigcup_{a \in G} Ha) = \sum_{a \in G} \#(Ha) .$$

Assuming there are n such right cosets, and since $\#(Ha) = \#(H)$, we have

$$\sum_{i=1}^{n} \#(Ha) = n\#(H) .$$

Therefore,

$$\#(G) = n\#(H) .$$

Hence $\#(G)/\#(H)$ equals the integer n where n is the number of right cosets of H in G. □

Note that Lagrange's Theorem does not require that H be a normal subgroup. Theorem T4.28 presented a method for constructing improper crystallographic point groups from a given proper crystallographic point group G by using the halving groups of G. Lagrange's Theorem tells us that if $\#(G)$ is even, then there may be subgroups of G of order $\#(G)/2$. However, Lagrange's Theorem does not guarantee the existence of such a subgroup. In fact, in Chapter 5 we will find that the tetrahedral point group *23* has order 12 but no subgroups of order 6. Likewise, the icosahedral group *235* of order 60 has no subgroup of order 30.

Factor Groups: Let G denote a group and let H denote a subgroup of G. Under some conditions the right cosets of H in G can be made into a group under the operation

$$KL = \{\mathbf{k}\boldsymbol{\ell} \mid \mathbf{k} \in K \text{ and } \boldsymbol{\ell} \in L\} \tag{A7.1}$$

where K and L are right cosets of H in G. When the right cosets form a group, it is called a *factor group* and is denoted G/H (read "G mod H").

(EA7.14) Example: Let $G = 322$ and $H = 3$. Show that the right cosets of H under the operation described in (A7.1) is a group.

Solution: The right cosets of *3* are $\{1, 3, 3^{-1}\}$ and $\{[100]_2, [010]_2, [110]_2\}$. Forming the products we have

$$\{1,3,3^{-1}\}\{{}^{[100]}2, {}^{[010]}2, {}^{[110]}2\}$$

$$= \{1^{[100]}2, 1^{[010]}2, 1^{[110]}2, 3^{[100]}2, 3^{[010]}2, 3^{[110]}2, 3^{-1[100]}2,$$

$$3^{-1[010]}2, 3^{-1[110]}2\}$$

$$= \{{}^{[100]}2, {}^{[010]}2, {}^{[110]}2\} .$$

Similarly,

$$\{1,3,3^{-1}\}\{1,3,3^{-1}\} = \{1,3,3^{-1}\}$$

$$\{{}^{[100]}2, {}^{[010]}2, {}^{[110]}2\}\{1,3,3^{-1}\} = \{{}^{[100]}2, {}^{[010]}2, {}^{[110]}2\}$$

$$\{{}^{[100]}2, {}^{[010]}2, {}^{[110]}2\}\{{}^{[100]}2, {}^{[010]}2, {}^{[110]}2\} = \{1,3,3^{-1}\} .$$

Denoting the two right cosets by *3* and $3^{[100]}2$, we can use the results above to form the multiplication table shown in Figure A7.1. We can see from the table that these right cosets form the factor group *322/3*. Note that *3* is the identity element of *322/3*.

322/3	*3*	$3^{[100]}2$
3	*3*	$3^{[100]}2$
$3^{[100]}2$	$3^{[100]}2$	*3*

Figure A7.1

(PA7.15) Problem: Let G = *422* and H = *4*. Show that the right cosets of H under the operation described in (A7.1) is a group.

(EA7.15) Example: Let G = *322* and $H = {}^{[100]}2$. Show that the set of right cosets of H fail to form a group under the operation described in (A7.1).

Solution: Consider the right cosets $\{3, {}^{[010]}2\}$ and $\{3^{-1}, {}^{[110]}2\}$. Then

$$\{3, {}^{[010]}2\}\{3^{-1}, {}^{[110]}2\} = \{1, {}^{[010]}2, {}^{[100]}2, 3\}$$

which is not even a right coset! Therefore, the operation described in (A7.1) is not a binary operation on this set of right cosets. □

(PA7.16) Problem: Let $G = 422$ and $H = [100]_2$. Show that the set of right cosets of H fail to form a group under the operation described in (A7.1).

The condition that guarantees that the set of right cosets of H will form a group under the operation described in (A7.1) is that H be a normal subgroup of G. Suppose H is a normal subgroup of G. Then which coset must $H\mathbf{a}H\mathbf{b}$ be? Since $\mathbf{1}$, the identity element, is in H, $\mathbf{1a1b} = \mathbf{ab} \in H\mathbf{a}H\mathbf{b}$. Therefore, $H\mathbf{a}H\mathbf{b}$ must be the coset of $H\mathbf{ab}$. Consequently, we get the following theorem.

(TA7.16) Theorem: Let H denote a normal subgroup of G. Then the set of right cosets of H in G form a group, denoted G/H, called the *factor group* of G and H, under the operation

$$H\mathbf{a}H\mathbf{b} = H\mathbf{ab} .$$

If G is finite, then $\#(G/H) = \#(G)/\#(H)$.

(EA7.17) Example: Let $G = \bar{3}$ and $H = \bar{1} = \{\mathbf{1}, i\}$. Find the right and left cosets of $\bar{1}$ in $\bar{3}$ and show that $\bar{1}$ is a normal subgroup of $\bar{3}$. Using the operation defined in TA7.16, show that $\bar{3}/\bar{1}$ is a group and form its multiplication table.

Solution: The right and left cosets are as follows:

$$\bar{1}\mathbf{1} = \bar{1}i = \mathbf{1}\bar{1} = i\bar{1} = \{\mathbf{1}, i\}$$
$$\bar{1}\mathbf{3} = \bar{1}\bar{\mathbf{3}} = \mathbf{3}\bar{1} = \bar{\mathbf{3}}\bar{1} = \{\mathbf{3}, \bar{\mathbf{3}}\}$$
$$\bar{1}\mathbf{3}^{-1} = \bar{1}\mathbf{3}^{-1} = \mathbf{3}^{-1}\bar{1} = \bar{\mathbf{3}}\bar{1} = \{\mathbf{3}^{-1}, \bar{\mathbf{3}}^{-1}\} .$$

Since we can see that each of the right costs of $\bar{1}$ is also a left coset of $\bar{1}$, we conclude that $\bar{1}$ is a normal subgroup of $\bar{3}$. To construct the group $\bar{3}/\bar{1}$ we need a list of the right cosets. Each of the right cosets has two equally valid names corresponding to the two elements (representations) in each coset. For example, we could call the coset $\{3,\bar{3}\}$ either $\bar{1}3$ or $\bar{1}\bar{3}$. It makes no difference which of the equivalent names is chosen. To form our table we choose $\bar{1}1$, $\bar{1}\bar{3}$, $\bar{1}\bar{3}^{-1}$ to represent the cosets. We apply the binary operation $HaHb = Hab$ to form the multiplication. A problem occurs, however, since

$$(\bar{1}\bar{3})(\bar{1}\bar{3}^{-1}) = \bar{1}\bar{3}\bar{3}^{-1} = \bar{1}i$$

and $\bar{1}i$ is not in our chosen list. However, $\bar{1}i$ equals $\bar{1}1$ which is in our list. Hence we record $\bar{1}1$ in the table corresponding to the product $(\bar{1}\bar{3})(\bar{1}\bar{3}^{-1})$. The table is shown in Figure A7.2. By inspection we can see that $\bar{3}/\bar{1}$ is a group.

$\bar{3}/\bar{1}$	$\bar{1}1$	$\bar{1}\bar{3}$	$\bar{1}\bar{3}^{-1}$
$\bar{1}1$	$\bar{1}1$	$\bar{1}\bar{3}$	$\bar{1}\bar{3}^{-1}$
$\bar{1}\bar{3}$	$\bar{1}\bar{3}$	$\bar{1}\bar{3}^{-1}$	$\bar{1}1$
$\bar{1}\bar{3}^{-1}$	$\bar{1}\bar{3}^{-1}$	$\bar{1}1$	$\bar{1}\bar{3}$

Figure A7.2

In the course of the last example, we found that it is convenient to exchange one representative of a right coset for another. This is similar to the way in which we add fractions. For example, if we were asked to calculate the sum 2/3 + 4/5 , we would exchange each of these fractions for other equivalent fractions so as to facilitate the addition. In this case, if we exchange 10/15 for 2/3 and 12/15 for 4/5, we obtain 10/15 + 12/15 which can be easily calculated to be 22/15. Note that the set of fractions equivalent to any given fraction is an equivalence class under the equivalence relation on $\{a/b \mid a,b \in Z,\ b \neq 0\}$ defined by

$$a/b \sim c/d \iff ad = bc \ . \tag{A7.2}$$

The fact that the addition of rational numbers is such that the substitution of equivalent fractions is an addition problem yields equivalent sums is described by "rational addition is well defined with respect to the equivalence relation (A7.2)." Hence TA7.16 says that coset multiplication is well defined on G with respect to the equivalence relation of EA7.7.

(PA7.17) Problem: Let $G = 4/m$ and $H = \bar{1} = \{1, i\}$. Show that $\bar{1}$ is a normal subgroup of $4/m$ and that $(4m)/\bar{1}$ is a group and construct its multiplication table using the representations $H1$, $H\bar{4}$, $H2$ and $H\bar{4}^{-1}$.

APPENDIX 8

ISOMORPHISMS

In Chapter 3 we observed that *322* and $M_D(322)$ are isomorphic and that the mapping such that $\alpha \to M_D(\alpha)$ for each $\alpha \in 322$ is an isomorphism. In this Appendix we shall give definitions of these notions and explore their implications.

(DA8.1) Definition: Let G and H denote groups and let θ denote a mapping from G to H. If θ is one-to-one and onto and if

$$\theta(\mathbf{g}_1\mathbf{g}_2) = \theta(\mathbf{g}_1)\theta(\mathbf{g}_2) \quad \text{for all } \mathbf{g}_1, \mathbf{g}_2 \in G ,$$

then θ is called an *isomorphism.*

Note that in the expression $\theta(\mathbf{g}_1\mathbf{g}_2) = \theta(\mathbf{g}_1)\theta(\mathbf{g}_2)$, the binary operation on the left between $\mathbf{g}_1$ and $\mathbf{g}_2$ is that of G while the binary operation on the right between $\theta(\mathbf{g}_1)$ and $\theta(\mathbf{g}_2)$ is that of H.

(EA8.2) Example: Let G denote a group of isometries and let D denote a basis. Show that $\theta: G \to M_D(G)$ defined by

$$\theta(\alpha) = M_D(\alpha)$$

is an isomorphism from G to $M_D(G)$.

Solution: In (3.9) we showed that

$$M_D(\alpha\beta) = M_D(\alpha)M_D(\beta) .$$

Hence

$$\theta(\alpha\beta) = \theta(\alpha)\theta(\beta) .$$

Furthermore, since each isometry has exactly one matrix representation with respect to a basis D, θ is one-to-one. Since $M_D(G)$ contains only the matrix representations of elements of G, θ is onto. Hence G and $M_D(G)$ are isomorphic groups. □

(PA8.1) Problem: Independently verify the result given in Example EA8.2 for *322* and *422* by comparing the multiplication tables of G and $M_D(G)$ for each as discussed in Chapter 3.

The proof of the following theorem can be found in Durbin (1979, p. 81).

(TA8.3) Theorem: The relation ~ defined on the set of all groups of isometries by

$$G \sim H \iff G \text{ is isomorphic to } H$$

is an equivalence relation.

In many developments of algebra, any two groups that are isomorphic are considered to be indistinguishable. That is, the equivalence relation of TA8.3 is taken to be an equality. However, in our application of group theory to crystallography it will be necessary to distinguish between certain isomorphic groups. For example, *2* and *m* are isomorphic groups, yet the way in which their elements transform space is quite different. Moreover, the shape of an object with symmetry *2* is different from one of symmetry *m*.

(PA8.2) Problem: Show that 4 and $\bar{4}$ are isomorphic groups.

Let θ denote an isomorphism from the group G to the group H and let $\mathbf{g} \in G$. Then, by DA8.1,

$$\theta(\mathbf{g}^2) = \theta(\mathbf{gg}) = \theta(\mathbf{g})\theta(\mathbf{g}) = [\theta(\mathbf{g})]^2 .$$

Applying this idea n times we obtain

$$\theta(\mathbf{g}^n) = [\theta(\mathbf{g})]^n .$$

This fact and some other important properties are stated in the following theorem.

(TA8.4) Theorem: Let G and H denote isomorphic groups and let θ denote the isomorphism from G to H. Then the following statements hold where $\mathbf{g} \in G$.

(1) $\theta(\mathbf{e}_G) = \mathbf{e}_H$ where $\mathbf{e}_G$ and $\mathbf{e}_H$ are the identities of G and H, respectively.

(2) $\theta(\mathbf{g}^{-1}) = [\theta(\mathbf{g})]^{-1}$

(3) If $G = \langle \mathbf{g} \rangle$ (the cyclic group generated by $\mathbf{g}$), then H is cyclic and $H = \langle \theta(\mathbf{g}) \rangle$.

(4) If G is finite, then so is H and $\#(G) = \#(H)$.

We shall not give a formal proof of this theorem. See Durbin (1979) for a discussion of these properties. Note that part (4) follows from the fact that θ is one-to-one and onto and remarks made in Appendix 1.

(EA8.5) Example: Let H denote a subgroup of G, let $\mathbf{g} \in G$ and $T = \{\mathbf{ghg}^{-1} | \mathbf{h} \in H\}$. Show that T is a subgroup of G and that H is isomorphic to T.

Solution: Since H is a subgroup, it is nonempty and hence T is nonempty. Let $\mathbf{t}_1, \mathbf{t}_2 \in T$. Then there exist elements $\mathbf{h}_1, \mathbf{h}_2 \in H$ such that $\mathbf{t}_1 = \mathbf{gh}_1\mathbf{g}^{-1}$ and $\mathbf{t}_2 = \mathbf{gh}_2\mathbf{g}^{-1}$. Hence

$$\begin{aligned} \mathbf{t}_1\mathbf{t}_2 &= (\mathbf{gh}_1\mathbf{g}^{-1})(\mathbf{gh}_2\mathbf{g}^{-1}) \\ &= \mathbf{gh}_1(\mathbf{g}^{-1}\mathbf{g})\mathbf{h}_2\mathbf{g}^{-1} \\ &= \mathbf{gh}_1\mathbf{h}_2\mathbf{g}^{-1} . \end{aligned}$$

Since H is a subgroup, it is closed and so $\mathbf{h}_1\mathbf{h}_2 \in H$. Therefore, $\mathbf{t}_1\mathbf{t}_2 \in T$. Let $\mathbf{t} \in T$. Then there exists $\mathbf{h} \in H$ such that $\mathbf{ghg}^{-1} = \mathbf{t}$. Then $\mathbf{s} = \mathbf{gh}^{-1}\mathbf{g}^{-1} \in T$, since $\mathbf{h}^{-1} \in H$, and

$$\begin{aligned} \mathbf{ts} &= (\mathbf{ghg}^{-1})(\mathbf{gh}^{-1}\mathbf{g}^{-1}) \\ &= \mathbf{e} , \end{aligned}$$

where $\mathbf{e}$ is the identity of G. Hence $\mathbf{t}^{-1} = \mathbf{s} \in T$. Therefore, T is a subgroup of G. To show that H is isomorphic to T, we must find a candidate for the isomorphism. Looking at the way T is defined, it is natural to try the mapping θ defined by

$$\theta(\mathbf{h}) = \mathbf{ghg}^{-1} .$$

We first must show that θ is one-to-one. Suppose h_1 and h_2 are elements of H such that $\theta(h_1) = \theta(h_2)$. Then

$$gh_1g^{-1} = gh_2g^{-1} .$$

By the right and left cancellation laws, $h_1 = h_2$ Hence θ is one-to-one. Note that, by the definition of T, θ is onto. Finally,

$$\begin{aligned}\theta(h_1h_2) &= gh_1h_2g^{-1} \\ &= gh_1eh_2g^{-1} \quad \text{(where } e \text{ is the identify of } G\text{)} \\ &= gh_1(g^{-1}g)h_2g^{-1} \\ &= (gh_1g^{-1})(gh_2g^{-1}) \\ &= \theta(h_1)\theta(h_2)\end{aligned}$$

Hence θ is an isomorphism. □

An application of the result discussed in EA8.5 is to T5.5 where we prove that G_p is isomorphic to G_q when p and q are G-equivalent. In particular, if p and q are G-equivalent, then $\#(G_p) = \#(G_q)$ and each is isomorphic to a cyclic group of the same order.

REFERENCES

Bloss, F.D. (1971) Crystallography and Crystal Chemistry. Holt, Rinehart & Winston, Inc., New York, 545 p.

Boisen, M.B., Jr. and G.V. Gibbs (1976) A derivation of the 32 crystallographic point groups using elementary theory. Amer. Mineral., 61, 145-165.

Boisen, M.B., Jr. and G.V. Gibbs (1978) A method for constructing and interpreting matrix representations of space-group operations. Canadian Mineral., 16, 293-300.

Bravais, A. (1850) On the systems formed by points regularly distributed on a plane or in space. Translated by A.J. Shaler, Monograph No. 4, Amer. Crystallogr. Assoc., J. Ecole Polytechnique, Canier 33, Tome XIX, 1-128.

Bronowski, J. (1973) The Ascent of Man. Little, Brown & Co., Boston.

Brown, H., R. Bulow, J. Neubuser, H. Wondratschek, and H. Zassenhaus (1978) Crystallographic Groups of Four-dimensional Space. John Wiley & Sons, New York, 443 p.

Buerger, M.J. (1942) X-ray Crystallography. John Wiley & Sons, New York, 531 p.

Buerger, M.J. (1963) Elementary Crystallography. John Wiley & Sons, New York, 528 p.

Burnham, C.W. (1971) The crystal structure of pyroxferroite from Mari Tranquillitis. Proc. Second Lunar Sci. Conf., Vol. 1, 47-57, M.I.T. Press, Cambridge, Massachusetts.

Durbin, J.R. (1979) Modern Algebra. John Wiley & Sons, New York, 329 p.

Farkas, D.R. (1981) Crystallographic groups and their mathematics. Rocky Mountain J. Math., 11, 511-551.

Fisher, D.J. (1952) Lattice constants of synthetic chalcanthite by the x-ray precession technique using a single mounting of the crystal. Amer. Mineral., 37, 95-114.

Gibbs, G.V. (1969) The crystal structure of protoamphibole. Mineral. Soc. Amer. Spec. Paper 2, 101-110.

Gibbs, G.V. (1982) Molecules as models for bonding in silicates. Amer. Mineral., 67, 421-450.

Gibbs, G.V., C.T. Prewitt, and K.J. Baldwin (1977) A study of the structural chemistry of coesite. Z. Kristallogr., 145, 108-123.

Gibbs, G.V., E.P. Meagher, M.D. Newton, and D.K. Swanson (1981) A comparison of experimental and theoretical bond length and angle variations for minerals, inorganic solids and molecules. In Structure and Bonding in Crystals, Vol. 1, M. O'Keeffe and A. Navrotsky, Eds., Ch. 9, 195-225.

Grossman, I. and W. Magnus (1964) Groups and their graphs. Random House, The L.W. Singer Co., New York, 191 p.

Hahn, T. (1983) International Tables for Crystallography, Volume A, Space-group Symmetry. D. Reidel Publishing Co., Dordrecht, Holland, 854 p.

Hazen, R.M. and L.W. Finger (1982) Comparative Crystal Chemistry. John Wiley & Sons, New York, 231 p.

Hehre, W.J., L. Radon, R. Schleyer, and J.A. Pople (in press) *Ab initio* Molecular Orbital Theory. John Wiley & Sons, New York.

Henry, N.F.M. and K. Lonsdale, eds. (1952) International Table for X-ray Crystallography, Vol. 1, Symmetry Groups. The Kynoch Press, Birmingham, England, 558 p.

Herstein, I.N. (1975) Topics in Algebra. John Wiley & Sons, New York, 388 p.

Higgins, J.B. and P.H. Ribbe (1979) Sapphirine II. A neutron and X-ray diffraction study of (Mg-Al) and (Si-Al) ordering in monoclinic sapphirine. Contrib. Mineral. Petrol., 68, 357-368.

Horne, R.W. and P. Wildy (1961) Symmetry in virus architecture. Virology, 15, 348-359.

Johnson, C.K. (1976) OR TEP-11: A Fortran thermal-ellipsoid plot program for crystal-structure illustrations. U.S. Dept. Commerce, Nat'l Tech. Information Serv., Springfield, Virginia, 125 p.

Johnson, L.W. and R.D. Reiss (1980) Introduction to Linear Algebra. Addison-Wesley Publishing Co., Reading, Massachusetts, 358 p.

Lasaga, A.C. and R.J. Kirkpatrick (1981) Kinetics of Geochemical Processes. Reviews in Mineralogy, Vol. 8, Mineral. Soc. Amer., Washington, D.C., 398 p.

Levien, L., C.T. Prewitt, and D.J. Weidner (1980) Structure and elastic properties of quartz at pressure. Amer. Mineral., 65, 920-930.

Lopes-Vierra, A. and J. Zussman (1969) Further detail on the crystal structure of zussmanite. Mineral. Mag., 37, 49-60.

Megaw, H.D. (1970) Structural relationship between coesite and felspar. Acta Crystallogr., B26, 261-265.

Moore, P.B. (1968) The crystal structure of sapphirine. Amer. Mineral., 54, 31-49.

Moore, P.B. (1976) The glaserite, $K_3Na[SO_4]_2$, structure type as a "super" dense-packed oxide: evidence for icosahedral geometry and cation-anion mixed layer packings. N. Jahrb. Mineral. Abh., 127,187-196.

Newman, M. (1972) Integral Matrices. Academic Press, New York, 224 p.

O'Keeffe, M. and G.V. Gibbs (1984) Defects in amorphous silica; *Ab initio* MO calculations. J. Chem. Physics, 81, 876-879.

Peacor, D.R. and M.J. Buerger (1962) The determination and refinement of the structure of narsarsukite, $Na_2TiOSi_4O_{10}$. Amer. Mineral., 47, 539-556.

Pluth, J.J., J.V. Smith, and J. Faber, Jr. (1985) Crystal structure of low cristobalite at 10, 293, and 73 K: Variation of framework geometry with temperature. J. Appl. Physics, 57, 1045-1049.

Prewitt, C.T. and C.W. Burnham (1966) The crystal structure of jadeite, $NaAlSi_2O_6$. Amer. Mineral., 51, 956-975.

Ross, M., J.J. Papike, and K.W. Shaw (1969) Exsolution textures in amphiboles as indicators of subsolidus thermal histories. Mineral. Soc. Amer. Spec. Paper 2, 275-299.

Rotman, J.J. (1976) Theory of Groups. Allyn & Bacon, Inc., Boston, Massachusetts, 342 p.

Schectmann, D., I. Blech, D. Gratias, and J.W. Cahn (1984) Metallic phase with language orientational order and no translational symmetry. Phys. Rev. Letters, 53, 1951-1953.

Schoenflies, A. (1923) Theorie der Kristallstruktur. Berlin.

Seitz, F. (1935) A matrix-algebraic development of the crystallographic groups, III. Z. Kristallogr., 91, 135-165.

Senechal, M., G. Fleck, and A. Ludman, eds. (1974) Patterns of Symmetry. University of Massachusetts Press, Amherst, Massachusetts.

Slater, J.C. (1972) Symmetry and energy bonds in crystals. Dover Publications, Inc., New York, 563 p.

Smith, J.V. (1982) Geometrical and Structural Crystallography. John Wiley & Sons, New York, 450 p.

Taylor, W.H. and W.W. Jackson (1928) The structure of cyanite, Al_2SiO_5. Proc. Roy. Soc. London, A, 119, 132-146.

Villiers, J.P.R. (1971) Crystal structures of aragonite, strontianite, and witherite. Amer. Mineral., 56, 758-767.

Wainwright, J.E. and J. Starkey (1971) A refinement of the structure of anorthite. Z. Kristallogr., 133, 75-84.

Warren, B.E. (1930) The structure of tremolite. Z. Kristallogr., 72, 42-57.

Weyl, H. (1952) Symmetry. Princeton University Press, Princeton, New Jersey, 168 p.

Wilson, E.B.W. (1960) Vector Analysis. Dover Publications, Inc., New York, 436 p.

Winter, J.K. and S. Ghose (1979) Thermal expansion and high-temperature crystal chemistry of the Al_2SiO_5 polymorphs. Amer. Mineral., 64, 573-586.

Whittaker, E.J.W. and J. Zussman (1961) The choice of axes in amphiboles. Acta Crystallogr., 14, 54-55.

Wondratschek, H. and J. Neubuser (1967) Determination of the symmetry elements of a space group from the "general positions" listed in International Tables for X-ray Crystallography, Vol. 1. Acta Crystallogr., 23, 349-352.

Zachariasen, W.H. (1945) Theory of x-ray diffraction in crystals. John Wiley & Sons, New York, 225 p.

Zoltai, T. and M.J. Buerger (1959) The crystal structure of coesite, the dense, high pressure form of silica. Z. Kristallogr., 41, 129-141.

INDEX